Holography - Recent Advances and Applications

Edited by Joseph Rosen

Published in London, United Kingdom

Holography - Recent Advances and Applications
http://dx.doi.org/10.5772/intechopen.102216
Edited by Joseph Rosen

Contributors
Gangi Reddy Salla, Vanitha Patnala, Ravindra Pratap Singh, Cindy X. Xue Wen Chen, Adam Wax, Hillel B. Price, Kavan Ahmadi, Anton Goncharsky, Svyatoslav Durlevich, Joseph Rosen, Angika Bulbul, Nathaniel Hai, Mani R. Rai, Vijayakumar Anand, Soon Hock Ng, Tomas Katkus, Daniel Smith, Vinoth Balasubramani, Denver P. Linklater, Pierre J. Magistretti, Christian Depeursinge, Elena P. Ivanova, Saulius Juodkazis, Frank Dubois, Catherine Yourassowsky, Jérôme Dohet-Eraly, Karim Zouaoui Boudjeltia, Kedar Khare, Randy A. Bartels, Lang Wang, Gabriel Murray, Jeffrey Field, Anett Mehler-Bicher, Dirk Weitzel, Lothar Steiger, Mariana Potcoava, Christopher Mann, Simon Alford, Jonathan Art, Tina Sabel-Grau, Marga C. Lensen, Rahima Rahman, Zhenfang Zhang, Cigdem Yesildag, Arina Tyushina, Wanbin Zhang, Baosheng Li, Jianquan Li, Nandan S. Bisht, Vipin Tiwari, Vinu Raveendran Pillai Vasantha Kumari, Ziyang Chen, Rakesh Singh, Jixiong Pu, Krzysztof A. Cyran, Kamil Wereszczyński, Haijun Wu, Weikang Jiang, Varun Kumar, Chandra Shakher

First published in London, United Kingdom, 2023 by IntechOpen
IntechOpen is the global imprint of INTECHOPEN LIMITED, registered in England and Wales, registration number: 11086078, 5 Princes Gate Court, London, SW7 2QJ, United Kingdom

British Library Cataloguing-in-Publication Data
A catalogue record for this book is available from the British Library

Additional hard and PDF copies can be obtained from orders@intechopen.com

Holography - Recent Advances and Applications
Edited by Joseph Rosen
p. cm.
Print ISBN 978-1-80356-563-7
Online ISBN 978-1-80356-564-4
eBook (PDF) ISBN 978-1-80356-565-1

For EU product safety concerns:
IN TECH d.o.o., Prolaz Marije Krucifikse Kozulić 3, 51000 Rijeka, Croatia,
info@intechopen.com or visit our website at intechopen.com.

Meet the editor

Joseph Rosen is a Benjamin H. Swig Professor of Optoelectronics in the School of Electrical and Computer Engineering, Ben-Gurion University, Negev, Israel. He received his BSc, MSc, and DSc in Electrical Engineering from the Technion - Israel Institute of Technology. He is a fellow of Optica (formerly the Optical Society of America) and the International Society for Optical Engineering (SPIE). His research interests include holography, image processing, optical microscopy, diffractive optics, interferometry, biomedical optics, statistical optics, optical computing, and pattern recognition. He has co-authored more than 250 scientific journal papers, book chapters, and conference publications.

Contents

Preface

A good preface of a book about holography is to put the systems and methods described herein in their historical context. The early holograms and some of the holograms described in this book are recorded as the result of interference between two light waves. The phenomenon of optical two-wave interference has been well known since the first decade of the 19th century when Thomas Young published his famous double-slit experiment. In Young's experiment, the interference happens between two waves where none of them carries any image. Therefore, Young's experiment produces an interference pattern between two light waves but not a hologram. The revolutionary transition from a simple interference pattern to a hologram happened in 1948 in Dennis Gabor's pioneering work presenting for the first time what is known today as the Gabor hologram. The methods described in Chapters 1–4 in this book are conceptually close in the sense that the holograms are recorded by two-wave interference between a wave carrying the object information and another wave called the reference wave, which does not contain any object information.

The holograms described in Chapters 5 and 6 are additional technological steps in the evolution of holography beyond the holograms recorded with a non-image reference beam. These holograms of incoherently illuminated objects are recorded using the principle of self-interference. Self-interference holograms are a new stage in the evolutionary chain in which both interfering waves carry the object's image. However, the image information is never the same in both interferometer channels. In holograms described in Chapters 5 and 6, the images are in-focus at different distances from the aperture. The hologram described in Chapter 7, COACH, is an additional evolutionary stage in which one of the two images passes through a coded scattering mask. The other image is in focus at an infinite distance from the aperture.

From the stage of COACH, the technology could be evolved to a system in which both interfering waves pass through scattering masks and/or to a system without two-wave interference at all. At the time of this writing, the second option has been chosen, mainly because the interference-less COACH is simpler and more efficient than the COACH with two-wave interference. The acoustic holograms of Chapter 8 also belong to the type of holograms recorded without two-wave interference because the entire phase and amplitude information of the scene can be detected directly from a single acoustic wave without the need to record interference with any reference wave. What are the future systems in this evolutionary chain? The answer is that only those systems providing any advantage over other existing systems at a reasonable cost will probably justify their existence.

The holography of today is in transition from using traditional photo materials to the use of digital cameras and computers in what are called digital holograms. However, not all applications of holography can be implemented by digital holograms. Chapters 9 and 10 describe methods of recording holograms on non-traditional photo materials.

The unique applications of holographic display in Chapter 9 and of the holographic lens in Chapter 10 dictate the photo material in these holograms. Regarding other holography applications, various chapters in the second section of the book discuss diverse applications, such as holographic microscopy in Chapters 11–13, heat transfer measurements in Chapter 14, quantum computing in Chapter 15, augmented reality in Chapter 16, beam-shaping in Chapter 17, and holographic encryption in Chapter 18.

Holography is a broad field developed in the meeting between optics, and material sciences, and contains more or different information on the observed scene than a regular image of the same scene. To extract the required information from raw holograms, a digital algorithm is applied according to the specific application of the system. The development of the field has been accelerated lately due to the improvement of digital cameras, computers, and spatial light modulators.

As a multidisciplinary area, holography connects experts in electro-optical engineering, image processing, computer algorithms, and sometimes in material engineering.

More experts are needed when holography is utilized in various applications such as microscopy, industrial inspection, biomedicine, entertainment, and others. This book provides an overview of the world of holography from the aspect of concepts, system architectures, and applications. The various chapters deal with ongoing research of many current developments in the field of holography. Thus, the present book is only an interim summary of a research field that aims to encourage better technology for current and new applications. Many thanks and deep appreciation to all the authors for contributing excellent chapters to this book. I sincerely thank Author Service Managers Zrinka Tomicic and Romina Rovan and the staff of IntechOpen for their valuable support. We hope that the scientific community dealing with these and related topics will find this book interesting, helpful, and inspiring.

Finally, I would like to mention here my two dear colleagues and teachers, Gabriel Popescu (1971–2022) and John T. Sheridan (1964–2022), both of whom prematurely passed away while preparing this book. These two scientists contributed many original ideas to the science of holography and have greatly influenced numerous researchers in the field. May they rest in peace.

Joseph Rosen
School of Electrical and Computer Engineering,
Ben-Gurion University of the Negev,
Beer-Sheva, Israel

Section 1

Theoretical Concepts and Methods

Chapter 1

Spatial Light Modulators and Their Applications in Polarization Holography

Vipin Tiwari and Nandan S. Bisht

Abstract

Liquid crystal spatial light modulators (LC-SLMs) have gained substantial interest of the research fraternity due to their remarkable light modulation characteristics in modern imaging applications. Replacing the conventional optical elements from the SLM-based computer-generated holograms (CGHs) is a trending approach in modern digital holographic applications due to the optimized phase shift depending on the phase modulation features of SLMs. Apparently; SLMs serve a crucial role in the experimental implementation of digital holographic techniques. However, the resolution of the CGHs are sometimes limited by the structural discrepancies (fill factor, spatial anomalies, refresh rate, etc.) of SLM. Therefore, it is recommended to calibrate the modulation characteristics of SLMs prior to their implementation for imaging applications. This chapter provides comprehensive literature (review) of the LC-SLMs along with their major calibration methods. In addition, recent interesting applications of LC-SLMs have been discussed thoroughly within the framework of polarization holography.

Keywords: spatial light modulator, calibration methods, spatial anomalies, polarimetry, digital holography, polarization holography

1. Introduction

Spatial light modulators (SLMs) are electro-optical devices, pertaining to manipulating the fundamental characteristics, viz., amplitude, phase, and polarization state of light. SLMs have gained significant attention of the research fraternity due to their versatile applicability in various optical imaging applications, including digital holography [1, 2], adaptive optics [3], computational imaging [4–6], holographic display [7]. In fact, SLM driven CGHs have the potential to eliminate the undesired artifacts (low resolution, noise, etc.) in holographic applications resulting in dynamic modulation response and high resolution.

The structure of SLM is primarily based on either the optical micro-electro-mechanical system (MEMS) [8] or liquid crystal on silicon (LCOS) technology [9, 10]. Digital micrometer devices (DMDs) are commonly used MEMS-based SLMs, which consist of pixel-dependent micro-mirrors. The rapid switching between fixed states of micro-mirrors makes DMDs suitable for amplitude modulation of light with a high frame rate (up to 40 kHz) [9]. However, corresponding amplitude modulation is binary (depends on two fixed states of micro-mirrors) and therefore DMDs are not sufficient for customized phase and polarization modulation of light. On the other hand, LCOS SLMs have the potential to modulate the amplitude, phase, and polarization of light due to the electrically controlled birefringence (ECB) properties of liquid crystals (LCs). The desired phase shift can be obtained by addressing different grayscales (pixel voltages) of the SLM. In practice, the phase response of the SLM can vary from the ideal grayscale-phase response [provided in look-up-table (LUT) of SLM model] due to various exogenous factors, viz., manufacturing defects [11, 12], non-linear optical response of LC, spatial anomalies [10, 13, 14], etc. Therefore, calibration of LC-SLM for grayscale-phase is recommended for its precise utilization in various holographic applications.

Digital holography is a conventional optical technique that enables the simultaneous recording of amplitude and phase of light (encoded in hologram) using the complex electric field modulation of light [15]. However, polarization is another crucial degree of freedom of light along with its amplitude and phase. The inclusion of polarization encoding in conventional digital holograms paves the route for the development of polarization holography [16]. Conventional polarization holography requires polarization-sensitive holograms (PSH) for reconstructing polarization multiplexed holographic images. In general, these PSHs are encoded with the help of polarization-dependent birefringent materials [17]. However, recent developments in polarization holography use LC-SLM for encoding the polarization-sensitive holograms [18–21]. In addition, LC-SLM serves key role in the experimental demonstration of advanced digital holographic techniques, viz. coded aperture correlation holography (COACH) [22–24], Fresnel incoherent correlation holography (FINCH) [2, 24], Stokes correlation holography [21, 25, 26], polarization holography [18, 19, 27–29], etc. This chapter presents comprehensive literature (review) on the LC-SLM and its applicability in digital holography. Subsequent sections of the chapter illustrate the construction, working principle, and the major calibration methods for LC-SLM, followed by some crucial applications of LC-SLMs in the context of polarization holography.

2. Liquid crystal on silicon (LCOS) spatial light modulator

2.1 Structure and working principle of LCOS SLM

LCOS-based SLMs are programmed such that the orientation of LC molecules can be changed depending on their applied pixel voltage (gray scales). The use of LC materials in SLMs depends on their optical and electrical anisotropic properties. Typically, a thin layer of LC material can be described as a birefringent material with two refractive indices. The orientation of the index ellipsoid depends on the direction of the director of LC molecules. Ferroelectric and nematic alignments are the most commonly deployed LC mesophases for SLM fabrication [30]. The ferroelectric LC SLMs exhibit a faster refresh rate but suffer from the zeroth order diffraction artifact and are limited up to binary phase modulation only [31]. However, the nematic LC SLMs can provide customized multi-grayscale phase modulation. Further, the commonly used nematic LC alignments in SLMs are parallel aligned nematic (PAN) or homogeneous, vertically aligned nematic (VAN) or homeotropic cells. **Figure 1** represents the schematic of the LC alignments. In homogeneous and homeotropic configuration, the alignment layers are parallel to each other, so the LC molecules have

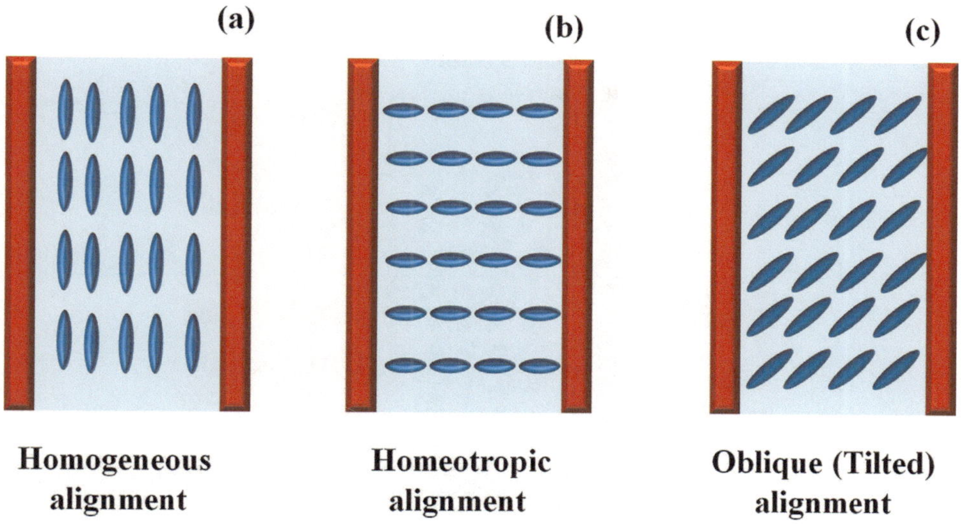

Figure 1.
Schematic of liquid crystal alignments in LC-SLM (a) Homogeneous, (b) homeotropic, and (c) tilted.

the same orientation and are generally used to obtain amplitude modulation of light. On the other hand, in the twisted nematic liquid crystal (TNLC) alignment, the orientation of the molecules can be adjusted such that the uppermost and bottom layers are differed by a fixed angle (0°, 45°, 90°).

SLMs can be used either in the transmission or reflection arm of an optical system, depending upon their structure and specifications. A schematic of the fundamental structure of the LC-SLM is shown in **Figure 2**. The inner structure of SLMs is comprised of a typical alignment of LC molecules (nematic phase), sandwiched between transparent electrodes and glass substrates. The active matrix circuit is formed on the silicon substrate to control each pixel electrode by the applied electrical potential individually using semiconductor technology. The pixel electrodes are typically made of reflective aluminum mirrors with pixel circuitry, which contains gate lines and buried transistors for a maximum fill factor with high reflectivity. Two alignment layers with rubbing direction could be used to initialize the orientation of the modulated LC molecules. The transparent electrode is deployed to work with the pixel electrode for generating the electric field across the LC cell.

The typical alignment of LC molecules of SLMs allows the change only along their director axis while the other orthogonal direction remains invariant. It results in optical anisotropy (phase retardation) due to SLM and is responsible for the optimized phase shift at any location of the SLM display. The optimized phase shift for each pixel of the SLM is programmed as a function of 256 gray scales ranging from 0 to 255. Therefore the typical anisotropic properties of SLM (phase shift and birefringence) can also be interpreted as a function of its gray scales.

2.2 Electro-optic effect in LC-SLM

LC-SLM can exhibit anisotropic nature due to the relative phase delay (birefringence) for the orthogonal polarization components of its LC array. Let us consider the

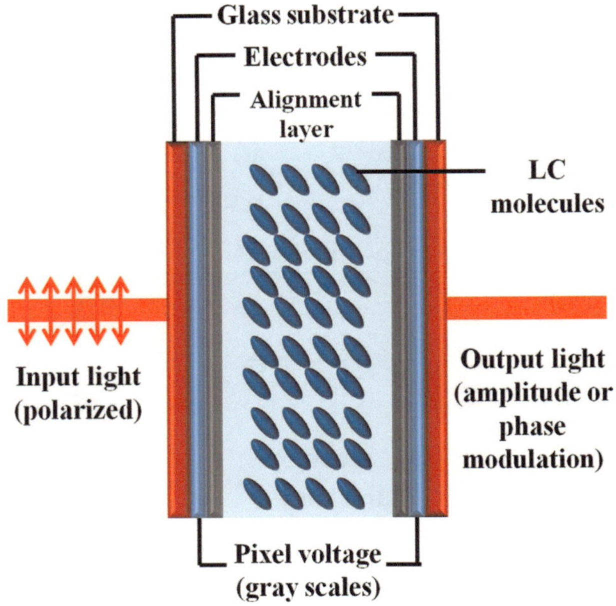

Figure 2.
Schematic of the structure of LC-SLM.

SLM is designed to modulate the phase of light along the vertically polarized (y-polarized) light. The phase delay experienced by the SLM is given as [32]

$$\varphi_x(x,y) = \frac{4\pi n_o}{\lambda}.d \tag{1}$$

$$\varphi_y(x,y) = \frac{4\pi n_e}{\lambda}.d \tag{2}$$

Here, n_o *and* n_e are the refractive indices for the slow and fast axis of the LC-SLM respectively and d is the thickness of the LC cell. The relative rotation of LC molecules of SLM for applied pixel voltage (grayscale) is shown in **Figure 3**. The LC molecules of SLM exhibit the tendency to align themselves along the direction of an external applied electric field. This effect is known as electrically controlled birefringence (ECB).

When pixel voltage (grayscale) is applied to the SLM, it provides an additional phase to the y-polarized light, whereas x-component of light remains unaffected.

The resultant phase of the x-polarized component of light is

$$\delta_x(x,y) = \varphi_x(x,y) \tag{3}$$

The resultant phase of the y-polarized component of light is

$$\delta_y(x,y) = \varphi_y(x,y) + \gamma\,(x,y) \tag{4}$$

Here, $\gamma(x,y)$ is the additional phase due to the applied grayscale of the SLM.

The retardance of the SLM can be defined as

$$R\,(x,y) = \delta_y(x,y) - \delta_x(x,y) = \left(1 - \frac{n_o}{n_e}\right)\varphi_y(x,y) + \gamma\,(x,y) \tag{5}$$

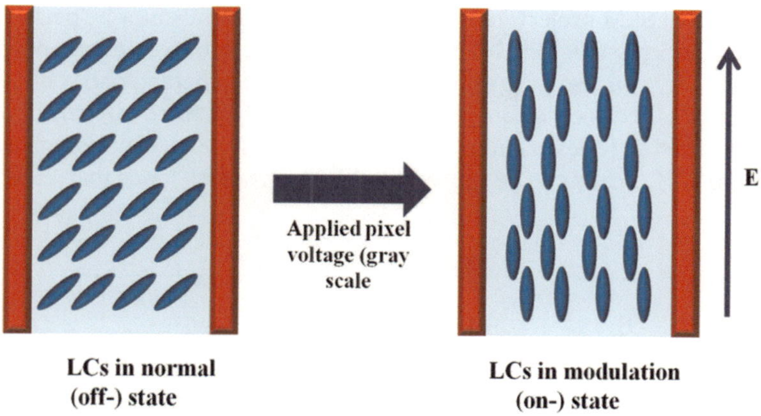

Figure 3.
Electo-optic effect in TNLC-SLM.

2.3 Light modulation characteristics of LC-SLM

It is already discussed that LC-SLM is known for customized light modulation characteristics in terms of its amplitude, phase, and polarization with respect to their grayscale values. The amplitude modulation can be obtained by illuminating twisted PAN or VAN-based SLMs with linearly polarized light. It should be transmitted through a polarizer with orthogonal direction with respect to the initial polarization of light. The level of attenuation by the second polarizer can be tuned by applying an electric field to the cell, which leads to a change of the birefringence β [9]. The phase-only modulation of light can be conveniently achieved using PAN LC cell-based SLMs. The reason is that PAN LC cells are most responsive to the ECB effect, i.e., light passing through the extraordinary axis of PAN cells exhibits phase retardation as a function of voltage-controlled birefringence (gray scales). On the other hand, TNLC SLMs are very complicated to handle for phase-only modulation as there exists amplitude modulation as well in that case [33, 34]. Therefore, TNLC-SLMs exhibit complex field (amplitude and phase) modulation or phase mostly modulation of light.

Polarization modulation is another crucial feature of LC-SLMs, which lead to very interesting applications of LC-SLMs in polarization holography. It is evident that when SLM is placed between two quarter-wave plates (QWPs) with their fast axis rotated by +45° and −45° with respect to the incident light beam. The Jones matrix of this combined optical configuration is typically a rotation matrix, which represents the phase shift of SLM. Therefore, it is possible to convert a phase-only SLM into a pixel-dependent phase-shifting device by inserting it between two QWPs. For example, a linearly polarized light can be converted into azimuthally polarized light by exposing SLM with a vortex structure under polarization modulation configuration.

3. Characterization of LC-SLM

SLMs are pixelated programmable devices. Exogenous parameters like pixel size, and fill factor are too crucial to investigate in order to obtain the desired phase response of LC-SLM. For instance, smaller pixel size leads to better resolution and larger diffraction angles whereas a higher fill factor is preferred as it can eliminate the zeroth order diffraction issue in SLM projection. However, the said issue cannot be perfectly eliminated due to pixel crosstalk in SLM. Moreover, other factors (frame rate, power supply, incident angle, etc) can also perturb the ideal modulation characteristics of LC-SLM.

The effect of LC cells on a coherent light source with a fixed polarization state can be described by using Jones calculus [35]. In particular, the modulation characteristics of light due to TNLC cells can be represented in terms of its Jones matrix as [36]

$$J_{TNLC} = \exp(i\varphi)\begin{pmatrix} a - ib & c - id \\ -c - id & a + ib \end{pmatrix} \tag{6}$$

Where, a, b, c, and d are known as anisotropic coefficients, which typically depend on the LC physical parameters (twist angle, director orientation angle, etc.). These physical parameters are usually not provided by manufacturers of LC-SLM and therefore it is very challenging to determine precise modulation features of TNLC-SLM.in a straightforward manner. However, the Jones matrix of LC-SLM can be measured using optical characterization techniques (interferometry, digital holography, cryptography, etc).

3.1 LC-SLM calibration methods

3.1.1 Interferometric methods

The phase modulation characteristics of LC-SLM can be calibrated using interferometric measurements [14, 27, 37–42]. The interferometry enables to record the intensity and phase of light simultaneously in the form of fringes (interference pattern). Interferometry relies on the principle of the reference beam (unperturbed light beam) and object beam (beam with phase shifting), where the phase difference between these two beams results in the fringe-shift depending upon the provided phase to the object beam. In the context of phase calibration of LC-SLM, the interference pattern is obtained from the superposition of the SLM modulated beam (object beam) and the reference beam. The fringe shift is obtained by changing the gray scales on SLM and the phase modulation of SLM can be retrieved from the relative displacement of fringes. However, a good contrast is required between the reference beam and object beam in order to obtain sharp fringes. Therefore, generating the reference arm of the interferometer is a critical factor to be optimized in these calibration methods. In particular, the reference beam can be generated either from the LC-SLM itself (self-reference) or by splitting the wavefront into two parts before interacting with SLM (out-reference). **Figure 4** depicts the schematic of the self-referenced interferometric method and reference calibration using a splitted SLM screen scheme [43]. In this technique, two different phase masks were imported into LC-SLM [PLUTO-VIS, Holoeye] such that half of the screen was addressed with uniform varying gray values while the other half portion was exposed with vertical binary diffraction grating (**Figure 4b**).

In a self-referenced method, the active area of SLM is divided into two parts. In one part different grayscales are provided (modulating part) meanwhile a fixed gray scale '0' is provided to the other half portion of the SLM screen (reference part) [27]. The beam passing through the modulating part undergoes a phase-shift depending on the addressed gray values on the SLM pixels. Hence, the phase-shift due to SLM modulation can be calibrated by investigating the fringe displacement (deformation) of the recorded interference pattern. Although the self-referenced interferometric calibration

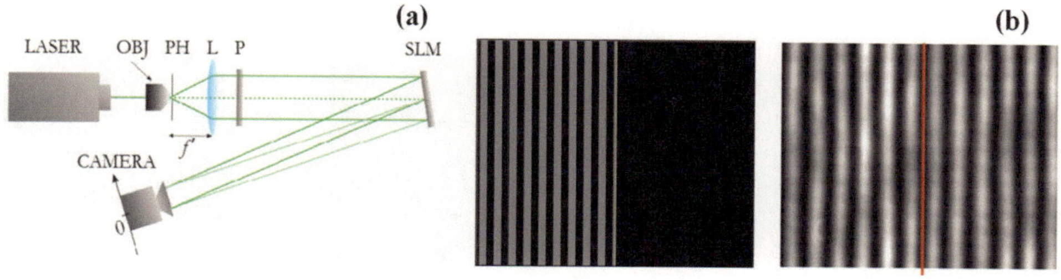

Figure 4.
(a) Schematic of self-referenced interferometric technique for phase calibration of reflective LC-SLM [PLUTO-VIS, Holoeye] calibration, and (b) imported pattern on LC-SLM (left) and recorded fringe pattern (right). [OBJ: Microscopic objective, L: Lens, P: Polarizer] (adapted from [43]).

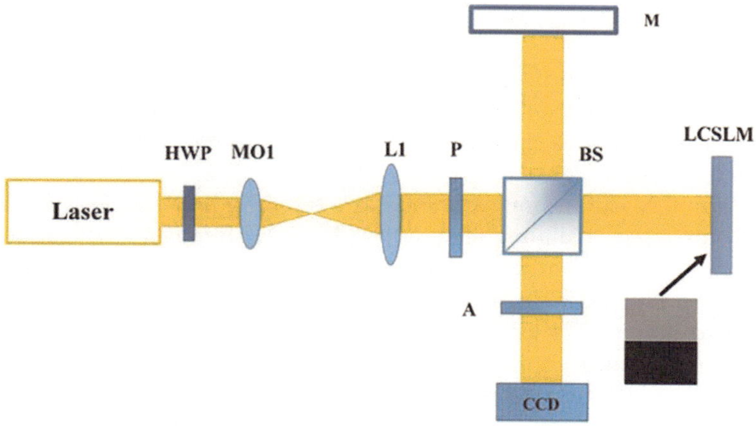

Figure 5.
Schematic of off-axis-referenced Twyman-green interferometer for phase calibration of LC-SLM [HWP: Half wave plate, MO1: Micro-objective, L1: Lens, P: Polarizer, BS: Beam splitter, M: Mirror, A: Analyzer, CCD: Charge-coupled device] (adapted from [44]).

methods are compact and stable but these methods are not compatible to determine the spatially varying (pixel addressable) modulation characteristics of LC-SLM.

On the other hand, an external reference beam (independent of LC-SLM) is used in the off-axis-referenced method. The simplest experimental configuration of these types of methods is the Twyman-Green interferometer, where the light beam is splitted into two orthogonal directions with the help of a beam-splitter [41]. In one arm, a plane-mirror is inserted whereas the SLM is placed on the other arm of the interferometer. The interference pattern formed due to the superposition of these two light beams is recorded at the camera plane (**Figure 5**). Although off-axis-referenced calibration methods require complicated experimental set-ups but these methods are very useful for precise measurements of spatial uniformity in SLM.

3.1.2 Polarization-sensitive digital holography

SLMs are polarization-sensitive devices and therefore the polarization properties of light are expected to be modulated after interacting with an LC-SLM. It is evident that the polarization properties [diattenuation, degree of polarization (DOP), retardance, etc.] cannot be determined using conventional interferometric methods.

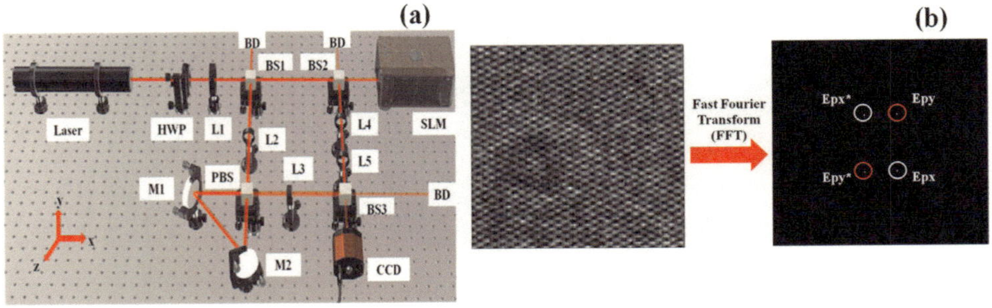

Figure 6.
(a) Schematic of polarization-sensitive digital holography for reflective LC-SLM (LC-R 720, holoeye), and (b) recorded fringe pattern and corresponding Fourier spectrum. [HWP: Half-wave plate, L: Lens, BS: Beam splitter, PBS: Polarization beam splitter, CCD: Charge coupled device, BD: Beam dump, M: Mirrors] (adapted from [32]).

Therefore, polarization-sensitive characterization methods are developed to account for the polarimetric calibration of SLM. In this context, spatially resolved Jones matrix imaging has the potential to extract the phase information of LC-SLM from the complex electric field of light.

The modern SLM calibration techniques are typically based on Polarimetric methods accompanied by digital holography [45, 46]. Moreover, the consolidation of digital holography with polarimetry has paved the convenient approach to SLM calibration. **Figure 6** illustrates the polarization-sensitive digital holography scheme for Jones matrix imaging of a reflective type LC-SLM (Holoeye, LC-R 720). The proposed experimental technique is equipped with a combination of Mach-Zehnder interferometer and Sagnac interferometer, which provide angular multiplexing to recover orthogonal polarization components. Moreover, the triangular Sagnac interferometer in the reference arm of the experimental setup is capable to tune the carrier frequency of reference orthogonal components according to object beam requirements.

Using the proposed experimental set-up, different interference patterns have been recorded at various grayscales of LC-SLM in two shots for input states of polarization (SOPs) +45° and −45° respectively. The phase shift can be measured by observing fringe-shift for the recorded interference patterns. For instance, **Figure 7** illustrates the recorded fringe shifts for gray values 0, 180, and 255 for input SOPs +45° (a-c) and −45° (d–f) respectively. The observed fringe shift with respect to the gray values indicates the SOP modulating characteristics of the LC-SLM.

Mathematically, the interference pattern for input +45° polarized ($I_p(x,y)$) and −45° polarized ($I_m(x,y)$) input light beam can be given as

$$I_p(x,y) = \left|R_{px}(x,\ y) + R_{py}(x,\ y) + JE_p(x,\ y)\right|^2 \tag{7}$$

$$I_m(x,y) = \left|R_{mx}(x,\ y) + R_{my}(x,\ y) + JE_m(x,\ y)\right|^2 \tag{8}$$

Here, $E_p(x,y)$,$E_m(x,y)$ denotes the modulated electric field, $R_{px}(x,y)$,$R_{py}(x,y)$ and $R_{mx}(x,y)$,$R_{my}(x,y)$ denotes the reference beams for x and y components for input light field corresponding to +45° and −45° polarized input light beam respectively. J is the Jones matrix of the LC-SLM.

Jones matrix calculus establishes a connection between the electric field components of emerging light from the object (E'_x,E'_y) and electric field components of input light beam (E_x,E_y) as

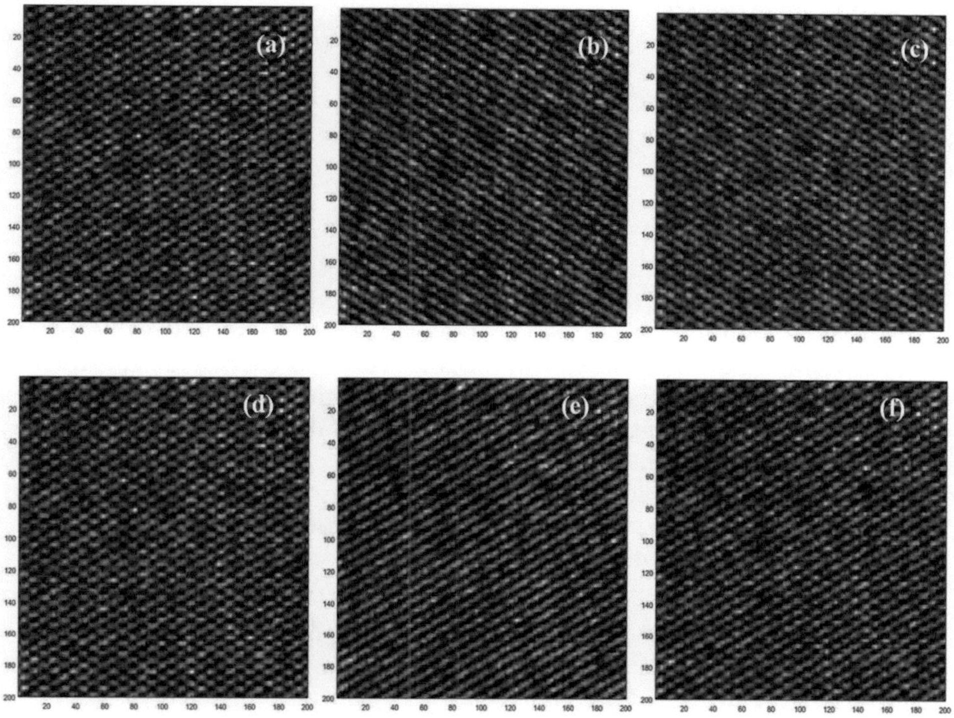

Figure 7.
Recorded interference patterns for +45° polarized incident light (a-c), for −45° polarized incident light (d–f) at grayscales 5 (a, d), 180 (b, e), and 255 (c, f) of LC-SLM respectively (adapted from [32]).

$$\begin{pmatrix} E'_x \\ E'_y \end{pmatrix} = \begin{pmatrix} J_{xx} & J_{xy} \\ J_{yx} & J_{yy} \end{pmatrix} \begin{pmatrix} E_x \\ E_y \end{pmatrix} \tag{9}$$

Here, $J = \begin{pmatrix} J_{xx} & J_{xy} \\ J_{yx} & J_{yy} \end{pmatrix}$ is the Jones matrix of the object.

Taking reference calibration of SLM display into account, the electric field components of the modulated light can be defined as [27]

$$E'_{calibrated} = \frac{E'modulated}{E'reference} \tag{10}$$

Here, $E'_{modulated}$ and $E'_{reference}$ is the electric field component corresponding to modulating part and reference part (grayscale '0') of the SLM display respectively.

For an illuminating object with +45° and −45° polarized input light, Eq. (9) turns into

$$\begin{pmatrix} E'_{px} \\ E'_{py} \end{pmatrix} = \begin{pmatrix} J_{xx} & J_{xy} \\ J_{yx} & J_{yy} \end{pmatrix} \begin{pmatrix} 1 \\ 1 \end{pmatrix} \tag{11}$$

$$\begin{pmatrix} E'_{mx} \\ E'_{my} \end{pmatrix} = \begin{pmatrix} J_{xx} & J_{xy} \\ J_{yx} & J_{yy} \end{pmatrix} \begin{pmatrix} 1 \\ -1 \end{pmatrix} \tag{12}$$

The Jones matrix of the object from the modulated field components can be retrieved as

$$
\left.\begin{aligned}
J_{xx} &= \frac{1}{2}\left[E_{px} + E_{mx}\right] \\
J_{xy} &= \frac{1}{2}\left[E_{px} - E_{mx}\right] \\
J_{yx} &= \frac{1}{2}\left[E_{py} + E_{my}\right] \\
J_{yy} &= \frac{1}{2}\left[E_{py} - E_{my}\right]
\end{aligned}\right\} \tag{13}
$$

The measured Jones matrix of the LC-SLM with respect to the gray values of the LC-SLM is depicted in **Figure 8**. The Jones matrix elements contain the vital information of the SLM, including the amplitude, phase, and anisotropic properties (birefringence, dichroism, retardance, etc.), which can be easily decomposed using the Jones decomposition methods [32, 47]. However, Jones matrix cannot provide the polarization information of the medium, containing depolarization. In this case, other polarimetric methods (Mueller matrix imaging, etc.) can be used for polarimetric calibration of LC-SLM [48, 49].

3.1.3 Determination of spatial anomalies in LC-SLM using Mueller-Stokes polarimetry

The inner structure of SLMs is comprised of a systematic alignment of Liquid crystal (LC) cells in specific patterns. It permits SLM to utilize the ability of LC cells to align themselves with respect to applied voltage i.e. gray values of SLM. Technically, the required modulation characteristics of light originate from the relative rotation of liquid-crystal (LC) cells about their optical axis within the inner structure of SLM. In general, SLMs are considered as homogenous in structure but they are not perfectly homogeneous in practice due to manufacturing discrepancies [10, 50]. The curvature of cover glass and silicon backplane polishing of SLM may also cause a significant non-uniformity in SLM display. This non-homogeneity in the SLM display gives rise to a

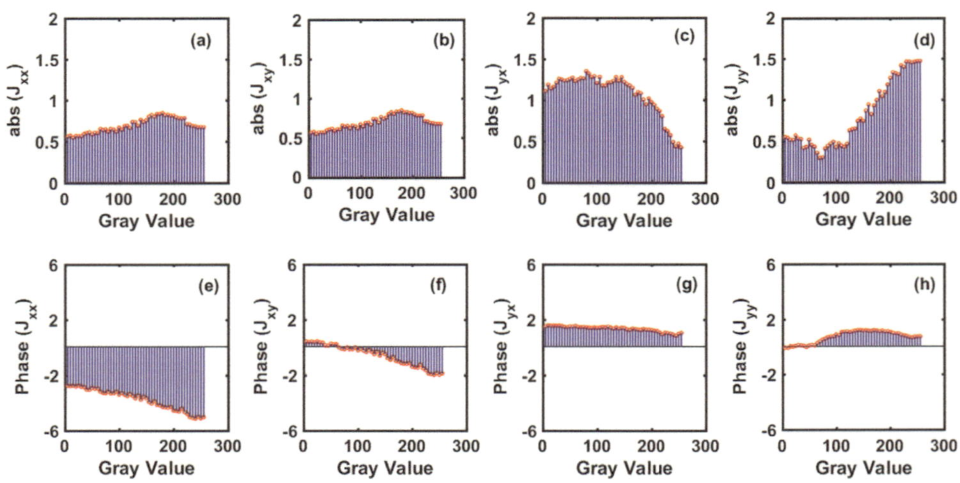

Figure 8.
Jones matrix components as a function of gray scales of the SLM. Amplitude (a–d) and phase (e–h) of Jones matrix components respectively (adapted from [32]).

higher light modulation capability at the edges than at the centre of the SLM display [51]. In addition, limited fill factor, the existence of non-active area, optical efficiency, and refresh rate of SLM display are other crucial parameters that are responsible for the experimentally observable non-uniformity of SLM [52]. It is noteworthy that this spatial non-uniformity may yield distorted wavefront and irregular modulation characteristics at different spatial parts of the SLM display. With a limited image array portion of the SLM display, it can be possible that these factors can generate discrepancies in optimized light modulation characteristics of SLM, especially for the applications where a broader laser beam is required.

The spatial anomalies of LC-SLM can be calibrated using Mueller matrix imaging. **Figure 9** illustrates the schematic of the improvised Mueller matrix imaging polarimeter (MMIP), which is used for rapid Mueller matrix measurement of a reflective type LC-SLM (Holoeye, LC-R720) with respect to its grayscales. A partially polarized light beam of wavelength 532 nm is emerging from a green-diode laser is spatially filtered (SF) and collimated by a collimating lens of the focal length of 20 cm. It is then allowed to pass through an MMIP, which consists of the polarization components i.e. linear polarizers (P_1 and P_2), quarter-wave plates (Q_1 and Q_2), a Half Wave Plate (HWP), and a Beam Splitter (BS). Theoretically, MMIP has two arms i.e. Polarization State Generator (PSG) and Polarization State Analyzer (PSA). Four SOPs (H, V, P, R) are generated in both PSG and PSA arms with the help of polarization components, and corresponding 16 intensity images (different combinations of SOPs) are recorded in a CCD camera (Procilica GT 2750, 2752 $\times$ 2200 pixels and pixel size of 4.54 μm), placed at the image plane. A total of 288 intensity images have been recorded for 18 gray values of SLM at the interval of 15 (0, 15, 30, up to 255) respectively at standard room temperature. Polarization properties (diattenuation, retardance, polarizance, and depolarization) have been retrieved from the measured Mueller matrices using the polar decomposition method of Mueller matrices [53].

To allocate spatial anomalies in polarization modulation produced due to SLM display, we have selected five spatial coordinates (pixel values) from different parts i.e. central part, right edge, left edge, upper part, and lower part of the light beam over SLM display (**Figure 9b**). Spatially addressable polarization properties are plotted for different grayscales of the LC-SLM and are represented in **Figure 10**. Interestingly, the determined polarization properties exhibit spatial fluctuations (particularly at the edges of the SLM display) and thus indicate the spatial non-uniformity in the LC-SLM.

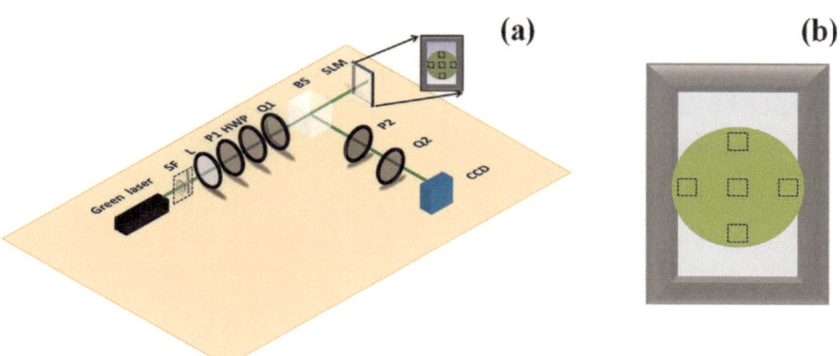

Figure 9.
(a) Schematic of Mueller matrix imaging Polarimeter (MMIP) for Mueller matrix imaging of reflective LC-SLM (LC-R 720, holoeye), and (b) selection of spatial locations of SLM display for measurements of spatial anomalies (adapted from [12]).

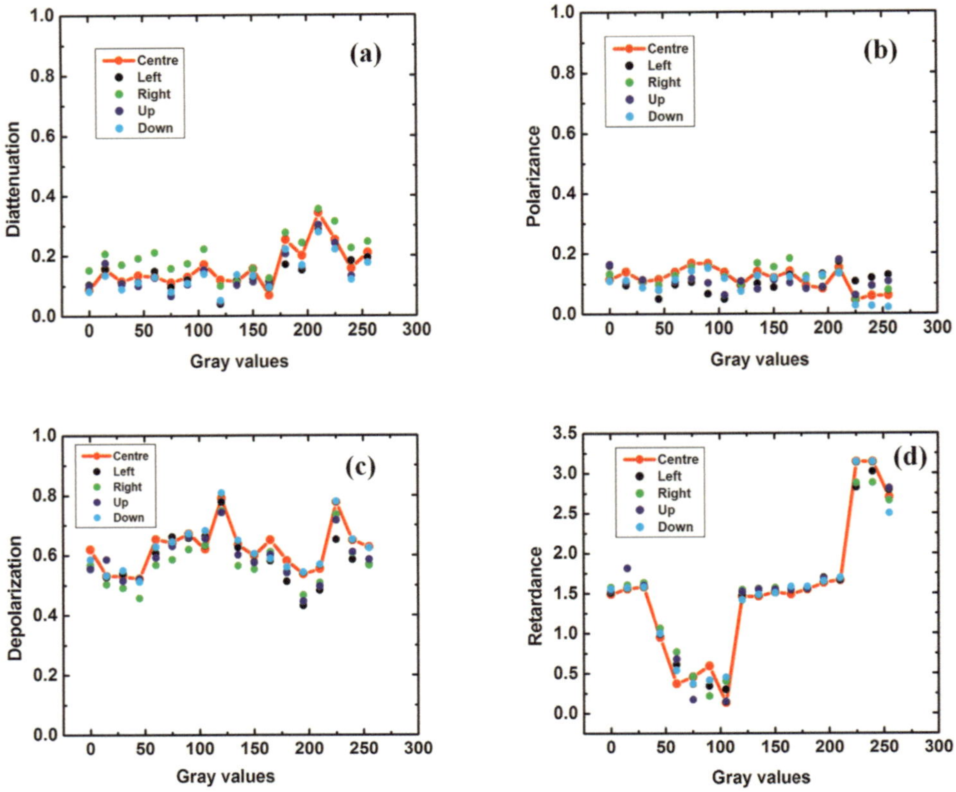

Figure 10.
Polarization characteristics [diattenuation (a), polarizance (b), depolarization (c), and retardance (d)] of the LC-SLM as function of gray values of SLM at different spatial locations (adapted from [12]).

4. Applications of LC-SLMs in digital holography

4.1 Phase-shifting digital holography

SLMs are widely employed in phase-shifting digital holography (PSDH), where multiple digital holograms are imported with known phase shifts in the reference beam. In such cases, LC-SLM offers a convenient means to import CGHs with optimized phase shifts for the experimental implementation of PSDH.

The reference beams with four pre-determined phase-shifts $\left(0, \frac{\pi}{2}, \ \pi, and \ \frac{3\pi}{2}\right)$ can be mathematically represented as

$$H_0 = |R|^2 + |O|^2 + 2|R||O|\cos(\varphi_O - \varphi_R) \tag{14}$$

$$H_{\pi/2} = |R|^2 + |O|^2 - 2|R||O|\sin(\varphi_O - \varphi_R) \tag{15}$$

$$H_\pi = |R|^2 + |O|^2 - 2|R||O|\cos(\varphi_O - \varphi_R) \tag{16}$$

$$H_{3\pi/2} = |R|^2 + |O|^2 + 2|R||O|\sin(\varphi_O - \varphi_R) \tag{17}$$

Here, φ_O and φ_R denote the 2-D phase functions corresponding to the object and reference beams respectively.

For a known reference phase $[\varphi_R(x, y)]$, the object beam phase can be given as

$$\varphi_O(x, y) = \varphi_R(x, y) + arctan\left(\frac{H_{3\pi/2} - H_{\pi/2}}{H_0 - H_\pi}\right) \tag{18}$$

Phase retrieval of a hologram (vortex of topological charge 1) using the PSDH technique is represented in **Figure 11**. Four digital phase shifts have been provided to a hologram computationally and the corresponding phase is retrieved from the phase-shifted holograms.

4.2 Polarization holography

Polarization holography can be interpreted as an advanced version of conventional digital holography, which enables to simultaneous recording of the SOPs of light beam along with its amplitude, and phase. Polarization holography is an emerging domain in digital holography leveraging several interesting applications such as holographic storage technology [16], multichannel polarization multiplexing [54], vector beam generation [19, 27, 55, 56], and optical functional devices [57], etc.

In polarization holography, a periodic change in the polarization state of light is required in terms of polarization-sensitive holograms. These polarization-sensitive holograms can be obtained with the help of SLMs using their pixel-dependent phase-shift. In this context, SLMs can be employed as key components for the experimental implementation of polarization holography. In a recent work, Nabadda et al. have demonstrated a phase-shifting interferometry (PSI) to evaluate the reconstruction of complex-valued holograms displayed onto a phase-only SLM (Hamamatsu X10,468-01, with 800 × 600 square pixels, 20 μ pixel pitch and 98% fill factor) and

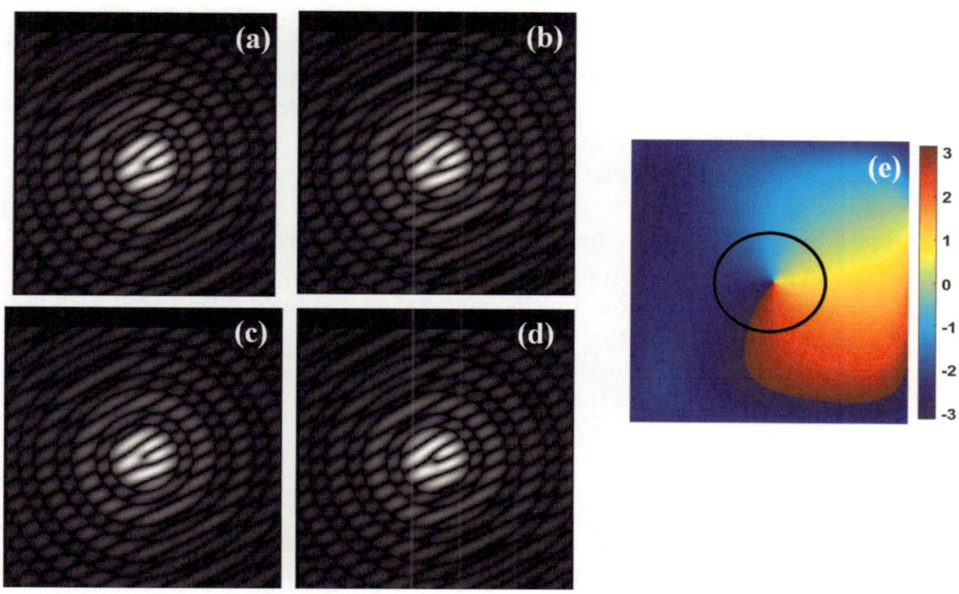

Figure 11.
Phase reconstruction of hologram (vortex, topological charge 1) using PSDH technique. (a–d) Holograms with phase shift 0, π/2, π, and 3π/2 respectively, and (e) reconstructed phase of a vortex (charge 1). [PSDH: Polarization sensitive digital holography].

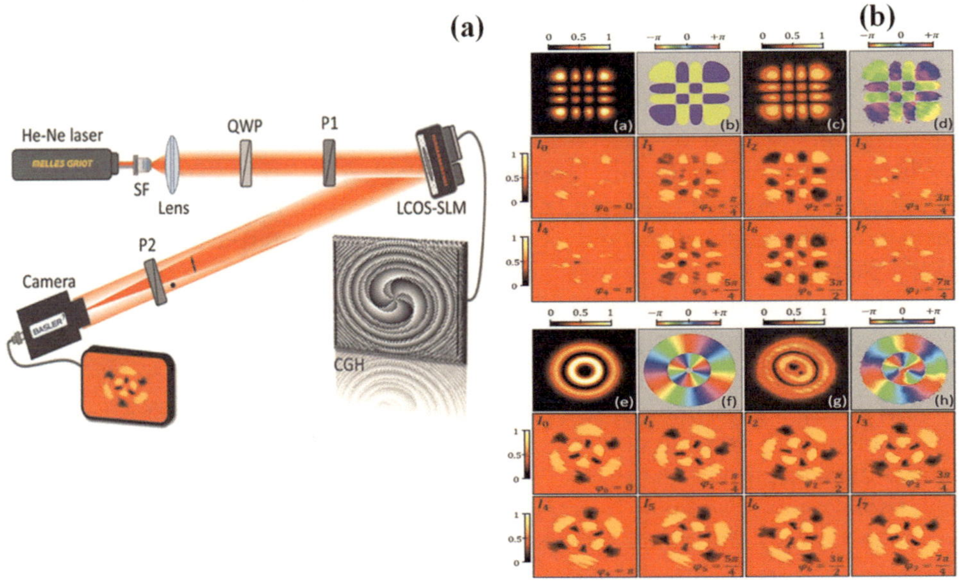

Figure 12.
(a) Scheme of the optical system. SF: Spatial filter; QWP: Quarter-wave plate;P: Linear polarizer; CGH: Displayed phase-only computer-generated hologram, and (b) realization of a HG_{33} *(A–D) and* LG_{13} *(E–H) beam. Theoretical intensity (A, E) and phase (B, F) distributions. Experimental intensity (C, G) and phase (D, H) distributions. The phase distribution of each beam is retrieved from the interferograms shown below each case with eight consecutive phase shifts of step* $\Delta\varphi = \pi/4$ *(adapted from [58]).*

corresponding experimental scheme is presented in **Figure 12a**. **Figure 12b** shows the experimental observations for the encoded Hermite-Gauss (HG) and a Laguerre-Gauss (LG) beam. It was observed that the theoretical intensity and phase distributions exhibit its characteristic double ring and spiral pattern. In this case, the experimental interferograms show three bright and three dark lobes inside each of the two intensity rings. The bright and dark lobes are opposite from one ring to the other, as corresponds to the π-phase shift between them, and they progressively rotate as phase-shifts are added in successive interferograms [58] .

In another development under polarization holography, vectorial wavefront holography is proposed using the wavefront shaping of a structured vector beam [59]. The study shows that the phase hologram can be used to tailor the polarization interference of a vector beam in momentum space, resulting in arbitrary polarization states. The designed polarization-multiplexing holograms are implemented with the help of LC-SLM and an azimuthal polarization converter was used to convert an incident linear polarization into azimuthally polarized vector beams (APVB).

5. Conclusion

LC-SLMs are essential optical components for the experimental implementation of modern optical imaging techniques, viz., computational imaging [4, 60], ghost imaging [61, 62], structured light-based imaging [63, 64], orbital angular momentum manipulation [65, 66], etc. The complicated structure and subtle working principle require the SLM calibration prior to its utilization for various imaging applications. SLM calibration methods can be broadly divided into two categories, i.e., interferometric techniques, and polarimetric techniques. In interferometric calibration techniques, phase modulation characteristics of LC-SLM can be calibrated by quantifying the relative phase-shift of the recorded interference patterns, which can be formed using traditional interferometry. In such methods, reference beams can be either considered off-axis (out-referenced) or generated from the LC-SLM itself (self-referenced). Self-referenced interferometric methods are easy to implement as they offer compactness and stability to the experimental set-up meanwhile off-axis referenced interferometric methods require tedious experimental set-ups as compared to the earlier case but offer wide applicability in the form of spatially resolved modulation characteristics, pixel-crosstalk, etc. On the other hand, SLMs are polarization-sensitive devices and hence polarization of light plays a vital role in SLM operation. In this context, polarization modulation characteristics of SLM can be accounted for using the polarimetric calibration methods. The polarimetric methods revolve around two polarimetric techniques, i.e. Jones matrix imaging and Mueller matrix imaging. In general, Jones matrix imaging is suitable for carryout polarimetric measurements of LC-SLM due to their homogeneous structure. However, SLMs are not perfectly homogenous and hence can exhibit depolarization as well [12, 49]. In such a scenario, Mueller matrix imaging can be employed to investigate the polarimetric characteristics of LC-SLM.

In summary, a comprehensive literature study of LC-SLMs (including their construction, working principle, calibration methods, and applications) has been revisited in this chapter. We have reviewed the detailed polarimetric characteristics of LC-SLM (Holoeye, LC-R 720) concerning its grayscales using polarization-sensitive digital holographic technique followed by recent investigations on the spatial anomalies in the LC-SLM using Mueller-Stokes polarimetry. Further, a few crucial applications of LC-SLMs have been quickly discussed within the framework of polarization holography. The present work is an attempt to consolidate the major SLM calibration methods and is expected to serve as a manual guide for LC-SLMs along with their applications in advanced imaging domains.

Acknowledgements

Vipin Tiwari acknowledges support from DST-INSPIRE (IF-170861).

Author details

Vipin Tiwari[1] and Nandan S. Bisht[1,2]*

1 Applied Optics and Spectroscopy Laboratory, Department of Physics, Kumaun University, SSJ Campus, Almora, Uttarakhand, India

2 Department of Physics, Soban Singh Jeena (SSJ) University, Almora, Uttarakhand, India

*Address all correspondence to: bisht.nandan@gmail.com

References

[1] Lin M, Nitta K, Matoba O, Awatsuji Y. Parallel phase-shifting digital holography with adaptive function using phase-mode spatial light modulator. Applied Optics. 2012;**51**: 2633-2637. DOI: 10.1364/AO.51.002633

[2] Rosen J, Kelner R, Kashter Y. Incoherent digital holography with phase-only spatial light modulators. Journal of Micro/Nanolithography, MEMS, MOEMS. 2015;**51**(14):2633-2637. DOI: 10.1117/1.jmm.14.4.041307

[3] Mu Q, Cao Z, Hu L, Li D, Xuan L. Adaptive optics imaging system based on a high-resolution liquid crystal on silicon device. Optics Express. 2006;**14**:8013. DOI: 10.1364/oe.14.008013

[4] King SV, Doblas A, Patwary N, Saavedra G, Martínez-Corral M, Preza C. Spatial light modulator phase mask implementation of wavefront encoded 3D computational-optical microscopy. Applied Optics. 2015;**54**:8587. DOI: 10.1364/ao.54.008587

[5] Klauss A, Conrad F, Hille C. Binary phase masks for easy system alignment and basic aberration sensing with spatial light modulators in STED microscopy. Scientific Reports. 2017;7:1-11. DOI: 10.1038/s41598-017-15967-5

[6] Beversluis MR, Novotny L, Stranick SJ. Programmable vector point-spread function engineering. Optics Express. 2006;**14**:2650. DOI: 10.1364/ oe.14.002650

[7] Kozacki T. Holographic display with tilted spatial light modulator. Applied Optics. 2011;**50**:3579-3588. DOI: 10.1364/ AO.50.003579

[8] Dunayevsky J, Marom DM. MEMS spatial light modulator for phase and amplitude modulation of spectrally dispersed light. Journal of Microelectromechanical Systems. 2013; **22**:1213-1221. DOI: 10.1109/JMEMS. 2013.2262600

[9] Lazarev G, Hermerschmidt A, Krüger S, Osten S. LCOS spatial light modulators: Trends and Applications. In: Opt. Imaging Metrol. Adv. Technol. Wiley Online Library. 2012. pp. 1-29. DOI: 10.1002/9783527648443.ch1. ISBN: 9783527648443

[10] Otón J, Millán MS, Ambs P, Pérez-Cabré E. Advances in LCoS SLM characterization for improved optical performance in image processing. In: Opt. Digit. Image Process. SPIE Photonics Europe, Strasbourg, France: SPIE digital library. 2008. p. 70001V. DOI: 10.1117/12.783587

[11] Li R, Cao L. Progress in phase calibration for liquid crystal spatial light modulators. Applied Sciences. 2019;**9**:1-13. DOI: 10.3390/app9102012

[12] Tiwari V, Pandey Y, Bisht NS. Spatially addressable polarimetric calibration of reflective-type spatial light modulator using Mueller–Stokes polarimetry. Frontiers of Physics. 2021;**9** (709192):1-9. DOI: 10.3389/ fphy.2021.709192

[13] Bilgeri LM, Salazar Bloise F, Lu M, Wang S, Jakobi M, Koch AW. Intensity distortions due to phase-only spatial light modulation: Characterization for applications in electronic speckle-pattern interferometry. The Review of Scientific Instruments. 2018;**89**(8). DOI: 10.1063/ 1.5029914

[14] Xia J, Chang C, Chen Z, Zhu Z, Zeng T, Liang PY, et al.

Pixel-addressable phase calibration of spatial light modulators: A common-path phase-shifting interferometric microscopy approach. Journal of Optics (United Kingdom). 2017;**19**(12):125701. DOI: 10.1088/2040-8986/aa8fbc

[15] Hariharan P. Basics of Holography. Cambridge University Press. 2002. DOI: 10.1017/cbo9780511755569. ISBN: 9780511755569

[16] Hong YF, Zang JL, Liu Y, Fan FL, Wu AA, Shao L, et al. Review and prospect of polarization holography. Chinese Optics. 2017;**10**:588-602. DOI: 10.3788/CO.20171005.0588

[17] Zhai Y, Cao L, Liu Y, Tan X. A review of polarization-sensitive materials for polarization holography. Materials (Basel). 2020;**13**:1-15. DOI: 10.3390/ma13235562

[18] Kohler C, Haist T, Schwab X, Osten W. Hologram optimization for SLM-based reconstruction with regard to polarization effects. Optics Express. 2008; **16**:14853. DOI: 10.1364/oe.16.014853

[19] Barada D, Ochiai T, Fukuda T, Kawata S, Kuroda K, Yatagai T. Dual-channel polarization holography: A technique for recording two complex amplitude components of a vector wave. Optics Letters. 2012;**37**(21):4528-4530. DOI: 10.1364/ol.37.004528

[20] Wang J, Tan X, Qi P, Wu C, Huang L, Xu X, et al. Linear polarization holography. Opto-Electronic Science. 2022;**1**:210009-210009. DOI: 10.29026/oes.2022.210009

[21] Singh RK, Naik DN, Itou H, Miyamoto Y, Takeda M. Stokes holography. Optics Letters. 2012;**37**(5): 966-968. DOI: 10.1364/ol.37.000966

[22] Rosen J, Anand V, Rai MR, Mukherjee S, Bulbul A. Review of 3D imaging by coded aperture correlation holography (COACH). Applied Sciences. 2019;**9**:1-25. DOI: 10.3390/app9030605

[23] Yu X, Wang K, Xiao J, Li X, Sun Y, Chen H. Recording point spread functions by wavefront modulation for interferenceless coded aperture correlation holography. Optics Letters. 2022;**47**:409. DOI: 10.1364/ol.439615

[24] Rosen J, Hai N, Rai MR. Recent progress in digital holography with dynamic diffractive phase apertures [invited]. Applied Optics. 2022;**61**:B171. DOI: 10.1364/ao.442364

[25] Sarkar T, Parvin R, Brundavanam MM, Singh RK. Higher-order Stokes-parameter correlation to restore the twisted wave front propagating through a scattering medium. Physical Review A. 2021;**104**. DOI: 10.1103/PhysRevA.104.013525

[26] Parvin R, Roy A, Varshney SK, Brundavanam MM. Modulation of coherence-polarization property of speckles using a birefringent scatterer. Applied Optics. 2021;**60**:7259. DOI: 10.1364/ao.430164

[27] Tiwari V, Gautam SK, Naik DN, Singh RK, Bisht NS. Characterization of a spatial light modulator using polarization-sensitive digital holography. Applied Optics. 2020;**59**(7):2024-2030. DOI: 10.1364/ao.380572

[28] Kim Y, Jeong J, Jang J, Kim MW, Park Y. Polarization holographic microscopy for extracting spatio-temporally resolved Jones matrix. Optics Express. 2012;**20**:9948. DOI: 10.1364/oe.20.009948

[29] Blanche P-AJ. Field Guide to Holography. F. Guid. to Hologr. 2014. DOI: 10.1117/3.1002325

[30] Carlescu I. Introductory chapter: Nematic liquid crystals. Liquid Crystals and Display Technology. 2020. DOI: 10.5772/INTECHOPEN.92726

[31] Goodby JW, Toyne KJ, Hird M, Styring P, Lewis RA, Beer A, et al. Chiral liquid crystals for ferroelectric, electroclinic, and antiferroelectric displays and photonic devices. In: Liq. Cryst. Mater. Devices, Flat Panel Displays. Electronic Imaging. San Jose, CA, United States: SPIE Digital Library; 2000. DOI: 10.1117/12.379967

[32] Tiwari V, Bisht NS. Combined Jones–Stokes polarimetry and its decomposition into associated anisotropic characteristics of spatial light modulator. Photonics. 2022;**9**:195. DOI: 10.3390/PHOTONICS9030195

[33] Davis JA, Moreno I, Tsai P. Polarization eigenstates for twisted-nematic liquid-crystal displays. Applied Optics. 1998;**37**:937. DOI: 10.1364/ao.37.000937

[34] Clemente P, Durán V, Martínez-León L, Climent V, Tajahuerce E, Lancis J. Use of polar decomposition of Mueller matrices for optimizing the phase response of a liquid-crystal-on-silicon display. Optics Express. 2008;**16**: 1965. DOI: 10.1364/oe.16.001965

[35] Goldstein DH. Polarized Light. Third ed2017. DOI: 10.1201/b10436

[36] Dainty JC. Optical waves in crystals. Optics and Laser Technology. 1985;**17**: 217-218. DOI: 10.1016/0030-3992(85) 90094-5

[37] Dai Y, Antonello J, Booth MJ. Calibration of a phase-only spatial light modulator for both phase and retardance modulation. Optics Express. 2019; **27**(13):17912-17926. DOI: 10.1364/oe.27.017912

[38] Zhao Z, Xiao Z, Zhuang Y, Zhang H, Zhao H. An interferometric method for local phase modulation calibration of LC-SLM using self-generated phase grating. The Review of Scientific Instruments. 2018;**89**. DOI: 10.1063/1.5031938

[39] DeMars LA, Mikuła-Zdańkowska M, Falaggis K, Porras-Aguilar R. Single-shot phase calibration of a spatial light modulator using geometric phase interferometry. Applied Optics. 2020; **59**(13):D125-D130. DOI: 10.1364/ao.383610

[40] Engström D, Milewski G, Bengtsson J, Galt S. Diffraction-based determination of the phase modulation for general spatial light modulators. Applied Optics. 2006;**45**(28):7195-7204. DOI: 10.1364/AO.45.007195

[41] Villalobos-Mendoza B, Granados-Agustín FS, Aguirre-Aguirre D, Cornejo-Rodríguez A. Phase shifting interferometry using a spatial light modulator to measure optical thin films. Applied Optics. 2015;**54**(26):7997-8003. DOI: 10.1364/ao.54.007997

[42] Villalobos-Mendoza B, Granados-Agustín FS, Aguirre-Aguirre D, Cornejo-Rodríguez A. Obtaining the curve "phase shift vs gray level" of a spatial light modulator Holoeye LC2012. Journal of Physics Conference Series. 2015. DOI: 10.1088/1742-6596/605/1/012016

[43] Fuentes JLM, Fernández EJ, Prieto PM, Artal P. Interferometric method for phase calibration in liquid crystal spatial light modulators using a self-generated diffraction-grating. Optics Express. 2016;**24**:14159. DOI: 10.1364/oe.24.014159

[44] Zhao Z. High precision optical Wavefront generation using liquid crystal spatial light modulator (LC-SLM). In: Liq. Cryst. [Working Title].

Intech-Open. 2021. DOI: 10.5772/intechopen.100379

[45] Zeng Z, Li Z, Fang F, Zhang X. Phase compensation of the non-uniformity of the liquid crystal on silicon spatial light modulator at pixel level. Sensors (Switzerland). 2021;**21**:1-15. DOI: 10.3390/s21030967

[46] Engström D, Persson M, Bengtsson J, Goksör M. Calibration of spatial light modulators suffering from spatially varying phase response. Optics Express. 2013;**21**(13):16086-16103. DOI: 10.1364/oe.21.016086

[47] Nee S-MF. Decomposition of Jones and Mueller matrices in terms of four basic polarization responses. Journal of the Optical Society of America. A. 2014; **31**(11):2518-2528. DOI: 10.1364/josaa.31.002518

[48] Tiwari V, Bisht NS. Statistical interpretation of Mueller matrix images of spatial light modulator. In: 2019 Work. Recent Adv. Photonics, WRAP. 2019 Workshop on Recent Advances in Photonics (WRAP). Guwahati, India: IEEE Xplore; 2019. p. 2019. DOI: 10.1109/WRAP47485.2019.9013707

[49] Dev K, Asundi A. Mueller-Stokes polarimetric characterization of transmissive liquid crystal spatial light modulator. Optics and Lasers in Engineering. 2012;**50**(4):599-607. DOI: 10.1016/j.optlaseng.2011.10.004

[50] Marco D, Vargas A, Sánchez-López d M, Moreno I. Measuring the spatial deformation of a liquid-crystal on silicon display with a self-interference effect. Optics Letters. 2020;**45**:4480. DOI: 10.1364/ol.396105

[51] Lingel C, Haist T, Osten W. Optimizing the diffraction efficiency of SLM-based holography with respect to the fringing field effect. Applied Optics. 2013;**52**:6877-6883. DOI: 10.1364/AO.52.006877

[52] Veronica C, Elena A, Katkovnik V, Shevkunov I, Egiazarian K. Pixel-Wise Calibration of the Spatial Light Modulator. Frontiers in Optics / Laser Science-2020, Optica Publishing Group, 2020, paper JTu1B.31. 2021. pp. 1-2. DOI: 10.1364/fio.2020.jtu1b.31

[53] Lu S-Y, Chipman RA. Interpretation of Mueller matrices based on polar decomposition. Journal of the Optical Society of America. A. 1996;**13**(5): 1106-1113. DOI: 10.1364/josaa.13.001106

[54] Zhou H, Sain B, Wang Y, Schlickriede C, Zhao R, Zhang X, et al. Polarization-encrypted orbital angular momentum multiplexed Metasurface holography. ACS Nano. 2020;**14**: 5553-5559. DOI: 10.1021/acsnano.9b09814

[55] Wang T, Fu S, Zhang S, Gao C, He F. A Sagnac-like interferometer for the generation of vector beams. Applied Physics B: Lasers and Optics. 2016;**122**. DOI: 10.1007/s00340-016-6507-8

[56] Lohmann AW. Reconstruction of Vectorial Wavefronts. Applied Optics. 1965;**4**(12):1667-1668. DOI: 10.1364/ao.4.001667

[57] Deng ZL, Wang ZQ, Li FJ, Hu MX, Li X. Multi-freedom metasurface empowered vectorial holography. Nano. 2022;**11**:1725-1739. DOI: 10.1515/nanoph-2021-0662

[58] Nabadda E, García-Martínez P, Sánchez-López d M, Moreno I. Phase-shifting common-path polarization self-interferometry for evaluating the reconstruction of holograms displayed

on a phase-only display. Frontiers of Physics. 2022;**0**:627. DOI: 10.3389/FPHY.2022.920111

[59] Ren H. Vectorial wavefront holography based on a polarisation-insensitive hologram. Journal of Optics. 2022;**24**:64008. DOI: 10.1088/2040-8986/AC6882

[60] Tiwari V, Bisht NS. Extended depth of field of diffraction limited imaging system using spatial light modulator based intensity compensated polarization coded aperture. 2022. DOI: 10.48550/arxiv.2204.04005. (preprint)

[61] Manisha V, Tiwari NS, Bisht B, Das RKS. Ghost diffraction: A spatial statistical approach2022. DOI: 10.48550/arxiv.2204.03860

[62] Wang Y, Liu Y, Suo J, Situ G, Qiao C, Dai Q. High speed computational ghost imaging via spatial sweeping. Scientific Reports. 2017;**7**:1-8. DOI: 10.1038/srep45325

[63] Otte E, Tekce K, Denz C. Spatial multiplexing for tailored fully-structured light. Journal of Optics (United Kingdom). 2018;**20**. DOI: 10.1088/2040-8986/aadef3

[64] Li R, Cao L. Complex wavefront sensing based on alternative structured phase modulation. Applied Optics. 2021;**60**:A48. DOI: 10.1364/ao.405630

[65] Andersen JM, Alperin SN, Voitiv AA, Holtzmann WG, Gopinath JT, Siemens ME. Characterizing vortex beams from a spatial light modulator with collinear phase-shifting holography. Applied Optics. 2019;**58**:404. DOI: 10.1364/ao.58.000404

[66] Wan C, Chen J, Zhan Q. Compact and high-resolution optical orbital angular momentum sorter. APL Photonics. 2017;**2**:1-6. DOI: 10.1063/1.4974824

Chapter 2

Ghost Diffraction Holography: A Correlation Assisted Quantitative Tool for Complex Field Imaging and Characterization

Vinu Raveendran Pillai Vasantha Kumari, Ziyang Chen, Rakesh Kumar Singh and Jixiong Pu

Abstract

The fascinating domain of ghost imaging has been a subject of interest in the fundamental and applied research for the last two decades with its promising applications in various imaging and characterization scenarios. In this chapter, we discuss the recently developed ghost diffraction holography (GDH) system with due emphasis on the capability of quantitative complex-field imaging in the ghost framework. The development of the unconventional correlation-assisted GDH technique by adopting the holography concept in ghost diffraction scheme is described, and the quantitative phase imaging capability is demonstrated in the microscopy. In addition, the technique exploits the spatial statistics of time-frozen recorded speckle intensity with snapshot detection in ghost framework, which could broaden the applications of the developed microscopy to real-time imaging of two- and three-dimensional biological samples with high resolution. Furthermore, we discuss demonstrated applications of the technique in the imaging various spatially varying complex-valued macroscopic and microscopic samples and the potential application of the technique in the recovery and characterization of orbital angular momentum modes encoded in spatially incoherent speckle field.

Keywords: ghost imaging, ghost diffraction, digital holography, digital holographic microscopy, speckles, quantitative phase imaging, orbital angular momentum

1. Introduction

Optical imaging relying on ghost approaches has been emerged as a novel imaging modality with the superior capability of high-resolution complex-valued image restoration that the traditional approaches cannot match even in the imaging scenarios with chaotic or turbulent environment [1–4]. Last two decades witnessed the theoretical and applied advancements in ghost imaging (GI) modalities from quantum to classical to computational approaches by utilizing the cross-correlation of intensity measurements in spatial, temporal or spectroscopy domains [5–11]. The counterintuitive imaging modality of ghost schemes acquire the high-resolution ghost image by exploiting the cross-correlation of intensities of object interacted scattered field detected by a single pixel detector and a field that never interacted with the object detected by a multi pixel detector. Since the first observance of ghost effect by Belenskii and Klyshko in 1994 [12] and later the experimental demonstration by Pittmann in 1995 [13] with entangled light from a spontaneous parametric down conversion source, numerous advancements have been reported by making use of various light sources and different correlation protocols. Besides the initial discussion over the quantum versus classical nature of the GI [14], the demonstration of pseudo-thermal light source in place of the parametric down conversion source opens the door for the development of more applied imaging techniques [15, 16]. Later the progress in the computational techniques and the introduction of computational GI [9] paves the path for the development of several techniques exploiting the advanced algorithms such as compressive sensing [17], deep learning [18, 19], differential measurement [20, 21], sequential deviation [22], etc. In recent years, researchers exploited the fascinating features of ghost schemes in other realms of applied imaging with cutting edge technological demonstrations of ghost tomography [23], ghost spectroscopy [8], X-ray ghost imaging [24, 25], temporal GI [7], electron GI [26], GI with atoms [27], etc.

Alongside the significant developments of GI techniques in high-resolution image restoration, the classical framework of GI schemes encounters the challenges of pure phase object imaging, which significantly limits the applications in biomedical imaging. Owing to the phase imaging challenge in GI framework, a few approaches utilizing ghost diffraction (GD) [28, 29], phase retrieval algorithms [30], interferometry schemes [31–34], Fourier-space filtering [35], and phase-shifting digital holography [34], etc. has been demonstrated in recent years. Most of the demonstrated phase imaging techniques in ghost schemes emphasis on low spatially varying phase objects, and demands the sophisticated measuring setups, time sequential detection of large set of sampling data, high quality reconstruction algorithm, etc. for the faithful phase recovery. In addition, the real-time quantitative phase imaging applications with ghost schemes demands the requirement of phase recovery from an instantaneous or time frozen intensity recordings. In this chapter we describe our recently invented technique of ghost diffraction holography (GDH) [36–38], where the quantitative phase recovery in a ghost system is made possible by adopting the concept of off-axis holography in combination with the conventional ghost diffraction system. The technique utilizes a snapshot recording scheme with synchronized detectors and exploits the spatial statistics of time frozen speckle field from a pseudo-thermal light source for the correlation of intensity fluctuations. In the following sections we describe different features of GDH system by giving emphasis on the snapshot detection scheme in GD system, recovery of complex correlation function, extension of GDH to a ghost diffraction holographic microscopy (GDHM) system, and finally on the recovery and characterization of orbital angular momentum modes (OAM) using a GDH system.

2. Classical framework of ghost diffraction system

In the last two decades GI and GD techniques has been emerged as well-established optical imaging techniques, which exploits the intensity correlations to obtain an image or diffraction pattern of an object, respectively. The realization of classical counterpart of the ghost schemes with entangled source open the door for the utilization of various light sources and thereby by the execution of the approach in various applied imaging scenarios where the conventional imaging techniques limits in the performance [14–16]. The classical ghost schemes perform coherent imaging with spatially incoherent light by make use of the intensity correlations of identical spatially correlated beams, one of them interact with the object, while the other has never interacted with the object. To understand the classical framework of ghost diffraction scheme, let us consider the conceptual schematic of a ghost diffraction system as shown in **Figure 1**. The stochastic field emanating from the spatially incoherent source is divided in to two identical copies by a beam splitter, where one of the fields reaches the detector, D1after illuminating an object, while the other field reaches detector, D2 without interacting with the object. A conventional GD system uses a point detector as D1 in the object arm and a multipixel detector as D2 in the reference arm. For a stochastic process that obeys Gaussian statistics, the cross-correlation of spatial intensity fluctuations of object field and reference field retrieves the diffraction pattern of the object [16], which is expressed as

$$\boldsymbol{G} = \langle \Delta I_o(\boldsymbol{r}_1) \Delta I_r(\boldsymbol{r}_2) \rangle = \left| \boldsymbol{g}^{(G)}(\boldsymbol{r}_1, \ \boldsymbol{r}_2) \right|^2 \tag{1}$$

where $\langle \dots \rangle$ represents the ensemble averaging, $I_o(\boldsymbol{r}) = |u_o(\boldsymbol{r})|^2$ and $I_r(\boldsymbol{r}) = |u_r(\boldsymbol{r})|^2$ are the intensity values at the detectors of object field $(u_o(\boldsymbol{r}))$ and reference field $(u_r(\boldsymbol{r}))$, respectively with $\Delta I(\boldsymbol{r}) = I(\boldsymbol{r}) - \langle I(\boldsymbol{r}) \rangle$ representing the spatial intensity fluctuation with respect to its average value, $\boldsymbol{r}$ the position coordinate in the detector plane and $\boldsymbol{g}^{(G)}(\boldsymbol{r}_1, \boldsymbol{r}_2) = \langle u_o^*(\boldsymbol{r}_1) u_r(\boldsymbol{r}_2) \rangle$ is the ghost diffraction correlation function that represents the diffraction pattern of the object.

The cross-correlation of intensity fluctuations implemented in the GD scheme directly retrieves only the modulus of the cross-correlation function or diffraction pattern of the object. This limits the direct and simultaneous recovery of phase along with the amplitude of the object, which restricts the execution of ghost schemes in complex-valued object imaging. A few GI schemes are demonstrated in recent years on considering the significance of optical phase imaging by utilizing the phase retrieval techniques or by using other indirect experimental schemes. Owing to the phase recovery challenges in ghost schemes, we have recently developed a GDH system by adopting the off-axis holography scheme in combination with the

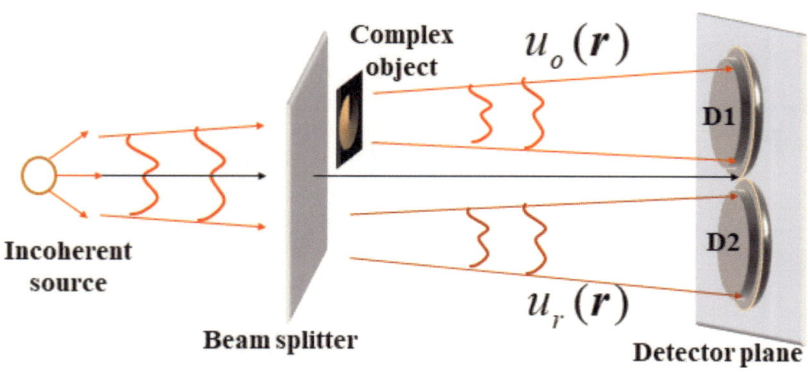

Figure 1.
Conceptual schematic of classical ghost diffraction system. Adapted with permission from [36] © The Optical Society.

conventional GD system. Section 3 describes in detail the GDH system and its extension to the microscopy for quantitative complex field imaging.

3. Ghost diffraction holography

The GDH combines the holographic concept with the conventional ghost diffraction scheme for the simultaneous quantitative phase and amplitude imaging. A conceptual schematic of the GDH scheme is shown in **Figure 2**, where a reference random field generated from an independent source is mixed with the fields emanating from the conventional GD system. The cross-correlation of spatial intensity fluctuations at the respective detectors retrieves a correlation hologram in the intensity correlation function. The digital processing of the correlation hologram recovers the complex correlation function, and thereby make possible the complex-valued object imaging. In comparison to conventional GD approaches, the GDH utilizes snapshot detection scheme by utilizing the time-frozen speckle field from a pseudothermal light source by considering the ergodicity in space.

3.1 Snapshot GDH based on spatial averaging

The GDH system utilizes the holographic addition of the reference random field $u_R(\mathbf{r})$ generated from an independent source with the object field $u_o(\mathbf{r})$ and reference field $u_r(\mathbf{r})$ of the conventional ghost diffraction system [36–38]. Therefore, the resultant intensity distributions at the respective detectors are given by,

$$\begin{aligned} I_o(r) &= (u_o(r) + u_R(r))^* (u_o(r) + u_R(r)) \\ I_r(r) &= (u_r(r) + u_R(r))^* (u_r(r) + u_R(r)) \end{aligned} \tag{2}$$

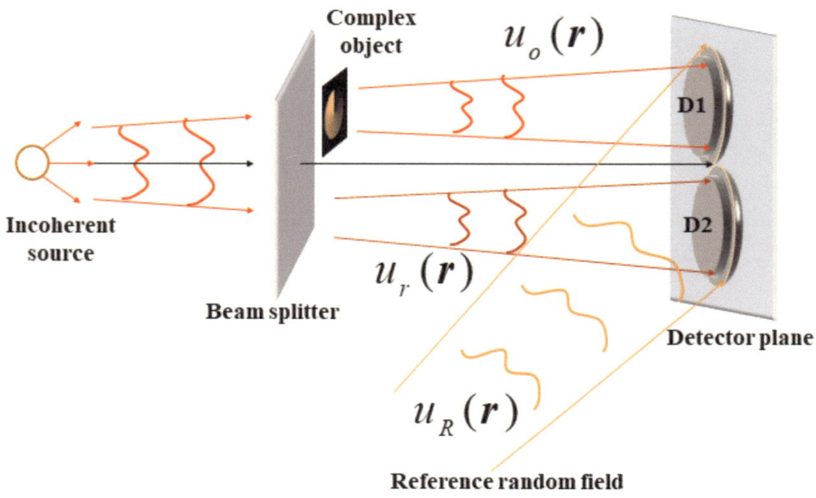

Figure 2.
Conceptual schematic of ghost diffraction holography (GDH) system. The GDH system utilizes a rotating ground glass diffuser as a spatially incoherent source. Adapted with permission from [36] © The Optical Society.

where the field distributions at the respective detectors are expressed as

$$\begin{aligned} u_o(r) &= \frac{\exp(\mathrm{ikz})}{\mathrm{i}\lambda \mathrm{z}} \int u_0(\hat{r}) \exp\left(i\varphi_g(\hat{r})\right) T(\hat{r}) \exp\left[\frac{\mathrm{ik}}{2\mathrm{z}}\left(|r|^2 - 2r.\hat{r} + |\hat{r}|^2\right)\right] \mathrm{d}\hat{r} \\ u_r(r) &= \frac{\exp(\mathrm{ikz})}{\mathrm{i}\lambda \mathrm{z}} \int u_0(\hat{r}) \exp\left(i\varphi_g(\hat{r})\right) \exp\left[\frac{\mathrm{ik}}{2\mathrm{z}}\left(|r|^2 - 2r.\hat{r} + |\hat{r}|^2\right)\right] \mathrm{d}\hat{r} \\ u_R(r) &= \frac{\exp(\mathrm{ikz})}{\mathrm{i}\lambda \mathrm{z}} \int u_R(\hat{r}) \exp\left(i\varphi_R(\hat{r})\right) \exp\left[\frac{\mathrm{ik}}{2\mathrm{z}}\left(|r|^2 - 2r.\hat{r} + |\hat{r}|^2\right)\right] \mathrm{d}\hat{r} \end{aligned} \tag{3}$$

with $'\lambda'$ the wavelength of light source, $\varphi_g(\hat{r})$ and $\varphi_R(\hat{r})$ are the random phase introduced by the rotating ground glass diffuser in the GD system and the reference ground glass diffuser, respectively, $'k = 2\pi/\lambda'$ the wave number, $'z'$ the propagation distances in the respective arms, $T(\hat{r})$ is the is the transmittance function of the object, and $\hat{r}$ and $\boldsymbol{r}$ are the position coordinate in the diffuser plane and detector plane, respectively.

The ghost schemes utilize the cross-correlation of spatial intensity fluctuations at the respective detector plane, which is expressed as:

$$G = \langle \Delta I_o(\boldsymbol{r}_1) \Delta I_r(\boldsymbol{r}_2) \rangle \tag{4}$$

where $\langle \dots \rangle$ represents the ensemble averaging, $\Delta I_o(\boldsymbol{r}_1) = I_o(\boldsymbol{r}_1) - \langle I_o(\boldsymbol{r}_1) \rangle$ and $\Delta I_r(\boldsymbol{r}_2) = I_r(\boldsymbol{r}_2) - \langle I_r(\boldsymbol{r}_2) \rangle$ represents the fluctuation of the intensity value with respect to its average value for the respective intensity distributions at the detectors. On assuming the scattered field from the diffusers obeys Gaussian statistics and by utilizing the fourth order moment of the field at the detectors, the cross-correlation of intensity fluctuations in Eq. (4) is related to the respective second order correlation function as [39, 40]:

$$\begin{aligned} \langle \Delta I_o(\boldsymbol{r}_1) \Delta I_r(\boldsymbol{r}_2) \rangle &= |\langle (u_o(\boldsymbol{r}_1) + u_R(\boldsymbol{r}_1))^* (u_r(\boldsymbol{r}_2) + u_R(\boldsymbol{r}_2)) \rangle|^2 \\ &= \left|\langle u_o^*(\boldsymbol{r}_1) u_r(\boldsymbol{r}_2) + u_o^*(\boldsymbol{r}_1) u_R(\boldsymbol{r}_2) + u_R^*(\boldsymbol{r}_1) u_r(\boldsymbol{r}_2) + u_R^*(\boldsymbol{r}_1) u_R(\boldsymbol{r}_2) \rangle\right|^2 \\ &= \left|\langle u_{o^*}(\boldsymbol{r}_1) u_r(\boldsymbol{r}_2) \rangle + \langle u_R^*(\boldsymbol{r}_1) u_R(\boldsymbol{r}_2) \rangle\right|^2 \\ &= |g^G(\boldsymbol{r}_1, \boldsymbol{r}_2) + g^R(\boldsymbol{r}_1, \boldsymbol{r}_2)|^2 \end{aligned} \tag{5}$$

where $g^G(\boldsymbol{r}_1, \boldsymbol{r}_2) = \langle u_o^*(\boldsymbol{r}_1) u_r(\boldsymbol{r}_2) \rangle$, and $g^R(\boldsymbol{r}_1, \boldsymbol{r}_2) = \langle u_R^*(\boldsymbol{r}_1) u_R(\boldsymbol{r}_2) \rangle$ are the second order correlation functions corresponding to the ghost diffraction field and the reference random field from the holographic arm, respectively. Here, the contribution from the cross terms is taken as zero. i.e., $\langle u_o^*(\boldsymbol{r}_1) u_R(\boldsymbol{r}_2) \rangle = \langle u_R^*(\boldsymbol{r}_1) u_r(\boldsymbol{r}_2) \rangle = 0$, since independent diffusers are used to generate the respective random fields.

The execution of ensemble averaging in the correlation process can be performed either with the temporal averaging or with the spatial averaging. In view of the real-time implementation of the ghost schemes with quantitative phase imaging potential, the GDH system rely on spatial averaging by considering the time frozen speckle fields at the detector plane under the assumption of spatial stationarity [41, 42]. A snapshot detection scheme in GDH is implemented by replacing the point detector of a conventional GD system with a multi pixel detector D1, which is synchronized with another multi pixel detector D2. The spatial averaging operation on the detected intensity distributions is carried out by fixing the distance $\Delta \boldsymbol{r} = \boldsymbol{r}_1 - \boldsymbol{r}_2$ equivalent to the pixel size at the detector plane, and by moving a specific window size in the intensity distribution $I_o(r)$ over the entire intensity image $I_r(r)$. On applying a change of variables $\boldsymbol{r}_1 = \boldsymbol{r} + \Delta\boldsymbol{r}$ and $\boldsymbol{r}_2 = \boldsymbol{r}$ with $\Delta\boldsymbol{r} = \boldsymbol{r}_1 - \boldsymbol{r}_2$, the Eq. (5) modifies to

$$\begin{aligned} G &= \langle \Delta I_o(\boldsymbol{r} + \Delta\boldsymbol{r}) \Delta I_r(\boldsymbol{r}) \rangle \\ &= \left| g^G(\boldsymbol{r} + \Delta\boldsymbol{r},\ \boldsymbol{r}) + g^R(\boldsymbol{r} + \Delta\boldsymbol{r}, \boldsymbol{r}) \right|^2 \end{aligned} \tag{6}$$

where $g^G(\boldsymbol{r} + \Delta\boldsymbol{r}, \boldsymbol{r})$ represents the complex field correlation function that encodes the diffraction pattern of the object, which can be further expressed as.

$$\begin{aligned} g^G(\boldsymbol{r} + \Delta\boldsymbol{r}, \boldsymbol{r}) &= \left\langle u_o^*(\boldsymbol{r} + \Delta\boldsymbol{r}) u_r(\boldsymbol{r}) \right\rangle \\ &= \frac{1}{\lambda^2 z^2} \iint u_0^*(\hat{\boldsymbol{r}}_1) T(\hat{\boldsymbol{r}}_1) u_0(\hat{\boldsymbol{r}}_2) \delta(\hat{\boldsymbol{r}}_1 - \hat{\boldsymbol{r}}_2)\, exp\left(-i\left(\varphi_g(\hat{\boldsymbol{r}}_1) - \varphi_g(\hat{\boldsymbol{r}}_2)\right)\right) \\ &\quad exp\left(-i\frac{k}{2z}\left(|\hat{\boldsymbol{r}}_1|^2 - |\hat{\boldsymbol{r}}_2|^2\right)\right) \times exp\left(-i\frac{2\pi}{\lambda z}\Delta\boldsymbol{r}.\hat{\boldsymbol{r}}_1\right) d\hat{\boldsymbol{r}}_1 d\hat{\boldsymbol{r}}_2 \end{aligned} \tag{7}$$

with $\delta(\hat{\boldsymbol{r}}_1 - \hat{\boldsymbol{r}}_2) \propto \int \exp\left(-i\frac{2\pi}{\lambda z}(\hat{\boldsymbol{r}}_1 - \hat{\boldsymbol{r}}_2).\boldsymbol{r}\right) d\boldsymbol{r}$ the delta function. The contributions corresponding to the $\boldsymbol{r}$ dependent phase factors $exp\left(ik\frac{|\boldsymbol{r}|^2}{2z}\right)$ and $exp\left(ik\frac{|\boldsymbol{r}+\Delta\boldsymbol{r}|^2}{2z}\right)$ were canceled out while estimating the cross correlation of intensity fluctuations as in Eq. (6). On considering $\hat{\boldsymbol{r}}_2 = \hat{\boldsymbol{r}}_1 = \hat{\boldsymbol{r}}$ and $u_0^*(\hat{\boldsymbol{r}}) u_0(\hat{\boldsymbol{r}}) = I_0(\hat{\boldsymbol{r}})$, the Eq. (7) modifies to

$$g^G(\Delta\boldsymbol{r}) = \frac{1}{\lambda^2 z^2} \int I_0(\hat{\boldsymbol{r}}) T(\hat{\boldsymbol{r}})\, exp\left(-i\frac{2\pi}{\lambda z}\Delta\boldsymbol{r}.\hat{\boldsymbol{r}}\right) d\hat{\boldsymbol{r}} \tag{8}$$

Similarly, the reference correlation function $g^R(\boldsymbol{r} + \Delta\boldsymbol{r}, \boldsymbol{r})$ in the Eq. (6) results into

$$\begin{aligned} & g^R(\boldsymbol{r} + \Delta\boldsymbol{r}, \boldsymbol{r}) = \left\langle u_R^*(\boldsymbol{r} + \Delta\boldsymbol{r}) u_R(\boldsymbol{r}) \right\rangle \\ & = \frac{1}{\lambda^2 z^2} \iint u_R^*(\hat{\boldsymbol{r}}_1) u_R(\hat{\boldsymbol{r}}_2) \delta(\hat{\boldsymbol{r}}_1 - \hat{\boldsymbol{r}}_2)\, exp\left(-i(\varphi_R(\hat{\boldsymbol{r}}_1) - \varphi_R(\hat{\boldsymbol{r}}_2))\right) exp\left(-i\frac{k}{2z}\left(|\hat{\boldsymbol{r}}_1|^2 - |\hat{\boldsymbol{r}}_2|^2\right)\right) \\ & \times exp\left(-i\frac{2\pi}{\lambda z}\Delta\boldsymbol{r}.\hat{\boldsymbol{r}}_1\right) d\hat{\boldsymbol{r}}_1 d\hat{\boldsymbol{r}}_2 \end{aligned} \tag{9}$$

On considering $\hat{\boldsymbol{r}}_2 = \hat{\boldsymbol{r}}_1 = \hat{\boldsymbol{r}}$ and $u_R^*(\hat{\boldsymbol{r}}) u_R(\hat{\boldsymbol{r}}) = I_R(\hat{\boldsymbol{r}})$ at the reference diffuser plane, Eq. (9) modifies to

$$g^R(\Delta\boldsymbol{r}) = \frac{1}{\lambda^2 z^2} \int I_R(\hat{\boldsymbol{r}})\ exp\left(-i\frac{2\pi}{\lambda z}\Delta\boldsymbol{r}.\hat{\boldsymbol{r}}\right) d\hat{\boldsymbol{r}} \tag{10}$$

By substituting Eqs. (8) and (10) in Eq. (4), the intensity correlation function in terms of the position vector $\Delta\boldsymbol{r}$ is expressed as

$$\begin{aligned} G &= \langle \Delta I_o(\boldsymbol{r} + \Delta\boldsymbol{r}) \Delta I_r(\boldsymbol{r}) \rangle = \left| g^G(\Delta\boldsymbol{r}) + g^R(\Delta\boldsymbol{r}) \right|^2 \\ &= \left| \frac{1}{\lambda^2 z^2} \left\{ \int I_0(\hat{\boldsymbol{r}}) T(\hat{\boldsymbol{r}})\, exp\left(-i\frac{2\pi}{\lambda z}\Delta\boldsymbol{r}.\hat{\boldsymbol{r}}\right) d\hat{\boldsymbol{r}} + \int I_R(\hat{\boldsymbol{r}})\, exp\left(-i\frac{2\pi}{\lambda z}\Delta\boldsymbol{r}.\hat{\boldsymbol{r}}\right) d\hat{\boldsymbol{r}} \right\} \right|^2 \end{aligned} \tag{11}$$

Eq. (11) describes the cross-correlation of intensity fluctuations in the GDH scheme, which results into the generation of the ghost correlation hologram. In the GDH scheme, the reference correlation function is generated by an off-axis point source illumination on an independent ground glass diffuser with the intensity at the source structure, $I_R(\hat{r}) = circ\left(\frac{\hat{r}-r_s}{a}\right)$, which is controlled by the position r_s and size a of the aperture used to illuminate the independent reference ground glass diffuser. The $g^R(\Delta r)$ provides a linear phase to generate the ghost correlation hologram, and the size of the aperture is selected in such a way that it covers the extent of $g^G(\Delta r)$. In the GDH system, the recording and reconstruction of the object information is carried out in terms of the coherence functions rather than in terms of the optical field as in the conventional holography system [36, 43–45]. A digital processing in the correlation hologram by making use of Fourier transform method [46], retrieves the complex correlation function $g^G(\Delta r)$ from other redundant terms of the correlation hologram. The complex correlation function retrieval provides the flexible opportunity to recover the complex-valued object information at any specific plane by making use of digital propagation methods [47].

3.2 GDH design

A schematic diagram of the GDH system with snapshot recording scheme is shown in **Figure 3**. The design consists of a conventional ghost diffraction part (Part I), an outer holographic reference arm with an independent random field (Part II) and the snapshot recording and digital processing module (Part III). A coherent light source splitting into two beams by a non-polarizing beam splitter (BS1) act as the source beams for Part I and Part II of the system. The coherent beam transmitting through a rotating ground glass diffuser (GG1) generates the speckle field, which act as the pseudothermal source for the ghost diffraction system. The design utilizes the orthogonal polarized components of light beams by using a polarization beam splitter (PBS1) for the development of a common path snapshot recording of the ghost object and reference fields. The reflected beam from BS1 generates a reference random field by allowing the beam to pass through the combination of an off-axis microscope objective and a ground glass diffuser GG2. This reference random field from Part II propagates and combines with the common path propagating orthogonally polarized scattered fields from the Part I of the system using a beam splitter BS2. A snapshot recording scheme is developed by using the combination of two synchronized charge coupled device (CCD) cameras, CCD1 and CCD2 and polarizers P1 and P2 at specific orientations. The polarizers are used for the projection and superposition of corresponding polarization components from Part I and part II of the system with maximum visibility. At a specific instant of time the CCD's records the intensity distributions of superposed object and reference fields.

3.3 Recovery of complex correlation function

The performance of the GDH system (shown in **Figure 3**) for complex correlation function recovery is theoretically and experimentally demonstrated with various pure phase objects and real-valued objects. The pure phase of helical mode with topological charge, $l = 1$ (shown in **Figure 4a**) from a vortex phase plate (VPP) is used as a pure phase object in the experimental design shown in **Figure 3**.

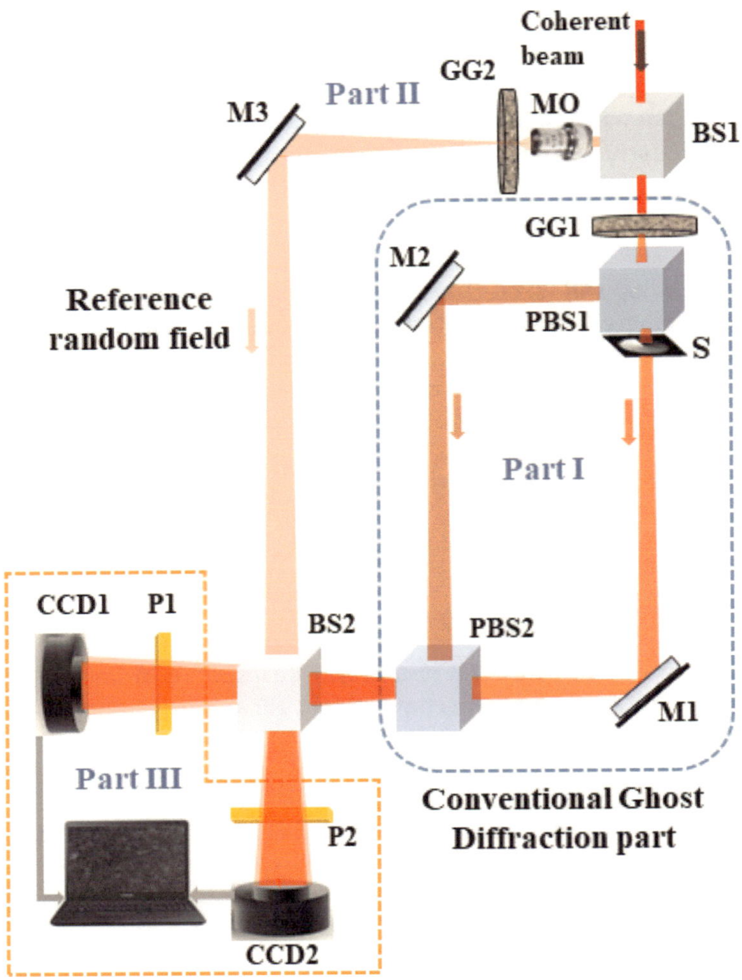

Figure 3.
Schematic of the GDH system; part I: Conventional ghost diffraction part, part II: Holographic arm with reference random field, part III: Snapshot recording and digital processing module, BS: Beam splitter, GG: Ground glass diffuser, S: Sample, M: Mirror, PBS: Polarization beam splitter, MO: Microscope objective, P: Polarizer, and CCD: Charge coupled device. Adapted with permission from [36] © The Optical Society.

The snapshot detection scheme records the intensity distribution corresponding to the object and reference fields, which are shown in **Figure 4b** and **c**, respectively. The cross-correlation of intensity fluctuations from the two detectors results in a ghost correlation hologram as expressed by Eq. (6) and is shown in **Figure 4d**. The existence of a fork pattern in the ghost correlation hologram clearly indicates the encoding of the phase information in the correlation hologram. A digital analysis based on Fourier transform method [46] of fringe analysis recovers the complex correlation function from the ghost correlation hologram. The recovered amplitude and phase distribution at the detector plane corresponding to the helical mode with $l = 1$ is shown in **Figure 4e** and **f**. Moreover, to illustrate the validity of the system, a GDH system is simulated with helical wavefront $exp\ (il\phi)$, where l is the topological charge and ϕ the angular coordinate of the pure phase sample. The simulation of the GDH system is implemented by considering a light source of wavelength 632.8 nm and the beam of specific size like the experimental conditions to illuminate the rotating diffuser.

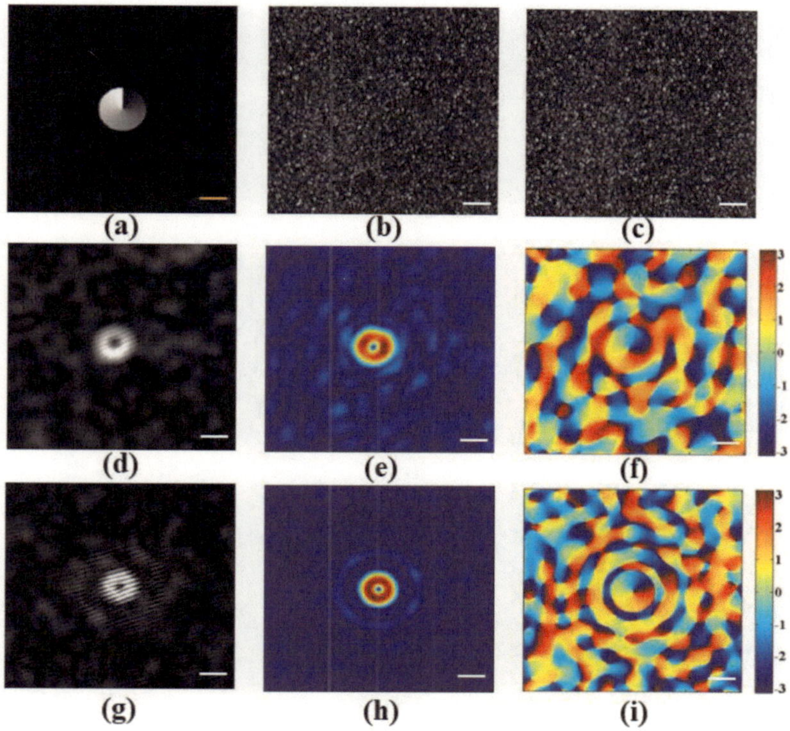

Figure 4.
Recovery of complex correlation function for a phase object; (a) pure phase of a helical mode with topological charge 1, (b) & (c) recorded intensity distributions corresponding to the object field and reference fields, (d)–(f) experimental results of ghost correlation hologram, amplitude, and phase distributions of complex correlation functions, respectively, (g)–(i) simulation results of ghost correlation hologram, amplitude, and phase distributions of complex correlation functions, respectively, scale bar: (a)2.5 mm, (b) & (c) 1.32 mm, and (d)–(i) 138 μm.

The reference random field is generated by an off-axis point source illumination on an independent diffuser. The simulation results corresponding to ghost correlation hologram, recovered amplitude and phase distributions are shown in **Figure 4g–i**, respectively.

Furthermore, the ability of the approach in real-valued case is demonstrated with a triangular pattern of aperture (shown in **Figure 5a**) as an object for the experimental design shown in **Figure 3**. The triangular aperture generates off-axis propagating random fields on illuminating the surface with a speckle field from the rotating diffuser. The superposition of three off axis propagating random light beams creates an array of vortices in the complex correlation function [48, 49]. The recorded intensity distributions corresponding to the object field and reference field are shown in **Figure 5b** and **c**, respectively. The cross-correlation of intensity fluctuations of the object field and reference field at the detector plane retrieves the ghost correlation hologram as shown in **Figure 5d**, which clearly has an array of fork fringes in the intensity distribution. A digital processing on the hologram retrieves the complex correlation function. The amplitude and phase distribution of complex correlation function is shown in **Figure 5e** and **f**, which shows an array of dark core and the helical phase distribution corresponding to the dark core in its amplitude and phase distribution, respectively. The experimental results were compared with the simulation results corresponding to a triangular aperture as sample, and the respective

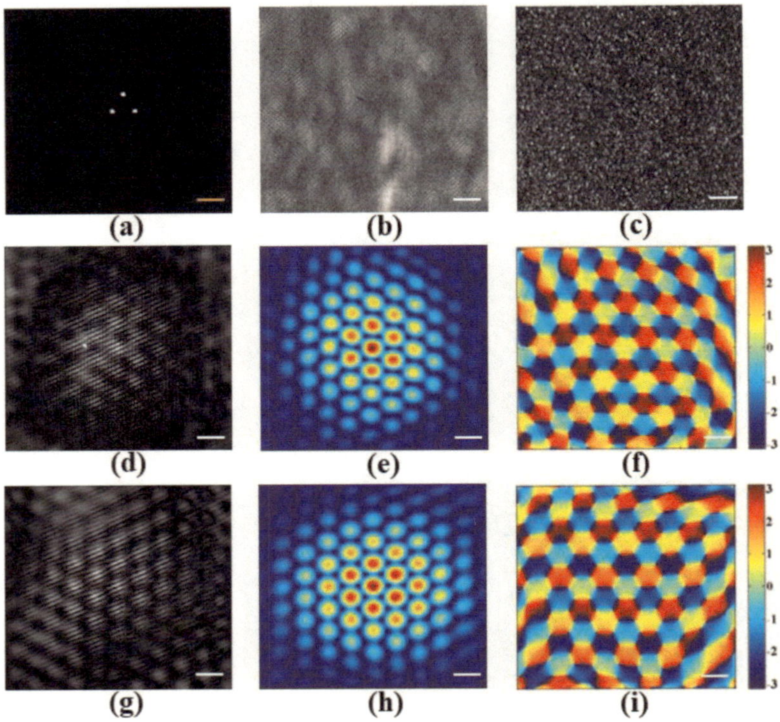

Figure 5.
Recovery of complex correlation function for a triangular aperture; (a) pure phase of a helical mode with topological charge 1, (b) & (c) recorded intensity distributions corresponding to the object field and reference fields, (d)–(f) experimental results of ghost correlation hologram, amplitude, and phase distributions of complex correlation functions, respectively, (g)–(i) simulation results of ghost correlation hologram, amplitude, and phase distributions of complex correlation functions, respectively, scale bar: (a) 2.5 mm, (b) & (c) 1.32 mm, and (d)–(i) 138 μm.

results corresponding to ghost correlation hologram, amplitude and phase distributions are shown in **Figure 5g–i**. A good agreement between simulation and experimental results in the recovery of complex correlation function for various objects using GDH provides the potential provision to realize the technique in quantitative imaging of complex-valued object.

3.4 Quantitative phase and amplitude imaging

The complex field correlation function recovery from snapshot detected intensities make the provision to recover the complex-valued image of the object at specific plane by employing a digital beam propagation approach based on angular spectrum method [47]. The digital analysis procedure employed in the technique for the simultaneous quantitative phase and amplitude image recovery of an object is shown in **Figure 6**. The synchronized CCD's records the intensity distribution corresponding to the ghost object field and reference field at a fixed time. A spatial averaging assisted cross-correlation of intensity fluctuations retrieves the ghost correlation hologram at the detector plane. By applying a Fourier transform operation on the ghost correlation hologram, the spectra and its conjugate spectra are separated from the central dc term. The complex correlation function is retrieved by performing an inverse Fourier transform operation on the filtered and centrally shifted spectrum. The simultaneous

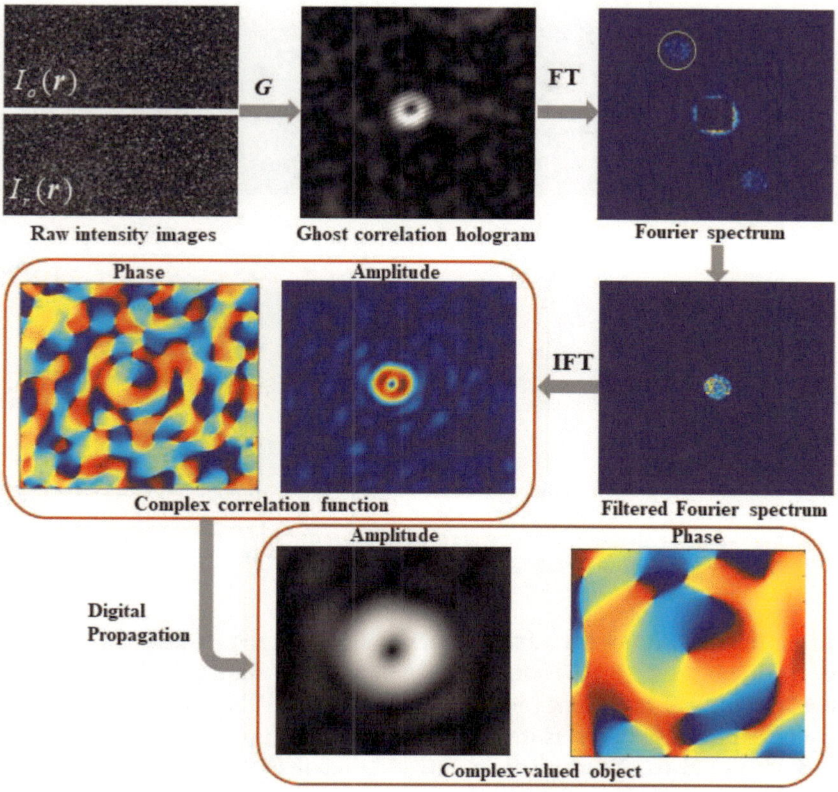

Figure 6.
Digital analysis scheme of quantitative phase and amplitude imaging with GDH system; $I_o(\mathbf{r})$ and $I_r(\mathbf{r})$ are the recorded intensity distributions corresponding to the object and reference fields; G: Cross-correlation of intensity fluctuations; FT: Fourier transform, and IFT: Inverse Fourier transform. Adapted with permission from [36] © The Optical Society.

retrieval of amplitude and phase distribution of correlation function make the provision to recover the complex-valued object at the desired plane using angular spectrum method of digital propagation. The performance of the GDH system in complex-valued object imaging is demonstrated with various objects like pure phase object, planar transparency, 1951 USAF resolution test target, etc. The quantitative phase and amplitude imaging of various objects used in the GDH system are shown in **Figure 7**.

3.5 Ghost diffraction holographic microscopy

In view of the significance of the digital holographic microscopy in biomedical imaging, the imaging potential of the GDH system is extended to the domain of microscopy, the GDH microscopy (GDHM) system. An experimental design of the GDHM system is shown in **Figure 8**. The system utilizes identical microscopy configuration in the ghost object and reference field as represented in part I of the **Figure 8**. An adjustable combination of different microscope objectives and tube lenses are utilized in the system according to the dimensions of the sample. The random fields from microscopy assisted ghost object and reference fields superpose with the independent reference field from Part II of the system. The combination of polarizers and CCDs records the intensity distribution corresponding to the polarization components

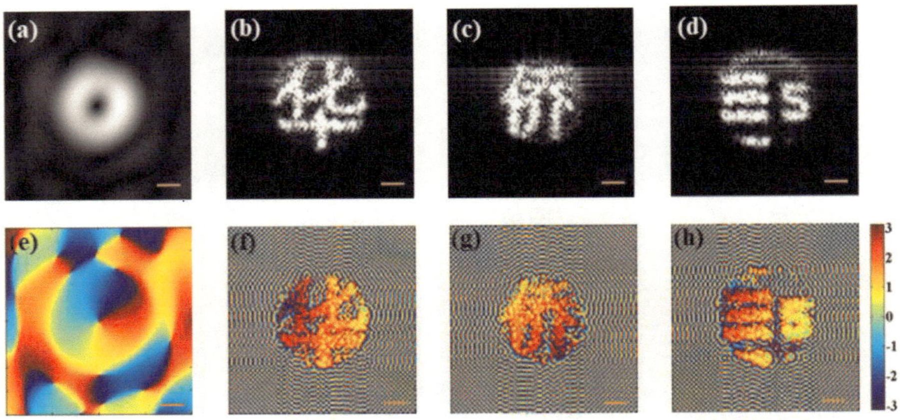

Figure 7.
Quantitative phase and amplitude imaging of various objects with GDH system; first row: Recovered amplitude distributions; second row: Recovered phase distributions; (a) & (e) VPP with $l = 1$; (b) & (f) Chinese character "HUA"; (c) & (g) Chinese character "QIAO"; and (d) & (h) negative 1951 USAF resolution test target. Scale bar:1.15 mm. Adapted with permission from [36] © The Optical Society.

from different arms. A quantitative comparison of imaging results of USAF 1951 resolution test target with GDH and GDHM system are shown in **Figure 9**. The first two rows in **Figure 9** represent the imaging results of GDH system, where the reconstruction quality deteriorates as it reaches Group-0, Element-4 of the test target. The last two rows in **Figure 9** represent the imaging results corresponding to GDHM system with various configurations of microscope objectives and tube lenses. A detailed quantitative comparison of imaging results is given in **Table 1**. The GDHM system have a good resolving ability up to Group-7, Element-1 with the given experimental design, which corresponds to 128-line pairs/mm with a line width of 3.9 μm (**Table 1**).

4. Recovery and characterization of orbital angular momentum modes

The characteristic feature of the complex correlation function recovery of GDH system is recently exploited to develop a technique for the recovery and characterization of OAM modes [50]. The OAM associated with a vortex electromagnetic wave is an independent physical dimension of light having the characteristics of orthogonality and infinite-dimensional basis, which have remarkable applications in imaging, optical communication, quantum and classical information processing, multidimensional coding, etc. [51–53]. The optical vortex beams carries an OAM equivalent to $l\hbar$ per photon ($\hbar$ is the Planck's constant divided by 2π) with an azimuthal phase of $exp\,(il\phi)$, where l is the topological charge and ϕ the angular coordinate. Last two decades witnessed a tremendous development of various techniques for the generation of vortex beams, and in recent years the research efforts shifting from coherent vortex beams to low coherent light with helical phase modes in partially coherent beams [54–56].

Alongside the progress in the developments of vortex beam generation techniques, a significant number of techniques were introduced for the detection and sorting of various OAM modes [51, 52]. However, most of the conventional detection and

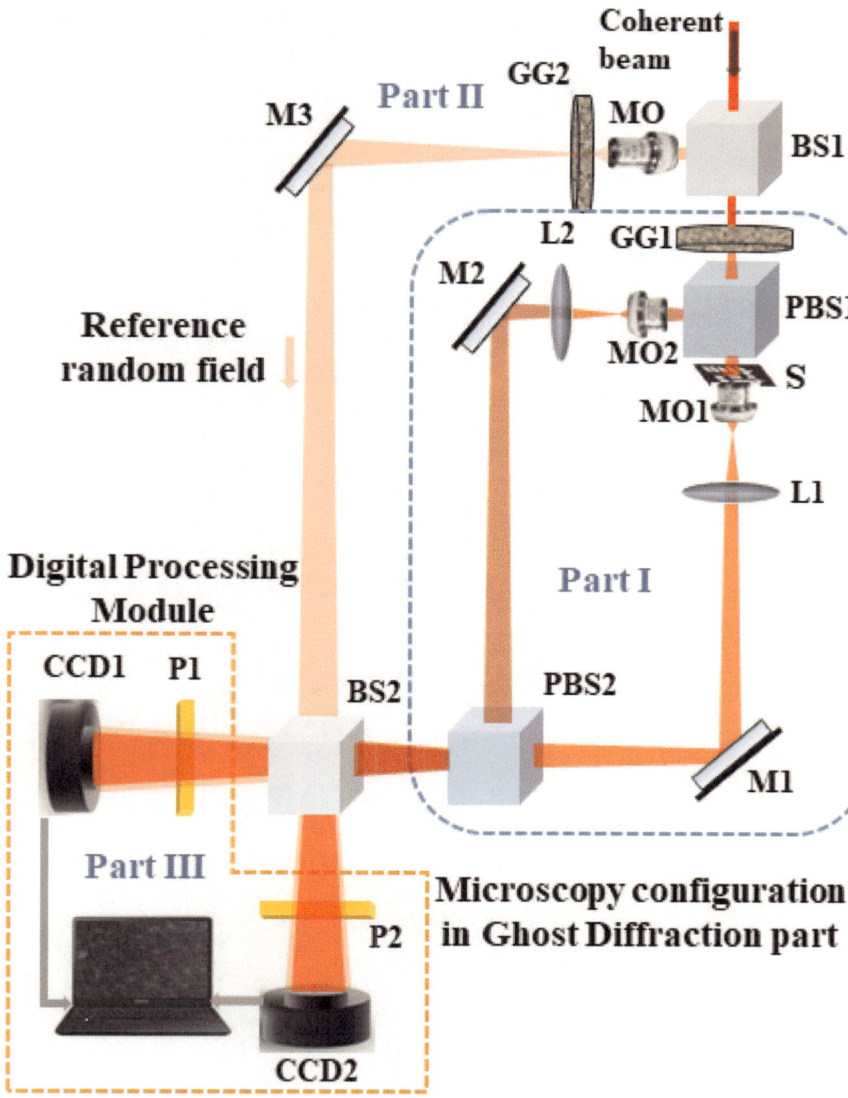

Figure 8.
Schematic of the GDHM system; Part I: Conventional ghost diffraction part with microscopy configuration, Part II: Holographic arm with reference random field, part III: Snapshot recording and digital processing module, BS: Beam splitter, GG: Ground glass diffuser, S: Sample, M: Mirror, PBS: Polarization beam splitter, MO: Microscope objective, P: Polarizer, and CCD: Charge coupled device. Adapted with permission from [36] © The Optical Society.

sorting techniques limits its application in scenarios where the propagating medium is inhomogeneous or turbulent in nature, which scrambles the helical wavefront of the vortex beam and destroys the orthogonality. A few research efforts were put forwarded in recent years to tackle the challenges of vortex beam propagation through scattering medium by exploiting the information carrying potential of speckle fields [45, 49, 57–59]. In addition, a few techniques were demonstrated for helicity and topological charge (TC) determination of partially coherent vortex beams [60, 61], robustness investigation of low coherent vortex beam propagation through noisy and turbulent media [62, 63], discrimination of incoherent vortex states of light [64], and

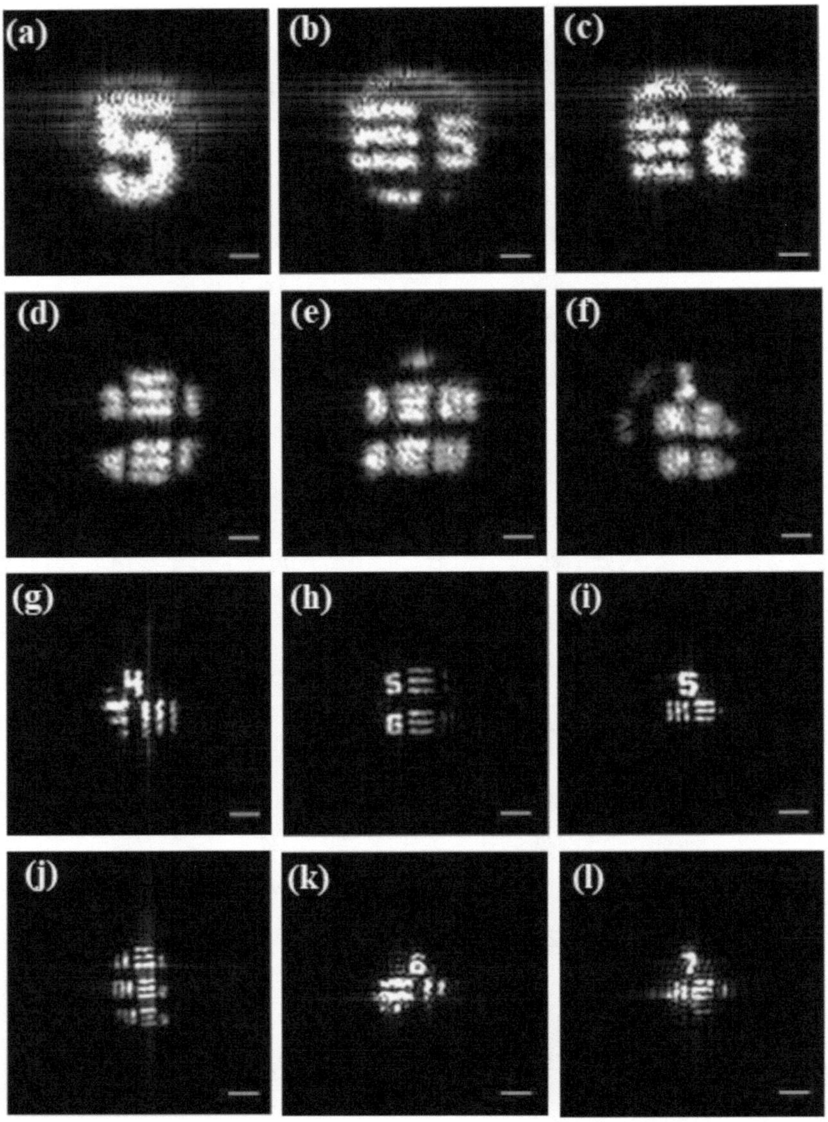

Figure 9.
Quantitative imaging results of GDH and GDHM system with negative 1951 USAF resolution test target as sample. Recovered amplitude distributions: (a) group −2 element 5, (b) group −1 element 5, (c) group −1 element 6, (d) group 0 element 3–4, (e) group 0 element 5–6, & (f) group 1 element 1; GDHM system: (g) group 4 element 1, (h) group 4 element 5–6, (i) group 5 element 1, (j) group 5 element 1–3, (k) group 6 element 1 & (l) group 7 element 1. Scale bar: (a)–(f) 1.15 mm, (g)–(j) 57.5 μm, (k) 23.0 μm, and (l) 11.5 μm. Adapted with permission from [36] © The Optical Society.

sorting of incoherent optical vortex modes [65]. This section describes the recently developed technique for OAM mode detection and characterization by exploiting the potential features of GDH system. The technique exploits the properties of speckle field to encode and transmit the OAM modes. The holography assisted GD scheme is utilized for simultaneous detection of amplitude and phase of the OAM modes, and thereby provides an additional feature of determining the helicity of the various modes.

Imaging system	System specifications			Reconstruction area in USAF resolution target	Reconstruction limit	
	MO	L (mm)	M		Line pairs/mm	Line width (μm)
GDH			1	G0–E4	1.41	355
GDHM	40X/ 0.65NA	100	20	G4–E6	28.50	17.5
	60X/ 0.85NA	100	30	G5–E3	40.30	12.4
	100X/ 1.25NA	100	50	G6–E2	71.80	7.0
	100X/ 1.25NA	200	100	G7–E1	128.0	3.9

M: Magnification; MO: Microscope Objective; L: Tube Lens; G: Group; and E: Element.
Adapted with permission from [36] © The Optical Society.

Table 1.
Quantitative analysis on the imaging system.

4.1 Generation and recovery of OAM modes with GDH

The applicability of the GDH technique in the generation and recovery of OAM modes is demonstrated with the experimental system shown in **Figure 10**. The light from a coherent light source splits into two parts by a non-polarized beam splitter (BS1), where the transmitted part passes through a rotating ground glass diffuser (GG1) controlled by an aperture (A) of specific size and generates a spatially incoherent light beam with random speckle field at each instant of time. The random field from the rotating ground glass illuminates the reflecting type of phase sensitive SLM, which is used to encode different vortex modes with specific l values. The vortex

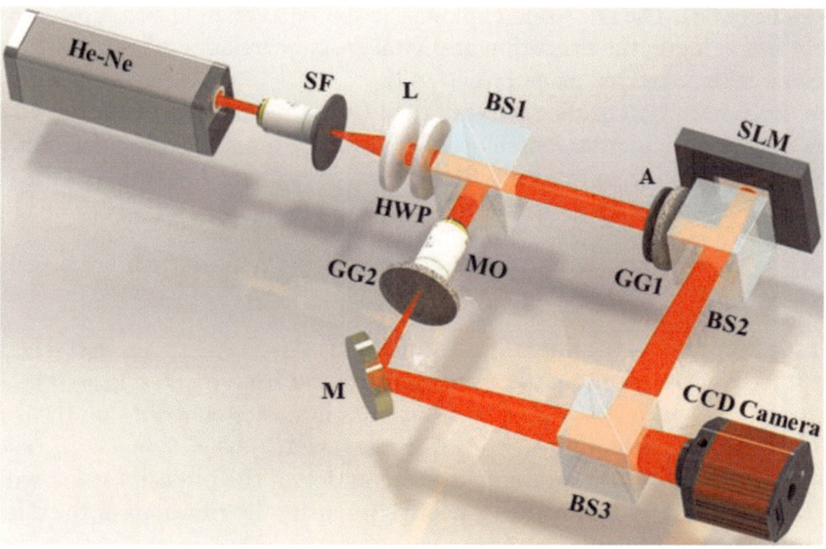

Figure 10.
Schematic of the experimental design of the GDH system for the generation and recovery of OAM modes; He-Ne: helium-neon laser source; M: mirror; SF: spatial filter assembly; L: lens; HWP: half wave plate; A: aperture; GG: ground glass diffuser; BS: beam splitter; SLM: spatial light modulator; MO: microscope objective; and CCD: charge coupled device camera. Adapted from [50].

mode encodes in the speckle field generated from the diffuser, and the field distribution in a transverse plane at a distance z from the diffuser plane is given by,

$$\psi_1(r) = \frac{\exp(\mathrm{ikz})}{\mathrm{i}\lambda \mathrm{z}} \int \psi_0(\hat{r}) \exp\left(i\varphi_g(\hat{r})\right) \mathrm{V}_0^{\varepsilon l}(\hat{r}) \exp\left[\frac{\mathrm{ik}}{2\mathrm{z}}\left(|r|^2 - 2r.\hat{r} + |\hat{r}|^2\right)\right] \mathrm{d}\hat{r} \quad (12)$$

where $\psi_0(\hat{r})$ is the incident light field at the exit plane of the diffuser, $\varphi_g(\hat{r})$ the random phase introduced by the ground glass diffuser, $V_0^{\varepsilon l}(\hat{r}) = exp\,(i\varepsilon l\phi)$ the vortex mode with zero radial index and l azimuthal index, where ε and l represent the sign and magnitude of the TC of the specific OAM mode, λ the wavelength of the light source, $k = 2\pi/\lambda$ the wave number of light, z the propagation distance, and $\hat{r}$ and r are the position coordinates at the source plane and detector plane, respectively.

The holographic random field for the GD scheme is achieved by making an off axis point source illumination on an independent diffuser (GG2) in the reflected beam from BS1. The reference random field generated from the diffuser in the holographic arm is given by,

$$\psi_R(r) = \frac{\exp(\mathrm{ikz})}{\mathrm{i}\lambda \mathrm{z}} \int \psi_R(\hat{r}) \exp\left(i\varphi_R(\hat{r})\right) \exp\left[\frac{\mathrm{ik}}{2\mathrm{z}}\left(|r|^2 - 2r.\hat{r} + |\hat{r}|^2\right)\right] \mathrm{d}\hat{r} \quad (13)$$

where $\psi_\mathrm{R}(\hat{r})$ is the incident light field at the exit plane of the diffuser, $\varphi_\mathrm{R}(\hat{r})$ the random phase introduced by the diffuser in the holographic arm. This reference random field propagates and combines independently with the specific vortex encoded spatially incoherent field from the SLM arm. The technique utilizes a sequential snapshot recording of the time frozen speckle fields corresponding to various modes and the recording is implemented using a CCD, which is synchronized with GG1 and SLM. The approach exploits the spatial averaging as a replacement of ensemble averaging in the execution of digital cross-correlation of recorded intensity fluctuations at the detector plane [36, 37, 42].

With the assumption of the scattered field from the diffusers obeys the Gaussian statistics and by utilizing the fourth order moment of the field at the detector plane [39, 40], the correlation of intensity fluctuations is expressed in terms of the respective second order correlation function as,

$$\langle \Delta I_m(\boldsymbol{r}_1)\Delta I_0(\boldsymbol{r}_2)\rangle = \left|W(\boldsymbol{r}_1,\ \boldsymbol{r}_2) + W^R(\boldsymbol{r}_1, \boldsymbol{r}_2)\right|^2 \quad (14)$$

with $\langle \ldots \rangle$ represents the ensemble averaging, $\mathrm{I_m}(r_1) = (\psi_\mathrm{m}(r_1) + \psi_\mathrm{R}(r_1))^*$ $(\psi_\mathrm{m}(r_1) + \psi_\mathrm{R}(r_1))$ the resultant intensity contribution from $\psi_\mathrm{m}(r)$ and $\psi_\mathrm{R}(r)$, $\mathrm{I_0}(r_2) = (\psi_0(r_2) + \psi_\mathrm{R}(r_2))^* (\psi_0(r_2) + \psi_\mathrm{R}(r_2))$ the resultant intensity contribution from $\psi_0(r)$ and $\psi_\mathrm{R}(r)$ with m representing any positive integer, $\Delta I_m(\boldsymbol{r}_1) = I_m(\boldsymbol{r}_1) - \langle I_m(\boldsymbol{r}_1)\rangle$, $\Delta I_0(\boldsymbol{r}_2) = I_0(\boldsymbol{r}_2) - \langle I_0(\boldsymbol{r}_2)\rangle$ represents the fluctuation of the intensity value with respect to its average value for the respective intensity distributions at the detector plane, $W(\boldsymbol{r}_1, \boldsymbol{r}_2) = \langle \psi_m^*(\boldsymbol{r}_1)\psi_0(\boldsymbol{r}_2)\rangle$, and $W^R(\boldsymbol{r}_1, \boldsymbol{r}_2) = \langle \psi_R^*(\boldsymbol{r}_1)\psi_R(\boldsymbol{r}_2)\rangle$ are the second order correlation functions corresponding to vortex encoded field and a reference random field, respectively. Here, we are justified in taking the contribution from the cross terms zero. i.e., $\langle \psi_m^*(\boldsymbol{r}_1)\psi_R(\boldsymbol{r}_2)\rangle = \langle \psi_R^*(\boldsymbol{r}_1)\psi_0(\boldsymbol{r}_2)\rangle = 0$, since independent diffusers are used to generate the respective random fields. Eq. (14) describes that the intensity correlation at the detector plane results in to a 'correlation hologram', which is the result of the superposition of second order complex field correlation functions [36, 43, 44].

For a fixed wavelength λ and at a fixed propagation distance z, by considering $\hat{r}_2 = \hat{r}_1 = \hat{r}$, $r_1 = r + \Delta r$, $r_2 = r$, and $\Delta r = r_1 - r_2$; the recovered complex cross-correlation function is expressed as,

$$\boldsymbol{W}(\Delta r) \propto \int \boldsymbol{I}_0(\hat{r}) \mathrm{V}_0^{el}(\hat{r}) \ \boldsymbol{exp}\left[-\boldsymbol{i}\frac{2\pi}{\lambda z}\Delta r.\hat{r}\right] d\hat{r} \tag{15}$$

where $\boldsymbol{I}_0(\hat{r}) = \psi_0^*(\hat{r})\psi_0(\hat{r})$ represents the non-stochastic intensity distribution at the diffuser plane. The correlation function $\boldsymbol{W}(\Delta r)$ is directly related to the intensity distribution at the diffuser plane $\boldsymbol{I}_0(\hat{r})$ and the encoded vortex phase information $\mathrm{V}_0^{el}(\hat{r})$ at the exit plane of the diffuser. In a similar way, the reference correlation function $W^R(\Delta \boldsymbol{r})$ is considered as,

$$\boldsymbol{W}^R(\Delta r) \propto \int \boldsymbol{I}_R(\hat{r}) \ \boldsymbol{exp}\left[-\boldsymbol{i}\frac{2\pi}{\lambda z}\Delta r.\hat{r}\right] d\hat{r} \tag{16}$$

where $I_R(\hat{r}) = circ\left(\frac{\hat{r}-r_s}{a}\right)$ is the intensity distribution at the reference diffuser GG2, which is utilized in the GDH system for the generation of the correlation hologram. i.e., the intensity of the reference field at off-axis position $'\boldsymbol{r}_s'$ and size $'a'$ of the circular aperture. The reference random field is created by an off axis point source illumination on GG2 with an aperture size $'a'$ and position $'\boldsymbol{r}_s'$, which generates a linear phase that support the recording of the correlation hologram.

A digital recovery process utilizing Fourier domain filtering approach [46] retrieves the complex correlation function $\boldsymbol{W}(\Delta r)$ from other redundant terms in Eq. (14). The retrieval of the complex correlation function offers a new detection scheme for the OAM modes scrambled in the speckle pattern. The retrieval of phase distribution along with the amplitude of the OAM modes with the GDH system delivers a direct scheme for the simultaneous determination of sign and magnitude of the topological charge of specific helicity. An anticlockwise helical structure direction represents the positive sign of the TC, and a clockwise helical structure direction represents a negative TC for the recovered OAM modes.

4.2 Recovered OAM modes

The GDH system assisted with SLM is utilized to encode phase masks corresponding to various TC. The sequential encoding of specific phase mask in the SLM and illumination with the random light from the rotating diffuser generate the vortex encoded speckle fields with various OAM modes. The corresponding intensity distributions of speckle patterns were recorded with CCD for various position shifts of the rotating diffuser. The intensity distribution of resultant speckle patterns obtained with specific topological charges are shown in **Figure 11a–d** and the respective reference speckle patterns are shown in **Figure 11e–h**. The inset of **Figure 11** shows the enlarged view of the marked specific region, where in **Figure 11a–d** clearly represents the speckle pattern modulation with interference fringes corresponding to various vortex phase masks in comparison to the inset images of **Figure 11e–h** of reference speckle intensity distributions.

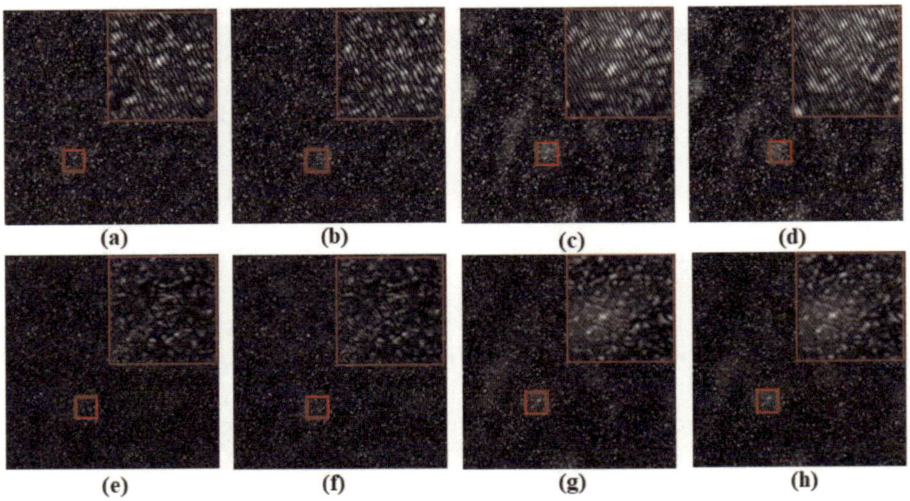

Figure 11.
(a–d) CCD recorded resultant speckle intensity distribution for OAM modes with $l = 1,4,-1\&-4$, *respectively; (e–h) CCD recorded resultant speckle intensity distribution for respective reference speckle field.*

The cross-correlation of CCD recorded intensity fluctuations based on spatial averaging retrieves the correlation hologram at the detector plane. The retrieved correlation hologram corresponding to $\varepsilon \boldsymbol{l} = +1\&{-1}$ and $\varepsilon \boldsymbol{l} = +4\&{-4}$ are shown in **Figure 12a–d**. A digital recovery scheme utilizing Fourier transform method of fringe analysis [46] recovers the complex field distribution corresponding to various OAM modes. The recovered amplitude and phase distributions of the vortex modes for respective topological charges are shown in **Figure 12e–l**. The performance of the technique is validated by simulation for various vortex modes with specific helicity and TC by generating a spatially incoherent light source by illuminating a random diffuser with a collimated coherent light source of wavelength 632.8 nm and a beam size of 6.5 mm. The respective simulation results corresponding to $\varepsilon \boldsymbol{l} = +1\&{-1}$ and $\varepsilon \boldsymbol{l} = +4\&{-4}$ are shown in **Figure 13a–l**. The simultaneous recovery of both phase and amplitude distributions corresponding to specific OAM modes provides direct advantage of the quantitative determination of magnitude and sign of the topological charge of the generated OAM modes on evaluation of the recovered phase. The recovered phase distributions shown in **Figure 12i–l** in experiment and **Figure 13i–l** in simulation exhibit a phase variation around the singularity in the order of $2l\pi$ with the number of phase jumps equal to the value of l. In addition, the quantitative analysis on the recovered phase distribution allows the direct estimation of the sign of the TC of OAM modes based on the direction of rotation of helicity of the respective modes. An anticlockwise rotation of phase distribution as shown in **Figure 12i–k** and **Figure 13i** and **k** corresponds to positive TC and a clockwise rotation of phase distribution as shown in **Figure 12j** and **l** and **Figure 13j** and **l** corresponds to negative TC of OAM modes.

4.3 Characterization of OAM modes

Furthermore, the recovered OAM modes were characterized by performing a quantitative analysis by making use of the orthogonal projection method [66–68].

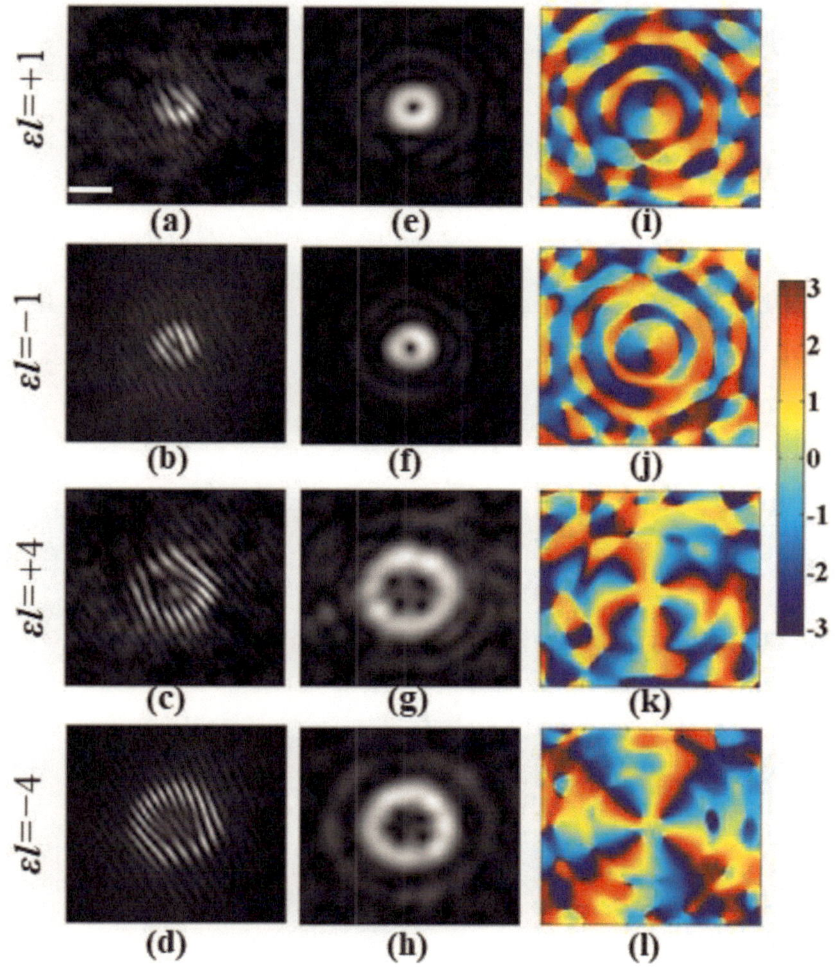

Figure 12.
Experimental results: (a–d) retrieved correlation holograms; (e–h) recovered amplitude distributions of the OAM modes; and (i–l) recovered phase distributions of the OAM modes; scale bar: 110 μm. Adapted from [50].

The projection techniques were utilized to examine the angular momentum associated the field distribution of the light beam. The OAM modes encoded in the recovered correlation function is examined by projecting into respective spiral harmonics $exp\,(il\phi)$. The determination of OAM power spectrum gives the provision of decomposition of each OAM components in terms of azimuthal modes. The simulation and experimental results corresponding to the orthogonal projection analysis shown in **Figure 14a–h**, respectively, which exhibits a good agreement. Moreover, an experimental recovery and characterization is performed with the GDH system for OAM modes with different TCs $|\varepsilon \boldsymbol{l}| = 1$ to 11, and the quantitative characterization results of respective OAM modes are represented in the plot shown in **Figure 14**. The plot illustrates the experimentally generated OAM modes of $|\varepsilon \boldsymbol{l}| = 1$ to 11 and their respective simulation results. The inner and outer radii of OAM modes were retrieved from the respective amplitude distribution of correlation function corresponding to various TCs recovered using the GDH system. The direct dependence of doughnut structure of the amplitude distribution of OAM mode is demonstrated in the plot. The variation of inner and outer radii of the recovered

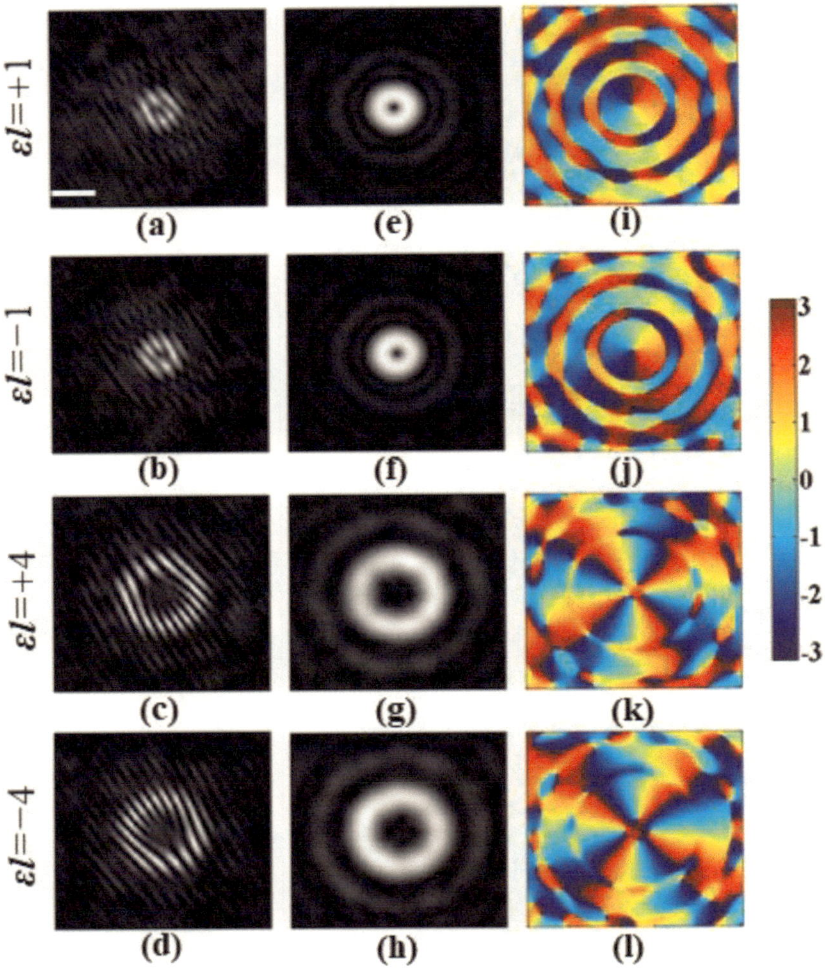

Figure 13.
Simulation results: (a–d) retrieved correlation holograms; (e–h) recovered amplitude distributions of the OAM modes; and (i–l) recovered phase distributions of the OAM modes; scale bar: 110 μm. Adapted from [50].

modes with various TC are shown in **Figure 15a**, and the variation of areas of dark core and the annular bright rings of the doughnut structure with the TC are shown in **Figure 15b**. The dark core and the annular bright ring have a direct relation with the order of the OAM mode, where an increase in the order of the vortices makes a corresponding increase in the areas of the recovered OAM mode distribution. A good agreement is observed in the quantitative comparison of experimental and theoretical results for all the recovered modes with various l values, which highlight the potential of the technique in recovery and characterization of any OAM modes.

5. Discussion and summary

The development of unconventional correlation assisted GDH design capable of simultaneous detection of ghost diffraction intensity distribution by synchronizing

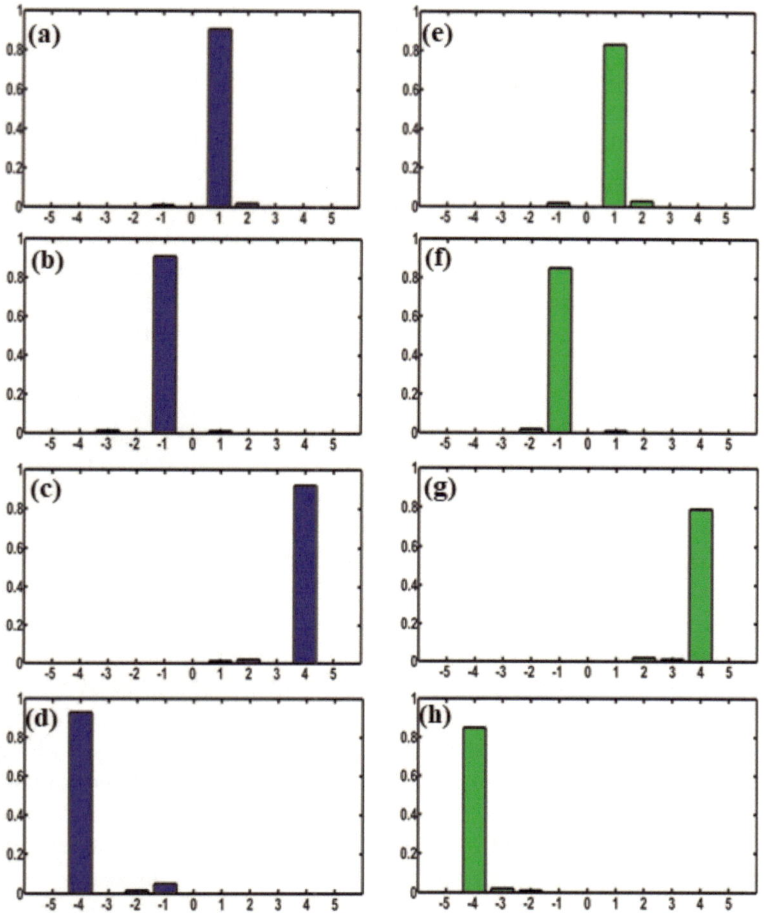

Figure 14.
OAM distribution with X-axis the topological charge (εl) *and Y-axis the OAM power spectrum* (P(l))*; (a)–(d) simulation results; and (e)–(h) experimental results.*

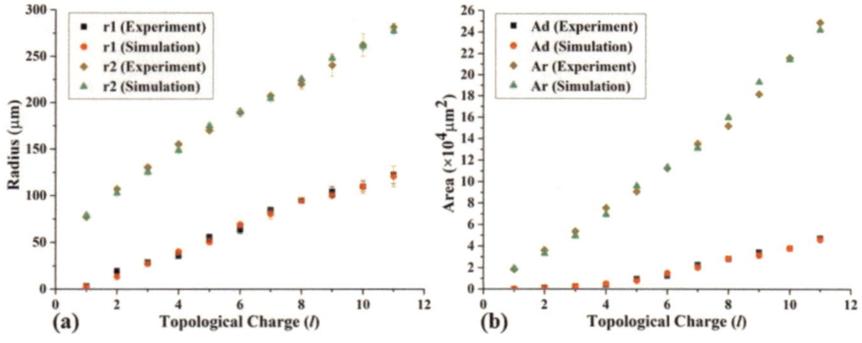

Figure 15.
Quantitative comparison of experimental and simulation results of various OAM modes recovered using GDH system: (a) variation of inner and outer radii with the topological charge (r1, inner radius and r2, outer radius); (b) variation of area of dark core (Ad, dark core) and area of bright annular ring (Ar, annular ring) of the recovered OAM modes. Adapted from [50].

the detectors and recording at an instant of time expands the applied domain of ghost schemes to quantitative phase imaging. The GDH system exploits the wide-sense spatial stationarity of the scattered fields at the detector plane and implements the spatial averaging approach as a replacement of ensemble averaging in the execution of correlation of intensity fluctuations. In recent times, a significant number of research efforts in correlation-based imaging and characterization techniques utilizes the potential features of spatial statistical optics [41, 43, 69–74], but the approach has only very few executions in ghost scenarios [29]. The utilization of spatial averaging provides the potential opportunity to develop the snapshot GDH system, where the single shot detection of two intensity distributions corresponding to the ghost diffraction fields are sufficient to retrieve the ghost correlation hologram, thereby the simultaneous quantitative phase and amplitude imaging.

The extension of the GDH design to the domain of microscopy for imaging of spatially varying complex-valued objects with high-speed and high spatial resolution configurations enlarge the application of the system to quantitative phase microscopy with potential implications in biomedical imaging. Moreover, the spatial resolution of the microscopy system is not only governed by the numerical aperture of the microscopy configuration but also influence on the size of the speckle grains illuminating the object. The design of GDHM system consisting of a variable aperture placed at the rotating diffuser plane controls the beam size illuminating the diffuser, and thereby ensure the proper speckle grain size to improve the resolution of the system. Furthermore, the complex-correlation function recovery feature of the GDH system is exploited to develop a new basis for the recovery of OAM modes encoded in a spatially incoherent random field. The OAM mode propagation and recovery challenge in inhomogeneous or turbulent atmosphere is tackled by exploiting the robust propagation characteristics of spatially incoherent speckle field generated from the pseudo-thermal source, and designing a holography assisted GD scheme with simultaneous determination of amplitude and phase of the encoded vortex modes.

In this chapter, we have presented the recent advancements in the ghost scenarios with the adoption of an unconventional correlation assisted holography scheme for complex-valued object imaging and characterization. The quantitative imaging feature of the GDH technique in comparison to other ghost approaches is established both theoretically and experimentally by imaging of various spatially varying macroscopic and microscopic samples. The implementation of spatial averaging with time-frozen intensity patterns to retrieve the ghost correlation hologram provides the advantage of the real-time imaging with new application dimensions in holography, microscopy, and tomography. Finally, the utilization of GDH scheme for the quantitative recovery of scrambled OAM modes encoded in the random speckle pattern is expected to find significant applications in sorting of superposed optical vortex modes, optical communication, multi-dimensional imaging, quantum and classical information processing, imaging through turbid media, etc.

Acknowledgements

Support from the National Natural Science Foundation of China (NSFC) under grant 12150410318.

Author details

Vinu Raveendran Pillai Vasantha Kumari[1*], Ziyang Chen[1], Rakesh Kumar Singh[2] and Jixiong Pu[1]

1 College of Information Science and Engineering, Fujian Provincial Key Laboratory of Light Propagation and Transformation, Huaqiao University, Xiamen, China

2 Laboratory of Information Photonics and Optical Metrology, Department of Physics, Indian Institute of Technology (BHU), Varana, India

*Address all correspondence to: vinurv@hqu.edu.cn

References

[1] Padgett M, Aspden R, Gibson G, et al. Ghost imaging. OPN. 2016;**27**:38-45

[2] Hoenders BJ. Review of a bewildering classical-quantum phenomenon: Ghost imaging. In: Advances in Imaging and Electron Physics. Vol. 208. San Diego: Academic Press Inc.; 2018. pp. 1-41

[3] Shapiro JH, Boyd RW. The physics of ghost imaging. Quantum Information Processing. 2012;**11**:949-993

[4] Shih Y. The physics of turbulence-free ghost imaging. Technologies. 2016; **4**:39

[5] Padgett MJ, Boyd RW. An introduction to ghost imaging: Quantum and classical. Philosophical Transactions of the Royal Society A: Mathematical, Physical and Engineering Sciences. 2017; **375**:20160233

[6] Moreau P-A, Toninelli E, Gregory T, et al. Ghost imaging using optical correlations. Laser Photonics Reviews. 2018;**12**:1700143

[7] Ryczkowski P, Barbier M, Friberg AT, et al. Ghost imaging in the time domain. Nature Photonics. 2016;**10**: 167-170

[8] Janassek P, Blumenstein S, Elsäßer W. Ghost spectroscopy with classical thermal light emitted by a Superluminescent diode. Physical Review Applied. 2018;**9**:1-6

[9] Shapiro JH. Computational ghost imaging. Physical Review A. 2008;**78**: 61802

[10] Devaux F, Moreau P-A, Denis S, et al. Computational temporal ghost imaging. Optica. 2016;**3**:698

[11] Erkmen BI, Shapiro JH. Ghost imaging: From quantum to classical to computational. Advances in Optics and Photonics. 2010;**2**:405-450

[12] Belinskii A, v, Klyshko DN. Two-photon optics: Diffraction, holography, and transformation of two-dimensional signals. Journal of Experimental and Theoretical Physics. 1994;**78**:259-262

[13] Pittman TB, Shih YH, Strekalov D v, et al. Optical imaging by means of two-photon quantum entanglement. Physical Review A 1995; 52: R3429-R3432.

[14] Gatti A, Brambilla E, Bache M, et al. Ghost imaging with thermal light: Comparing entanglement and classical correlation. Physical Review Letters. 2004;**93**:93602

[15] Ferri F, Magatti D, Gatti A, et al. High-resolution ghost image and ghost diffraction experiments with thermal light. Physical Review Letters. 2005;**94**: 183602

[16] Borghi R, Gori F, Santarsiero M. Phase and amplitude retrieval in ghost diffraction from field-correlation measurements. Physical Review Letters. 2006;**96**:183901

[17] Duarte MF, Davenport MA, Takhar D, et al. Single-pixel imaging via compressive sampling. IEEE Signal Processing Magazine. 2008;**25**:83-91

[18] He Y, Wang G, Dong G, et al. Ghost imaging based on deep learning. Scientific Reports. 2018;**8**:6469

[19] Rizvi S, Cao J, Zhang K, et al. DeepGhost: Real-time computational ghost imaging via deep learning. Scientific Reports. 2020;**10**:11400

[20] Ferri F, Magatti D, Lugiato LA, et al. Differential ghost imaging. Physical Review Letters. 2010;**104**:253603

[21] Losero E, Ruo-Berchera I, Meda A, et al. Quantum differential ghost microscopy. Physical Review A. 2019; **100**:63818

[22] Li Y-X, Yu W-K, Leng J, et al. Pseudo-thermal imaging by using sequential-deviations for real-time image reconstruction. Optics Express. 2019;**27**:35166-35181

[23] Kingston Andrew M, Pelliccia D, Rack A, et al. Ghost tomography. Optica. 2018;**5**:1516-1520

[24] Pelliccia D, Rack A, Scheel M, et al. Experimental X-ray ghost imaging. Physical Review Letters. 2016; **117**:113902

[25] Olbinado MP, Paganin DM, Cheng Y, et al. Phase-sensitive x-ray ghost imaging. 2019

[26] Li S, Cropp F, Kabra K, et al. Electron ghost imaging. Physical Review Letters. 2018;**121**:114801

[27] Khakimov RI, Henson BM, Shin DK, et al. Ghost imaging with atoms. Nature. 2016;**540**:100-103

[28] Abouraddy AF, Stone PR, Sergienko AV, et al. Entangled-photon imaging of a pure phase object. Physical Review Letters. 2004;**93**:213903

[29] Bache M, Magatti D, Ferri F, et al. Coherent imaging of a pure phase object with classical incoherent light. Physical Review A. 2006;**73**:53802

[30] Zhang M, Wei Q, Shen X, et al. Lensless Fourier-transform ghost imaging with classical incoherent light. Physical Review A. 2007;**75**:21803

[31] Gong W, Han S. Phase-retrieval ghost imaging of complex-valued objects. Physical Review A. 2010;**82**: 23828

[32] Zhang DJ, Tang Q, Wu TF, et al. Lensless ghost imaging of a phase object with pseudo-thermal light. Applied Physics Letters. 2014;**104**:121113

[33] Shirai T, Friberg AT. Ghost Mach-Zehnder interferometry for phase measurement with spatially incoherent light. Journal of Optics. 2020;**22**:045604

[34] Clemente P, Durán V, Tajahuerce E, et al. Single-pixel digital ghost holography. Physical Review A. 2012;**86**: 41803

[35] Shirai T, Setälä T, Friberg AT. Ghost imaging of phase objects with classical incoherent light. Physical Review A. 2011;**84**:41801

[36] Vinu RV, Chen Z, Singh RK, et al. Ghost diffraction holographic microscopy. Optica. 2020;**7**:1697-1704

[37] Huang Y, Chen Z, et al. Snapshot ghost diffraction imaging based on spatial averaging. Proceedings of SPIE. Advanced Optical Imaging Technologies IV. 2021;**11896**:16-21. DOI: 101117/ 122600803

[38] Singh RK, Vinu RV, Chen Z, et al. Quantitative phase recovery in ghost imaging. In: 2021 IEEE Photonics Conference, IPC Proceedings, Vancouver. 2021. DOI: 10.1109/ IPC48725.2021.9592877

[39] Goodman JW. Speckle Phenomena in Optics: Theory and Applications. Englewood: Coberts &Co; 2007

[40] Goodman JW. Statistical Optics. NJ: Wiley-Interscience; 2000. DOI: 10.1017/ CBO9781107415324.004

[41] Takeda M, Wang W, Naik DN, et al. Spatial statistical optics and spatial correlation holography: A review. Optical Review. 2014;**21**:849-861

[42] Takeda M. Spatial stationarity of statistical optical fields for coherence holography and photon correlation holography. Optics Letters. 2013;**38**: 3452-3455

[43] Singh RK, Vyas S, Miyamoto Y. Lensless Fourier transform holography for coherence waves. Journal of Optics. 2017;**19**:115705

[44] Singh RK, Vinu RV, Sharma MA. Recovery of complex valued objects from two-point intensity correlation measurement. Applied Physics Letters. 2014;**104**:111108

[45] Vinu RV, Singh RK. Determining helicity and topological structure of coherent vortex beam from laser speckle. Applied Physics Letters. 2016;**109**: 111108

[46] Takeda M, Ina H, Kobayashi S. Fourier-transform method of fringe-pattern analysis for computer-based topography and interferometry. Journal of the Optical Society of America. 1982; **72**:156-160

[47] Goodman JW. Introduction to Fourier Optics. New York: McGraw-Hill; 1996

[48] Vyas S, Senthilkumaran P. Interferometric optical vortex array generator. Applied Optics. 2007;**46**: 2893-2898

[49] Singh RK, Sharma AM, Senthilkumaran P. Vortex array embedded in a partially coherent beam. Optics Letters. 2015;**40**: 2751-2754

[50] Huang Y, Vinu RV, Chen Z, et al. Recovery and characterization of orbital angular momentum modes with ghost diffraction holography. Applied Sciences. 2021;**11**:12167

[51] Yao AM, Padgett MJ. Orbital angular momentum: Origins, behavior and applications. Advances in Optics and Photonics. 2011;**3**:161

[52] Chen R, Zhou H, Moretti M, et al. Orbital angular momentum waves: Generation, detection, and emerging applications. IEEE Communications Surveys and Tutorials. 2020;**22**:840-868

[53] Willner AE, Ren Y, Xie G, et al. Recent advances in high-capacity free-space optical and radio-frequency communications using orbital angular momentum multiplexing. Philosophical Transactions of the Royal Society A: Mathematical, Physical and Engineering Sciences. 2017;**375**. DOI: 10.1098/RSTA.2015.0439

[54] Gori F, Santarsiero M, Borghi R, et al. Partially coherent sources with helicoidal modes. Journal of Modern Optics. 1998;**45**:539-554

[55] Gbur G, Visser TD. Phase singularities and coherence vortices in linear optical systems. Optics Communications. 2006;**259**:428-435

[56] Zeng J, Lin R, Liu X, et al. Review on partially coherent vortex beams. Frontiers of Optoelectronics. 2019;**12**:229-248

[57] Jesus-Silva AJ, Hickmann JM, Fonseca EJS. Strong correlations between incoherent vortices. Optics Express. 2012;**20**:19708

[58] Reddy SG, Prabhakar S, Kumar A, et al. Higher order optical vortices and formation of speckles. Optics Letters. 2014;**39**:4364

[59] Salla GR, Perumangattu C, Prabhakar S, et al. Recovering the vorticity of a light beam after scattering. Applied Physics Letters. 2015;**107**: 021104

[60] Yang Y, Mazilu M, Dholakia K. Measuring the orbital angular momentum of partially coherent optical vortices through singularities in their cross-spectral density functions. Optics Letters. 2012;**37**:4949

[61] Lu X, Zhao C, Shao Y, et al. Phase detection of coherence singularities and determination of the topological charge of a partially coherent vortex beam. Applied Physics Letters. 2019;**114**: 201106

[62] Wang T, Pu J, Chen Z. Propagation of partially coherent vortex beams in a turbulent atmosphere. Optical Engineering. 2008;**47**:036002

[63] Jesus-Silva AJ, Alves CR, Fonseca EJS. Robustness of a coherence vortex. Applied Optics. 2016;**55**:7544-7549

[64] Chen J, Li Y. Discrimination of incoherent vortex states of light. Optics Letters. 2018;**43**:5595

[65] Bezerra DO, Amaral JP, Fonseca EJS, et al. Sorting of spatially incoherent optical vortex modes. Scientific Reports. 2020;**10**:10

[66] Sarkar T, Parvin R, Brundavanam MM, et al. Higher-order stokes-parameter correlation to restore the twisted wave front propagating through a scattering medium. Physical Review A. 2021;**104**:013525

[67] D'Errico A, Piccirillo B, Cardano F, et al. Measuring the complex orbital angular momentum spectrum and spatial mode decomposition of structured light beams. Optica. 2017;**4**:1350-1357

[68] Molina-Terriza G, Torres JP, Torner L. Management of the Angular Momentum of light: Preparation of photons in multidimensional vector states of angular momentum. Physical Review Letters. 2001;**88**:013601

[69] Naik DN, Singh RK, Ezawa T, et al. Photon correlation holography. Optics Express. 2011;**19**:1408-1421

[70] Takeda M, Singh AK, Naik DN, et al. Holographic Correloscopy-unconventional holographic techniques for imaging a three-dimensional object through an opaque diffuser or via a Scattering Wall: A review. IEEE Transactions on Industrial Informatics. 2016;**12**:1631-1640

[71] Kim K, Somkuwar AS, et al. Imaging through scattering media using digital holography. Optics Communications. 2019;**439**:218-223

[72] Chen L, Chen Z, Singh RK, et al. Increasing field of view and signal to noise ratio in the quantitative phase imaging with phase shifting holography based on the Hanbury Brown-Twiss approach. Optics and Lasers in Engineering. 2022;**148**:106771

[73] Vinu RV, Singh RK. Experimental determination of generalized stokes parameters. Optics Letters. 2015;**40**: 1227-1230

[74] Vinu RV, Singh RK. Synthesis of statistical properties of a randomly fluctuating polarized field. Applied Optics. 2015;**54**:6491-6497

Chapter 3

Standardization Techniques for Single-Shot Digital Holographic Microscopy

Kedar Khare

Abstract

Digital holographic microscopy (DHM) is a mature technology for quantitative phase imaging. Thousands of articles have been published on this topic over the last couple of decades. Our goal in this article is to emphasize that single-shot holographic microscopy systems offer several practical advantages and in principle capture the full diffraction-limited information of interest. Since phase cannot be measured directly, phase reconstruction is inherently a computational problem. In this context, we describe some traditional algorithmic ideas as well as newer sparse optimization-based methodologies for phase reconstruction from single-shot holograms. Robust operation of a DHM system additionally requires a number of auxiliary algorithms associated with fractional fringe detection, phase unwrapping, detection of focus plane, etc., that will be discussed in some detail. With the data-driven nature of applications of DHM being developed currently, the standardization or benchmarking of algorithmic ideas for DHM systems is important so that same sample imaged by different DHM systems provides the same numerical phase maps. Such uniformity is also key to establishing effective communication between DHM developers and potential users and thereby increasing the reach of the DHM technology.

Keywords: quantitative phase imaging, microscopy, sparse optimization, phase unwrapping, autofocus methods for phase objects

1. Introduction

Digital holography (DH) is an interferometric imaging modality [1–4] which enables recording and numerical reconstruction of phase map of a coherent light beam that has interacted with the object of interest. In DH, the interference pattern is recorded on an array sensor. As a result, unlike traditional film-based holography, the interference fringe data is now available as a matrix of numbers. This numerical nature of the data enables one to apply a number of algorithmic ideas for quantitative phase estimation and 3D image formation. While film-based holograms were typically reconstructed by re-illumination with the conjugate reference beam, DH does not necessarily have to mimic this physical hologram replay process. As we will see in this article, novel algorithmic ideas can in principle take DH beyond film based holographic imaging performance. Digital form of holography was first demonstrated in early work [5] where recording of holograms was performed using a vidicon tube array detector and a fast Fourier transform (FFT)-based numerical reconstruction was demonstrated. The DH modality gained much popularity starting early 1990s when charge-coupled device (CCD) and complementary metal oxide semiconductor (CMOS) array sensors became readily available with simultaneous increase in computational power at low cost. Since then, DH has has received a lot of attention in the Science and Engineering literature. A quick web search for the term "digital holography" yields over 150,000 documents, and interestingly almost half of them are related to "digital holographic microscopy" (DHM) suggesting the attention received by DHM as an imaging modality. Compared to intensity or amplitude-based microscopic imaging modalities commonly in use in Life Sciences, quantitative phase is known to be more sensitive modality and provides a much more complete information about a sample object under investigation. Quantitative phase imaging further allows the possibility of label-free live cell imaging. The minimal wet-lab processing required for label-free imaging also makes quantitative phase a potentially cost-effective imaging modality. A large number of publications on DHM continue to appear (mainly in the physics, optics, and engineering journals); however, it is still not used widely as a preferred imaging modality in the Life Sciences community. For example, if one walks into a typical bioscience research laboratory in a university setting or a pathology clinic in a hospital, the researchers and technicians there are most likely not using quantitative phase as an imaging modality in their work. While DHM can be considered as an emerging modality at present, its widespread use in future will require standardization and benchmarking of algorithmic methodologies as well as systems. In particular with a number of system configurations and algorithmic combinations being used, it is not clear if we will get identical numerical phase maps if the same sample is imaged by two different DHM systems. This aspect is very important in the light of current trend of data-driven imaging application development [6]. Otherwise, individual applications will likely become system or algorithm-specific, thus limiting their usage. This article summarizes a number of ideas from our work over the last several years that are relevant to standardization of algorithms for single-shot DHM systems that record image plane digital holograms.

The organization of this article is as follows. In Section 2, we discuss the importance of employing a single-shot DHM system, first from the perspective of simplicity of deploying such a system in practical settings. We additionally argue that the single-shot holographic image record actually contains all the diffraction-limited information about the object field that is of interest. In Section 3, we present algorithmic methodologies that may be used to demodulate a single-shot hologram record with emphasis on the sparse-optimization-based approaches and their noise and resolution advantage. Beyond the core requirement of good-quality phase reconstruction, additional auxiliary methodologies are required to make the DHM imaging performance robust. In Section 4, we discuss a few algorithmic methods for this purpose that make a DHM instrument more user-friendly. Section 5 describes some interesting issues related to computational 3D image formation from holographic data which is an important research topic in itself. We conclude the discussion in Section 6 by providing further insights gained from our experience of working with clinicians and emphasize the need for a DHM user consortium which will help popularization of DHM technologies within the Life Sciences community.

2. Single-shot DHM systems: motivation

A DHM system in bare-bones form is shown in **Figure 1**. The system as shown here mainly comprises of a coherent plane wave illuminating a sample of interest, which is then imaged using an afocal imaging system consisting of a microscope objective (MO) and a tube lens. The array sensor is placed in the image plane so that an image field $O(x, y)$ corresponding to the magnified version of the exit wave in the sample plane is incident on the sensor. Since DHM is an interferometric modality, a part of the illuminating beam is split beforehand and is recombined at the sensor. The reference beam field at the sensor may be denoted as $R(x, y)$. If sufficient temporal coherence is available and the path lengths in the interferometer are matched well, an interference pattern with good contrast may be recorded on the sensor, which may be denoted as:

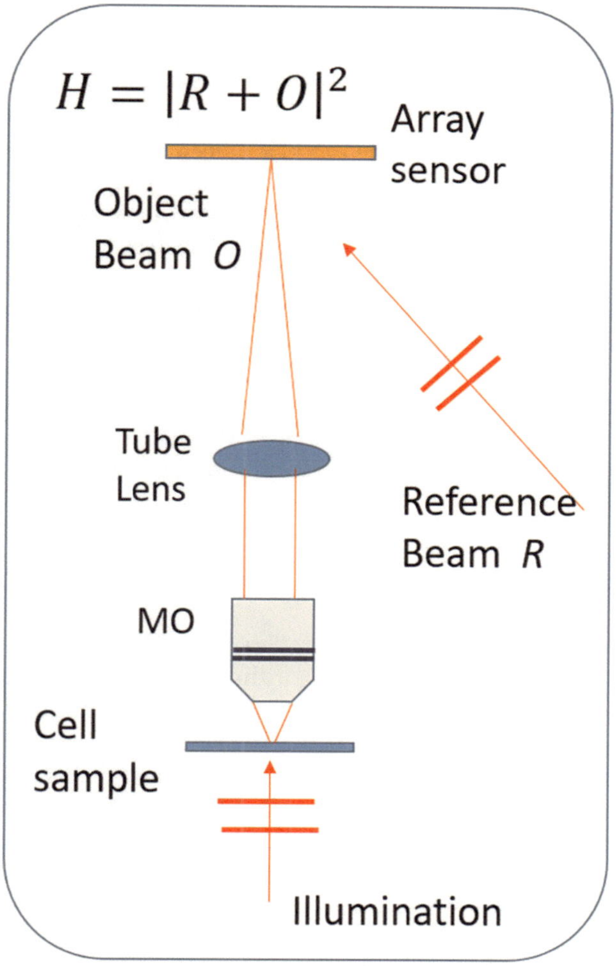

Figure 1.
Single-shot digital holographic microscopy (DHM) in a bare-bones form. The basic imaging system is an afocal infinity-corrected microscope which forms image field $O(x, y)$ at the sensor plane containing full diffraction-limited information as permitted by the microscope objective (MO). The reference beam $R(x, y)$ is derived from the same illuminating source. The signal detected at the sensor consists of off-axis interference fringes.

$$H(x,y) = |R(x,y) + O(x,y)|^2. \tag{1}$$

If future DHM systems are to be deployed in field, e.g. at a clinic, one may not always have access to research laboratory like environment with vibration isolation platforms. Luckily, the CCD or CMOS arrays available off-the-shelf at reasonable costs today are quite sensitive. As a result, a DHM system employing a few mW laser source requires just a few milliseconds of exposure time in order to record good-quality interference fringes with high contrast. Such short exposure times (much shorter compared to the timescale of typical ambient vibration) imply that vibration isolation is not required for recording a single frame of hologram data. The configuration in **Figure 1** further suggests that the image field $O(x,y)$ contains full diffraction-limited information that one may wish to recover. So in principle, a single recorded data frame $H(x,y)$ contains interference fringes with good contrast and also embeds full diffraction-limited information in it. Multi-shot DHM systems, e.g. employing the phase shifting technique, on the other hand require stringent vibration isolation platforms. Additionally, there is a need of deploying piezo-transducers for introducing three or four phase shits in a repeatable manner for generation of a single phase image. The vibration isolation and phase shifting hardware can increase the system costs beyond what is affordable in a place like a primary healthcare center. The necessity of multiple interference records for a phase-shifting-based DHM system also suggests that it cannot perform phase imaging of dynamic samples as may be of interest to researchers engaged in live cell imaging. As described in the next section, traditionally the discussion regarding use of single-shot vs. multi-shot (or phase shifting) DHM systems has concentrated on achievable image resolution in the two methodologies. However, the possibility of full diffraction-limited imaging performance in single-shot DHM records via sparse optimization algorithms as we will present here addresses this debate to some extent. We can now therefore envision DHM systems that offer all practical advantages of a single-shot operation and at the same time do not suffer in terms of loss of resolution or phase accuracy.

3. Single-shot phase reconstruction algorithms

Having made a case for single-shot operation of DHM systems, we now turn to DHM configurations and corresponding algorithms for phase reconstruction. In recording of a digital hologram, two geometrical configurations are typically encountered. The off-axis configuration involves a plane reference beam $R(x,y)$ which nominally makes an angle with respect to the object beam $O(x,y)$ as shown in **Figure 1**. The second one is the in-line hologram recording configuration where the reference and object beam are nominally colinear. The in-line configuration is typically employed when we wish to image weak scattering objects like particulates or microorganisms floating in a fluid medium with a DHM system. Colinear nature of such systems makes them more compact which in itself offers several practical advantages. When morphology of more complex structures is to be imaged, the weak scattering approximation may not be fully valid, and in such cases, the off-axis configuration is preferred. The off-axis configuration has received a lot of attention in the literature. Let us for instance assume a reference beam for the off-axis case to be of the form: $R(x,y) = R_0 \exp\left(i2\pi f_{0x}x\right)$. The interference signal as in Eq. (1) can then be written more explicitly as:

$$H(x, y) = |O|^2 + |R_0|^2 + 2|O||R_0| \cos\left[2\pi f_{0x}x - \phi_O(x,\ y)\right]. \tag{2}$$

In the aforementioned relation, $\phi_O(x, y)$ represents the object beam phase. The special form of the reference beam means that the Fourier transform structure of the hologram signal is quite simple. In particular, the first two terms of the hologram are located at the center of the Fourier space and the cosine term leads to two lobes in Fourier space centered on the carrier fringe frequency as shown in **Figure 2**. Note that if the spatial bandwidth of the function $O(x, y)$ is given by $2B_x$ in the $x-$ direction, the bandwidth of the $|O|^2$ is $4B_x$. The bandwidth of $|R_0|^2$ is practically well within the bandwidth of $|O|^2$ as it most contains some low-frequency variations. In the traditional film-based holography, the replay of the off-axis hologram would involve re-illuminating the developed film by conjugate reference beam. The digital processing equivalent of this procedure if the filter out the cross-term represented by the side lobe, place it in the center of the Fourier space, and evaluate inverse Fourier transform in order to get an estimation of the object beam $O(x, y)$ at the array sensor plane [7]. As seen from **Figure 2**, we however note that in order to recover the full information about the object wave, the three terms cannot overlap in the Fourier space. If the hologram can be considered to be sampled well on the digital array sensor, then the non-overlap requirement on the three energy lobes in Fourier space is highly restrictive as it implies that the allowable bandwidth of the object wave is much lower compared to the bandwidth of the array sensor itself. In other words, the filtering methodology inherently leads to lower-resolution object wave recovery, despite the fact that a higher resolution information is present in the hologram data. With the hologram being sampled on digital array sensors with typical pixel sizes of 2–5 μm already restricts the reference beam angle in order to get well-sampled fringes on the sensor and the non-overlap requirement on Fourier bands further worsens the situation. A relaxation in the overlap condition to enable higher resolution is achieved by two notable approaches [8, 9]. If the afocal microscope system as in **Figure 1** was used with typical incoherent bright-field illumination (e.g. using LED or lamp), the user is expected to get full diffraction-limited image resolution. However, as soon as one switches to the digital holography mode with coherent illumination and introduction of off-axis reference beam, the recovered resolution is worse compared to the diffraction-limited resolution when linear filtering methodologies are used. This

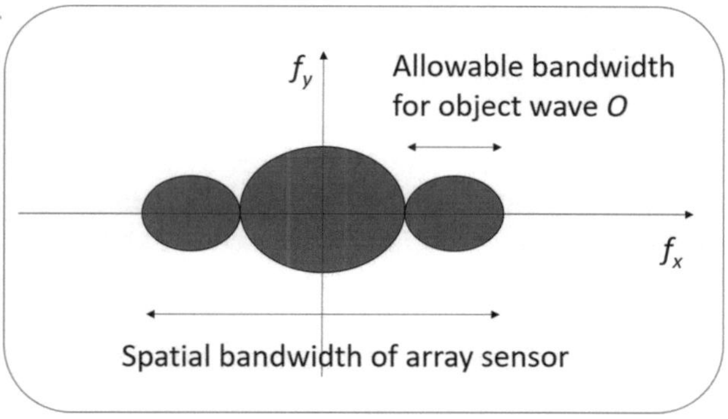

Figure 2.
Fourier space representation of the off-axis hologram. Note that if the hologram is processed by filtering of the cross-term, then the allowable bandwidth for object wave is much lower compared to the bandwidth of the detector array.

situation is highlighted in image plane holography as poor resolution implies loss of edge information which is critical to visual perception of the image field by users. We emphasize here that the loss of image resolution as discussed here is an artifact of the Fourier filtering-based algorithmic framework used for object field recovery which prevents us to recover the full pixel resolution. The single-shot object field recovery problem therefore needs to be examined further. We observe that the recorded hologram represented by Eq. (2) is a quadratic equation in terms of the amplitude $|O|$ of the object beam. A solution for the object beam amplitude can thus be expressed as:

$$|O| = -|R_0|\cos\left(2\pi f_{0x}x - \phi_O\right) \pm \sqrt{H - |R_0|^2 \sin^2\left(2\pi f_{0x}x - \phi_O\right)}. \quad (3)$$

This relation is curious as it suggests that the magnitude $|O|$ explicitly depends on the phase ϕ_O of the object beam itself. One may now choose any arbitrary phase map ϕ_O, compute $|O|$ as per Eq. (3), and the corresponding complex-valued object field will exactly satisfy the hologram data represented by $H(x, y)$. The solution of the single-shot holography problem is therefore ambiguous when we consider the problem in this numerical form. More specifically, since a single hologram frame $H(x, y)$ is to be used to estimate the complex-valued object wave function $O(x, y)$ (assuming the prior knowledge of the reference beam $R(x, y)$), the numerical problem of estimating $O(x, y)$ from a single hologram frame $H(x, y)$ can be considered to be an incomplete data problem. The particular form of the reference beam played no role in this incomplete data argument, and so this solution ambiguity exists for in-line holography case as well. In the in-line case, various terms of the hologram signal actually overlap completely in the Fourier space, and there is no possibility of obtaining a solution for $O(x, y)$ by means of a simple linear filtering operation. The off-axis case at least provides an approximate lower-resolution estimate of the unknown object wave $O(x, y)$ due to the particular Fourier structure of the hologram signal. The size of Fourier space window to be used to filter out the cross-term, however, still remains a subjective choice. In **Figure 3(a)**, we show an off-axis image plane hologram for a red blood cell (RBC) sample. The recovered phase map of the object beam is shown in **Figure 3(b)–(d)** for three choices of filter windows of sizes of sizes $0.2\rho_0$, $0.5\rho_0$, and $0.7\rho_0$, respectively, where ρ_0 refers to the distance between the zero frequency and the carrier frequency peak positions.

The three phase reconstructions differ in terms of their resolution and background artifacts which may have arisen due to the contribution of the dc or zero-frequency terms to the reconstruction. The choice of filter sizes is generally left to the user; however, this may lead to variability of numerical phase maps between different users which will generally not be acceptable to the user community. Overall, the aforementioned discussion suggests that the phase reconstruction problem for single-shot image plane holograms is not quite simple. A solution maybe designed for one setup for a particular sample by a given group of researchers. But translating this solution to a robust system to be used by third-party users is a nontrivial proposition. Solving such variability issues and obtaining a full pixel resolution seem to be possible with the phase shifting methodology which is however a multi-shot approach not suitable for practical deployment of DHM systems. Since the single-shot phase reconstruction is an incomplete data problem, the use of image sparsity ideas popularized by the compressive sensing community can be handy. This, however, requires one to change the reconstruction framework to an optimization procedure as we will explain in detail in the following subsection.

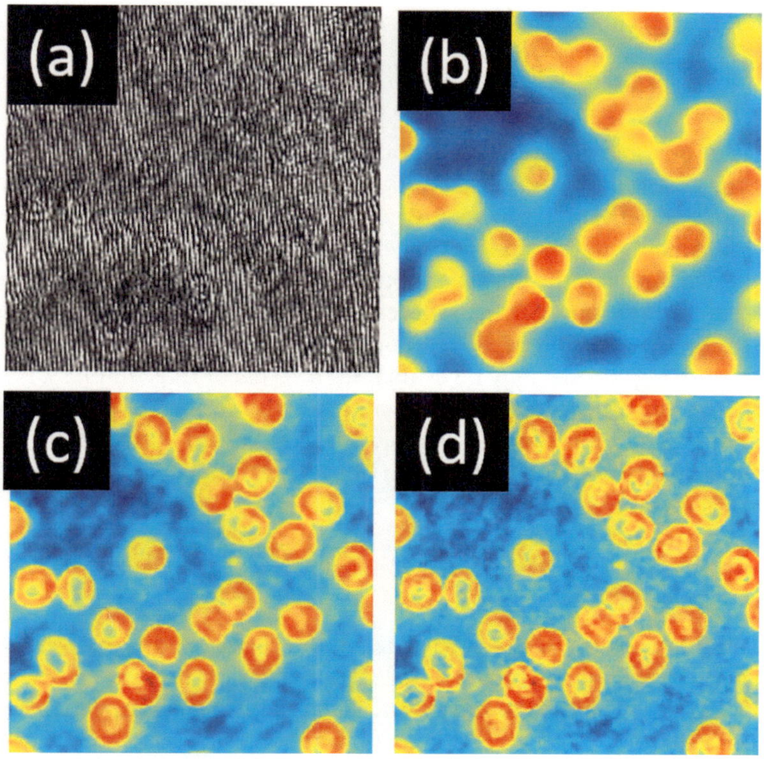

Figure 3.
(a) Image plane digital hologram of RBCs. (b)–(d) Phase of object beam determined by Fourier filtering of the cross-term with filter window sizes $0.2\rho_0$, $0.5\rho_0$, and $0.7\rho_0$, where ρ_0 is the distance between the zero frequency and the carrier frequency peak positions. The variability of resolution and background artifacts makes this methodology highly subjective.

3.1 Single-shot phase reconstruction as an optimization problem

It is now a widely accepted fact in signal and image processing community that natural images have number of degrees of freedom that are much smaller than the number of visual pixels used to represent them. In fact, a number of image and video compression standards like JPEG or MPEG regularly exploit this redundancy in natural images that arise due to structures of the objects present in the typical images. The same ideas must apply to phase reconstruction as well. In particular, when observing objects like biological cells using a DHM, one should be able to exploit the representational redundancy of the desired complex-valued field $O(x, y)$ at the sensor plane corresponding to the objects being imaged. As explained before, the Fourier filtering solution for the off-axis holography essentially has origins in the way holograms used to be replayed in film-based holography. A different framework for complex object wave recovery is, however, possible in digital holography as the interference record is now available in the numerical form, and there is no need to mimic film-based holography in the numerical processing of this numerical data. In the following discussion, we describe an optimization framework [10] for recovering the object wave $O(x, y)$ from an image plane hologram $H(x, y)$ that may help in addressing the difficulties associated with the single-shot phase reconstruction problem. In particular, we seek to minimize the functional:

$$C(O, O^*) = C_1 + C_2 \\ = \left\| H - \left(|R|^2 + |O|^2 + R^* O + R O^* \right) \right\|^2 + \alpha \ \psi(O, O^*). \tag{4}$$

Here, $\| \dots \|^2$ denotes the squared L2-norm of the quantity inside the norm. The first term C_1 of the cost function is, therefore, a least-square data fit to the interference model. The second term refers to a constraint or penalty term which encourages solution with some desired properties. The positive constant α decides the weight between the two terms of the cost function. Note that the overall cost function in Eq. (4) is real- and positive-valued, whereas the solution we are seeking is complex-valued. In such cases, the steepest descent direction needed for the purpose of iterative optimization algorithms may be evaluated in terms of the complex or Wirtinger derivatives. The Wirtinger derivative is evaluated with respect to the variable O^* and makes the implementation almost as simple as a real-valued optimization problem. The Wirtinger derivative for the first term of the cost function is straightforward to evaluate and is given by:

$$\nabla_{O^*} C_1 = -2\left[H - |R + O|^2\right] \cdot (R + O), \tag{5}$$

A few suitable choices for the penalty term, the corresponding expressions for the Wirtinger derivatives, and the characteristic properties of the penalty functions are provided in **Table 1**. The first penalty is the squared gradient sum which encourages locally smooth solutions. The second penalty function is the total variation (TV) penalty which is known to be edge-preserving and encourages piece-wise constant solutions. The third is the modified Huber penalty which locally acts like the squared gradient sum or TV penalties depending on the parameter δ. This parameter can be selected based on the statistics of gradients in a given image. For example, δ may be made proportional to the median of the gradient magnitudes. The Huber penalty will then act like the edge-preserving TV penalty at pixels where $|\nabla O| >> \delta$ and act like the smoothing quadratic penalty for pixels where $|\nabla O| << \delta$. The expressions for the Wirtinger derivatives for various choices of the penalty function are provided in the second column of **Table 1** for convenience of the readers. Implementation of the optimization solution can proceed with any of the gradient-based iterative schemes, the simplest being the gradient descent scheme. The $(n+1)-$th iteration of this scheme may be described as:

$$O_{n+1} = O_n - \tau[\nabla_{O^*} C]_{O_n}. \tag{6}$$

Penalty term	Wirtinger derivative	Remarks
$\sum_k \|\nabla O_k\|^2$	$-[\nabla^2 O]_k$	Smoothness penalty
$\sum_k \|\nabla O_k\|$	$-\left[\nabla \cdot \left(\frac{\nabla O}{\|\nabla O\|}\right)\right]_k$	Total variation (edge-preserving)
$\sum_k \left[\sqrt{1 + \frac{\|\nabla O_k\|^2}{\delta^2}} - 1\right]$	$-\left[\nabla \cdot \left(\frac{\nabla O}{\sqrt{1 + \frac{\|\nabla O\|^2}{\delta^2}}}\right)\right]_k$	Huber penalty (locally adaptive)

Table 1.
Suitable of penalty functions for optimization-based image recovery in single-shot digital holography along with the Wirtinger derivatives.

The parameter τ represents a step size which may be derived using backtracking line search [11]. The gradient descent scheme can be replaced by alternative iterative methods such as conjugate gradient or Nesterov accelerated gradient. While several algorithmic choices are available, one of the important practical problems is that the solutions obtained by this procedure depend on the regularization parameter α. Tuning of α for individual hologram data sets can become tedious. In **Figure 4**, we show object wave recovery for an image plane hologram corresponding to a step object. The TV penalty has been used in this case. It is observed that the quality of phase solution can change significantly depending on the value of α. For a low numerical value of α, the phase reconstruction shows fringe-like artifacts. On the other hand for a large numerical value of α, the phase solution shows over-smoothing. A "good" value of α proves that, in principle, an excellent recovery of the edge (and thus full pixel resolution) is possible via the optimization approach. A methodology that does not require any free parameter like α is thus practically very useful. The optimization problem for this purpose may be restated as follows [12]. We wish to determine a solution $O(x, y)$ satisfying

$$||H - |R + O|^2||_2^2 < \varepsilon, \tag{7}$$

such that among all solutions satisfying the aforementioned criterion, the desired solution has the lowest numerical value of the penalty function. This objective may be achieved by progressing the iterative solution in the following manner:

- At the beginning of (n + 1)th iteration, the solution O_n is updated so as to reduce the error objective C_1.
- The resulting intermediate solution may be denoted as $O_{n,int}$.
- The TV (or other penalty) for this iterative solution may be reduced with recursive steepest descent iterations of the form:

$$O_{n,k+1} = O_{n,k} - \tau[\nabla_{O^*}\psi(O,\ O^*)]_{O_{n,k}}. \tag{8}$$

At the end of N_{TV} sub-iterations of this form starting with $O_{n,0} = O_{n,int}$, we set $O_{n+1} = O_{n,N_{TV}}$.

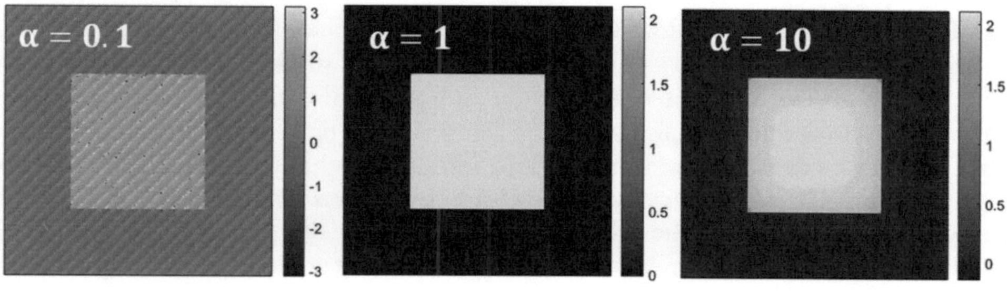

Figure 4.
Phase recoveries from a single-shot hologram corresponding to a step phase object. The optimization method with TV penalty is used, and the solution is shown for three values of the free parameter $\alpha = 0.1, 1, 10$ in the cost function in Eq. (4).

- Further, the reduction of C_1 and ψ is performed in an adaptive such that the distances $d_1 = \|O_n - O_{n,int}\|_2$ and $d_2 = \|O_{n+1} - O_{n,int}\|_2$ are made approximately equal.

The aforementioned strategy is inspired by the Adaptive Steepest Descent Projection onto Convex sets (ASD-POCS) algorithm [13]. The main thought process behind the scheme aforementioned is that the solution is driven away from minima in C_1, where the numerical value of the penalty function C_2 may be large. Eventually, the solution is driven to an equilibrium point where both the reduction in C_1 and C_2 oppose each other. While the scheme aforementioned needs multiple sub-iterations in each iteration, another scheme that naturally achieves the goal of $d_1 \approx d_2$ is the mean gradient descent (MGD) [14, 15]. In MGD, the solution is moved along the direction $\hat{u}$ that bisects the steepest descent directions corresponding to the two objectives C_1 and C_2. The solution therefore proceeds as:

$$O_{n+1} = O_n - \tau[\hat{u}_{O_n}], \tag{9}$$

where,

$$\hat{u} = \frac{\hat{u}_1 + \hat{u}_2}{2}, \tag{10}$$

and $\hat{u}_{1,2}$ are the unit vectors defined as:

$$\hat{u}_{1,2} = \frac{\nabla_{O^*} C_{1,2}}{\|\nabla_{O^*} C_{1,2}\|_2}. \tag{11}$$

Moving in the bisector direction typically increases the angle between the unit vectors $\hat{u}_1$ and $\hat{u}_2$ till the angle settles to a large obtuse value. At this point, simultaneous reduction of both the objectives C_1 and C_2 is not possible, and the solution thus reaches a balance point. Nominally, the two schemes aforementioned lead to a similar solution. Note that the goal of the optimization here is not to achieve a minimum of an overall cost function but rather to achieve an equilibrium between various objective functions (like C_1 and C_2) that represent the desirable properties of the solution $O(x, y)$. **Figure 5** shows two illustrations of the adaptive optimization process applied to the single-shot image plane hologram data corresponding to a step phase object. The recovery of phase step is excellent independent of the off-axis or on-axis configuration. As long as the reference beam is known, the two cases are operationally the same for the optimization algorithm. The holograms in **Figure 5** have been simulated with Poisson noise, and the RMS error achieved for the phase solution using the optimization procedure is observed to be better than the single-pixel-based shot noise limit. An experimental demonstration of this sub-shot noise phase recovery via the optimization procedure was also shown earlier with low-light-level interferograms. In particular, it was demonstrated that for interferograms with light level of the order of 10 photons/pixel, the RMS phase accuracy for a simple lens phase object was 5× better that what is expected from the shot noise limit [16]. Achieving such phase accuracy with classical light interferometry is not surprising as we are implicitly constraining the solutions by making a suitable choice of the penalty function. The important message here is that optimization-based phase recovery is expected to provide phase measurements accuracy better than array sensor's noise floor. The single-shot full pixel resolution capability along with improved phase accuracy makes this technique

Figure 5.
Adaptive optimization-based phase recoveries for image plane hologram of a step phase object. Both off-axis and on-axis configurations are shown.

suitable for employing with a quantitative phase microscopy system. In the illustration in **Figure 6**, we demonstrate optimization-based phase recovery using an image plane hologram of cervical cell nucleus. The cervical cell sample was obtained from a clinical collaborator (All India Institute of Medical Sciences) and is a typical pap-smear slide sample. Phase solutions using Fourier filtering as well as optimization method are displayed. The Fourier filtering solution used a circular filter window of radius 0.6 times the distance between the dc and cross-term peaks. An interesting aspect of the optimization procedure is that the iterations are fully in the image domain, and as a result, the iterative reconstruction may be performed over a selected region of interest (ROI) [17]. This makes it possible to implement reconstruction of individual cell regions in near real time and also allows for the possibility of parallelizing the phase reconstruction process. The iterative optimization here used the Huber penalty and clearly has a higher resolution compared to the Fourier filtering reconstruction. We emphasize here that the higher resolution information is already present in the image plane hologram and the optimization method that able to extract is much better compared to the Fourier filtering method. Improved resolution in single-shot holography operation offers multiple advantages. First of all, higher resolution enables one to observe fine textural features in a phase reconstruction, which

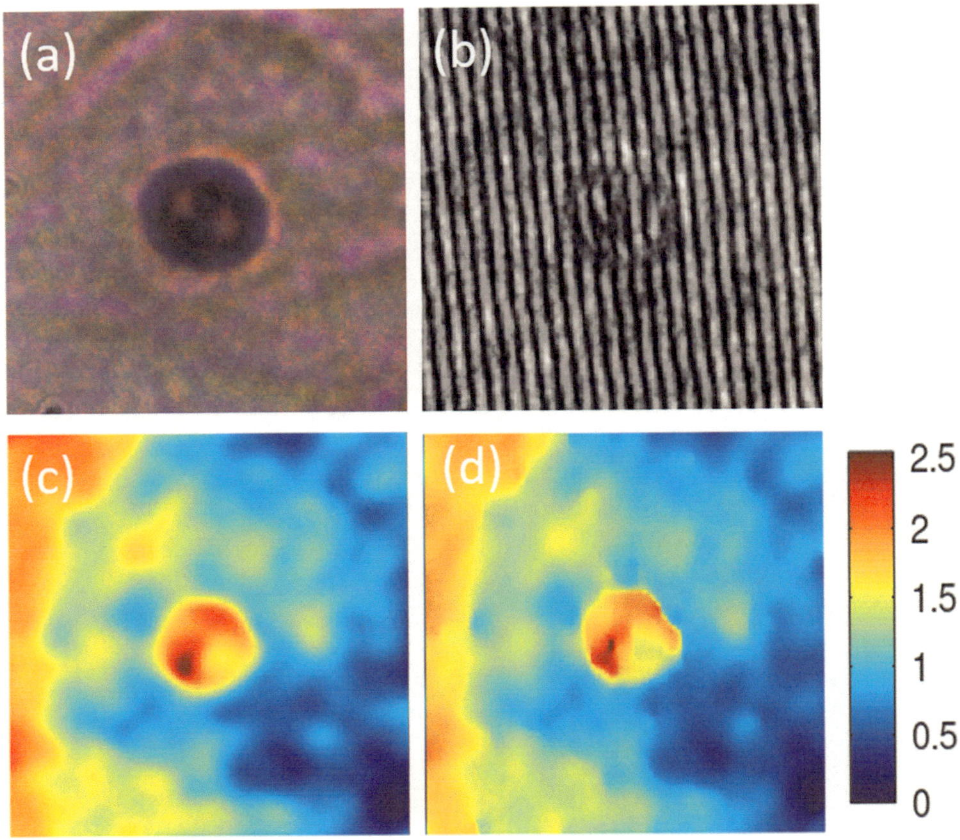

Figure 6.
Adaptive optimization-based phase recoveries for image plane hologram of a cervical cell nucleus, (a) bright-field image of cell nucleus, (b) 256 × 256 ROI of image plane hologram for the cell nucleus, phase reconstruction using (c) Fourier filtering method and (d) optimization method with Huber penalty. Optimization-based recovery clearly shows higher-resolution edge reconstruction. Fourier filtering-based reconstruction was performed with whole camera frame, and then the same ROI in phase reconstruction is cropped in (c).

may for example be important in diagnostics. For example, it is well known that cervical cell nuclei have higher "roughness" which show up in the phase profile [18, 19]. Secondly, the single-shot operation makes the DHM system simpler from hardware perspective, thus making it affordable for practical field deployment. The cost reduction here is not due to use of cheaper optics or other hardware components but due to the superior resolution and noise capabilities of the optimization approach to phase recovery which makes such a system possible.

4. Allied algorithmic requirements for robust phase reconstruction

Beyond the core phase reconstruction algorithm, the making of a robust DHM system requires a few other allied concepts so that the phase reconstructions are numerically repeatable for the same sample across different users and/or DHM systems. We discuss three such ideas in this section that are handy in building a robust DHM system.

4.1 Fractional fringe shift detection

The accurate knowledge of the reference beam $R(x, y)$ is essential for any phase reconstruction algorithm. While the amplitude $|R(x, y)|$ can be determined as by a calibration step by blocking the object beam, estimating the phase of $|R(x, y)|$ needs some additional discussion. Any error in phase estimate for $R(x, y)$ ends up as an additional phase factor in the object phase. If an afocal system is used for image formation in a DHM system, the aberrations are minimized, but the problem of accurate determination of carrier fringe frequency still remains an important problem. In off-axis holography configuration, the fringe frequency is typically estimated by locating the Fourier domain peaks corresponding to the cross-terms in the digital hologram. This step is commonly performed by using the fast Fourier transform (FFT) operation on the recorded off-axis hologram. For a hologram with $N \times N$ pixels, the spatial frequencies in the FFT domain are sampled at discrete intervals of $1/N$ pixel^{-1}. In a given setup, the true carrier fringe frequency peak may be located in between these discrete sample locations in the Fourier domain. If the fringe frequency determination has an error of (Δ_x, Δ_y) with $|\Delta_x|,|\Delta_y| \leq 0.5$ units in the x and y directions, this leads to a phase error at pixel location (m, n) that may be given by:

$$\Delta\phi(m, n) = \frac{2\pi}{N}\left(\Delta_x m + \Delta_y n\right). \quad (12)$$

This phase error causes a ramp phase background over the desired phase solution. For multiple DHM systems, it is not practically possible to ensure that the fringes will be oriented identically on an array sensor, and this will cause different ramp phase errors for the same sample. As demonstrated in [20], even a small fractional degree camera rotation can lead to different ramp phase backgrounds, making this fractional fringe problem quite sensitive for applications. Detection and elimination of this fractional fringe effects requires a two-step algorithmic approach as we describe here.

The main task at hand here is to get a higher-resolution representation of the cross-term peak in the Fourier domain compared to what is already available from the FFT of the hologram. The fractional fringe detection therefore needs the following steps:

- Compute 2D FFT of the hologram $H(x, y)$ and locate the integer pixel location (u_0, v_0) corresponding to the cross-term peak,
- A local discrete Fourier transform computation is then performed as a three matrix product in the following manner in the neighborhood of (u_0, v_0):

$$\tilde{H}(\mathbf{P}, \mathbf{Q}) = \exp\left(-i\frac{2\pi}{N}\mathbf{P}\mathbf{X}^T\right) H(\mathbf{X}, \mathbf{Y}) \exp\left(-i\frac{2\pi}{N}\mathbf{Y}\mathbf{Q}^T\right). \quad (13)$$

Here **X**,**Y** denote coordinate vectors $[-N/2, -N/2+1, \ldots, N/2-1]^T$. If an upsampling factor of α is to be employed for a region of (1.5×1.5) pixels surrounding (u_0, v_0), then the vectors **P**,**Q** are defined as:

$$(\mathbf{P}, \mathbf{Q}) = (u_0, v_0) + \left[-\frac{1.5}{2}, -\frac{1.5}{2} + \frac{1}{\alpha}, \ldots, \frac{1.5}{2} - \frac{1}{\alpha}\right]. \quad (14)$$

The three matrices in Eq. (13) are of size $(1.5\alpha \times N)$,$(N \times N)$,$(N \times 1.5\alpha)$, respectively, and their multiplication has marginal additional computational cost when the upsampling factor α is much less than hologram size N. Following this procedure, an oversampled local Fourier transform $|\tilde{H}(\mathbf{P}, \mathbf{Q})|$ is computed and its sub-pixel peak shift is determined. The ramp phase error $\Delta\phi$ in Eq. (12) can then be corrected.

In **Figure 7**, we show phase reconstruction for a red blood cell sample with and without the fractional fringe correction, which clearly shows the effectiveness of this simple algorithmic method in removing any residual phase ramp background. A 20-fold upsampling of a (1.5×1.5) pixel region centered on integer pixel peak location was performed in this case, making it possible to detect small fractional fringe shift. Such upsampling of Fourier transform of the hologram can also be performed by zero-padding of the hologram prior to computing its Fourier transform. However, a 20-fold upsampling will make the total image size after zero-padding impractical when high-resolution Fourier transform is required only in the neighborhood of the carrier-frequency peak. The methodology may be used routinely with off-axis holograms with marginal additional computational burden.

4.2 Focusing of unstained cells

The unique advantage of a DHM system is that it can image unstained cell samples, since the contrast mechanism for imaging is the natural refractive index of the cells relative to their surrounding medium. The second interesting aspect is that since the complex-valued object field in the image plane is being recovered from the hologram, it is possible to numerically propagate this field to achieve computational refocus. One of the difficulties with this computational refocusing is that the typical Life Sciences users are not accustomed to such a methodology and often want to physically focus the cells by moving the sample stage in z-direction. Further, since most users without prior Optics or Physics background cannot interpret the phase images, they usually prefer to first record a bright-field image in focused position to know what they are imaging. The quantitative phase image obtained at the same sample location makes it easier for them to interpret the morphology of the cell being imaged. For this purpose, a DHM system can be built with a dual illumination so that it can be easily switched between the bright-field and the quantitative phase modes of operation. Focusing of unstained samples is however not trivial in bright-field mode as there is minimal amplitude contrast. Computational refocus is also not possible in the bright-field mode as the recorded image does not contain any phase information. The focusing

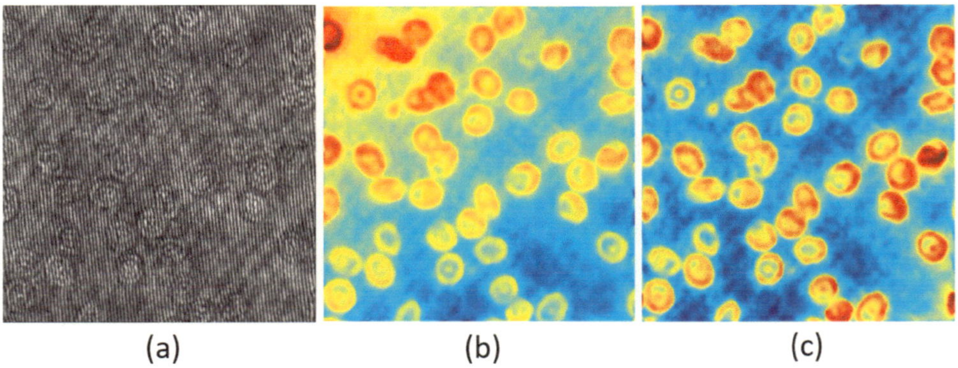

Figure 7.
Illustration of fractional fringe elimination technique, (a) image plane hologram of RBCs, (b), (c) phase reconstruction with and without sub-pixel localization of carrier-frequency peak in Fourier domain. The phase ramp along the diagonal direction in (b) has been removed in (c).

problem can be addressed in a clever manner by employing a focusing criterion directly in the hologram domain [21]. Unstained samples can be considered to be nearly pure phase objects in the focus plane, and this implies that the interference fringes in the focus-plane hologram will predominantly show phase modulation. In other words in the focus plane, fringes mainly show bending without any amplitude modulation at the location of the cell. It is well known that when the sample is defocused, the phase information gets transferred to amplitude which causes amplitude modulation of fringes in the form of dark or bright halo near the cell boundaries. This effect is clearly illustrated in **Figure 8** where we show three through-focus holograms of the same red blood cell as the microscope stage is translated in z-direction in small steps. The bottom row of **Figure 8** also shows the corresponding phase reconstructions for these sample positions. Interestingly, the total phase variation is highest for the hologram in **Figure 8(b)** where the amplitude modulation of fringes is minimal. One may define a hologram domain focusing criterion as follows: *The focus plane may be defined as the one with minimal amplitude modulation on the interference fringes.* This criterion is very important for standardization. Note that the phase profiles in the defocus planes in **Figure 8** are quite distorted in comparison with that in the focus plane. This in turn suggests that numerical phase map obtained by different users may be different for the same sample unless a uniform focusing criterion is used. Such a situation is not desirable considering that many imaging applications including DHM imaging are moving toward data-driven applications where numerical values of phase maps are very important in statistical learning, for example, for the purpose of cell classification applications. The minimum amplitude modulation criterion described here is easily implementable visually or by means of an autofocus algorithm in the hologram domain. The methodology also provides a seamless experience to potential Life Sciences users of DHM systems who may prefer to focus the sample physically.

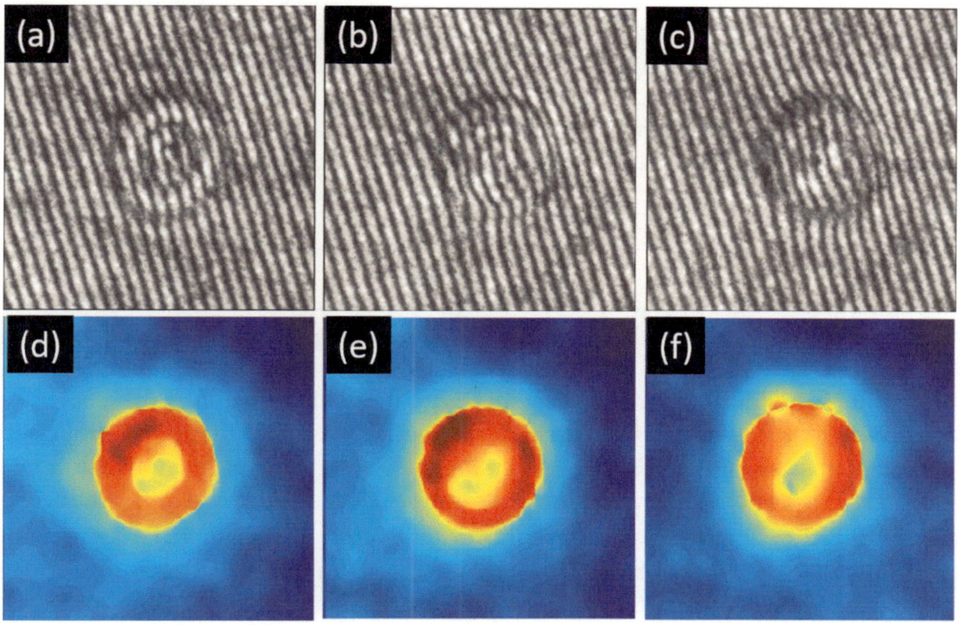

Figure 8.
Illustration of hologram domain focusing criterion, (a), (b), (c): through-focus holograms of the same cell; (d), (e), (f): their corresponding phase reconstructions.

4.3 Fast unwrapping of two-dimensional phase maps

The phase map $\phi_O(x, y)$ recovered from the single-shot image plane hologram may be interpreted approximately (at least for thin samples like mono-layer of cells) as:

$$\phi_O(x, y) = \frac{2\pi}{\lambda} \int dz \; n(x, y, z). \tag{15}$$

Here, λ is the illumination wavelength, z is the nominal propagation direction, and $n(x, y, z)$ stands for the relative refractive index of the sample with respect to the surrounding medium. Typically, the reconstruction algorithm provides the complex-valued object wave $O(x, y)$, and the phase $\phi_{O,w}(x, y)$ is defined as the arc-tangent of the ratio of imaginary and real parts of the complex-valued function:

$$\phi_{O,w}(x, y) = \arctan\left(\frac{Im[O(x, \; y)]}{Re\,[O(x, \; y)]}\right). \tag{16}$$

Since the arc-tangent function is defined over the interval $(-\pi, \pi]$, the phase map $\phi_{O,w}(x, y)$ in Eq. (16) has jump discontinuities that are not expected in ϕ_O as per Eq. (15). The $-\pi$ to ϕ phase jumps may occur as per the morphology of the sample in a complicated manner in the phase image. The phase jumps are however not physical and need to be removed. The unwrapped phase may therefore be expressed as:

$$\phi_O(x, y) = \phi_{O,w}(x, y) + 2m(x, y)\pi, \tag{17}$$

where $m(x, y)$ are integers that make the phase map continuous without the 2π jumps. Phase unwrapping in two dimensions is not a trivial problem, and multiple solutions have been proposed to address it. The popular solutions have come from radar literature, and their path following nature makes phase unwrapping a challenging image processing problem. A fast and robust solution to the two-dimensional phase unwrapping problem is possible [22] via an approach based on the transport of intensity equation (TIE). The TIE is a curious relation that relates the longitudinal intensity derivative of a propagating field (in Fresnel zone) to its transverse phase gradient. The TIE may be stated in the context of our problem as:

$$-k\frac{\partial I}{\partial z} = \nabla_{xy} \cdot \left(I\nabla_{xy}\phi_O\right). \tag{18}$$

Here, the gradient operator ∇_{xy} on the right-hand side denotes the transverse gradient and $I(x, y)$ denotes field intensity. This is probably the only relation in Optics which expresses the phase ϕ_O without the use of the arc-tangent function but rather as a differential equation. When the intensity $I(x, y)$ is nearly constant as for a pure phase object in focus plane, the TIE simplifies to the Poisson equation. We proposed a novel scheme based on the TIE for addressing the unwrapping problem which is fast and robust. In particular, one may use the wrapped phase map $\phi_{O,w}(x, y)$ obtained from $O(x, y)$ to generate an auxiliary field $u(x, y) = \exp\left[i\phi_{O,w}(x, \; y)\right]$. To estimate the longitudinal intensity derivative corresponding to this auxiliary field, it is propagated by small distances $\pm\Delta z$ and the intensity derivative is obtained as:

$$\frac{\partial I}{\partial z} \approx \frac{|u(x, \; y, \; \Delta z)|^2 - |u(x, y, -\Delta z)|^2}{2\Delta z} + O\left((\Delta z)^2\right). \tag{19}$$

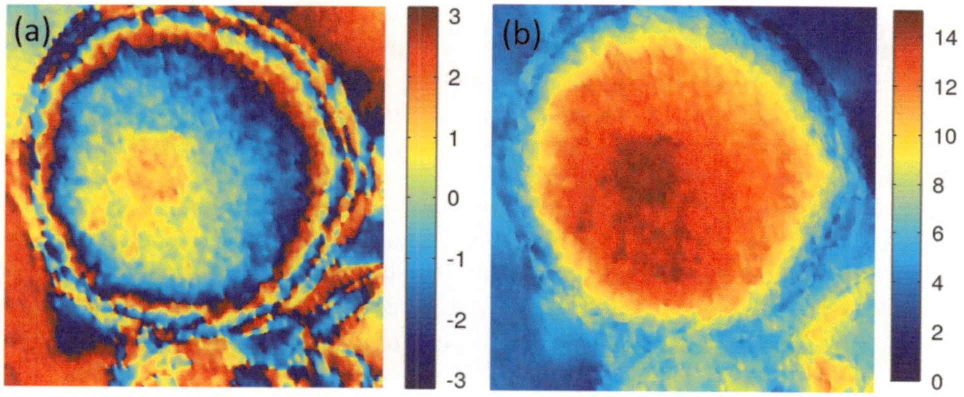

Figure 9.
Illustration of TIE-based phase unwrapping; (a) wrapped phase map recovered from image plane hologram of a pollen grain and (b) unwrapped version of (a).

Note that the distance Δz aforementioned is not determined by any experimental considerations like detector noise but is purely selected as a computational parameter which may be made small to improve the accuracy of the longitudinal intensity derivative. The TIE may be solved to get the unwrapped phase $\phi_O(x, y)$. A popular approach for solving the TIE involves the use of inverse Laplacian operator:

$$\phi_O(x, y) = -\frac{k}{I}\nabla_{xy}^{-2}\frac{\partial I}{\partial z}. \tag{20}$$

The inverse Laplacian may be implemented readily in Fourier domain as follows:

$$\nabla_{xy}^{-2} g(x, y) = \mathcal{F}^{-1}\left[\frac{D^2\ G\left(f_x, f_y\right)}{D^2 + \varepsilon^2}\right]. \tag{21}$$

Here, $D^2 = -4\pi^2\left(f_x^2 + f_y^2\right)$ and ε^2 is a small positive constant which avoids division by zero. This methodology offers a fast FFT-based solution to the phase unwrapping problem whose processing time is independent of the phase structure to be unwrapped. The phase unwrapping operation can thus be implemented with marginal additional computational efforts. In **Figure 9**, we illustrate unwrapping of a phase map obtained from image plane hologram of a pollen grain using the TIE-based solution. This technique is useful and works well except in cases where the transverse gradient of the phase map has a circulating (or curl) component.

5. Does digital holography truly provide 3D information?

The readers may have noticed that in this article so far, we have avoided showing the phase maps associated with various objects in a 3D surface-rendered form. This is because the recovered phase $\phi_O(x, y)$ using image plane digital holography is purely a 2D function and does not in itself contain any tomographic information. The 3D surface rendering is usually justified, since the phase map is related to the optical thickness of the sample under consideration. In this context, we provide a short discussion on the topic of whether phase imaging via DHM even provides a true 3D information. Holography if often naturally associated with “3D imaging.” In traditional film-based display holography, the holographic replay is performed by re-illumination of the recorded hologram by conjugate reference beam. When a human observer inspects the replayed wavefront visually, there is a perception of 3D objects which mainly arises because of the capability of eyes to focus onto high-contrast objects while ignoring any diffuse and blurred background. In digital holography, this ability of human eye to focus onto sharp objects has no important role to play, since the reconstruction is performed numerically. The problem of 3D reconstruction over $(N_x \times N_y \times N_z)$ voxels using object field $O(x, y)$ recovered from a digital hologram of size $(N_x \times N_y)$ also poses a problem of mismatch of degrees-of-freedom [23]. The 3D reconstruction problem has been attempted in the prior literature in two different ways. In one configuration, multiple (typically a few hundred) digital holograms of a transparent object are recorded from different views and the individual phase reconstructions are them combined using the Fourier diffraction theorem [24]. This method requires dedicated hardware where in order to change illumination angle or sample orientation or both. Such tomographic approaches also have relevance to the 3D reconstruction problems in cryo-electron microscopy of viruses or macromolecules [25]. In a second approach, a much smaller number of holograms are used and the reconstruction is performed via a sparse optimization method [26, 27]. The problem of dimensionality mismatch is attempted in such cases by exploiting the object sparsity. Tomographic imaging remains to be a challenging problem for digital holography, and much work needs to be done to model forward (object to hologram) and backward (hologram to object volume) propagation of light fields. The problem is particularly difficult when the sample is thick. Due to the complexity of the problem, some recent works have also explored machine learning-based approaches to tomographic phase (or refractive index) reconstruction [28]. This is likely to be a research problem of interest in future.

6. Popularization of DHM technology among life sciences researchers

Widespread usage of DHM technology among Life Sciences researchers and practitioners is somewhat lacking at present, and a lot needs to be done by DHM researchers to correct this status. The multiple advantages of the quantitative phase imaging modality suggest that it can be routinely used in practice just like bright-field or fluorescence microscopy. Here, I lay out a few steps that are needed going forward.

- **Education about quantitative phase imaging**: A quick look at biosciences and medical school curriculum suggests that quantitative phase as a modality is never introduced to them. This means that most Life Sciences researchers are not even aware of quantitative phase imaging. They might hear about it at advanced level as some exotic microscopic methodology but are then not inclined to study or use it for their own work. Even with very large number of publications by DHM researchers in Optics and Engineering journals, the situation may not change unless concrete efforts are made in this direction.

- **Standardization of protocols**: Life Sciences community is very particular about sample preparation and imaging protocols. The lack of standardized protocols for phase microscopy prevents multiple laboratories to communicate with each other effectively. The present article, in a limited sense, describes standardization issues associated with algorithmic methodologies. However, wide ranging efforts must be made in terms of benchmarking DHM hardware configurations, associated reconstruction algorithms, and their capabilities such as spatial resolution and phase accuracy. This will allow potential users to understand the modality much better before they decide to use it seriously. Such standardization is even more important if phase images are to be used with machine learning for designing diagnostic applications.

The popularization of DHM technology thus seems to be a task beyond what individual research groups can achieve in isolation. A DHM users' consortium will be needed in order to move toward the aforementioned goals.

7. Conclusions

It this article, we summarized algorithmic methodologies that enable robust phase reconstruction performance for a single-shot digital holographic microscope system. First, we presented a sparse optimization-based phase reconstruction approach that provides single-shot full-resolution imaging performance. Further allied methodologies for fractional fringe correction, hologram domain focusing of unstained samples, and transport-of-intensity-equation-based fast 2D-phase unwrapping were discussed. Finally, we highlighted the importance and need of a DHM users' consortium for benchmarking of DHM hardware and reconstruction algorithms, if this technology is to be popularized as a routine microscopy modality among Life Sciences researchers.

Conflict of interest

The authors declare no conflict of interest.

Author details

Kedar Khare
Optics and Photonics Centre, Indian Institute of Technology Delhi, New Delhi, India

*Address all correspondence to: kedark@physics.iitd.ac.in

References

[1] Javidi B, Carnicer A, Anand A, Barbastathis G, Chen W, Ferraro P, et al. Roadmap on digital holography [invited]. Optics Express. 2021;**29**:35078-35118

[2] Kim MK. Principles and techniques of digital holographic microscopy. SPIE Reviews. 2010;**1**:018005

[3] Park Y, Depeursinge C, Popescu G. Quantitative phase imaging in biomedicine. Nature Photonics. 2018;**12**: 578-589

[4] Kemper B, Illy E. Digital holographic microscopy. PhotonicsViews. 2020; **17**(1):32-35

[5] Goodman JW, Lawrence R. Digital image formation from electronically detected holograms. Applied Physics Letters. 1967;**11**(3):77-79

[6] Rivenson Y, Wu Y, Ozcan A. Deep learning in holography and coherent imaging. Light: Science and Applications. 2019;**8**(1):85-98

[7] Takeda M, Ina H, Kobayashi S. Fourier-transform method of fringe-pattern analysis for computer-based topography and interferometry. Journal of the Optical Society of America. A. 1982;**72**:156-160

[8] Trusiak M, Mico V, Garcia J, Patorski K. Quantitative phase imaging by single-shot Hilbert–Huang phase microscopy. Optics Letters. 2016;**41**: 4344-4347

[9] Baek YS, Lee KR, Shin S, Park YK. Kramers–Kronig holographic imaging for high-space-bandwidth product. Optica. 2019;**6**:45-51

[10] Khare K, Samsheerali PT, Joseph J. Single shot high resolution digital holography. Optics Express. 2013;**21**: 2581-2591

[11] Nocedal J, Wright SJ. Numerical Optimization. New York, NY: Springer New York; 1999

[12] Singh M, Khare K. Single-shot interferogram analysis for accurate reconstruction of step phase objects. Journal of the Optical Society of America. A. 2017;**34**:349-355

[13] Sidky EY, Pan X. Image reconstruction in circular cone-beam computed tomography by constrained, total-variation minimization. Physics in Medicine and Biology. 2008;**53**: 4777-4807

[14] Rajora S, Butola M, Khare K. Mean gradient descent: An optimization approach for single-shot interferogram analysis. Journal of the Optical Society of America. A. 2019;**36**:D7-D13

[15] Rajora S, Butola M, Khare K. Regularization-parameter-free optimization approach for image deconvolution. Applied Optics. 2021;**60**: 5669-5677

[16] Singh M, Khare K, Jha AK, Prabhakar S, Singh RP. Accurate multipixel phase measurement with classical-light interferometry. Physical Review A. 2015;**91**:021801

[17] Singh M, Khare K. Single-shot full resolution region-of-interest (ROI) reconstruction in image plane digital holographic microscopy. Journal of Modern Optics. 2018;**65**:1127-1134

[18] Mangal J, Monga R, Mathur SR, Ahlawat S, Khare K. Unsupervised organization of cervical cells using

bright-field and single-shot digital holographic microscopy. Journal of Biophotonics. 2019;**12**:e201800409

[19] Zink D, Fischer AH, Nickerson JA. Nuclear structure in cancer cells. Nature Reviews Cancer. 2004;**4**:677-687

[20] Singh M, Khare K. Accurate efficient carrier estimation for single shot digital holographic imaging. Optics Letters. 2016;**41**:4871-4874

[21] Malik R, Sharma P, Poulose S, Ahlawat S, Khare K. A practical criterion for focusing of unstained cell samples using a digital holographic microscope. Journal of Microscopy. 2020;**279**:114-122

[22] Pandey N, Ghosh A, Khare K. Two-dimensional phase unwrapping using the transport of intensity equation. Applied Optics. 2016;**55**:2418-2425

[23] Birdi J, Rajora S, Butola M, Khare K. True 3d reconstruction in digital holography. Journal of Physics: Photonics. 2020;**2**:044004

[24] Balasubramani V, KuÅ A, Tu H-Y, Cheng C-J, Baczewska M, Krauze W, et al. Holographic tomography: Techniques and biomedical applications. Applied Optics. 2021;**60**:B65-B80

[25] Frank J. Three-Dimensional Electron Microscopy of Macromolecular Assemblies. Oxford, UK: Oxford University Press; 2006

[26] Wang H, Tahir W, Zhu J, Tian L. Large-scale holographic particle 3d imaging with the beam propagation model. Optics Express. 2021;**29**: 17159-17172

[27] Denis L, Lorenz D, Thiébaut E, Fournier C, Trede D. Inline hologram reconstruction with sparsity constraints. Optics Letters. 2009;**34**:3475-3477

[28] Kamilov US, Papadopoulos IN, Shoreh MH, Goy A, Vonesch C, Unser M, et al. Learning approach to optical tomography. Optica. 2015;**2**: 517-522

Chapter 4

Three Dimensional Widefield Imaging with Coherent Nonlinear Scattering Optical Tomography

Lang Wang, Gabriel Murray, Jeff Field and Randy A. Bartels

Abstract

A full derivation of the recently introduced technique of Harmonic Optical Tomography (HOT), which is based on a sequence of nonlinear harmonic holographic field measurements, is presented. The rigorous theory of harmonic holography is developed and the image transfer theory used for HOT is demonstrated. A novel treatment of phase matching of homogeneous and in-homogeneous samples is presented. This approach provides a simple and intuitive interpretation of coherent nonlinear scattering. This detailed derivation is aimed at an introductory level to allow anyone with an optics background to be able to understand the details of coherent imaging of linear and nonlinear scattered fields, holographic image transfer models, and harmonic optical tomography.

Keywords: nonlinear optics, tomography, computational imaging, nonlinear scattering, nonlinear holography, optical holography, optical tomography, phasematching

1. Introduction

Optical microscopy permits the noninvasive acquisition of information that is revealed through light-matter interactions. These light-matter interactions are generally referred to as contrast mechanisms and come in many forms. The information carried by an optical contrast mechanism depends on the properties of the illumination light, the properties of the light produced by the contrast mechanism, and the details of the light detection. Most optical imaging systems rely on light that can be described approximately as classical, although there is a steadily growing body of work describing microscopy methods that exploit quantum correlations to enable the extraction of new information from objects.

The coherence properties of the light used for illumination and detection are also critical drivers of the properties of an optical imaging system. The classical theory of optical coherence is concerned with the statistical properties of classical fields that are treated as random variables. Although light is, in general, partially coherent, it is often suitable to describe light in the limiting case of either fully coherent or fully incoherent. Heuristically, we can describe coherent light as a field that is statistically similar across either temporal or spatial points on the field, whereas fully incoherent light lacks any correlation either along temporal or spatial displacements. Consideration of optical coherence is critical for understanding the broader context of optical microscopy and optical tomographic imaging.

While light propagation is not directly modified by optical coherence, field coherence strongly impacts the observed signal from a detector. For our purposes, a semi-classical model of light detection is suitable, where we consider generated photocurrents in a photodiode or photo-generated electrons detected by a camera chip. In all instances treated herein, we assume detector integration times are longer than the temporal coherence times of the light fluctuations. As such, detected light intensities are inferred from a long-time average of the incident optical field's instantaneous intensity. To provide a consistent framework for our discussion of coherent tomographic imaging, we briefly review optical imaging theory to ensure that the reader is familiar with the notation used in this treatment.

In this chapter, we focus on imaging systems that can be described in a classical optical formalism that uses coherent nonlinear scattering as a contrast mechanism. Coherent nonlinear scattering exploits the microscopic properties of materials that exhibit a nonlinear dipole in response to a sufficiently strong incident field [1–5]. While materials can produce higher-order responses than a nonlinear dipole, the high field strength required generally precludes the use of higher-order terms to prevent damage to the object under study. It is suitable to describe the sensitivity of the nonlinear response as a nonlinear susceptibility tensor that is obtained from a Taylor expansion of the dipole response to the applied electric field. While the tensorial nature of nonlinear response depends on the incident fields and the distribution, we will suppress the vectorial dependence of both the coherent nonlinear light-matter interactions and the light scattering [3]. Within this approximation, coherent nonlinear scattering is described by a scalar field. For the purposes of imaging and tomography, we must then build a model for the propagation of the scattered scalar field through an imaging system, the detection of that light, and the processing required to obtain a microscopic or tomographic image.

This chapter is organized as follows. Coherent imaging theory is outlined and the application to tomographic imaging using coherent scattering is described. This section will show that while the imaging system permits spatial magnification of the field to enable the observation of small features, this magnification comes at the cost of low-pass filtering of the spatial frequency span of the collected coherent scattered light. Next, the scalar model of the nonlinear scattered field is developed to produce the working equations for the contrast signal that is collected by the imaging system. Then, the physical implications of the scattering and image formation models are discussed. Finally, the implications of the image transfer model for holographic optical tomography (HOT) and the tomographic reconstruction algorithm for both second harmonic generation (SHG) and third harmonic generation (THG) are discussed. We conclude by discussing prospects for widefield HOT imaging.

2. Optical holography

Optical detectors respond to the incident optical intensity rather than the field. As a result, all phase information is lost when making any direct optical measurement of a field. However, we know that when the measured light has suitable coherence, we may convert phase differences into intensity modulations through optical interference. Generally, we may have the desired signal field, U_s, so that if we have a well-characterized reference field, U_r, we may then recover the desired signal field through suitable processing of the interference intensity. This imaging method that recovers the complex wave, i.e., amplitude and phase, is referred to as holography [6].

Holography was first described as a linear scattering model where the scattering object is much smaller than the extent of the incident wave [6]. In this in-line Gabor holography, the unscattered (ballistic) part of the incident wave is treated as the reference wave. The interference between the scattered and unscattered portions of the field constitutes the hologram. In-line holography generated limited excitement initially because the desired scattered field was contaminated by an unwanted conjugate (twin-image) field. This contamination significantly degraded the utility of early holograms. Around the same time that Gabor was working on in-line holography, Leith and co-workers were working in a secret US government program to process synthetic aperture radar films optically. Leith and Upatnieks independently discovered holography, but with a communications theory perspective that employed a spatial frequency carrier for off-axis holography [7–9]. In off-axis holography, with a suitably large incident reference beam angle, the complete complex signal field may be recovered, thus solving the twin-image problem. In 1997, Yamaguchi demonstrated another elegant solution to the twin-image problem by taking a sequence of holograms where the relative phase of the signal and reference field was shifted, allowing for unique extraction of the complex signal field from a series of in-line holograms [10]. While early holographic work made use of photographic plates, modern holography makes use of digital cameras and numerical processing algorithms [11, 12].

Off-axis holography is hailed as a 3D imaging technique. Indeed, when a hologram produced from an exposed photographic plate is illuminated by a duplicate of the reference wave, one will observe the signal wave as if the object were still present. An observer will see a 3D image of the object. Off-axis holography was first revealed in dramatic fashion with holograms of trains produced in Leith's laboratory. However, it must be remembered that human visual perception is stereoscopic and not truly 3D. Thus, an observer *perceives* depth, but does not truly resolve the 3D spatial structure of an object!

Emil Wolf analyzed optical holography from the perspective of linear scattering in the first Born approximation. He developed what is now referred to as the Fourier diffraction theorem, which shows that while a holographic field can be propagated or refocused, *the field does not include any axially localized (optically sectioned) information about the object under study* [13]. The 3D holography observations can be explained as surface scattering from an object with varied depth so that upon viewing, the observed perceives depth through stereoscopic processing.

The focus of this chapter is on nonlinear holographic imaging. In nonlinear microscopy, signals are recorded from coherent nonlinear scattering, which arises from a distortion of the induced oscillating dipole response of an atom or molecule subjected to a suitably strong illumination (fundamental) field. This nonlinear response can scatter light to new frequencies at harmonics of the incident fundamental field frequency, $\omega_m = m\omega_1$, where ω_1 is the fundamental frequency and $m > 1$ is an integer. Both second harmonic generation (SHG, $m = 2$) [3] and third harmonic generation (THG, $m = 3$) [14–16] are routinely used for optical microscopy [2]. The first SHG images demonstrated the approach, but routine use had to await the arrival of reliable ultrafast laser oscillators. Today, SHG microscopy is routinely used in biological imaging [4, 5] to look primarily at muscle fibers [17] and collagen [18–20]. Beyond the study of morphology [21], one may obtain macromolecular structure [22].

Due to the weak strength of the nonlinear optical susceptibility, coherent nonlinear holography had to await the development of more powerful ultrafast sources. Roughly 60 years after the first reports of linear holographic imaging, Demetri Psaltis' group described SHG holography of SHG-active nanoparticles using a 10 Hz laser amplifier system [23] in a special issue of applied optics that was dedicated to the memory of Emmett Leith [24]. Subsequently, Psaltis demonstrated focusing and imaging of point scatterers in biological tissues and phase conjugation to improve image quality [25–28]. Shortly after this initial demonstration of SHG holography, imaging in biological systems with oscillators was demonstrated [29–34]. By optimizing the experimental configuration, quasi-3D imaging at rates of nearly 1500 volumes per second was demonstrated [34]. This early holography work was still limited in the ability to produce detailed 3D imaging and this problem was only recently solved with the introduction of harmonic optical tomography (HOT) [35].

3. Optical diffraction tomography

A measured optical field that has scattered from an object because of a spatial inhomogeneity in either the linear or nonlinear optical susceptibility reveals information about the spatial distribution of the susceptibility. The goal of any coherent imaging system is to uncover quantitative data on the spatial distribution of susceptibility variations, $\delta\chi(\mathbf{r})$. The information that is transferred from the incident light field to the scattered field—whether this involves linear or nonlinear scattering—depends on both the properties of the incident light and the nature of the optical physics exploited for the contrast mechanism. In this chapter, we focus our discussion on tomographic imaging obtained through coherent nonlinear optical scattering holographic measurements. To put this technique in context, we will recite the key properties of tomographic imaging under the range of coherent properties of the illumination light and for various contrast mechanisms.

The concept of diffraction tomography was introduced by Emil Wolf in his seminal optics communications paper analyzing optical holography for the case of linear scattering [13]. Wolf's treatment is reproduced in this chapter and the extension to coherent nonlinear scattering [35] is developed.

In classic holography [7–9, 11, 12, 36], we consider the illumination of the object by a spatially coherent, monochromatic plane wave with a particular incident propagation wavevector $\mathbf{k}_i$. The key observation that follows from the Fourier diffraction theorem is that very limited information is transferred from the scattered field in the process of optical scattering of the incident light from the linear susceptibility perturbation. In a holographic imaging scenario, the transverse spatial frequency span is limited by the numerical aperture of the objective lens, which sets the transverse imaging resolution [8, 37, 38]. However, for the spatially coherent illumination case, the axial frequency support is identically zero. This means that a coherent scattered field recovered from holography does not permit optical sectioning in the imaging process. To fully resolve the object, all the spatial frequency information (or equivalently the spatial information) must be adequately sampled. The methods for fully capturing both the transverse and axial spatial frequency information, called optical diffraction tomography (ODT), make use of either object or beam rotation [37, 39, 40].

The opposite extreme of spatially coherent illumination is the case of spatially incoherent illumination [41, 42]. We may still assume a long coherence time for the case of quasi-monochromatic light. However, the temporally random variations of the illumination field sample the full range of possible incident spatial frequency that is supported by the numerical aperture of the condenser lens. The image transfer model for incoherent illumination exhibits a finite thickness at intermediate spatial frequencies, yet still exhibits no spatial frequency support near zero transverse spatial frequencies [43]. As a result, spatially incoherent imaging also lacks optical sectioning capabilities. However, when an object is placed within the depth of focus of a microscope with spatially incoherent illumination, an absorption tomographic image can be reconstructed [41] with only a slight modification to the standard computed tomography filtered back projection algorithm from a sequence of transillumination intensity images taken over a full rotation of the object [40]. This imaging modality is referred to as optical projection tomography.

Partially coherent illumination fits somewhere in between fully spatially coherent and incoherent imaging. The image transfer model was derived by Streibl for linear scattering under quasi-monochromatic, partially coherent illumination [43]. This transfer function falls between that found for fully coherent and fully incoherent illumination. While in the case of fully spatially coherent light, scattering is accumulated from all depths, the image transfer function of partially coherent light acts as a low-pass filter that rejects the defocused contributions [44]. Streibl showed that by acquiring a stack of images in axial steps of the depth of focus, the set of data can be deconvolved to obtain a three-dimensional image [45–47]. This same strategy—that of a 3D deconvolution of a stack of images—is a productive approach for 3D imaging of incoherent fluorescent emission [48, 49].

In recent years, Streibl's approach has been expanded to other partially coherent illumination sources. A method called white light diffraction tomography makes use of spatially coherent light with very broad bandwidth, and thus short temporal coherence [50]. When this very broad bandwidth light is used to illuminate a specimen with a very high NA objective, measurement of the complex field produced by linear scattering through a variant of holography broadens the imaging transverse function axially to permit optical sectioning. Three-dimensional images of the inhomogeneity of the linear susceptibility are then obtained through a 3D deconvolution from a sequence of images taken as the object is displaced in the in the axial direction.

Another strategy that avoids the use of interferometry for extracting the complex field is the use of asymmetric illumination apertures with partially coherent light [51]. When this illumination strategy is coupled with a rigorous model of the imaging transfer function, again a 3D deconvolution can be applied to an axial image stack to obtain a 3D image of the spatial variations in linear optical susceptibility.

The forms of tomography that we have discussed so far are primarily based on optical scattering. As we discuss in detail in the later section, the reliance on scattering with spatially coherent illumination allows for the measured field to relate the input and output scattering directions, which pinpoints the spatial frequency component of the object spatial susceptibility perturbation distribution. However, the short duration of the excited state lifetime and rapid dephasing of fluorescent emitters renders them spatially incoherent. While one might expect this spatial coherence to prevent interference, an individual fluorescent emitter will interfere with itself even though the lack of spatial coherence prevents interference between emitters. As a result, diffractive optical structures can be designed to enable depth-dependent interference intensity structures that allow for holography of incoherent emitters [52, 53]. An alternate strategy can be deployed for mimicking coherent scatting and holographic imaging with incoherent emitters based on the interference of spatially coherent illumination light, either between a plane wave and a point focus [54, 55] or with a pair of plane waves [56–61]. By using the interference between two spatially coherent illuminating plane waves, one may perform tomographic imaging with fluorescent emitters that exactly mimics ODT [62–65].

Widefield coherent nonlinear scattering enables the ability to form holograms when a coherent reference beam is directed to interfere with the light produced from the nonlinear scattering process [23, 29]. In the case of illumination with a plane wave fundamental beam, the scattering picture for nonlinear scattering is nearly equivalent to that of linear scattering, with a few modifications. These similarities and differences in the scattering picture will be discussed in later sections of this chapter. The key observation is that we will not obtain any optical sectioning with strictly plane wave illumination. However, due to the weak interaction strength for nonlinear scattering, the fundamental excitation beam is generally weakly focused to provide a balance between field of view activated in the nonlinear scattering process and signal strength that is driven by suitably large field strengths. In such a weak excitation case, moderate 3D imaging resolution is observed [34].

The fact that nonlinear scattering is driven by multiple input fields allows for a completely new form of optical tomography that we called holographic optical tomography (HOT) [35]. In HOT, we employ a high NA condenser to illuminate the object with a broad range of input fundamental spatial frequencies. To ensure widefield illumination, the object is illuminated at defocused plane where the beam is spread out spatially. Because the nonlinear scattering process draws from a broad distribution of illumination spatial frequencies from the full transverse spatial frequency support of the condenser NA, the coherent transfer function for this widefield coherent nonlinear scattering imaging process gains axial spatial frequency support, and thus allows for optical sectioning. Now the strategy first demonstrated by Streibl may be deployed so that 3D tomographic imaging can be obtained from the deconvolution of an axial image stack using a model of the HOT coherent transfer function.

4. Description of optical imaging systems

Optical microscopy can be modeled as a two-dimensional or three-dimensional image collection system. As our focus here is the treatment of tomographic imaging with coherently scattered light, we will provide a discussion of the imaging of spatially coherent light. In the case of coherent nonlinear scattering, however, the weak nonlinear light-matter interaction strength necessitates the use of pulsed light fields. Because the light propagation [66, 67] is linear and shift-invariant, after the coherent nonlinear scatter has occurred (as described in the following section), we may treat the light propagation for each temporal and spatial frequency independently, so that the total field may be obtained from the superposition of the imaged fields. A schematic of the optical imaging system is shown in **Figure 1**.

The spatio-temporal variation of the scalar field is denoted by

$$U_j^{\mathrm{sc}}(\mathbf{r}, t) = a_j(t) e^{-i\omega_j t} u_j(\mathbf{r}) \tag{1}$$

Here, j is the order of the incident pulse with a complex slowly-varying temporal envelope $a_j(t)$, that is centered on optical frequency $\omega_j = j\omega_1$. The object is illuminated by the incident fundamental pulsed field denoted with $j = 1$ having a field spectrum centered at ω_1. The full spectrum of the complex analytic scalar field is simply

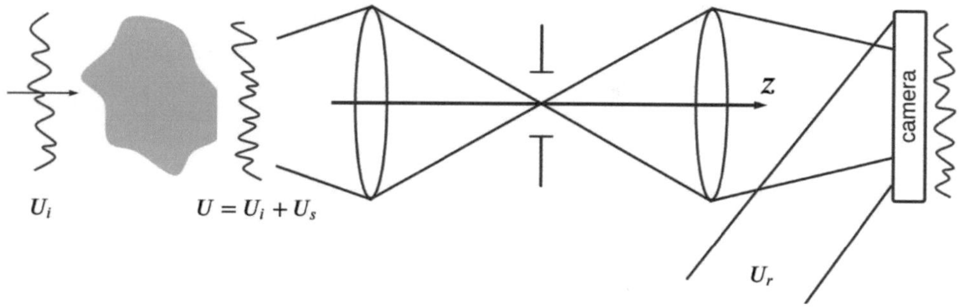

Figure 1.
The schematic of the optical 4-F imaging system. We use U^i to denote the incident field, U^t to denote the total field, U^s to denote the field scattered from the object that is imaged as the image field U^{im}, and U_r is the reference field that mixes with the images field to form the hologram.

$U_j(\mathbf{r}, \omega) = \mathcal{F}_t\left\{U_j^{\text{sc}}(\mathbf{r},\ t)\right\} = A_j(\Omega_j)\,u_j(\mathbf{r})$, where $\mathcal{F}\{\cdot\}$ is the Fourier transform operator with its subscript t denoting a transform with respect to the time variable t and the relative frequency is $\Omega_j = \omega - j\omega_1$. The complex temporal envelope is obtained through the inverse Fourier transform of the field spectral amplitude as $a_j(t) = \mathcal{F}_t^{-1}\{A_j(\Omega_j)\}$. The time, t, and optical frequency, variables are conjugate as are the three-dimensional spatial vector, $\mathbf{r} = (\mathbf{r}_\perp, z)$, and the spatial frequencies, $\mathbf{k} = 2\pi\left(\mathbf{f}_\perp, f_z\right)$. The transverse spatial coordinates are $\mathbf{r}_\perp = (x, y)$, with the corresponding transverse spatial frequencies $\mathbf{f}_\perp = \left(f_x, f_y\right)$.

When imaging a coherent field from an object plane to an imaging plane, we can use a simple shift-invariant model for each optical frequency component. The imaging system will be described by an ideal telecentric 4-F imaging system as shown in **Figure 1** [66, 68]. Green's function for a coherent 4-F imaging system is referred to as the amplitude (or coherent) spread function (CSF), $h(\mathbf{r}_\perp)$, which is expressed as the spatial convolution

$$u^{\text{im}}(\mathbf{r}_{\perp,\text{im}}, \omega) = \int_{\pm\infty} u^{\text{o}}(\mathbf{r}_{\perp,\text{o}}, \omega)\,h(\mathbf{r}_{\perp,\text{im}} - \mathbf{r}_{\perp,\text{o}}, \omega)\,d^2 r_{\perp,\text{o}}. \tag{2}$$

Here, $u^{\text{im}} = u(z = z_{\text{im}})$ is the field at the image plane, $u^{\text{o}} = u(z = z_{\text{o}})$ is the field at the object plane, $\mathbf{r}_{\perp,\text{im}}$ is at the image plane and $\mathbf{r}_{\perp,\text{o}}$ is at the object plane. The spatial frequency representation is quite compact and elucidates the low-pass transverse spatial filtering behavior of coherent optical imaging systems through the expression as follows:

$$u^{\text{im}}(\mathbf{f}_\perp, \omega) = H(\mathbf{f}_\perp, \omega)\,u^{\text{o}}(\mathbf{f}_\perp, \omega). \tag{3}$$

The coherent transfer function (CTF) is the Fourier transform of the CSF, $H(\mathbf{f}_\perp, \omega) = \mathcal{F}_{\mathbf{r}_\perp}\{h(\mathbf{r}_\perp,\ \omega)\}$. Here, we have assumed that the 4-F imaging system has unity magnification for simplicity of notation. The expressions are easily generalized to non-unity magnification [66].

As the propagation of coherent fields through a source-free region can be readily described using the angular spectral propagator, the object field can thus, be propagated from any reference plane to the conjugate object plane of the imaging system. Similarly, the field in the imaging region can also be propagated from one plane to another. The angular spectral propagator for fields propagating in the positive z direction, which we denote as our 4-F imaging system optical axis, is given simply by

$$u(\mathbf{f}_\perp, z + \Delta z, \omega) = \exp\left(i2\pi\gamma(\mathbf{f}_\perp, \ \omega)\Delta z\right)u(\mathbf{f}_\perp, z, \omega), \ z = z^{\text{im}}, z^{\text{o}}, \tag{4}$$

where the axial spatial frequency, $\gamma_j(\mathbf{f}_\perp, \omega_j) = \sqrt{(n\omega/2\pi c)^2 - \|\mathbf{f}_\perp\|^2}$, c is the speed of light in vacuum, and $n(\omega) \to n$ is the refractive index of the background medium at optical frequency ω. Only non-evanescent spatial frequencies, $\sqrt{\|\mathbf{f}\perp\|^2} < n\omega/2\pi c$, propagate to the far field to be detected.

Within the theory of semiclassical light detection, we may describe the signal recorded by a camera of an incident optical field as the time average of the zero-delay field autocorrelation,

$$I^{\text{det}}(\mathbf{r}_\perp, \tau) = \left\langle U_j^{\text{s}}(\mathbf{r}_\perp, \ z^{\text{im}}, \ t) U_j^{\text{r},*}(\mathbf{r}_\perp, \ z^{\text{im}}, \ t + \tau) \right\rangle_t. \tag{5}$$

The angle brackets, $< \cdot >$, denote a time average determined by the detector timescale (e.g., the camera integration time), which for practical nonlinear holographic imaging is orders of magnitude longer than the coherent time of the light fields. For $\tau = 0$, this signal can be equivalently represented by the weighted contributions by the cross-spectral density of the light $W^{\text{sr}}(\omega) = A^{\text{s}}(\omega)A^{\text{r},*}(\omega)$, leading to the expression

$$I^{\text{det}}(\mathbf{r}_\perp) = \int W^{\text{sr}}(\omega) I(\mathbf{r}_\perp, \omega) d\omega. \tag{6}$$

The intensity for spatially coherent fields, as we assume here, is defined as:

$$I(\mathbf{r}_\perp, \omega) = |u(\mathbf{f}_\perp, \ z^{\text{im}}, \ \omega)|^2. \tag{7}$$

Coherent tomographic imaging requires access to the field directly. This field can be approximately retrieved experimentally through holography that relies on interference with a reference field, which we denote as $u_{\text{r}} = A_{\text{r}} \exp(i\phi_{\text{r}})$. For simplicity, we have assumed that the reference field is unity amplitude and exhibits a relative phase shift ϕ_{r}. This phase shift can vary linearly as in off-axis holography [7, 8, 36] or relative phase shifts can be imparted in a series of measurements as is in phase shifting holography [10]. In either case, the field intensity for the total field given by the sum of the reference field and the images scattered field, $u^{\text{t}} = u^{\text{r}} + u^{\text{im}}$ leads to four terms in the intensity that read $I^{\text{t}} = I^{\text{r}} + I^{\text{im}} + u^{\text{r},*} u^{\text{im}} + u^{\text{r}} u^{\text{im},*}$. With suitable numerical processing, we may then isolate the scattered field from the measurement, leading to

$$I^{\text{holo}}(\mathbf{r}_\perp) = \int W^{\text{sr}}(\omega) u^{\text{r},*}(\mathbf{r}_\perp, z^{\text{im}}, \omega) u^{\text{im}}(\mathbf{r}_\perp, \omega) d\omega \tag{8}$$

for the case of a unity amplitude reference field. In the transverse spatial frequency domain, we may write this expression as:

$$I^{\text{holo}}(\mathbf{r}_\perp) = e^{i\beta_r z^{\text{im}}} \int W^{\text{sr}}(\omega) u^{\text{im}}(\mathbf{r}_\perp, \omega) d\omega. \tag{9}$$

Here, we note that the image of the scattered field, $u_s(\mathbf{f}_\perp, \omega)$, is low-pass filtered by the imaging system CTF.

With a model of coherent imaging of the scattered field, we now need a description of the scattered field to proceed. In the following section, we derive the scattered coherent nonlinear field the the m^{th}order harmonic driven by the fundamental field pulse with an incident fundamental center frequency of ω_1.

5. Coherent nonlinear scattering of scalar field

Our goal is to understand the imaging properties, capabilities, and limitations of coherent nonlinear optical holography and tomography. While a full description of coherent nonlinear scattering requires a vector treatment [17, 32, 34], we will restrict our discussion to scalar fields. Such a treatment may provide an understanding of holographic and imaging properties without loss of generality, as the measurement at the camera always involves a projection of the nonlinear scattered field polarization onto the reference field polarization [34]. Thus, we post-select a particular polarization component that can then be regarded as a scalar nonlinear field.

In the scalar description that follows, we begin with the wave equation, where we have made the usual assumptions for optical propagation. Explicitly, these assumptions are that we consider a region devoid of free charges and associated free-charge current densities. Moreover, we assume nonmagnetic media, so the magnetic permeability used is simply that of free space. To simplify the wave equation, we assume that both the linear, $\chi^{(1)}$, and nonlinear, $\chi^{(m)}$, optical susceptibilities are scalar quantities to facilitate a scalar treatment. Finally, we assume that any spatial variation in optical susceptibility is weak compared to the mean optical susceptibility, i.e., $\delta\chi^{(m)}(\mathbf{r}) \ll \overline{\chi}^{(m)} = \langle \chi^{(m)}(\mathbf{r}) \rangle_{\mathbf{r}}$, and the angle brackets denote a spatial average. This assumption means that to first order we may treat the medium as spatially homogeneous, which allows for simplification of the wave equation. The inverse scattering problem for imaging the spatial variations of optical susceptibility, $\delta\chi^{(m)}(\mathbf{r})$, are treated as a perturbation to the driven homogeneous wave equation.

5.1 Scattering model of scalar field

Making explicit use of the assumptions stated above, we may combine two of Maxwell's equations to obtain the optical wave equation:

$$\nabla \times \nabla \times \mathcal{U}(\mathbf{r}, t) + \mu_0 \frac{\partial^2 \mathcal{D}(\mathbf{r}, t)}{\partial t^2} = 0. \tag{10}$$

Here, $\mathcal{U}$ is the electric field which is assumed to be scalar. The right-hand side of the equation above is 0 because the external source of the field is excluded from the interested region and we assume there is no internal source, i.e., the material is not self-luminous. Here, $\mathcal{D}$ is the scalar displacement vector that describes the displacement current, including the linear and nonlinear response from bound charges in the material. The nonlinear contributions to this displacement constitute the quantity that we wish to image. Here, $\mathcal{U}$ and $\mathcal{D}$ denote real fields. Below, we will drive coupled wave equations for the propagation of the fundamental and nonlinear fields, where we assume that these fields may be described as complex analytic functions.

For isotropic media that can, to the first approximation, be treated as spatially homogeneous, in the absence of free charges, Gauss' law, $\nabla \cdot \mathcal{D} = 0$, allows us to make the approximation $\nabla \cdot \mathcal{U} \approx 0$. Making use of this simplification and a standard vector identity, the first term in Eq. (10) simplifies to

$$\nabla \times \nabla \times \mathcal{U} = \nabla(\nabla \cdot \mathcal{U}) - \nabla^2 \mathcal{U} = -\nabla^2 \mathcal{U}. \tag{11}$$

Now, Eq. (10) becomes

$$\nabla^2 \mathcal{U} - \mu_0 \frac{\partial^2 \mathcal{D}}{\partial t^2} = 0. \tag{12}$$

Our interest lies in the nonlinear response, which is encapsulated in the real displacement field, which is written as

$$\mathcal{D}(\mathbf{r}, t) = \varepsilon_0 \mathcal{U}(\mathbf{r}, t) + \mathcal{P}(\mathbf{r}, t). \tag{13}$$

The total real polarization density of the form of a superposition of the linear and the nonlinear response is given by

$$\mathcal{P}(\mathbf{r}, t) = \mathcal{P}^{\mathrm{L}}(\mathbf{r}, t) + \mathcal{P}^{\mathrm{NL}}(\mathbf{r}, t). \tag{14}$$

The linear polarization density follows a convolution of the linear optical response of the medium

$$\mathcal{P}^{\mathrm{L}}(\mathbf{r}, t) = \varepsilon_0 \int_{-\infty}^{t} R^{(1)}(\mathbf{r}, \tau) \mathcal{U}(\mathbf{r}, \tau) d\tau. \tag{15}$$

The linear, causal optical response function of the medium is $R^{(1)}(\mathbf{r}, t)$ and is related to the frequency-dependent optical susceptibility through the temporal Fourier transform relationship $R^{(1)}(\mathbf{r}, t) = \mathcal{F}_\omega\{\chi^{(1)}(\mathbf{r}, \omega)\}$. Here, $\mathcal{F}_\omega\{\cdot\} = \mathcal{F}_t^{-1}\{\cdot\} = (2\pi)^{-1} \int F(\omega) \exp(-i\omega t) d\omega$ denotes an inverse Fourier transform.

For nonresonant interactions, the nonlinear polarization density may generally be expanded as a power series of the form

$$\mathcal{P}^{\mathrm{NL}}(\mathbf{r}, t) = \mathcal{P}^{(2)}(\mathbf{r}, t) + \mathcal{P}^{(3)}(\mathbf{r}, t) + \cdots + \mathcal{P}^{(m)}(\mathbf{r}, t) + \cdots. \tag{16}$$

Details of the nonlinear polarization density will be deferred to a later section. These nonlinear polarization density terms drive a wide range of nonlinear optical processes. For our purposes, we will focus on $m = 2,3$ and only consider the processes that drive second and third harmonic generation (SHG and THG).

Combining all of these expressions, we arrive at our wave equation for coherent nonlinear scattering that reads the equation:

$$\nabla^2 \mathcal{U}(\mathbf{r}, t) - \mu_0 \varepsilon_0 \frac{\partial^2 \mathcal{U}(\mathbf{r}, t)}{\partial t^2} - \mu_0 \frac{\partial^2 \mathcal{P}^{\mathrm{L}}(\mathbf{r}, t)}{\partial t^2} - \mu_0 \frac{\partial^2 \mathcal{P}^{\mathrm{NL}}(\mathbf{r}, t)}{\partial t^2} = 0. \tag{17}$$

5.2 Wave equation in the frequency domain

The time-domain equation may easily be represented in the frequency domain by noting that the fields can be represented through an inverse Fourier transform as

$\mathcal{U}(\mathbf{r}, t) = \mathcal{F}_\omega\{\mathcal{U}(\mathbf{r}, \omega)\}$, $\mathcal{P}^{\mathrm{L}}(\mathbf{r}, t) = \mathcal{F}_\omega\{\mathcal{P}^{\mathrm{L}}(\mathbf{r}, \omega)\}$, and $\mathcal{P}^{\mathrm{NL}}(\mathbf{r}, t) = \mathcal{F}_\omega\{\mathcal{P}^{\mathrm{NL}}(\mathbf{r}, \omega)\}$. The spectra of the real fields are denoted with the argument ω.

Applying the second-order temporal partial derivatives to the inverse Fourier transform in Eq. (17) produces the frequency-domain wave equation:

$$\nabla^2\mathcal{U}(\mathbf{r}, \omega) + \omega^2\mu_0\varepsilon_0\mathcal{U}(\mathbf{r}, \omega) + \omega^2\mu_0\mathcal{P}^{\mathrm{L}}(\mathbf{r}, \omega) + \omega^2\mu_0\mathcal{P}^{\mathrm{NL}}(\mathbf{r}, \omega) = 0. \tag{18}$$

Making use of the time-domain linear response function in Eq. (15), we may write the equation:

$$\mathcal{P}^{\mathrm{L}}(\mathbf{r}, \omega) = \varepsilon_0\chi^{(1)}(\mathbf{r}, \omega)\mathcal{U}(\mathbf{r}, \omega). \tag{19}$$

Defining the refractive index in the usual way as $n^2(\mathbf{r}, \omega) = 1 + \chi^{(1)}(\mathbf{r}, \omega)$ and the wavenumber as $\beta(\mathbf{r}, \omega) = \omega n(\mathbf{r}, \omega)/c$ and where the phase velocity of light in a vacuum is $c = (\mu_0\varepsilon_0)^{-1/2}$, then the wave equation may be written as:

$$\nabla^2\mathcal{U}(\mathbf{r}, \omega) + \beta^2(\mathbf{r}, \omega)\mathcal{U}(\mathbf{r}, \omega) = -\omega^2\mu_0\mathcal{P}^{\mathrm{NL}}(\mathbf{r}, \omega). \tag{20}$$

This equation is now a forced Helmholtz equation, where the LHS describes linear scattering and the RHS is the nonlinear forcing function.

5.3 Slowly varying envelope approximation

The wave equation in Eq. (20) contains the spectrum of the real fields and polarization densities, which includes the complex conjugate of the positive frequencies in the negative frequency region. In addition, these spectra include all optical frequencies, including the fundamental and the nonlinear scattered fields. To simplify these expressions, we assume that each spectral region can be written as a separate spectral envelope, so that we consider, in general, a set of optical pulses (or cw fields) with frequencies centered at ω_j so that we may decompose the total (real) field as the superposition

$$\mathcal{U}(\mathbf{r}, t) = \sum_j \mathcal{U}_j(\mathbf{r}, t) \tag{21}$$

with the distinct spectral bands centered about ω_j. We now write the complex analytic scalar field U for the jth frequency band as:

$$\mathcal{U}_j(\mathbf{r}, t) = \frac{1}{2}U(\mathbf{r}, t) + \frac{1}{2}U^*(\mathbf{r}, t), \tag{22}$$

where * denotes the complex conjugate.

We assume that we have pulses well described by a slowly varying envelope in time relative to the rapid oscillations of a carrier (center) frequency, ω_j. Thus, the total field can be written as a slowly varying temporal envelope, $a_j(t)$, and spatial envelope, $u_j(\mathbf{r})$, multiplied by a rapidly varying carrier that is nominally propagating along the direction z, giving

$$U_j(\mathbf{r}, t) = a_j(t)u_j(\mathbf{r})\exp\left(i\beta_j z - i\omega_j t\right). \tag{23}$$

The wavenumber at frequency ω_j is defined by $\beta_j = \omega_j n_j / c$, where $n_j = \langle n(\mathbf{r},\ \omega_j) \rangle_{\mathbf{r}}$. The positive frequency (complex analytic field), $U_j(\mathbf{r}, \omega) = \mathcal{F}_\omega\{U_j(\mathbf{r},\ t)\}$, then reads the following equation:

$$U_j(\mathbf{r}, \omega) = A_j(\omega - \omega_j) u_j(\mathbf{r}) e^{i\beta_j z}. \tag{24}$$

For the convenience of notation, by defining $\Omega_j = \omega - \omega_j$, we may write $a_j(t) = \mathcal{F}_{\Omega j}\{A_j(\Omega_j)\}$. With this definition, we use the standard approach of describing a pulse in terms of a center of mass (of the power spectral density of the light field), ω_j, and a slowly varying envelope in time, $a_j(t)$. We assume that these fields have a temporal envelope that varies slowly with respect to the oscillation of the carrier, ω_j. This is the standard slowly varying envelope approximation.

5.4 Harmonic generation

The nonlinear polarization density, e.g., for SHG scattering with $m = 2$, (assuming scalar interactions for simplicity) is given by

$$\mathcal{P}^{\mathrm{NL}}(\mathbf{r}, t) = \varepsilon_0 \chi^{(m)}(\mathbf{r}) \mathcal{U}^m(\mathbf{r}, t). \tag{25}$$

We have explicitly ignored spectral dispersion of the second-order nonlinear coefficient in this expression, and as such we do not need the second-order time response integral nor the second-order response function. Physically, these assumptions equate to assuming that the second-order polarization density responds instantly. Note also that the field and polarization density are described by real quantities in this expression. While many nonlinear interactions can occur, we focus our discussion on coherent nonlinear scattering where we scatter to new frequencies at $\omega_m = m\omega_1$ due to a nonlinear dipole that oscillates at m times the input fundamental center frequency ω_1.

The many interaction terms are considered by taking the mth power of the total field that we consider as a superposition of the fundamental, $\mathcal{U}_1$, and the mth harmonic field, $\mathcal{U}_m$. This expansion provides two complex analytic terms leading to a polarization density that generates the mth harmonic frequency ω_m, given by

$$P^{(\mathrm{mHG})}(\mathbf{r}, t) = \frac{1}{2^{m-1}} \varepsilon_0 \chi^{(m)}(\mathbf{r}) U_1^m(\mathbf{r}, t). \tag{26}$$

This term drives coherent nonlinear scattering from the fundamental optical frequency centered at ω_1 to the harmonic frequency centered at ω_n. We may also consider the complementary process in which the nonlinear field is back-converted to the fundamental through the polarization density term that oscillates at the center frequency ω_1

$$P^{(\mathrm{bc})}(\mathbf{r}, t) = \varepsilon_0 \chi^{(m)}(\mathbf{r}) U_1^*(\mathbf{r}, t) U_m(\mathbf{r}, t). \tag{27}$$

In an imaging scenario, we may assume that very little coherent linearly scattered power is generated. Thus, we may assume that $U_m \ll U_1$, and that U_1 is constant throughout the interaction region. The mathematical approximation for this condition is referred to as the undepleted pump approximation. In this approximation, we may drop the back-conversion term. In addition, we will see that this approximation naturally leads to a nonlinear scattering equation that is homologous to linear scattering in the first Born approximation.

As a specific example, consider SHG, where the time-domain polarization density for the SHG source term reads

$$P^{(\mathrm{SHG})}(\mathbf{r},t)=\frac{1}{2}\varepsilon_0\chi^{(2)}(\mathbf{r})u_1^2(\mathbf{r})a_1^2(t)e^{i(2\beta_1 z-2\omega_1 t)}. \tag{28}$$

The polarization density for SHG oscillates with a center frequency of $\omega_2 = 2\omega_1$ and is described by the fundamental pulse spectral autocorrelation that appears from the Fourier transform of the square of the slowly varying fundamental pulse temporal envelope as given by

$$P^{(\mathrm{SHG})}(\mathbf{r},\Omega_2)=\frac{1}{2}\varepsilon_0\chi^{(2)}(\mathbf{r})e^{i2\beta_1 z}u_1^2(\mathbf{r})\mathcal{F}_\omega\{a_1^2(t)e^{-i2\omega_1 t}\}. \tag{29}$$

5.5 Coupled wave equations

Making use of the nonlinear scattering assumptions noted in the previous section we are now in a position to write the coupled wave equations for the coherent nonlinear scattering process.

$$\left[\nabla^2+\beta_{10}^2 n^2(\mathbf{r},\ \omega)\right]U_1(\mathbf{r},\omega)=0, \tag{30}$$

$$\left[\nabla^2+\beta_{m0}^2 n^2(\mathbf{r},\ \omega)\right]U_m(\mathbf{r},\omega)=-\omega^2\mu_0 P^{(\mathrm{mHG})}(\mathbf{r},\omega), \tag{31}$$

and where the nonlinear polarization density reads

$$P^{(\mathrm{mHG})}(\mathbf{r},\omega)=\frac{1}{2^{m-1}}\varepsilon_0\chi^{(m)}(\mathbf{r})u_1^m(\mathbf{r},\omega)e^{im\beta_1 z}\mathcal{F}_\omega\{a_1^m(t)e^{-im\omega_1 t}\}. \tag{32}$$

The first equation describes linear scattering, while the second is the nonlinear scattering at the mth harmonic. Critically here, we have used the undepleted pump approximation because the nonlinear scattered field is assumed to ever gain enough strength to drive the back conversion process. In addition, we assume zero input coherent nonlinear field at the input boundary. The free-space wavenumber for the jth frequency term is $\beta_{j0} = \omega_j/c$.

By defining the mth-order autocorrelation function as

$$\mathcal{A}_m(\Omega_m)=\mathcal{F}_\omega\{a_1^m(t)e^{-im\omega_1}\}, \tag{33}$$

we may write the forced equation governing linear and coherent nonlinear scattering as:

$$\left[\nabla^2+\beta_{10}^2 n^2(\mathbf{r},\ \omega)\right]u_1(\mathbf{r},\omega)\mathcal{A}_1(\Omega_1)=0, \tag{34}$$

$$\left[\nabla^2+\beta_{m0}^2 n^2(\mathbf{r},\ \omega)\right]u_m(\mathbf{r},\omega)\mathcal{A}_m(\Omega_m)=-\frac{\beta_{m0}^2}{2^{m-1}}\chi^{(m)}(\mathbf{r})u_1^m(\mathbf{r},\omega)\mathcal{A}_m(\Omega_m). \tag{35}$$

This form of the equations admits the construction of solutions from the free space Green's functions. Here, we have assumed that spectral width is sufficiently narrow than the multiplicative $\omega^2 \approx \omega_m^2$.

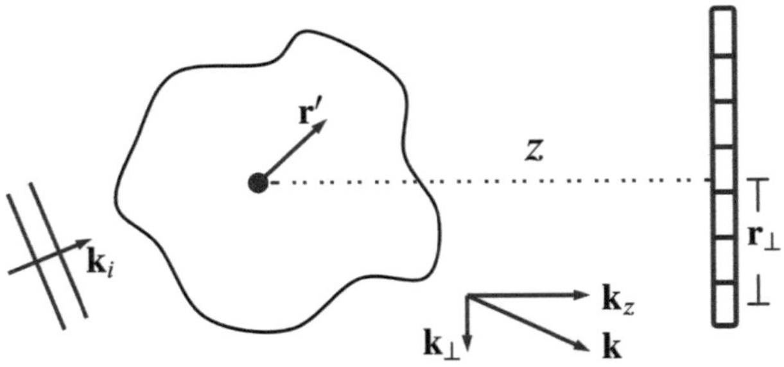

Figure 2.
The scattering model. The object is illuminated by a fundamental wave, for example, a plane wave with wavevector $\mathbf{k}_i$. The field scattered from the object, with a range of detected wavevectors $\mathbf{k}$, is measured at the plane perpendicular to z axis with a size limited by the objective. The origin of the coordinate is denoted by the black dot. For nonlinear scattering, we have $\mathbf{k}_i \to \mathbf{k}_1^i$, $k \to \mathbf{k}_m$.

The equations above allow for a general spectrally-varying treatment of coherent nonlinear holography and tomography. However, the effects of the spectral variation on propagation and on the interpretation of scattering are not strongly dependant on the pulse spectrum. In order to simplify the following interpretation of the imaging transfer function, we will assume that we have a narrow enough spectrum so we may make a continuous wave (cw) approximation, where $\mathcal{A}_m(\Omega_m) \approx A_m \delta(\Omega_m)$. Invoking this approximation and integrating over Ω_m and assuming unity amplitude fields leads to the simplified form of the coupled wave equations given by

$$\left[\nabla^2 + \beta_{10}^2 n^2(\mathbf{r},\ \omega_1)\right] u_1(\mathbf{r},\ \omega_1) = 0, \tag{36}$$

for the fundamental and

$$\left[\nabla^2 + \beta_{m0}^2 n^2(\mathbf{r},\ \omega)\right] u_m(\mathbf{r},\ \omega) = -\frac{\beta_{m0}^2}{2^{m-1}} \chi^{(m)}(\mathbf{r}) u_1^m(\mathbf{r},\ \omega) \tag{37}$$

for the nonlinear harmonic field (**Figure 2**).

5.6 Holography with a linear scattering

Inspection of the coupled wave equations in Eqs. (36) and (37) makes it clear that within the undepleted pump approximation the fundamental field solution is independent of the nonlinear scattering. Thus, it is fruitful to first obtain a solution to the linear field propagation, and we will consider the general case where the linear susceptibility varies in space. We may rewrite Eq. (36) in the form of the equation:

$$\left[\nabla^2 + \beta_{10}^2 n_1^2\right] u_1(\mathbf{r}) = -\beta_{10}^2 \delta\chi^{(1)}(\mathbf{r}) u_1(\mathbf{r}). \tag{38}$$

Our goal is to solve for $\delta\chi^{(1)}(\mathbf{r}) = \chi^{(1)}(\mathbf{r}) - \overline{\chi}^{(1)}$ to produce an image of image the susceptibility variation, which constitutes our object. The background linear optical susceptibility, $\overline{\chi}^{(1)}$, is chosen so that $\delta\chi^{(1)}(\mathbf{r})$ lies in a compact domain, i.e., so that it is contained within some volume V. The background refractive index at the fundamental frequency ω_1 is then given by $n_1^2 = 1 + \overline{\chi}^{(1)}$.

Solutions to Eq. (38) in the first Born approximation may be constructed using Green's theorem with the formula as follows:

$$u_1(\mathbf{r}) = u_1^{(i)}(\mathbf{r}) - \beta_{10}^2 \int_V g(\mathbf{r}, \mathbf{r}') \delta\chi^{(1)}(\mathbf{r}') u_1^{(i)}(\mathbf{r}') d^3\mathbf{r}'. \tag{39}$$

Such solutions in the domain outside of the compactly supported susceptibility perturbation, i.e., $\mathbf{r} \notin V$, can be constructed because we have defined the susceptibility perturbation so that it is contained inside of the volume V. The free-space Green's function is defined by

$$\nabla^2 g(\mathbf{r}, \mathbf{r}') + \beta_1^2 g(\mathbf{r}, \mathbf{r}') = \delta^{(3)}(\mathbf{r} - \mathbf{r}'), \tag{40}$$

where $\beta_1 = n_1 \beta_{10}$. The incident fundamental wave, $u_1^{(i)}(\mathbf{r})$, is a solution to the homogeneous wave equation, given by Eq. (38) when the susceptibility term vanishes with $\delta\chi^{(1)}(\mathbf{r}) = 0$.

Making use of a three-dimensional spatial frequency decomposition, where $\mathbf{k}_j = (k_{jx}, k_{jy}, k_{jz})$, and where the norm of the wavevector gives the wavenumber $\mathbf{k}_j \cdot \mathbf{k}_j = \beta_j^2$, Green's function is written as

$$g(\mathbf{r}, \mathbf{r}') = \frac{1}{(2\pi)^3} \int_{-\infty}^{\infty} g(\mathbf{k}_1) e^{i\mathbf{k}_1 \cdot (\mathbf{r} - \mathbf{r}')} d^3\mathbf{k}_1. \tag{41}$$

Making use of this expansion in Eq. (39) produces the spatial frequency spectrum of the free space Green's function as:

$$g(\mathbf{k}_1) = \frac{1}{(k_{1z} - \Gamma_1)(k_{1z} + \Gamma_1)}. \tag{42}$$

Here, we have defined the transverse spatial frequency vector $\mathbf{k}_{j\perp} = (k_{jx}, k_{jy})$ as well as $\Gamma_j = \sqrt{\beta_j^2 - \|\mathbf{k}_{j\perp}\|^2}$.

Computation of the inverse Fourier transform along the z direction by making use of contour integration leads to the Weyl expression for Green's function, that is:

$$g(\mathbf{r}_\perp, z, \mathbf{r}'_\perp, z') = \frac{1}{i4\pi} \int_{-\infty}^{\infty} \frac{e^{i\mathbf{k}_{1\perp} \cdot (\mathbf{r}_\perp - \mathbf{r}'_\perp)} e^{i\Gamma_1(\mathbf{k}_{1\perp})|z - z'|}}{\Gamma_1(\mathbf{k}_{1\perp})} d^2\mathbf{k}_{1\perp}. \tag{43}$$

By defining the source term in the RHS of Eq. (38) as $S_1(\mathbf{r}) = \delta\chi^{(1)}(\mathbf{r}) u_1^{(i)}(\mathbf{r})$ in the first Born approximation, we may write the solution for the scattered wave as:

$$u_1(\mathbf{k}_1) = u_1^{(i)}(\mathbf{k}_1) - \beta_{10}^2 g(\mathbf{k}_1) S_1(\mathbf{k}_1). \tag{44}$$

Here, the spectrum of the source term reads $S_1(\mathbf{k}_1) = \delta\chi^{(1)}(\mathbf{k}_1) \circledast \mathbf{k}_1 u_1^{(i)}(\mathbf{k}_1)$ and the spectrum of the input fundamental wave that is propagating along $+z$ is $u_1^{(i)}(\mathbf{k}_1) = H_i(\mathbf{k}_{1\perp}) \delta(k_{1z} - \Gamma_1(\mathbf{k}_{1\perp}))$. The operator $\circledast$ represents a convolution.

The forward and backward scattered fields are obtained by applying contour integration and selecting the suitable residue from the simple roots of Eq. (42). In the forward direction ($z > 0$), we can obtain the field as:

$$u_1^f(\mathbf{k}_{1\perp}, z) = u_1^{(i)}(\mathbf{k}_{1\perp}, z) - i \frac{\beta_{10}^2 e^{i\Gamma_1 z}}{2\Gamma_1(\mathbf{k}_{1\perp})} S_1(\mathbf{k}_{1\perp}, \Gamma_1). \tag{45}$$

In the backward direction, $z < 0$, we obtain

$$u_1^b(\mathbf{k}_{1\perp}, z) = i\frac{\beta_{10}^2 e^{-i\Gamma_1 z}}{2\Gamma_1(\mathbf{k}_{1\perp})} S_1(\mathbf{k}_{1\perp}, -\Gamma_1). \tag{46}$$

We will restrict our discussion to the forward-propagating wave collected by the imaging system with a coherent transfer function given by $H_o(\mathbf{k}_{1\perp})$. Thus, the imaged field reads as:

$$u_1^{\mathrm{im}}(\mathbf{k}_{1\perp}, z) = -iH_o(\mathbf{k}_{1\perp})\frac{\beta_{10}^2 e^{i\Gamma_1 z}}{2\Gamma_1(\mathbf{k}_{1\perp})} S_1(\mathbf{k}_{1\perp}, \Gamma_1). \tag{47}$$

Here, we have dropped the unscattered portion of the incident field to focus on the scattered field and simplify the discussion that follows.

In the process of recording a hologram, we multiply by a reference field, $u_1^{\mathrm{ref}}(\mathbf{k}_{1\perp}, z) = \delta^{(2)}(\mathbf{k}_{1\perp}) \exp(i\beta_1 z)$, so that our cw hologram signal, $I_1^{\mathrm{holo}}(\mathbf{r}_\perp, z) = u_1^{\mathrm{ref},*}(\mathbf{r}_\perp, z)u_1^{\mathrm{im}}(\mathbf{r}_\perp, z)$, which leads to the transverse spectrum of the hologram is given by the cross-correlation operation

$$I_1^{\mathrm{holo}}(\mathbf{k}_{1\perp}, z) = u_1^{\mathrm{im}}(\mathbf{k}_\perp, z) \otimes_{\mathbf{k}_{1\perp}} u_1^{\mathrm{ref}}(\mathbf{k}_{1\perp}, z). \tag{48}$$

The operator $\otimes$ represents a correlation. We have also assumed that the holographic interference term has been shifted to baseband. After evaluation of the cross-correlation integral, the hologram transverse spectrum now reads

$$I_1^{\mathrm{holo}}(\mathbf{k}_{1\perp}, z) = u_1^{\mathrm{im}}(\mathbf{k}_\perp, z)e^{-i\beta_1 z}. \tag{49}$$

To simplify our analysis of the hologram, we will first consider the special case of the fundamental incident wave as a plane wave incident along the direction $\mathbf{k}_1^i$, with amplitude $H_i(\mathbf{k}_{1\perp}^i)$, so that $u_1^i = H_i(\mathbf{k}_{1\perp}^i) \exp(i\mathbf{k}_1^i \cdot \mathbf{r})$ and the corresponding spectrum reads $u_1^i(\mathbf{k}_1) = (2\pi)^3 H_i(\mathbf{k}_{1\perp}^i)\delta^{(3)}(\mathbf{k}_1 - \mathbf{k}_1^i)$.

For the plane illumination case, we may specifically write out our source term convolution integral as follows:

$$S_1(\mathbf{k}_1) = \int \delta\chi^{(1)}(\mathbf{k}_1 - \mathbf{k}_{1'})u_1^i(\mathbf{k}_{1'})d^3\mathbf{k}_{1'} \tag{50}$$

as

$$S_1(\mathbf{k}_1) = H_i(\mathbf{k}_{1\perp}^i)e^{i\beta_1 z}\delta\chi^{(1)}(\mathbf{k}_1 - \mathbf{k}_1^i). \tag{51}$$

Now the imaged scattered field hologram for a single incident fundamental plane wave illumination reads as:

$$I_1^{\mathrm{holo}}(\mathbf{k}_{1\perp}, z) = -iH_o(\mathbf{k}_{1\perp})H_i(\mathbf{k}_{1\perp}^i)\frac{\beta_{10}^2 e^{i(\Gamma_1 - \beta_1)z}}{2\Gamma_1(\mathbf{k}_{1\perp})}\delta\chi^{(1)}(\mathbf{k}_{1\perp} - \mathbf{k}_{1\perp}^i, k_{1z} - \Gamma_1^i). \tag{52}$$

Note that $\Gamma_j^i(\mathbf{k}_{j\perp}^i) = \sqrt{\beta_j^2 - \left\|\mathbf{k}_{j\perp}^i\right\|^2}$ and the spatial vector is decomposed as $\mathbf{r} = (\mathbf{r}_\perp, z)$. We see that the 2D Fourier transform of the measured field is related to the 3D Fourier transform of the susceptibility distribution of the object.

Making a coordinate transform into the spatial frequency space of the object by defining the scattering vector $\mathbf{Q} = (\mathbf{Q}_\perp, Q_z) \equiv \mathbf{k}_1 - \mathbf{k}_1^i$, then we may rewrite the hologram transverse spectrum for a single plane wave illumination as:

$$I_1^{\text{holo}}(\mathbf{Q}_\perp, z) = -iH_o(\mathbf{Q}_\perp + \mathbf{k}_{1\perp}^i)H_i(\mathbf{k}_{1\perp}^i)\frac{\beta_{10}^2 e^{i(\Gamma_1(\mathbf{Q}_\perp+\mathbf{k}_{1\perp}^i)-\beta_1)z}}{2\Gamma_1(\mathbf{Q}_\perp + \mathbf{k}_{1\perp}^i)}\delta\chi^{(1)}(\mathbf{Q}). \tag{53}$$

Now, we may take the Fourier transform along z, giving us

$$I_1^{\text{holo}}(\mathbf{Q}_\perp, Q_z) = -i\beta_{10}^2 H_o(\mathbf{Q}_\perp + \mathbf{k}_{1\perp}^i)H_i(\mathbf{k}_{1\perp}^i)\frac{\delta[Q_z + \beta_1 - \Gamma_1(\mathbf{Q}_\perp + \mathbf{k}_{1\perp}^i)]}{2\Gamma_1(\mathbf{Q}_\perp + \mathbf{k}_{1\perp}^i)}\delta\chi^{(1)}(\mathbf{Q}). \tag{54}$$

The Dirac delta function has selected the portions of the Ewald sphere that are supported by the illumination and collection optical system transfer functions given as H_i and H_o, respectively. The hologram field may now be written as a simple linear shift-invariant model with

$$I_1^{\text{holo}}(\mathbf{Q}) = H_{\text{lin}}(\mathbf{Q})\delta\chi^{(1)}(\mathbf{Q}). \tag{55}$$

By inspection of Eq. (54), we may identify the transfer function for a single illumination plane wave, which is given by:

$$H_{\text{lin}}^i(\mathbf{Q}) \equiv -i\beta_{10}^2 H_o(\mathbf{Q}_\perp + \mathbf{k}_{1\perp}^i)H_i(\mathbf{k}_{1\perp}^i)\frac{\delta[Q_z + \beta_1 - \Gamma_1(\mathbf{Q}_\perp + \mathbf{k}_{1\perp}^i)]}{2\Gamma_1(\mathbf{Q}_\perp + \mathbf{k}_{1\perp}^i)}. \tag{56}$$

When using a non-negligible illumination condenser optic NA, then the super-position of all of the illumination k-vectors must be considered so that we can get:

$$H_{\text{lin}}(\mathbf{Q}) \equiv -i\beta_{10}^2 \int H_o(\mathbf{Q}_\perp + \mathbf{k}_{1\perp}^i)H_i(\mathbf{k}_{1\perp}^i)\frac{\delta[Q_z + \beta_1 - \Gamma_1(\mathbf{Q}_\perp + \mathbf{k}_{1\perp}^i)]}{2\Gamma_1(\mathbf{Q}_\perp + \mathbf{k}_{1\perp}^i)}d^2\mathbf{k}_{1\perp}^i. \tag{57}$$

Here, we have suppressed the explicit optical frequency dependence. For a short pulsed illumination, we would make use of an effective transfer function weighted by the cross-spectral density of the scattered and reference waves:

$$H_{\text{lin}}(\mathbf{Q}) = \int H_{\text{lin}}(\mathbf{Q})W_{\text{sr}}(\omega)d\omega. \tag{58}$$

Notice that only the terms β_1 and β_{10} exhibit spectral dependence, provided that we can neglect dispersion of $\delta\chi^{(1)}$.

5.7 Coherent nonlinear scattering, holography, and tomography

The linear scattering case can be easily extended to nonlinear scattering. For mth-order scattering, we first assume that we have zero-incident harmonic field, so that only the scattered fields appear in the expressions. We assume a mean refractive index at the harmonic of $\overline{n}_m^2 = \langle n_m^2(\mathbf{r})\rangle_{\mathbf{r}}$, so that we may define $\beta_m = \overline{n}_m \beta_{m0}$ and $\beta_{m0} = m\,\omega_1/c$. Furthermore, we define the magnitude of the phase mismatch at $\Delta\beta = \beta_m - m\beta_1 = m\beta_{10}\,(\overline{n}_m - \overline{n}_1)$. This parameter is defined to be positive for material that exhibits normal dispersion. This presents a scattering exclusion zone near the origin.

Consider Eq. (37), which is a driven Helmholtz equation analogous to Eq. (38), but where we make the substitutions $\beta_{10}^2 \to \beta_{m0}^2/2^{m-1}$, $\delta\chi^{(1)}(\mathbf{r}) \to \chi^{(m)}(\mathbf{r})$, and $u_1(\mathbf{r}) \to u_1^m(\mathbf{r})$. Now we obtain a nonlinear scattering version of Eq. (44), but where no incident field is present due to the zero boundary condition identified above and we make the further substitutions $\mathbf{k}_{1\perp} \to \mathbf{k}_{m\perp}$, $\Gamma_1 \to \Gamma_m$, and modify the source term to read

$$S_m(\mathbf{k}_m) = \chi^{(m)}(\mathbf{k}_m) \circledast_{\mathbf{k}_m} u_i^{(m)}(\mathbf{k}). \tag{59}$$

where $u_i^{(m)}(\mathbf{k}) = \mathcal{F}_{3D}\{u_1^m(\mathbf{r})\}$.

Defining a generalized nonlinear scattering vector $\mathbf{Q} = (\mathbf{Q}_\perp, Q_z) \equiv \mathbf{k}_m - \mathbf{k}_m^i$, where $\mathbf{k}_m^i = \sum_{j=1}^m \mathbf{k}_1^{(j)}$, and explicitly we have $\mathbf{k}_{m\perp}^i = \sum_{j=1}^m \mathbf{k}_{1\perp}^{(j)}$ and $\Gamma_m^i = \sum_{j=1}^m \Gamma_1^{(j)}$. Then, by following the arguments in the previous section, we may again obtain a linear shift-imaging imaging model given by

$$I_m^{\text{holo}}(\mathbf{Q}) = H_{\text{mHG}}(\mathbf{Q})\chi^{(m)}(\mathbf{Q}). \tag{60}$$

Here, the transfer function for the harmonic holography reads:

$$H_{\text{mHG}}(\mathbf{Q}) = \int H_{\text{mHG}}^i(\mathbf{Q})\, d^2\mathbf{k}_{m\perp}^i. \tag{61}$$

And where the integrand is given by:

$$H_{\text{mHG}}^i(\mathbf{Q}) \equiv -\frac{i}{2^{m-1}}\beta_{m0}^2 H_o\left(\mathbf{Q}_\perp + \mathbf{k}_{m\perp}^i\right) u_i^{(m)}\left(\mathbf{k}_{m\perp}^i\right) \frac{\delta\left[Q_z + \beta_m - \Gamma_m\left(\mathbf{Q}_\perp + \mathbf{k}_{m\perp}^i\right)\right]}{2\Gamma_m\left(\mathbf{Q}_\perp + \mathbf{k}_{m\perp}^i\right)}. \tag{62}$$

A similar extension to illumination with a short optical pulse can be applied to this transfer function as was applied in the linear scattering case (**Figure 3**).

5.8 Example: second harmonic generation holography

Coherent nonlinear holography offers new possibilities for expanded spatial frequency support due to the effect of noncollinear mixing of fundamental spatial frequencies in the nonlinear mixing process [35]. The key difference lies in the source term, which for SHG reads:

$$S_2(\mathbf{k}_2) = \int \chi^{(2)}(\mathbf{k}_2 - \mathbf{k}_{2'})\, u_i^{(2)}(\mathbf{k}_{2'})\, d^3\mathbf{k}_{2'}. \tag{63}$$

The spectrum of the square of the field is the autoconvolution of the spectrum, $u_2^i(\mathbf{k}_2) = u_1^i(\mathbf{k}_1) \circledast_{\mathbf{k}_1} u_1^i(\mathbf{k}_1)$, which is given by the integral

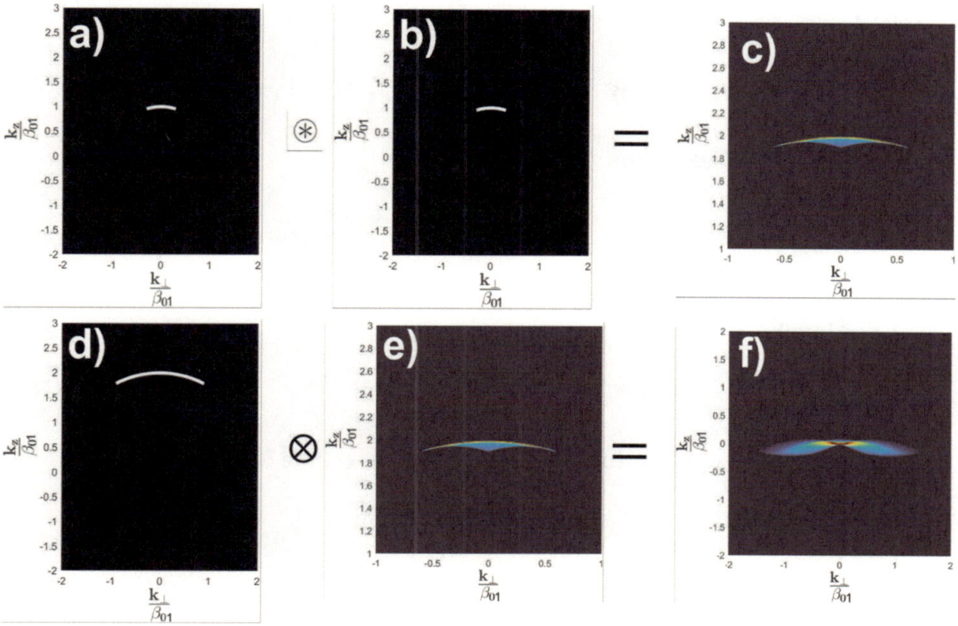

Figure 3.
Visualization of the construction of the SHG coherent transfer function (CTF) using Eqs. (64) and (65). Panel (a) and (b) show the fundamental pupil function of the condenser lens with an NA of .3. Panel (c) shows the resultant distribution for $u_i^{(2)}(\mathbf{k}_2)$. Panel (d) shows the pupil function for the objective lens collecting SHG light (NA = .9). Panel (e) shows panel (c) again to graphically illustrate the convolution between (d) and (e) that appears at the SHG CTF in panel (f). The final result in panel (f) is the total contribution of all possible plane wave angles and collected SHG allowed by the optics of the system as shown in Eq. (76).

$$u_i^{(2)}(\mathbf{k}_2) = \int u_1^i(\mathbf{k}_1) u_1^i(\mathbf{k}_2 - \mathbf{k}_1) d^3\mathbf{k}_1. \tag{64}$$

The autoconvolution of the fundamental field spectrum also appears in the problem of modeling reflectance confocal microscopy [69] and for an illumination objective with a half-opening angle α that is defined by $\sin\alpha = \mathrm{NA}/n$ and reads as:

$$u_i^{(2)}(\mathbf{k}_{2\perp}, k_{2z}) = \frac{4\beta_2}{\pi K} \sin^{-1}\left[\frac{1}{p}\left(1 - \frac{2\beta_2 \cos\alpha}{|k_{2z}|}\right)\right], \text{for } |k_{2z}| \geq 2\beta_2 \cos\alpha. \tag{65}$$

Here, we use the parameters $p = (2\beta_2 |k_{2\perp}|)\sqrt{1 - (K/2\beta_2)^2}/(K|k_{2z}|)$ and $K = \sqrt{k_{2\perp}^2 + k_{2z}^2}$.

Now, we consider the case where we have two input fundamental plane waves $u_{1'}(\mathbf{r}) = a' \exp(i\mathbf{k}_{1'} \cdot \mathbf{r})$ and $u_1{}''(\mathbf{r}) = a'' \exp\left(i\mathbf{k}_1{}'' \cdot \mathbf{r}\right)$. Now, the integral in Eq. (64)

$$u_i^{(2)}(\mathbf{k}_2) = a'a'' \int \delta^{(3)}(\mathbf{k}_1 - \mathbf{k}_{1'})\delta^{(3)}\left(\mathbf{k}_2 - \mathbf{k}_1 - \mathbf{k}_1{}''\right) d^3\mathbf{k}_1, \tag{66}$$

which simplifies to

$$u_i^{(2)}(\mathbf{k}_2) = a'a'' \delta^{(3)}\left(\mathbf{k}_2 - \mathbf{k}_{1'} - \mathbf{k}_1{}''\right). \tag{67}$$

So that now the source term evaluates to read

$$S_2(\mathbf{k}_2) = a'a''\chi^{(2)}\left(\mathbf{k}_{2\perp} - \mathbf{k}_{1\perp}' - \mathbf{k}_{1\perp}'', k_{2z} - \Gamma_{1'} - \Gamma_1''\right). \tag{68}$$

The imaged SHG field reads

$$u_2^{\text{im}}(\mathbf{k}_{2\perp}, z) = -iH_o(\mathbf{k}_{2\perp}) \frac{\beta_{20}^2 e^{i\Gamma_2(\mathbf{k}_{2\perp})z}}{4\Gamma_2(\mathbf{k}_{2\perp})} S_2(\mathbf{k}_{2\perp}, \Gamma_2). \tag{69}$$

The SHG hologram reads

$$I_2^{\text{holo}}(\mathbf{k}_{2\perp}, z) = u_2^{\text{im}}(\mathbf{k}_{2\perp}, z)e^{-i\beta_2 z}. \tag{70}$$

For the case with a pair of fundamental plane waves, the SHG hologram term

$$I_2^{\text{holo}}(\mathbf{k}_{2\perp}, z) = -iH_o(\mathbf{k}_{2\perp})u_i^{(2)}(\mathbf{k}_{2\perp}^i) \frac{\beta_{20}^2 e^{i(\Gamma_2(\mathbf{k}_{2\perp})-\beta_2)z}}{4\Gamma_2(\mathbf{k}_{2\perp})} \chi^{(2)}\left(\mathbf{k}_{2\perp} - \mathbf{k}_{2\perp}^i, k_{2z} - \Gamma_1^{i(2)}\right). \tag{71}$$

We have defined in incident SHG vector as the sum of the two incident fundamental k-vectors, $\mathbf{k}_2^i = \mathbf{k}_{1'} + \mathbf{k}_1''$, and explicitly we have $\mathbf{k}_{2\perp}^i = \mathbf{k}_{1\perp}' + \mathbf{k}_{1\perp}''$ and $\Gamma_1^{i(2)} = \Gamma_1' + \Gamma_1''$.

Defining a new SHG scattering vector $\mathbf{Q} = (\mathbf{Q}_\perp, Q_z) \equiv \mathbf{k}_2 - \mathbf{k}_2^i$, then we may rewrite the SHG hologram as

$$I_2^{\text{holo}}(\mathbf{Q}_\perp, z) = -iH_o(\mathbf{Q}_\perp + \mathbf{k}_{2\perp}^i)u_i^{(2)}(\mathbf{k}_{2\perp}^i) \frac{\beta_{20}^2 e^{i\left(\Gamma_2\left(\mathbf{Q}_\perp + \mathbf{k}_{2\perp}^i\right)-\beta_2\right)z}}{4\Gamma_2\left(\mathbf{Q}_\perp + \mathbf{k}_{2\perp}^i\right)} \chi^{(2)}(\mathbf{Q}). \tag{72}$$

Now we may take the Fourier transform along z, giving us (**Figure 4**)

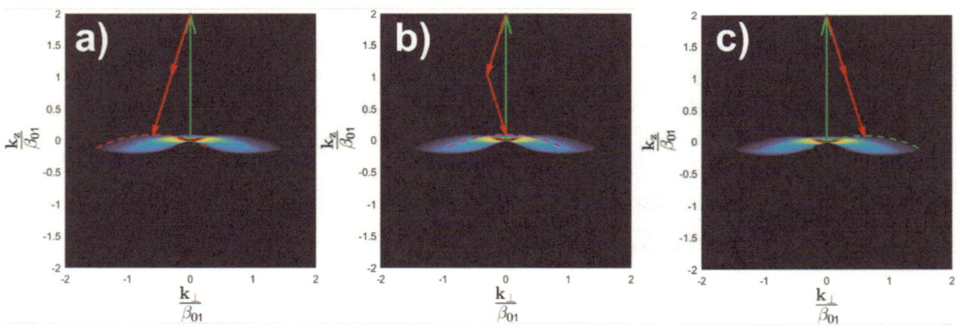

Figure 4.
An illustration of the contributions to the SHG CTF as described in Eq. (76). The green arrows represent the scattered SHG direction, the red arrows show the pair of incident fundamental plane wave direction that generate SHG scattering. The angle of the red arrows is limited by the condenser lens whereas the angle of the green arrow is limited by the objective lens. The color map is the contribution of all fundamental input plane waves given in Eq. (75). The dashed lines show the possible SHG scattering contributions of the detected scattering angle.

$$I_2^{\text{holo}}(\mathbf{Q}_\perp, z) = -i\beta_{20}^2 H_o(\mathbf{Q}_\perp + \mathbf{k}_{2\perp}^i) u_i^{(2)}(\mathbf{k}_{2\perp}^i) \frac{\delta[Q_z + \beta_2 - \Gamma_2(\mathbf{Q}_\perp + \mathbf{k}_{2\perp}^i)]}{4\Gamma_2(\mathbf{Q}_\perp + \mathbf{k}_{2\perp}^i)} \chi^{(2)}(\mathbf{Q}). \tag{73}$$

The SHG hologram field may now be written as a simple linear shift invariant model with

$$I_2^{\text{holo}}(\mathbf{Q}) = H_{\text{SHG}}(\mathbf{Q})\chi^{(2)}(\mathbf{Q}). \tag{74}$$

The transfer function for a pair of fundamental illumination plane waves, is given by

$$H_{\text{SHG}}^i(\mathbf{Q}) = -i\beta_{20}^2 H_o(\mathbf{Q}_\perp + \mathbf{k}_{2\perp}^i) u_i^{(2)}(\mathbf{k}_{2\perp}^i) \frac{\delta[Q_z + \beta_2 - \Gamma_2(\mathbf{Q}_\perp + \mathbf{k}_{2\perp}^i)]}{4\Gamma_2(\mathbf{Q}_\perp + \mathbf{k}_{2\perp}^i)}. \tag{75}$$

When using a non-negligible illumination condenser optic NA, then the superposition of all of the illumination k-vectors must be considered, so that

$$H_{\text{SHG}}(\mathbf{Q}) = -i\beta_{20}^2 \int H_o(\mathbf{Q}_\perp + \mathbf{k}_{2\perp}^i) u_i^{(2)}(\mathbf{k}_{2\perp}^i) \frac{\delta[Q_z + \beta_2 - \Gamma_2(\mathbf{Q}_\perp + \mathbf{k}_{2\perp}^i)]}{4\Gamma_2(\mathbf{Q}_\perp + \mathbf{k}_{2\perp}^i)} d^2\mathbf{k}_{2\perp}^i. \tag{76}$$

6. Harmonic optical tomography (HOT) conclusions

The conventional approach to harmonic holography, that is imaging of nonlinear scattering with holographic detection using a coherent harmonic reference beam, cannot provide depth information, known as optical sectioning. While one can rotate the illumination beam or the object, neither of these strategies are very practical. In the case of object rotation, the mechanical positioning errors introduced by the translation and rotation stages make high-resolution imaging all but impossible. While illumination beam scanning is easier, one is left with the classic missing cone problem, and thus estimation of the object is a difficult inverse problem that is prone to distortion.

A few years ago, we introduced a completely new strategy that takes advantage of the fact that coherent nonlinear scattering mixes all possible pairs of incident fundamental plane waves to produce a vast array of scattering directions. The result is that with a suitably large condenser NA for focusing the fundamental light, optical sectioning is admitted into the imaging process. Clearly, point scanning nonlinear scattering takes advantage of this very feature, but in that case, the total power of the scattered harmonic field is detected. As a result, one cannot obtain direct access to the desired nonlinear susceptibility, $\chi^{(m)}(\mathbf{r})$. Detection of the field in such a point scanning approach would allow for an identical information transfer from the object to the image as we demonstrate in HOT.

However, HOT is able to exploit cameras, which provides several advantages. First, we have increased speed because we capture a widefield holographic image from which the mHG field is extracted. Second, we benefit from heterodyne amplification of the field because we can bring a strong reference field to detect a

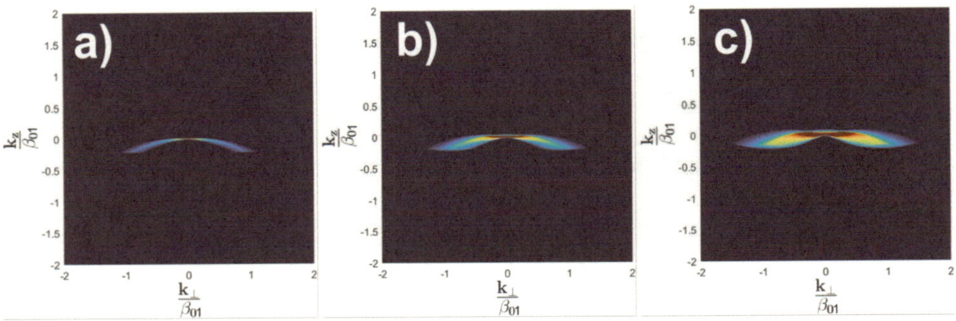

Figure 5.
SHG CTF for varying illumination condenser numerical aperture values of 0.1, 0.2 and 0.3 (a), (b) and (c) respectively). The imaging objective numerical aperture is fixed at 0.9.

weak harmonic field and push to very high imaging speeds [31]. Third, because the CTF exhibits broadening along the direction of propagation (k_{2z}), the SHG field image is in focus over a finite depth of field. This means that we may take an image stack by either translating the object axially (along z), or by imparting a defocus phase to scan the depth. To produce a 3D tomographic image of $\chi^{(2)}(\mathbf{r})$, the extracted field stack is deconvolved with the CTF (see **Figure 5**). While the low NA example in **Figure 5** shows negligible optical sectioning, increases in the condenser NA rapidly expand the axial spatial frequency support. There is ample opportunity to further optimize the resolution, speed, and sensitivity of HOT. In addition, there are opportunities to implement HOT for other coherent nonlinear scattering mechanisms, such as third harmonic generation (THG). Moreover, the polarization behavior can be exploited, and the fact that we detect the fields, rather than the intensity means that the sign, and thus the orientation of the susceptibility tensors are accessible.

This chapter focused on the background to introduce the uninitiated to the topics of coherent holography and tomography. We provided a full and rigorous scalar treatment of coherent nonlinear scattering for holographic and nonlinear imaging and tomography. Because most readers will be more familiar with linear scattering, we reviewed linear scattering and then demonstrated the homology to coherent nonlinear scattering through variable substitution to convert from the linear to the nonlinear scattering formulae. The critical difference between linear and nonlinear cases is that the source term in the nonlinear case provides vastly increased spatial frequency support. We demonstrated that this spatial frequency support could be related to a linear shift-invariant imaging model for coherent nonlinear scattering when holographic detection is used. As a result, the entire imaging process can be characterized by a coherent transfer function (e.g., **Figure 5**). Expressions for computing the CTFs for coherent nonlinear holographic imaging and tomography are derived. We hope that the theory introduced here will inspire new researchers to investigate the use of powerful coherent nonlinear holographic imaging and tomography tools.

Acknowledgements

We gratefully acknowledge funding from the Chan Zuckerberg Initiative's Deep Tissue Imaging Program.

Conflict of interest

The authors declare no conflict of interest.

Dedication

We dedicate this chapter to the memory of Gabriel ('Gabi') Poposecu with whom we collaborated on the original HOT paper. Gabi is a dear friend and colleague and he is sorely missed.

Author details

Lang Wang[1], Gabriel Murray[2], Jeff Field[1,3] and Randy A. Bartels[1]*

1 The Department of Electrical and Computer Engineering, Colorado State University, Fort Collins, USA

2 Department of Physics, Colorado State University, Fort Collins, USA

3 Center for Imaging and Surface Science, Analytical Resources Core Facility, Colorado State University, Fort Collins, USA

*Address all correspondence to: randy.bartels@colostate.edu

References

[1] Franken PA, Hill AE, Peters CW, Weinreich G. Generation of optical harmonics. Physical Review Letters. 1961;7:118-119

[2] Sheppard C, Gannaway J, Kompfner R, Walsh D. The scanning harmonic optical microscope. IEEE Journal of Quantum Electronics. 1977; **13**(9):912-912

[3] Campagnola PJ, de Wei M, Lewis A, Loew LM. High-resolution nonlinear optical imaging of live cells by second harmonic generation. Biophysical Journal. 1999;77(6):3341-3349

[4] Campagnola PJ, Millard AC, Terasaki M, Hoppe PE, Malone CJ, Mohler WA. Three-dimensional high-resolution second-harmonic generation imaging of endogenous structural proteins in biological tissues. Biophysical Journal. 2002;**82**(1):493-508

[5] Moreaux L, Sandre O, Charpak S, Blanchard-Desce M, Mertz J. Coherent scattering in multi-harmonic light microscopy. Biophysical Journal. 2001; **80**(3):1568-1574

[6] Gabor D. A new microscopic principle. Nature. 1948;**161**(4098): 777-778

[7] Leith EN, Upatnieks J. Reconstructed wavefronts and communication theory*. Journal of the Optical Society of America. 1962;**52**(10):1123-1130

[8] Leith EN, Upatnieks J. Microscopy by wavefront reconstruction. Journal of the Optical Society of America. 1965;**55**(5): 569-570

[9] Leith E, Chen C, Chen H, Chen Y, Dilworth D, Lopez J, et al. Imaging through scattering media with holography. Journal of the Optical Society of America. A. 1992;**9**(7): 1148-1153

[10] Yamaguchi I, Zhang T. Phase-shifting digital holography. Optics Letters. 1997;**22**(16):1268-1270

[11] Schnars U, Jüptner W. Direct recording of holograms by a ccd target and numerical reconstruction. Applied Optics. 1994;**33**(2):179-181

[12] Cuche E, Marquet P, Depeursinge C. Simultaneous amplitude-contrast and quantitative phase-contrast microscopy by numerical reconstruction of fresnel off-axis holograms. Applied Optics. 1999;**38**(34):6994-7001

[13] Wolf E. Three-dimensional structure determination of semi-transparent objects from holographic data. Optics Communications. 1969;**1**(4):153-156

[14] Squier JA, Müller M, Brakenhoff GJ, Wilson KR. Third harmonic generation microscopy. Optics Express. 1998;**3**(9): 315-324

[15] Yelin D, Silberberg Y. Laser scanning third-harmonic-generation microscopy in biology. Optics Express. 1999;**5**(8): 169-175

[16] Masihzadeh O, Schlup P, Bartels RA. Enhanced spatial resolution in third-harmonic microscopy through polarization switching. Optics Letters. 2009;**34**(8):1240-1242

[17] Plotnikov SV, Millard AC, Campagnola PJ, Mohler WA. Characterization of the myosin-based source for second-harmonic generation from muscle sarcomeres. Biophysical Journal. 2006;**90**(2):693-703

[18] Roth S, Freund I. Coherent optical harmonic generation in rat-tail tendon. Optics Communications. 1980;**33**(3): 292-296

[19] Freund I, Deutsch M, Sprecher A. Connective tissue polarity. Optical second-harmonic microscopy, crossed-beam summation, and small-angle scattering in rat-tail tendon. Biophysical Journal. 1986;**50**(4):693-712

[20] Cox G, Kable E, Jones A, Fraser I, Manconi F, Gorrell MD. 3-dimensional imaging of collagen using second harmonic generation. Journal of Structural Biology. 2003;**141**(1):53-62

[21] Yeh AT, Nassif N, Zoumi A, Tromberg BJ. Selective corneal imaging using combined second-harmonic generation and two-photon excited fluorescence. Optics Letters. 2002; **27**(23):2082-2084

[22] Tiaho F, Recher G, Rouède D. Estimation of helical angles of myosin and collagen by second harmonic generation imaging microscopy. Optics Express. 2007;**15**(19):12286-12295

[23] Ye P, Centurion M, Psaltis D. Harmonic holography: A new holographic principle. Applied Optics. 2008;**47**(4):A103-A110

[24] Bartels RA, Hoover BG, Zalevsky Z, John Caulfield H. Manipulating light waves: Introduction. Applied Optics. 2008;**47**(4):MLW1-MLW3

[25] Hsieh C-L, Grange R, Ye P, Psaltis D. Three-dimensional harmonic holographic microcopy using nanoparticles as probes for cell imaging. Optics Express. 2009;**17**(4):2880-2891

[26] Hsieh C-L, Ye P, Grange R, Psaltis D. Digital phase conjugation of second harmonic radiation emitted by nanoparticles in turbid media. Optics Express. 2010;**18**(12):12283-12290

[27] Hsieh C-L, Ye P, Grange R, Laporte G, Psaltis D. Imaging through turbid layers by scanning the phase conjugated second harmonic radiation from a nanoparticle. Optics Express. 2010;**18**(20):20723-20731

[28] Ye P, Psaltis D. Seeing through turbidity with harmonic holography invited. Applied Optics. 2013;**52**(4): 567-578

[29] Masihzadeh O, Schlup P, Bartels RA. Label-free second harmonic generation holographic microscopy of biological specimens. Optics Express. 2010;**18**(10): 9840-9851

[30] Shaffer E, Moratal C, Magistretti P, Marquet P, Depeursinge C. Label-free second-harmonic phase imaging of biological specimen by digital holographic microscopy. Optics Letters. 2010;**35**(24):4102-4104

[31] Smith DR, Winters DG, Schlup P, Bartels RA. Hilbert reconstruction of phase-shifted second-harmonic holographic images. Optics Letters. 2012; **37**(11):2052-2054

[32] Winters DG, Smith DR, Schlup P, Bartels RA. Measurement of orientation and susceptibility ratios using a polarization-resolved second-harmonic generation holographic microscope. Biomedical Optics Express. 2012;**3**(9): 2004-2011

[33] Rivard M, Popov K, Couture C-A, Laliberté M, Bertrand-Grenier A, Martin F, et al. Imaging the noncentrosymmetric structural organization of tendon with interferometric second harmonic generation microscopy. Journal of Biophotonics. 2014;**7**(8):638-646

[34] Smith DR, Winters DG, Bartels RA. Submillisecond second harmonic holographic imaging of biological specimens in three dimensions. Proceedings of the National Academy of Sciences. 2013;**110**(46):18391-18396

[35] Chenfei H, Field JJ, Kelkar V, Chiang B, Wernsing K, Toussaint KC, et al. Harmonic optical tomography of nonlinear structures. Nature Photonics. 2020;**14**(9):564-569

[36] Leith EN, Upatnieks J. Wavefront reconstruction with diffused illumination and three-dimensional objects*. Journal of the Optical Society of America. 1964;**54**(11):1295-1301

[37] Haeberlé O, Belkebir K, Giovaninni H, Sentenac A. Tomographic diffractive microscopy: Basics, techniques and perspectives. Journal of Modern Optics. 2010;**57**(9):686-699

[38] Kozacki T, Kujawińska M, Kniażewski P. Investigation of limitations of optical diffraction tomography. Opto-Electronics Review. 2007;**15**:102-109

[39] Devaney AJ. A filtered backpropagation algorithm for diffraction tomography. Ultrasonic Imaging. 1982;**4**(4):336-350

[40] Kak AC, Slaney M. Principles of Computerized Tomographic Imaging. Philadelphia, PA, USA: Society for Industrial and Applied Mathematics (SIAM); 2001

[41] Sharpe J, Ahlgren U, Perry P, Hill B, Ross A, Hecksher-Sørensen J, et al. Optical projection tomography as a tool for 3d microscopy and gene expression studies. Science. 2002;**296**(5567):541-545

[42] Sharpe J. Optical projection tomography. Annual Review of Biomedical Engineering. 2004;**6**(1):209-228. PMID: 15255768

[43] Streibl N. Three-dimensional imaging by a microscope. Journal of the Optical Society of America. A. 1985;**2**(2): 121-127

[44] Jenkins MH, Gaylord TK. Three-dimensional quantitative phase imaging via tomographic deconvolution phase microscopy. Applied Optics. 2015; **54**(31):9213-9227

[45] Bertero M, Boccacci P. Introduction to Inverse Problems in Imaging. Boca Raton, FL, USA: CRC Press; 1998

[46] Thiébaut E. Introduction toimage reconstruction and inverse problems. In: Foy R, Foy FC, editors. Optics in Astrophysics. Dordrecht: Springer Netherlands; 2005. pp. 397-422

[47] Bertero M, Piana M. Inverse Problems in Biomedical Imaging: Modeling and Methods of solution. Milano: Springer Milan; 2006

[48] Sarder P, Nehorai A. Deconvolution methods for 3-d fluorescence microscopy images. IEEE Signal Processing Magazine. 2006;**23**(3):32-45

[49] Darrell A, Meyer H, Marias K, Brady M, Ripoll J. Weighted filtered backprojection for quantitative fluorescence optical projection tomography. Physics in Medicine and Biology. 2008;**53**(14):3863-3881

[50] Kim T, Zhou R, Mustafa Mir S, Derin Babacan P, Carney S, Goddard LL, et al. White-light diffraction tomography of unlabelled live cells. Nature Photonics. 2014;**8**(3):256-263

[51] Chen M, Tian L, Waller L. 3d differential phase contrast microscopy.

Biomedical Optics Express. 2016;**7**(10): 3940-3950

[52] Linarès-Loyez J, Bon P, et al. Self-interference (selfi) microscopy for live super-resolution imaging and single particle tracking in 3d. Frontiers in Physics. 2019;**7**:68

[53] Nisan SBBGS, Vladimir L. High-magnification super-resolution finch microscopy using birefringent crystal lens interferometers. Nature Photonics. 2016;**10**(12):802-808

[54] Poon T-C. Scanning holography and two-dimensional image processing by acousto-optic two-pupil synthesis. Journal of the Optical Society of America. A. 1985;**2**(4):521-527

[55] Yoneda N, Saita Y, Nomura T. Motionless optical scanning holography. Optics Letters. 2020;**45**(12):3184-3187

[56] Futia G, Schlup P, Winters DG, Bartels RA. Spatially-chirped modulation imaging of absorbtion and fluorescent objects on single-element optical detector. Optics Express. 2011;**19**(2): 1626-1640

[57] Field JJ, Winters D, Bartels R. Plane wave analysis of coherent holographic image reconstruction by phase transfer (CHIRPT). Journal of the Optical Society of America. A. 2015; **32**(11):2156-2168

[58] Field JJ, Winters DG, Bartels RA. Single-pixel fluorescent imaging with temporally labeled illumination patterns. Optica. 2016;**3**(9):971-974

[59] Field JJ, Wernsing KW, Domingue SR, Allende Motz AM, DeLuca KF, et al. Super-resolved multimodal multiphoton microscopy with spatial frequency-modulated imaging. Proceedings of the National Academy of Sciences of the United States of America. 2016;**113**(24): 6605-6610

[60] Stockton P, Field J, Bartels R. Single pixel quantitative phase imaging with spatial frequency projections. Methods in Quantitative Phase Imaging in Life Science. 2018;**136**:24-34

[61] Field JJ, Wernsing KA, Squier JA, Bartels RA. Three-dimensional single-pixel imaging of incoherent light with spatiotemporally modulated illumination. Journal of the Optical Society of America. A. 2018;**35**(8): 1438-1449

[62] Stockton PA, Field JJ, Squier J, Pezeshki A, Bartels RA. Single-pixel fluorescent diffraction tomography. Optica. 2020;**7**(11):1617-1620

[63] Stockton P, Murray G, Field JJ, Squier J, Pezeshki A, Bartels RA. Tomographic single pixel spatial frequency projection imaging. Optics Communications. 2022;**520**:128401

[64] Schlup P, Futia G, Bartels RA. Lateral tomographic spatial frequency modulated imaging. Applied Physics Letters. 2011;**98**(21):211115

[65] Stockton PA, Wernsing KA, Field JJ, Squier J, Bartels RA. Fourier computed tomographic imaging of two dimensional fluorescent objects. APL Photonics. 2019;**4**(10):106102

[66] Mertz J. Introduction to Optical Microscopy. 2nd ed. Cambridge, United Kingdom: Cambridge University Press; 2019

[67] Novotny L, Hecht B. Principles of Nano-Optics. Cambridge, United Kingdom: Cambridge University Press; 2006

[68] Goodman JW. Introduction to Fourier optics. In: McGraw-Hill Physical and Quantum Electronics Series. New York City, USA: McGraw-Hill; 2005

[69] Sheppard CJR, Min G, Kawata Y, Kawata S. Three-dimensional transfer functions for high-aperture systems. Journal of the Optical Society of America. A. 1994;**11**(2):593-598

Chapter 5

Compact Incoherent Multidimensional Imaging Systems Using Static Diffractive Coded Apertures

Vijayakumar Anand, Soon Hock Ng, Tomas Katkus, Daniel Smith, Vinoth Balasubramani, Denver P. Linklater, Pierre J. Magistretti, Christian Depeursinge, Elena P. Ivanova and Saulius Juodkazis

Abstract

Incoherent holographic imaging technologies, in general, involve multiple optical components for beam splitting—combining and shaping—and in most cases, require an active optical device such as a spatial light modulator (SLM) for generating multiple phase-shifted holograms in time. The above requirements made the realization of holography-based products expensive, heavy, large, and slow. To successfully transfer the holography capabilities discussed in research articles to products, it is necessary to find methods to simplify holography architectures. In this book chapter, two important incoherent holography techniques, namely interference-based Fresnel incoherent correlation holography (FINCH) and interferenceless coded aperture correlation holography (I-COACH), have been successfully simplified in space and time using advanced manufacturing methods and nonlinear reconstruction, respectively. Both techniques have been realized in compact optical architectures using a single static diffractive optical element manufactured using lithography technologies. Randomly multiplexed diffractive lenses were manufactured using electron beam lithography for FINCH. A quasi-random lens and a mask containing a quasi-random array of pinholes were manufactured using electron beam lithography and photolithography, respectively, for I-COACH. In both cases, the compactification has been achieved without sacrificing the performances. The design, fabrication, and experiments of FINCH and I-COACH with static diffractive optical elements are presented in details.

Keywords: Fresnel incoherent correlation holography, coded aperture imaging, diffractive optics, electron beam lithography, photolithography, imaging, high-speed imaging, quasi-random lens, scattering, self-interference, incoherent imaging

1. Introduction

The invention of holography has revolutionized the field of imaging in numerous ways [1–3]. The capability to record complete three-dimensional (3D) amplitude and phase information of an object from a few camera recordings is a remarkable advantage for many applications. In coherent holography, the light diffracted from an object is interfered with a reference wave derived from the same source to form the hologram of the object. With a coherent source such as a laser, the above task can be achieved in any of the standard interferometry architectures [4]. In the recent years, there has been a shift in focus from using coherent light sources to incoherent ones due to many advantages such as low cost, high SNR and high resolution. In incoherent holography, due to lack of coherence, the hologram cannot be formed by interfering object wave with reference wave but between two differently modulated object waves [1]. The above interference is called self-interference. Achieving such a configuration was a challenging task in the past and many sophisticated architectures such as rotating shear interferometer [5], triangle interferometer [6], and conoscopic holography [7] were developed. With the advent of active devices, such as spatial light modulators, realization of incoherent holography systems became relatively easier. Some notable incoherent holography techniques with active devices include Fresnel incoherent correlation holography (FINCH) [8], Fourier incoherent single-channel holography (FISCH) [9], and self-interference digital holography (SIDH) [10]. Of all the above three methods, FINCH and FISCH can exhibit super resolution while SIDH in off-axis configuration cannot. FINCH requires the simplest and compact optical configuration of the above three methods with a high field of view. Due to the above reasons, FINCH became an attractive imaging system [11–15] and moved successfully to industry as powerful 3D fluorescence microscopy products: FINCH^{TM} and CINCH^{TM} [16].

Another research direction called coded aperture imaging (CAI) using incoherent illumination evolved alongside of FINCH [17, 18]. In CAI, the light from an object is modulated by a coded aperture and the intensity distribution is recorded. The recorded intensity distribution is reconstructed into object's image either by using the transmission function of the coded aperture or by using the recorded point spread function. The first reports of CAI came in 1968 using random array of pinholes to perform lensless imaging of X-rays and Gamma rays [19, 20]. However, the image reconstruction was noisy due to the computational reconstruction mechanism. So, CAI evolved in search of the coded aperture–computational reconstruction (CA-CR) compatible pair that can create object's images with a high quality. During this search, uniformly redundant array (URA)-based coded aperture was found to exhibit a high signal-to-noise ratio in comparison to random array of pinholes [21]. Later, modified URA (MURA) with two shots was found to perform better in signal-to-noise ratio [22]. In this direction, by designing an appropriate CA-CR pair, spatio-spectral imaging methods, such as coded aperture snapshot spectral imager (CASSI), were developed [23, 24]. The CASSI approach was later expanded by applying compressive sensing methods [25]. In 2011, optical imaging using a phase coded aperture was reported [26]. There are multiple patents awarded on CAI methods for developing compact imaging products and in some cases, CAI has been used as an assistive technology in a larger framework.

The FINCH technique has a relatively low axial resolution in comparison to lens-based imaging and so coded aperture correlation holography (COACH) was developed in 2016 from FINCH by modifying the beam modulations [27]. In FINCH, the light from an object was modulated by two quadratic phase masks with different focal distances and the resulting two different object waves were interfered to form the hologram. In COACH, the light from a point is modulated by a quasi-random phase mask and a constant phase mask and interfered to form the hologram. In FINCH, at least three camera shots were recorded with different phase-shifts between the two interfering beams and combined into a complex hologram in order to cancel the twin image and bias terms. This complex hologram was numerically propagated to one of the focal planes of the quadratic phase mask to reconstruct the object information. In COACH, the phase mask is random and therefore, there is no focal plane where the hologram can be propagated. So, in COACH, the complex hologram was reconstructed by cross-correlating it with the complex point spread hologram recorded for different depths. In COACH, the lateral and axial resolutions are therefore given by the lateral and axial correlation lengths governed by the numerical aperture as in the case of lens-based imaging. During the development of COACH, it was identified that two-beam interference was not needed to record 3D information of the object, which led to the development of non-interferometric or interferenceless COACH (I-COACH) [28]. The development of I-COACH linked the capabilities of incoherent holography with the simplicity of CAI. The connection became evident after the demonstration of a super-resolution I-COACH method using specially coded apertures that generate random array of sub-diffraction and a lateral resolution approaching that of FINCH [29].

As a matter of fact, the current state-of-the-art FINCH as well as I-COACH technology are not optimal. The FINCH™ and CINCH™ still require numerous optical components and active devices, which makes the implementation challenging and as a product, less attractive and expensive. In I-COACH, or in general CAI, the requirements are less but the challenges lie in the lower efficiency of coded aperture and limited applications. During the past years, at Nanolab of Swinburne University of Technology (SUT), we have been working on expanding FINCH and I-COACH beyond the current state-of-the-art technologies using static diffractive optical elements manufactured using advanced fabrication technologies [30–34]. In this book chapter, we present the design, fabrication, and implementation of static diffractive optical elements in FINCH and I-COACH. Furthermore, we also present a conceptual configuration using static diffractive optical elements for optical trapping and imaging.

2. Fresnel incoherent correlation holography (FINCH)

2.1 Methodology and simulation results

FINCH was developed by Rosen and Brooker in the year 2007 [8]. The optical configuration of FINCH is shown in **Figure 1a**. Light from an object point is split into two waves using an SLM in which a phase mask formed by a random multiplexing of two quadratic phase functions with different focal lengths was displayed. These two object waves were interfered to form the object hologram. As FINCH is in in-line configuration, at least three object holograms need to be recorded with relative phase shifts of $\Theta = 0$, $2\pi/3$, and $4\pi/3$ radians and combined in complex space to remove the twin image and bias terms during reconstruction by numerical backpropagation. Let us consider a point object with an amplitude of $\sqrt{I_o}$. The complex amplitude at a distance of z_1 from the point and entering the SLM is given as $C_1\sqrt{I_o}Q(1/z_1)$, where

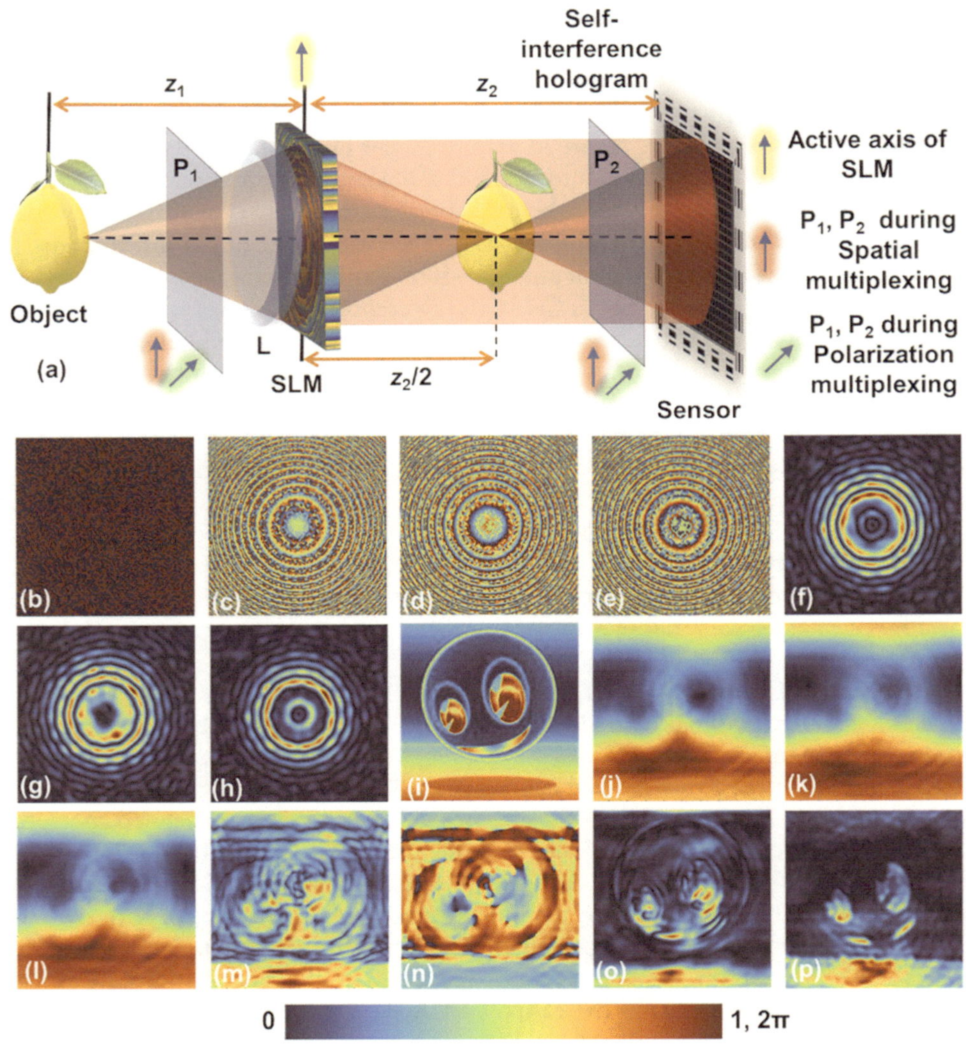

Figure 1.
(a) Optical configuration of FINCH. Two orientations of polarizers P1 and P2 are shown: red for spatial multiplexing and green for polarization multiplexing. The active axis of SLM is shown in yellow. Refractive lens L is not needed in the case of spatial multiplexing. (b) Random multiplexing matrix. Phase images of the randomly multiplexed bifocal lenses with (c) $\Theta = 0$, (d) $\Theta = 2\pi/3$, and (e) $\Theta = 4\pi/3$ radians. Simulated PSFs corresponding to (f) $\Theta = 0$, (g) $\Theta = 2\pi/3$, and (h) $\Theta = 4\pi/3$ radians. (i) Test object. Holograms simulated for (j) $\Theta = 0$, (k) $\Theta = 2\pi/3$, and (l) $\Theta = 4\pi/3$ radians. (m) Amplitude and (n) phase components of the complex hologram. Reconstructed image of the object by (o) numerical backpropagation and (p) NLR ($\alpha = 0$, $\beta = 0.6$).

$Q(b) = e^{\frac{j\pi b R^2}{\lambda}}$, $R = \sqrt{x^2+y^2}$ and C_1 is a complex constant. The phase function displayed on the SLM is given by $\Phi(x,y) = \left\{\left(\frac{\pi R^2}{\lambda f_1}\right)\right\}_{2\pi} M + \left\{\left(\frac{\pi R^2}{\lambda f_2}\right)\right\}_{2\pi}(1-M)$, where the first function collimates the incoming light when $f_1 = z_1$, while the second function focuses light at $z_2/2$, when $1/f_2 = 1/z_1 + 2/z_2$ and M is a binary random matrix {0,1}. The complex amplitude after the SLM is given as $C_2\sqrt{I_o}Q(1/z_1)e^{-j\Phi(x,y)}$, where C_2 is a complex constant. The self-interference point spread function is given as $I_{PSF} = \left|C_2\sqrt{I_o}Q(1/z_1)e^{-j\Phi(x,y)} \otimes Q(1/z_2)\right|^2$. The hologram H for an object O can be given as

$I_{\mathrm{PSF}} \otimes O$, where '$\otimes$' is a 2D convolutional operator. At least three holograms H_1, H_2, and H_3 corresponding to Θ = 0, 2π/3, and 4π/3 are recorded, and projected into the complex space as $H_C = H_1\left(e^{-j4\pi/3} - e^{-j2\pi/3}\right) + H_2\left(1 - e^{-j4\pi/3}\right) + H_3\left(e^{-j2\pi/3} - 1\right)$ to form a complex hologram. The image of the object is reconstructed as $I_R = H_C \otimes Q\left(\frac{1}{z_R}\right)$, where z_R is the reconstruction distance.

In the above case, FINCH hologram was reconstructed by a numerical backpropagation. Recalling the cases of COACH [27] and I-COACH [28], where there was no real or virtual image plane associated with the optical configuration and the phase masks, the reconstruction was carried out using cross-correlation with prerecorded point spread functions (PSFs). Since FINCH is a linear shift-invariant system, the same approach can be applied to FINCH as well. This idea was tested in 2020, which yielded notable results for FINCH [35, 36]. The past studies with scattering masks revealed that nonlinear reconstruction (NLR) [37] yielded better results than cross-correlation, Lucy–Richardson algorithm (LRA) [38, 39], and regularized filter algorithm [40]. The reconstructed image can be given as $I_{\mathrm{R}} = \left|\mathcal{F}^{-1}\left\{\left|\tilde{I}_{\mathrm{PSF}}\right|^{\alpha} exp\left[i\, arg\left(\tilde{I}_{\mathrm{PSF}}\right)\right]\left|\tilde{H}\right|^{\beta} exp\left[-i\, arg\left(\tilde{H}\right)\right]\right\}\right|$, where α and β are tuned between -1 and 1, to obtain the minimum entropy given as $-\sum\sum \phi(m,n)\log\left[\phi(m,n)\right]$, where $\phi(m,n) = |C(m,n)|/\sum_M\sum_N |C(m,n)|$ and (m,n) are the indexes of the correlation matrix, and $C(m,n)$ is the correlation distribution.

A simulation study was carried out using the above equations in MATLAB [41]. The matrix size was 500 × 500 pixels, pixel size Δ = 10 μm, λ = 650 nm, z_1 = 50 cm, z_2 = 1 m, and diameter of optical modulator was 3 mm. The image of the binary random function M used for multiplexing is shown in **Figure 1b**. The phase images for creating three different phase-shifted holograms with Θ = 0, 2π/3, and 4π/3 radians are shown in **Figure 1c–e**, respectively. The PSFs corresponding to **Figure 1c–e** are shown in **Figure 1f–h,** respectively. A test smiley object as shown in **Figure 1i** was used for simulation. The holograms H_1, H_2, and H_3 of the smiley object for Θ = 0, 2π/3 and 4π/3 are shown in **Figure 1j–l**, respectively. The amplitude and phase of H_c are shown in **Figure 1m** and **n**, respectively. The reconstructed image is shown in **Figure 1o**. The reconstruction results using NLR are shown in **Figure 1p**. By comparing **Figure 1o** and **p**, it is seen that the reconstruction using NLR was better. In the evolution of FINCH, the phase-shifting method removed the twin image and bias terms, but the noises associated with the random multiplexing affected the quality of reconstruction.

To avoid the reconstruction noises due to random multiplexing, a polarization multiplexing scheme was introduced [42]. In the first spatial random multiplexing scheme, the light from the object was polarized along the active axis of the SLM, whereas in the polarization multiplexing method, the light from an object was polarized at π/4 radians along the active axis of the SLM. As a result, only about 50% of the incoming light is modulated by the SLM, while the remaining part was not modulated. A second polarizer oriented at π/4 radians along the active axis of the SLM was mounted before the image sensor to create interference between the two object waves. An additional refractive lens was used to collimate the object wave before it is incident on the SLM. In the case of polarization multiplexing scheme, the PSF is given as $I_{\mathrm{PSF}} = \left| C_3\sqrt{I_o}Q\left(\frac{1}{z_1}\right)e^{-j\left(\frac{\pi R^2}{\lambda f_1}\right)} \otimes Q\left(\frac{1}{z_2}\right) + C_4\sqrt{I_o}Q\left(\frac{1}{z_1}\right)e^{-j\left(\frac{\pi R^2}{\lambda f_2}\right)} \otimes Q\left(\frac{1}{z_2}\right)\right|^2$, where C_3 and C_4 are complex constants. The phase images of the mask for Θ = 0, 2π/3 and 4π/3 are shown in **Figure 2a–c**, respectively. The images of the PSFs for Θ = 0, 2π/3, and 4π/3 are shown in **Figure 2d–f,** respectively and the images of object holograms for Θ = 0, 2π/3, and 4π/3 are shown in **Figure 2g–i,** respectively. The amplitude and phase of the complex hologram formed by projecting **Figure 2g–i** in complex space are shown in **Figure 2j** and **k,** respectively. The reconstructed image of the object is shown in **Figure 2l**. The reconstruction result using NLR is shown in **Figure 2m**.

Recently, a new reconstruction method called Lucy–Richardson–Rosen algorithm (LRRA) was developed by combining the well-known Lucy–Richardson algorithm (LRA) and NLR [43]. In LRA, the $(n+1)^{th}$ reconstructed image is given as $I_R^{n+1} = I_R^n \left\{ \frac{I_p}{I_R^n \otimes I_{PSF}} \otimes I_{PSF}' \right\}$, where I_{PSF}' is the complex conjugate of I_{PSF}. This process is iterated until an optimal reconstruction is obtained. The LRA has a forward convolution $I_R^n \otimes I_{PSF}$ and the ratio between this and I_p is correlated with I_{PSF}, which is replaced by the NLR to achieve a better estimation. However, a recent study showed the high sensitivity of LRRA to spatial shifts with a high-resolution image sensor and so may not be suitable for experiments [44]. The reconstruction result ($\alpha = 0$, $\beta = 0.7$, iterations = 10) of LRRA is shown in **Figure 2n** and the direct imaging result is shown in **Figure 2o**.

The above two spatial multiplexing and polarization multiplexing schemes remain the two main configurations of FINCH. Comparing the performances of the above two schemes, it is seen that spatial multiplexing has a higher light throughput than polarization multiplexing while polarization multiplexing has a higher signal-to-noise ratio than spatial multiplexing. As a matter of fact, polarization multiplexing was widely adapted for further developments of FINCH. A secondary multiplexing method was used to reduce the number of camera shots of FINCH from three to one.

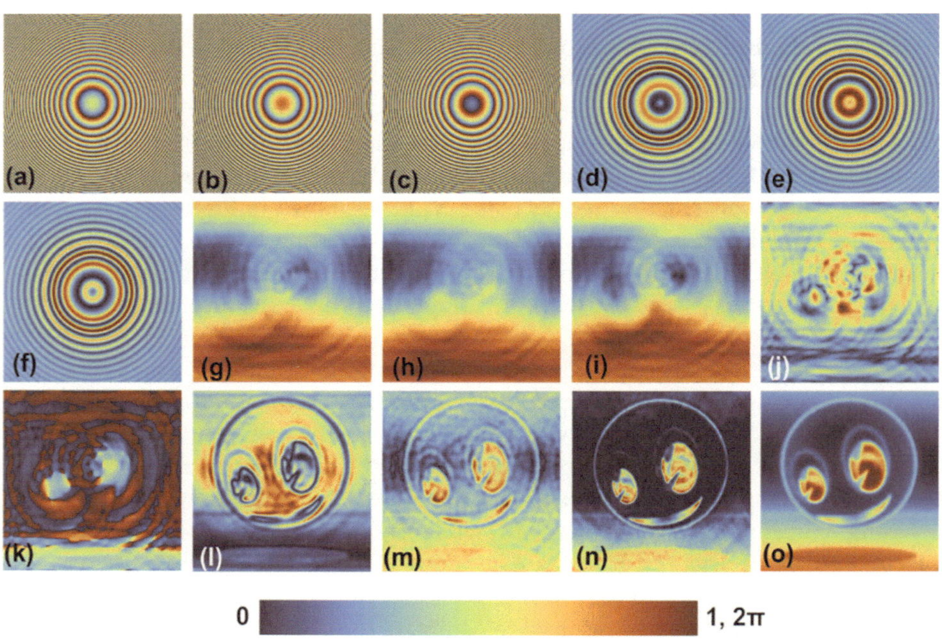

Figure 2.
Phase images of the quadratic phase masks for (a) Θ = 0, (b) Θ = 2π/3, and (c) Θ = 4π/3 radians. PSFs for (d) Θ = 0, (e) Θ = 2π/3, and (f) Θ = 4π/3 radians. Object holograms for (g) Θ = 0, (h) Θ = 2π/3, and (i) Θ = 4π/3 radians. (j) Amplitude and (k) phase of the complex hologram. Reconstruction by (l) numerical backpropagation, (m) nonlinear reconstruction, and (n) Lucy-Richardson-Rosen algorithm. (o) Direct imaging.

In the polarization multiplexing scheme, either spatial multiplexing [45, 46] or a polarization multiplexing with a 4-pol camera [47] was used to record multiple phase-shifted holograms from a single recording. The spatial multiplexing sacrificed the field of view while the polarization multiplexing sacrificed the signal-to-noise ratio. In the above secondary multiplexing schemes, the number of shots required remained the same but it was collected simultaneously distributed in space and polarization states. Recently, a two-step phase-shifting method was developed, which reduced the required number of camera shots in FINCH to two [48]. The evolution of FINCH over the years has been described chronologically in [1, 49].

2.2 Experimental results

To compactify FINCH, from a manufacturing point of view, the spatial multiplexing scheme is easier to realize as the diffractive optical elements involved have features that are super-wavelength. To realize FINCH in polarization multiplexing scheme, metasurfaces with subwavelength features are needed, which will increase the cost of the manufacturing process and often difficult to achieve a large area. As a matter of fact, there have been attempts to realize FINCH in polarization multiplexing scheme using metasurfaces and geometric phase lenses, which resulted in a high cost, low performance, and suffered from low light throughput due to polarizing of the incoming light [50, 51]. At Nanolab of SUT, two different randomly multiplexed diffractive optical elements were designed. The first element consisted of two binary Fresnel zone lenses with focal lengths of 2.5 cm and 5 cm such that the object is at 5 cm and the sensor is at 10 cm from the diffractive element. The second element consisted of a Fresnel zone lens with a focal length of 5 cm and an axicon. The Fresnel zone lenses were designed to include the thickness of glass plate to avoid spherical aberrations [52].

The randomly multiplexed Fresnel zone lenses were manufactured on Indium-Tin-Oxide (ITO)-coated commercial glass plates with a thickness of ~1 mm, index of refraction of ~1.5, and a high transmittivity (>95%). The substrate was cleaned in an ultrasonic bath in acetone and iso-propyl alcohol (IPA) for 10 minutes and baked on a hotplate at 180°C for 5 minutes to remove the residual solvents. When the substrate was cooled to room temperature, it was spin-coated with PMMA 950K A7 positive resist at 2000 RPM for a minute to achieve a resist thickness of λ. During spin coating, an edge of the substrate was masked with a tape, which was removed after spin coating. This masked area was connected to the metal clip of substrate holder in the electron beam lithography system RAITH150^{TWO}. An acceleration voltage of 10kV, aperture of 120 μm, and a beam current of ~3 nA was used with a working distance of 10 mm and write field of 100 μm. The diffractive element designed with a diameter of 5 mm was fabricated (~6 hours) without any stitching errors. The fabricated device was developed in methyl isobutyl ketone (MIBK) and IPA in the ratio of 1:3 for 1 minute. The optical microscope image of the randomly multiplexed Fresnel zone lenses is shown in **Figure 3a**. A second diffractive element consisting of a Fresnel zone lens and an axicon with a period of 100 μm was fabricated under the same fabrication conditions. The optical microscope image of the fabricated device is shown in **Figure 3b**. The magnified sections of the 3(a) and 3(b) are shown in **Figure 3c** and **d,** respectively.

Optical experiments were carried out using the fabricated devices and a spatially incoherent source (Thorlabs, M617L3, λ_c = 617 nm, FWHM = 18 nm). The diffractive element was located at 5 cm from the object and the sensor (Thorlabs DCU223M,

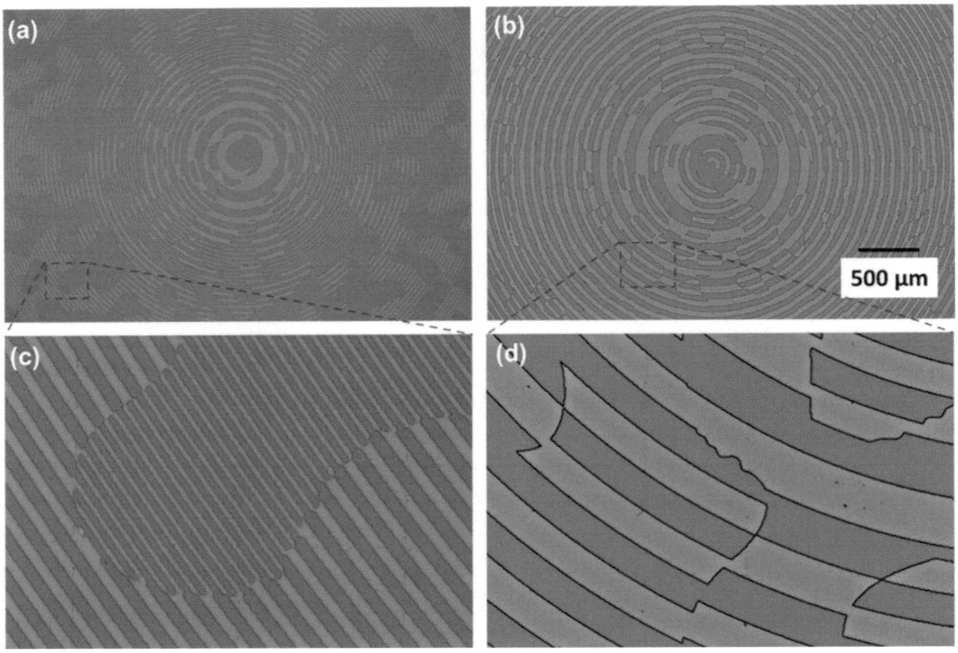

Figure 3.
Optical microscope images of randomly multiplexed (a) Fresnel zone lenses and (b) Fresnel zone lens and axicon. Magnified images of sections of (a) and (b) are shown in (c) and (d), respectively.

1024 × 768 pixels, pixel size = 4.65 μm) was located at 10 cm from the diffractive lens. In the first experiment, the randomly multiplexed Fresnel zone lenses were mounted. In the first step, the PSF was recorded using a pinhole with a diameter of 20 μm as shown in **Figure 4a**. The element 1 of Group 4 (16 lp/mm, grating period = 62.5 μm) of the United States Air Force (USAF) resolution target replaced the pinhole in the next step. The direct imaging result recorded at a distance of 5 cm from the diffractive element is shown in **Figure 4b**. The image of the recorded object hologram at 10 cm from the diffractive element is shown in **Figure 4c**. The reconstruction result using NLR (α = 0.2, β = 0.6) is shown in **Figure 4d**. The experiment was repeated to image a dragonfly larva wing, whose hologram is shown in **Figure 4e**. The reconstruction result and direct imaging result are shown in **Figure 4f** and **g,** respectively. The biological sample was reconstructed using PSFs recorded at different depths of the wing. The computational refocusing at different planes of the sample is shown in **Figure 4h**.

The experiment was repeated using the second diffractive element – randomly multiplexed Fresnel zone lens and axicon. The PSFs recorded for λ = 617 nm and λ = 530 nm are shown in **Figure 5a** and **b,** respectively. Two objects (Group – 2, Element – 6, 7.13 *lp/mm*) of USAF and National Bureau of Standards (NBS) resolution target (7.1 *lp/mm*) were mounted one after the other, and their holograms were captured and summed as shown in **Figure 5c**. The reconstructed results of USAF and NBS objects using the corresponding PSFs are shown in **Figure 5d** and **e,** respectively.

2.3 Discussion

FINCH has been implemented in a compact optical configuration using static diffractive optical elements manufactured using electron beam lithography

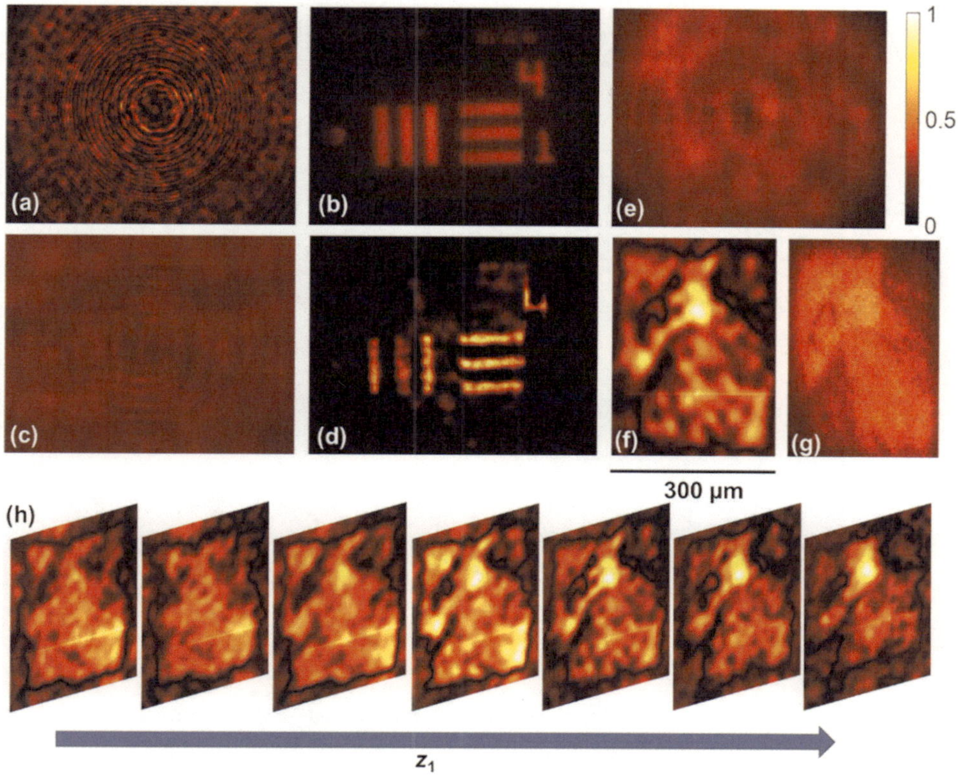

Figure 4.
(a) PSF. (b) Direct imaging result recorded at the focal plane of one of the Fresnel zone lenses, (c) object hologram and (d) reconstruction result for randomly multiplexed Fresnel zone lenses. (e) Hologram of a section of dragonfly larvae wing. (f) Reconstruction result using NLR (α = 0, β = 0.58) and (g) direct imaging result. (h) Results of 3D computational refocusing.

techniques. The imaging distances were only 5 cm between object to diffractive element and 10 cm between diffractive element and sensor, which is close to palm-size systems [53]. The diffraction efficiency was about 40%, which is higher than polarization multiplexing configurations. Both 3D spatial and spectral imaging have been demonstrated. The preliminary results are promising.

3. Interferenceless-coded aperture correlation holography (I-COACH)

3.1 Methodology and simulation results

The optical configuration of I-COACH or CAI is simpler than FINCH, as shown in **Figure 6,** consisting of only one optical element between the object and image sensor. In this study, only chaotic coded apertures are considered [54]. The use of URA [21], MURA [22], and masks for spectral imaging methods involve many optical components [23] unlike I-COACH with chaotic coded apertures. The intensity distribution for an object O is given as $I_O = I_{PSF} \otimes O$ and the reconstructed image is given as $I_R = I_O * I_{PSF} = I_{PSF} * I_{PSF} \otimes O$, where '$*$' is a 2D correlation operator. The above cross-correlation procedure is not optimal and NLR is the currently known best reconstruction method. So, the above procedure will be implemented using NLR. The simulation

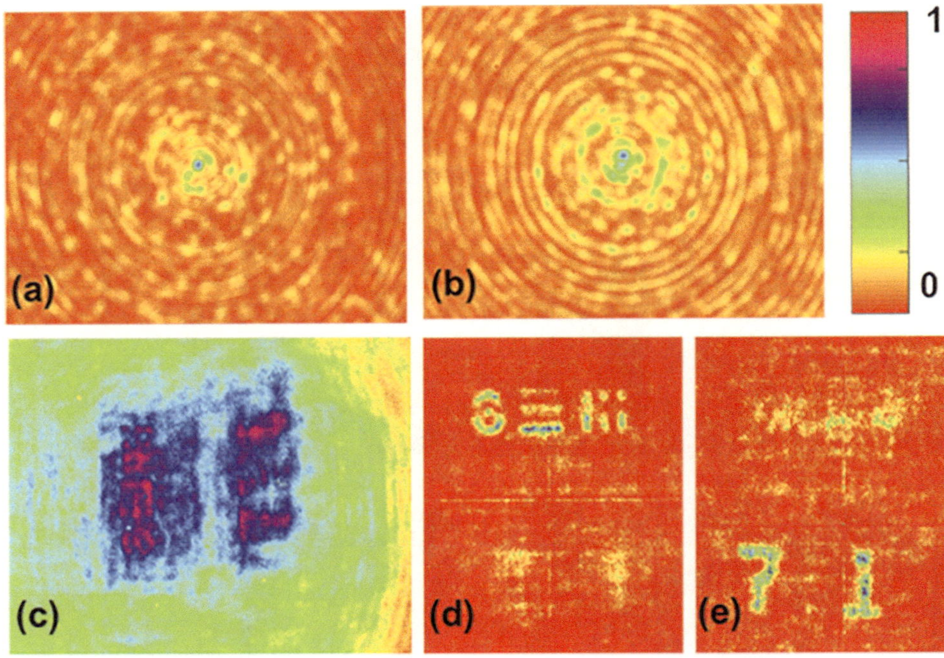

Figure 5.
PSFs recorded for (a) λ = 617 nm and (b) 530 nm. Holograms of USAF and NBS illuminated by λ = 617 nm and λ = 530 nm, respectively, but mounted at the same distance of 5 cm. The reconstruction results using NLR (α = 0, β = 0.63), (d) PSF (λ = 617 nm), and (e) PSF (λ = 530 nm).

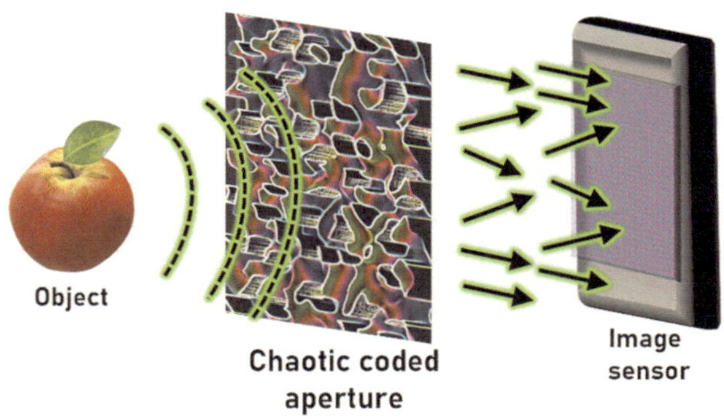

Figure 6.
Optical configuration of I-COACH or CAI with chaotic coded aperture.

conditions were as described in Section 2 with the object distance at 50 cm from the diffractive element, and the distance between the diffractive element and the image sensor is 1 m. The images of an amplitude type mask with a fill factor of 0.5 and chaotic coded aperture phase masks with scattering ratios σ = 1, 0.5, 0.2, 0.1, and 0.02 are shown in the first row of **Figure 7**. The simulated PSFs, object intensity distributions, and reconstruction results of the test object using NLR are shown in the second, third, and fourth rows, respectively.

The amplitude mask was designed by rounding off a normalized (0,1) 2D random function. The phase masks were engineered using Gerchberg–Saxton algorithm [55] by imposing a constraint on the size of the window in the spectrum domain as

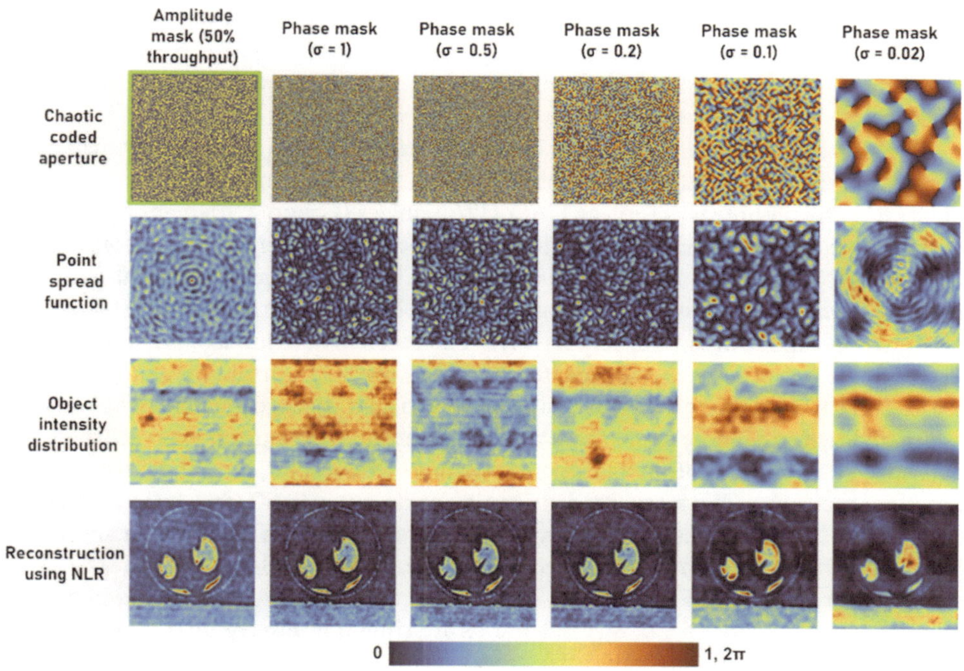

Figure 7.
Simulation results of PSF, object intensity distribution and reconstruction results using NLR for an amplitude mask with 50% light throughput (green window), phase masks with σ = 1, 0.5, 0.2, 0.1, and 0.02.

discussed in [54], where $\sigma = p/P$, where P is the size of the matrix and p is the size of the window. The speckle sizes decrease with an increase in the scattering degree and consequently, the resolution of the reconstruction improves with an increase in the scattering degree. The reconstruction result of the amplitude mask had a similar resolution as that of the phase mask with maximum scattering degree but additional background noise. In addition to the above, there are other scattering masks that generate a random array of points [56] and ring patterns [57], but multiplane imaging requires a multiplexing approach.

3.2 Experimental results

The chaotic coded apertures were realized as a pinhole array [58, 59] as well as a quasi-random phase mask manufactured using lithography procedures [60]. It must be noted that the first demonstrations of I-COACH without lens required an SLM as at least two camera shots were needed in order to create a bipolar matrix [28, 61]. Due to the expected background noise in amplitude masks, the pinhole array mask was designed using a two-step optimization procedure. The number of pinholes was selected as 2000. In the first step, the 2000 pinholes were spatially randomly arranged in 1000 ways and the reconstruction noise was simulated for every case. The optimal case of random arrangement was selected for the next round of optimization. In the next step, the location of every pinhole was shifted in steps of 5 pixels every time the background noise was estimated. When the noise decreased, the new position was retained and if the noise increased, the pinhole was shifted back to the previous position. The above two optimization procedures involved a total of about 5000 iterations. The final mask pattern demonstrated an SNR improvement of $\sim$60% in comparison to a case with a minimum SNR. The final mask design was transferred to a chromium-coated mask plate using Intelligent micropatterning SF100 XPRESS. The size of the QRAP was 8 mm and the diameter of the pinholes was 80 μm after fabrication.

The PSF was recorded using a pinhole of diameter of 20 μm for an object distance of 10 cm from the pinhole array and sensor at 10 cm from the pinhole array as shown in **Figure 8a**. The pinhole was replaced by a USAF object (Group 2, Elements 4, 5, and 6) and the intensity distribution was recorded as shown in **Figure 8b**. The reconstruction result using NLR for (α = 0, β = 0.6) is shown in **Figure 8c**. The experiment was repeated for a section of wing of an insect. The PSF and object intensity distributions are shown in **Figure 8d** and **e,** respectively. The reconstructed image is shown in **Figure 8f**. The experiment was repeated to record the 5D information along x, y, z, λ, and t. A spark was generated by creating an electrical discharge and the light was modulated by the pinhole array mask and the intensity distribution was recorded. A high-speed camera (Phantom v2512, monochrome, 800×1280, $\Delta = 28\,\mu m$) was used for recording the event. The intensity distribution at times t = 0, 80, and 160 μs are shown in **Figure 9a**, **d**, and **g**, respectively, which were reconstructed using green and red PSFs using NLR, and the reconstructed images for green wavelength at times t = 0, 80, and 160 μs are shown in **Figure 9b**, **e** and **f**, respectively. The reconstructed images for red wavelength at times t = 0, 80, and 160 μs are shown in **Figure 9c**, **f**, and **i**, respectively. The object intensity pattern was reconstructed using PSFs synthesized for different depths using the scaling approach and tomography of the spark pattern at time t = 0 was generated as shown in **Figure 9j**.

In the experiments using pinhole array, the results were significant, but the main challenge was the low light throughput. The above mask could not be used for imaging in low-intensity conditions such as fluorescence and astronomical imaging. To expand the applicability of the method, improvement to design was made at multiple levels. In the first step, the amplitude mask was replaced by a phase mask. A ground glass diffuser is a good choice but offers less control over the intensity distribution [62–64].

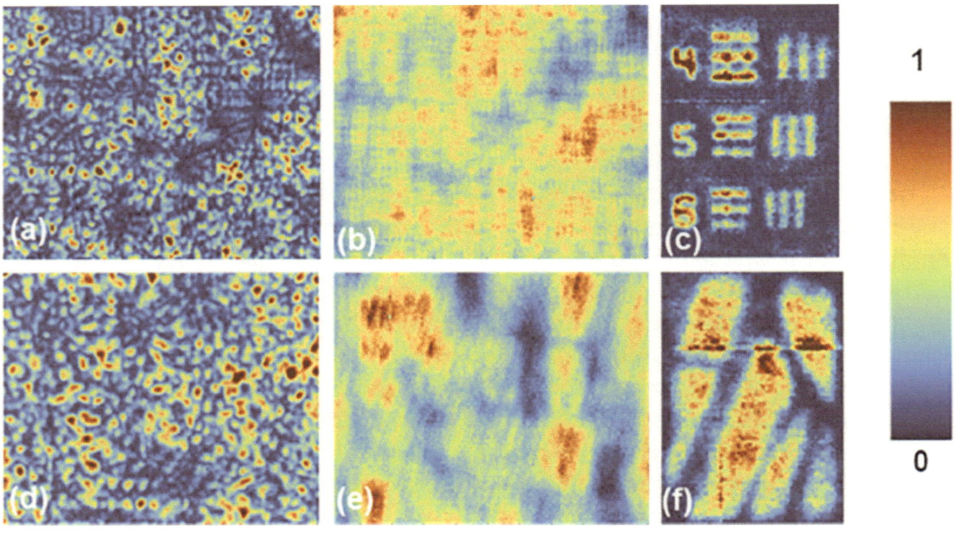

Figure 8.
Images of the (a) PSF and (b) object intensity distribution for USAF object, and (c) reconstruction result by NLR. (d) PSF and (e) object intensity distribution for wing of an insect, and (f) reconstruction result by NLR.

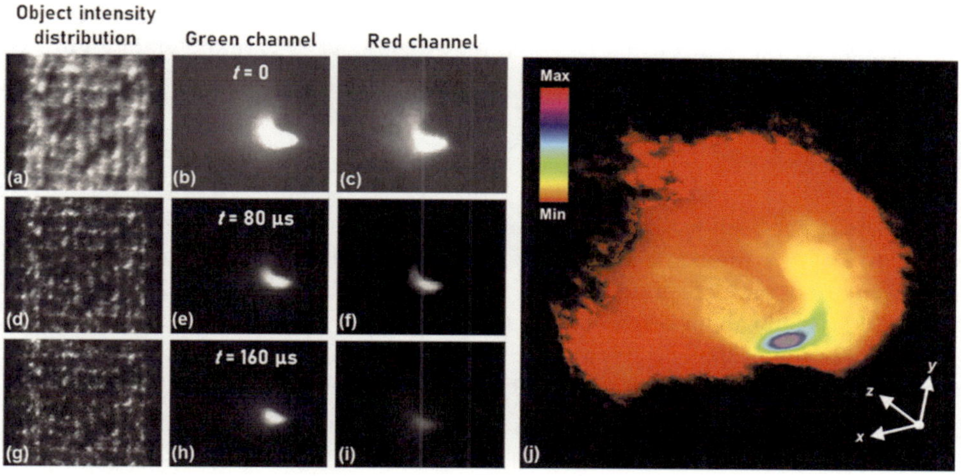

Figure 9.
(a) Recorded object intensity pattern and reconstruction result for (b) green channel and (c) red channel at t = 0. *(d) Recorded object intensity pattern and reconstruction result for (e) green channel and (f) red channel at* t = 80 *μs. (g) Recorded object intensity pattern and reconstruction result for (h) green channel and (i) red channel at* t = 160 *μs. (j) Tomography of the spark pattern.*

So, the phase mask was designed in a specific way by combining a chaotic coded phase mask and a quadratic phase function. The resulting phase function is called a quasi-random lens (QRL). This special lens behaves like a Fresnel zone lens, but instead of collecting light and focusing it on a point, it focuses on an area and within the area, light is scattered. In other words, the QRL creates controlled scattering similar to the mask designed in lensless I-COACH [61]. The generated phase of the QRL was converted into two levels by a standard rounding-off procedure. A QRL with a diameter of 5 mm and focal length of 5 cm was fabricated using electron beam lithography (RAITH150^{TWO}) using the same fabrication settings as described in Section 2. The optical microscope image of the fabricated device is shown in **Figure 10a**. The PSFs recorded using a pinhole with a diameter of 100 μm and using red (617 nm) and green wavelengths (530 nm) as shown in **Figure 10b** and **c**, respectively. The object hologram obtained by summing of the object intensity distributions recorded for two objects NBS (10 lp/mm) and USAF (Group 2, Element 2) separated by 1 cm and

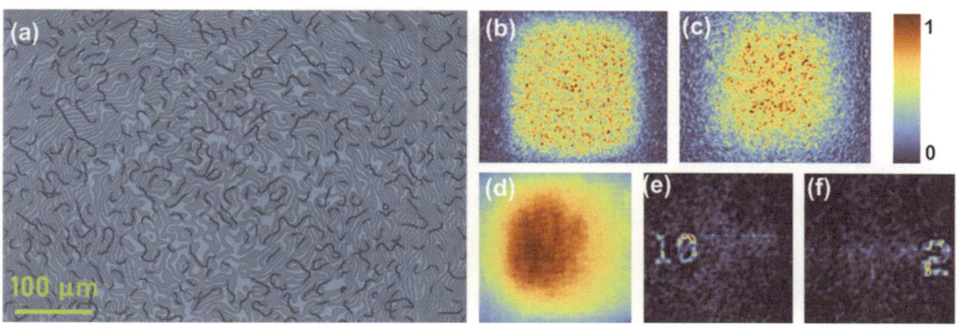

Figure 10.
(a) Optical microscope image of the QRL. Recorded PSFs for (b) 617 nm and (c) 530 nm. (d) Object hologram. Reconstruction results using PSFs recorded using (e) red and (f) green LEDs.

illuminated by red and green wavelengths, respectively, is shown in **Figure 10d**. The reconstruction results using NLR and red and green PSFs are shown in **Figure 10e** and **f**, respectively.

3.3 Discussion

I-COACH has been realized in a compact optical configuration with a total length of 20 cm. This is quite compact in comparison to CASSI, MURA, and also other incoherent holography techniques. The maximum efficiency achievable with the binary QRL is only about 40%, which can be improved to >90% by manufacturing a greyscale version of the QRL. With the latest advancements in rapid fabrication technologies and nanoimprint techniques, mass production of greyscale QRL is not expensive in comparison to using active devices such as SLM [65–68].

4. Optical tweezer perspectives inspired by FINCH configuration

Optical Tweezer (OT) is a powerful tool to manipulate micron-sized objects by creating an optical trap [69, 70]. In 1970, Arthur Ashkin first reported that the gradient forces from the laser can trap micron-sized particles [71]. Years later, Ashkin and his colleagues invented optical tweezers and demonstrated the ability of the single-beam optical trap to attract particles intended for analysis [72]. The advancements in the technologies such as invention of SLMs and diffractive optical elements led to the development of holographic optical tweezers (HOT) [73]. In general, a conventional tweezer system comprises trapping architecture and imaging architecture as shown in **Figure 11**. In HOT, computer-generated holograms (CGH) are designed and displayed in the SLM to modulate the trapping beam for sample control and manipulation. The trapping beam can be single or multiple based on the design adapted in the CGH and can be dynamically manipulated. The HOT quickly emerged and became indispensable in the field of biology [74–76].

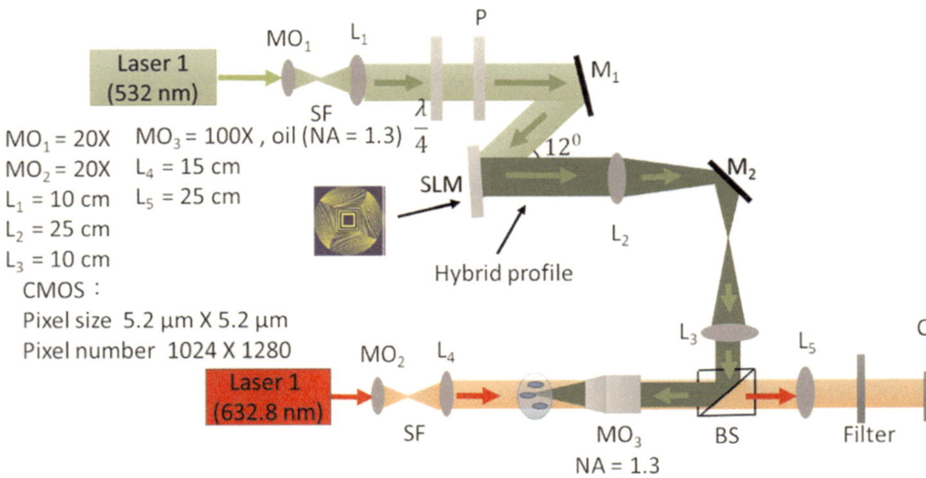

Figure 11.
Schematic of the typical holographic optical tweezer. Red and green represent the trapping and the imaging system architectures.

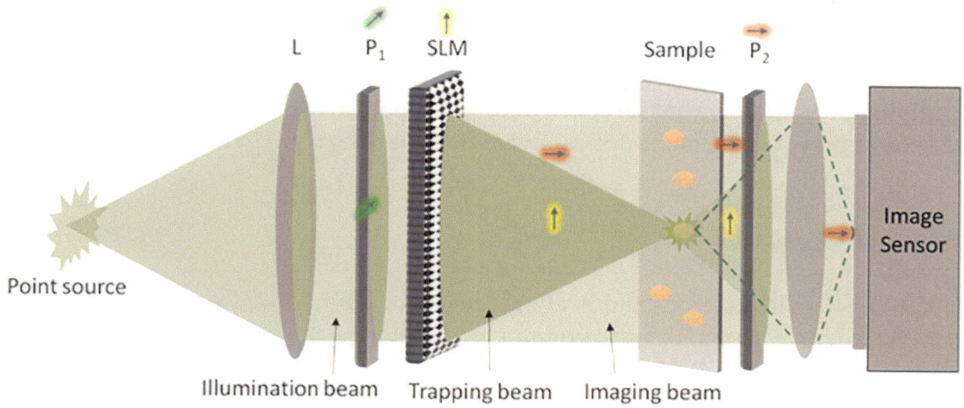

Figure 12.
Optical trapping and imaging configuration inspired for FINCH's polarization multiplexing scheme.

An alternate architecture inspired by FINCH polarization multiplexing scheme to the conventional trapping architecture for single beam compact trapping and imaging is shown in **Figure 12**. The proposed optical configuration uses a polarization multiplexing scheme. A polarizer P_1 at 45° orientation with respect to the active axis of the SLM was set before the SLM and a diffractive lens is displayed on the SLM. At the SLM, only part of the optical beam is modulated and so two beams namely, trapping beam and imaging beam, are generated. The modulated trapping beam creates a tightly focused spot to trap the sample and the unmodulated imaging beam is utilized for the imaging. Before the image sensor, a second polarizer P_2 oriented at 90° with respect to the active axis of the SLM was introduced to avoid the trapping beam, affecting the imaging. The optical power of the trapping beam can be controlled by adjusting the polarization degree of P_1. To achieve tight focusing, a projection system might be needed, which can increase the strength of the trap and imaging beams.

5. Summary and conclusion

In this chapter, two widely used incoherent holography systems, namely FINCH and I-COACH, are discussed with a focus on compactification using static diffractive optical elements. In FINCH, the required three camera shots and active optical device have been avoided by applying NLR. The diffractive optical element proposed in the first version of FINCH was designed and fabricated using electron beam lithography techniques in a compact optical configuration of 15 cm. I-COACH was realized using both an amplitude element and a phase element fabricated using photolithography and electron beam lithography systems, respectively, in a compact 20 cm configuration. The preliminary results are promising to begin research on building prototypes. There are still some challenges present in FINCH, as well as in I-COACH. From the manufacturing side, it is necessary to fabricate multilevel structures to improve the efficiency from the current 40% to >90%, which increases the cost and time of fabrication. With the latest developments in Femtosecond fabrication methods, it is possible to rapidly manufacture high-efficiency diffractive optical elements at a low cost [77–79]. The second challenge lies in the current performances of computational reconstruction methods. In both FINCH and I-COACH, the computational reconstruction generated background noise that could not be minimized to the level of lens-based imaging. With the advancements in deep learning techniques, we believe that it is possible to achieve a high SNR. The chapter concludes with an intriguing possibility of extending FINCH configuration for optical trapping experiments, which we believe may revolutionize optical tweezer and tomography research directions.

Acknowledgements

V.A thanks Tiia Lillemaa and Aravind Simon for their administrative support. D. S.; S. H. N.; T. K.; and S. J. are grateful for the financial support via ARC Linkage LP190100505 project. V. A. acknowledges the European Union's Horizon 2020 Research and Innovation Programme grant agreement No. 857627 (CIPHR). V.B.; P.J.M.; C.D. acknowledges King Abdullah University of Science and Technology (KAUST) for the funding.

Conflict of interest

Christian Depeursinge has financial interests in Lyncee Tec and Nanolive; Pierre J Magistretti has financial interests in Lyncee Tec.

Author details

Vijayakumar Anand[1*], Soon Hock Ng[2], Tomas Katkus[2], Daniel Smith[2], Vinoth Balasubramani[3], Denver P. Linklater[4], Pierre J. Magistretti[3], Christian Depeursinge[3], Elena P. Ivanova[4] and Saulius Juodkazis[2]

1 Institute of Physics, University of Tartu, Tartu, Estonia

2 Optical Sciences Centre, Swinburne University of Technology, Melbourne, Australia

3 Division of Biological and Environmental Sciences and Engineering, King Abdullah University of Science and Technology (KAUST), Saudi Arabia

4 School of Science, STEM College, RMIT University, Melbourne, Australia

*Address all correspondence to: vijayakumar.anand@ut.ee

References

[1] Rosen J, Vijayakumar A, Kumar M, Rai MR, Kelner R, Kashter Y, et al. Recent advances in self-interference incoherent digital holography. Advances in Optics and Photonics. 2019;**11**:1-66. DOI: 10.1364/AOP.11.000001

[2] Gabor D. Progress in holography. Reports on Progress in Physics. 1969;**32**: 395-404. DOI: 10.1088/0034-4885/32/1/308

[3] Osten W, Faridian A, Gao P, Körner K, Naik D, Pedrini G, et al. Recent advances in digital holography. Applied Optics. 2014;**53**:G44-G63. DOI: 10.1364/AO.53.000G44

[4] Goodman JW. Introduction to Fourier Optics. Englewood, Colorado: Roberts & Co.; 2005

[5] Roddier C, Roddier F, Demarcq J. Compact rotational shearing interferometer for astronomical applications. Optical Engineering. 1989; **28**:280166. DOI: 10.1117/12.7976903

[6] Cochran G. New method of making Fresnel transforms with incoherent light. JOSA. 1966;**56**:1513-1517. DOI: 10.1364/JOSA.56.001513

[7] Sirat G, Psaltis D. Conoscopic holography. Optics Letters. 1985;**10**:4-6. DOI: 10.1364/OL.10.000004

[8] Rosen J, Brooker G. Digital spatially incoherent Fresnel holography. Optics Letters. 2007;**32**:912-914. DOI: 10.1364/OL.32.000912

[9] Kelner R, Rosen J. Spatially incoherent single channel digital Fourier holography. Optics Letters. 2012;**37**: 3723-3725. DOI: 10.1364/OL.37.003723

[10] Hong J, Kim MK. Single-shot self-interference incoherent digital holography using off-axis configuration. Optics Letters. 2013;**38**:5196-5199. DOI: 10.1364/OL.38.005196

[11] Rosen J, Brooker G. Non-scanning motionless fluorescence three-dimensional holographic microscopy. Nature Photonics. 2008;**2**:190-195. DOI: 10.1038/nphoton.2007.300

[12] Wu P, Zhang D, Yuan J, Zeng S, Gong H, Luo Q, et al. Large depth-of-field fluorescence microscopy based on deep learning supported by Fresnel incoherent correlation holography. Optics Express. 2022;**30**:5177-5191. DOI: 10.1364/OE.451409

[13] Bu Y, Wang X, Li Y, Du Y, Gong Q, Zheng G, et al. Tunable edge enhancement by higher-order spiral Fresnel incoherent correlation holography system. Journal of Physics D: Applied Physics. 2021;**54**:125103. DOI: 10.1088/1361-6463

[14] Kashter Y, Vijayakumar A, Miyamoto Y, Rosen J. Enhanced super resolution using Fresnel incoherent correlation holography with structured illumination. Optics Letters. 2016;**41**: 1558-1561. DOI: 10.1364/OL.41.001558

[15] Anand V, Rosen J, Ng SH, Katkus T, Linklater DP, Ivanova EP, et al. Edge and contrast enhancement using spatially incoherent correlation holography techniques. Photonics. 2021;**8**:224. DOI: 10.3390/photonics8060224

[16] Siegel N, Brooker G. Single shot holographic super-resolution microscopy. Optics Express. 2021;**29**: 15953-15968. DOI: 10.1364/OE.424175

[17] Liang J. Punching holes in light: Recent progress in single-shot

coded-aperture optical imaging. Reports on Progress in Physics. 2020;**83**:116101. DOI: 10.1088/1361-6633/abaf43

[18] Caroli E, Stephen JB, Di Cocco G, Natalucci L, Spizzichino A. Coded aperture imaging in X-and gamma-ray astronomy. Space Science Reviews. 1987; **45**:349-403. DOI: 10.1007/BF00171998

[19] Ables JG. Fourier transform photography: A new method for X-ray astronomy. Publications of the Astronomical Society of Australia. 1968; **1**:172-173. DOI: 10.1017/S1323358000011292

[20] Dicke RH. Scatter-hole cameras for x-rays and gamma rays. The Astrophysical Journal. 1968;**153**:L101-L106. DOI: 10.1086/180230

[21] Fenimore EE, Cannon TM. Uniformly redundant arrays: Digital reconstruction methods. Applied Optics. 1981;**20**:1858-1864. DOI: 10.1364/AO.20.001858

[22] Olmos P, Cid C, Bru A, Oller JC, de Pablos JL, Perez JM. Design of a modified uniform redundant-array mask for portable gamma cameras. Applied Optics. 1992;**31**:4742-4750. DOI: 10.1364/AO.31.004742

[23] Wagadarikar A, John R, Willett R, Brady D. Single disperser design for coded aperture snapshot spectral imaging. Applied Optics. 2008;**47**:B44-B51. DOI: 10.1364/AO.47.000B44

[24] Kittle D, Choi K, Wagadarikar A, Brady DJ. Multiframe image estimation for coded aperture snapshot spectral imagers. Applied Optics. 2010;**49**: 6824-6833. DOI: 10.1364/AO.49.006824

[25] Arce GR, Brady DJ, Carin L, Arguello H, Kittle DS. Compressive coded aperture spectral imaging: An introduction. IEEE Signal Processing Magazine. 2013;**31**:105-115. DOI: 10.1109/MSP.2013.2278763

[26] Chi W, George N. Optical imaging with phase-coded aperture. Optics Express. 2011;**19**:4294-4300. DOI: 10.1364/OE.19.004294

[27] Vijayakumar A, Kashter Y, Kelner R, Rosen J. Coded aperture correlation holography–a new type of incoherent digital holograms. Optics Express. 2016; **24**:12430-12441. DOI: 10.1364/OE.24.012430

[28] Vijayakumar A, Rosen J. Interferenceless coded aperture correlation holography–a new technique for recording incoherent digital holograms without two-wave interference. Optics Express. 2017;**25**: 13883-11396. DOI: 10.1364/OE.25.013883

[29] Rai MR, Vijayakumar A, Ogura Y, Rosen J. Resolution enhancement in nonlinear interferenceless COACH with point response of subdiffraction limit patterns. Optics Express. 2019;**27**: 391-403. DOI: 10.1364/OE.27.000391

[30] Malinauskas M, Žukauskas A, Hasegawa S, Hayasaki Y, Mizeikis V, Buividas R, et al. Ultrafast laser processing of materials: From science to industry. Light: Science & Applications. 2016;**5**:e16133. DOI: 10.1038/lsa.2016.133

[31] Marcinkevičius A, Juodkazis S, Watanabe M, Miwa M, Matsuo S, Misawa H, et al. Femtosecond laser-assisted three-dimensional microfabrication in silica. Optics Letters. 2001;**26**:277-279. DOI: 10.1364/OL.26.000277

[32] Matsuo S, Juodkazis S, Misawa H. Femtosecond laser microfabrication of periodic structures using a microlens

array. Applied Physics A. 2005;**80**: 683-685. DOI: 10.1007/s00339-004-3108-x

[33] Juodkazis S, Rosa L, Bauerdick S, Peto L, El-Ganainy R, John S. Sculpturing of photonic crystals by ion beam lithography: Towards complete photonic bandgap at visible wavelengths. Optics Express. 2011; **19**:5802-5810. DOI: 10.1364/OE.19.005802

[34] Juodkazis S, Yamaguchi A, Ishii H, Matsuo S, Takagi H, Misawa H. Photo-electrochemical deposition of platinum on TiO_2 with resolution of twenty nanometers using a mask elaborated with electron-beam lithography. Japanese Journal of Applied Physics. 2001;**40**:4246. DOI: 10.1143/JJAP.40.4246

[35] Vijayakumar A, Katkus T, Lundgaard S, Linklater DP, Ivanova EP, Ng SH, et al. Fresnel incoherent correlation holography with single camera shot. Opto-Electronic Advances. 2020;**3**:08200004. DOI: 10.29026/oea.2020.200004

[36] Anand V, Katkus T, Juodkazis S. Randomly multiplexed diffractive lens and axicon for spatial and spectral imaging. Micromachines. 2020;**11**:437. DOI: 10.3390/mi11040437

[37] Rai MR, Vijayakumar A, Rosen J. Non-linear adaptive three-dimensional imaging with interferenceless coded aperture correlation holography (I-COACH). Optics Express. 2018;**26**: 18143-18154. DOI: 10.1364/OE.26.018143

[38] Richardson WH. Bayesian-based iterative method of image restoration. JoSA. 1972;**62**:55-59. DOI: 10.1364/JOSA.62.000055

[39] Lucy LB. An iterative technique for the rectification of observed distributions. The Astronomical Journal. 1974;**79**:745. DOI: 10.1086/111605

[40] Ng SH, Anand V, Katkus T, Juodkazis S. Invasive and non-invasive observation of occluded fast transient events: Computational tools. Photonics. 2021;**8**:253. DOI: 10.3390/photonics8070253

[41] Bhattacharya S, Vijayakumar A. Design and Fabrication Of Diffractive Optical Elements with MATLAB. Bellingham, Washington, USA: SPIE Press; 2017. DOI: 10.1117/3.2261461

[42] Rosen J, Siegel N, Brooker G. Theoretical and experimental demonstration of resolution beyond the Rayleigh limit by FINCH fluorescence microscopic imaging. Optics Express. 2011;**19**:26249-26268. DOI: 10.1364/OE.19.026249

[43] Vijayakumar A, Molong H, Jovan M, Hock NS, Tomas K, Annaleise K, et al. Single-shot mid-infrared incoherent holography using Lucy-Richardson-Rosen algorithm. Opto-Electronic Science. 2022;**1**:210006. DOI: 10.29026/oes.2022.210006

[44] Anand V, Khonina S, Kumar R, Dubey N, Reddy AN, Rosen J, et al. Three-dimensional incoherent imaging using spiral rotating point spread functions created by double-helix beams. Nanoscale Research Letters. 2022;**17**:1-3. DOI: 10.1186/s11671-022-03676-6

[45] Sakamaki S, Yoneda N, Nomura T. Single-shot in-line Fresnel incoherent holography using a dual-focus checkerboard lens. Applied Optics. 2020; **59**:6612-6618. DOI: 10.1364/AO.393176

[46] Nobukawa T, Muroi T, Katano Y, Kinoshita N, Ishii N. Single-shot

phase-shifting incoherent digital holography with multiplexed checkerboard phase gratings. Optics Letters. 2018;**43**: 1698-1701. DOI: 10.1364/OL.43.001698

[47] Tahara T, Kanno T, Arai Y, Ozawa T. Single-shot phase-shifting incoherent digital holography. Journal of Optics. 2017;**19**:065705. DOI: 10.1088/2040-8986/aa6e82

[48] Tahara T, Kozawa Y, Ishii A, Wakunami K, Ichihashi Y, Oi R. Two-step phase-shifting interferometry for self-interference digital holography. Optics Letters. 2021;**46**:669-672. DOI: 10.1364/OL.414083

[49] Rosen J, Alford S, Anand V, Art J, Bouchal P, Bouchal Z, et al. Roadmap on recent progress in FINCH technology. Journal of Imaging. 2021;**7**:197. DOI: 10.3390/jimaging7100197

[50] Zhou H, Huang L, Li X, Li X, Geng G, An K, et al. All-dielectric bifocal isotropic metalens for a single-shot hologram generation device. Optics Express. 2020;**28**:21549-21559. DOI: 10.1364/OE.396372

[51] Choi K, Yim J, Yoo S, Min SW. Self-interference digital holography with a geometric-phase hologram lens. Optics Letters. 2017;**42**:3940-3943. DOI: 10.1364/OL.42.003940

[52] Vijayakumar A, Bhattacharya S. Characterization and correction of spherical aberration due to glass substrate in the design and fabrication of Fresnel zone lenses. Applied Optics. 2013;**52**:5932-5940. DOI: 10.1364/AO.52.005932

[53] Tahara T, Oi R. Palm-sized single-shot phase-shifting incoherent digital holography system. OSA Continuum. 2021;**4**:2372-2380. DOI: 10.1364/OSAC.431930

[54] Anand V, Rosen J, Juodkazis S. Review of engineering techniques in chaotic coded aperture imagers. Light: Advanced Manufacturing. 2022;**3**:1-3. DOI: 10.37188/lam.2022.024

[55] Zalevsky Z, Mendlovic D, Dorsch RG. Gerchberg–Saxton algorithm applied in the fractional Fourier or the Fresnel domain. Optics Letters. 1996;**21**: 842-844. DOI: 10.1364/OL.21.000842

[56] Rai MR, Rosen J. Noise suppression by controlling the sparsity of the point spread function in interferenceless coded aperture correlation holography (I-COACH). Optics Express. 2019;**27**: 24311-24323

[57] Liu C, Man T, Wan Y. High-quality interferenceless coded aperture correlation holography with optimized high SNR holograms. Applied Optics. 2022;**61**(3):661-668. DOI: 10.1364/AO.444783

[58] Anand V, Ng SH, Maksimovic J, Linklater D, Katkus T, Ivanova EP, et al. Single shot multispectral multidimensional imaging using chaotic waves. Scientific Reports. 2020;**10**:1-3. DOI: 10.1038/s41598-020-70849-7

[59] Anand V, Ng SH, Katkus T, Juodkazis S. Spatio-spectral-temporal imaging of fast transient phenomena using a random array of pinholes. Advanced Photonics Research. 2021;**2**: 2000032. DOI: 10.1002/adpr.202000032

[60] Anand V, Ng SH, Katkus T, Juodkazis S. White light three-dimensional imaging using a quasi-random lens. Optics Express. 2021;**29**: 15551-15563. DOI: 10.1364/OE.426021

[61] Kumar M, Vijayakumar A, Rosen J. Incoherent digital holograms acquired by interferenceless coded aperture correlation holography system without

refractive lenses. Scientific Reports. 2017;7:1-11. DOI: 10.1038/s41598-017-11731-x

[62] Antipa N, Kuo G, Heckel R, Mildenhall B, Bostan E, Ng R, et al. DiffuserCam: Lensless single-exposure 3D imaging. Optica. 2018;**5**:1-9. DOI: 10.1364/OPTICA.5.000001

[63] Sahoo SK, Tang D, Dang C. Single-shot multispectral imaging with a monochromatic camera. Optica. 2017;**4**: 1209-1213. DOI: 10.1364/OPTICA.4.001209

[64] Lee K, Park Y. Exploiting the speckle-correlation scattering matrix for a compact reference-free holographic image sensor. Nature Communications. 2016;7:1-7. DOI: 10.1038/ncomms13359

[65] Sun YK, Wang L, Kamano M, Juodkazis S. Plasmonic nano-imprinting by photo-doping. Optics Letters. 2018; **43**:3786-3789. DOI: 10.1364/OL.43.003786

[66] Lu YM, Liu XQ, Zhu L, Chen QD, Juodkazis S, Sun HB. Vector scanning subtractive manufacturing technology for laser rapid fabrication. Optics Letters. 2021;**46**:1963-1966. DOI: 10.1364/OL.422455

[67] Hua JG, Tian ZN, Xu SJ, Lundgaard S, Juodkazis S. Fast fabrication of optical vortex generators by femtosecond laser ablation. Applied Surface Science. 2019;**475**:660-665. DOI: 10.1016/j.apsusc.2018.12.249

[68] Linklater DP, Juodkazis S, Ivanova EP. Nanofabrication of mechano-bactericidal surfaces. Nanoscale. 2017;**9**:16564-16585. DOI: 10.1039/C7NR05881K

[69] Nieminen TA, Knöner G, Heckenberg NR, Rubinsztein-Dunlop H. Physics of optical tweezers. Methods in Cell Biology. 2007;**82**:207-236. DOI: 10.1016/S0091-679X(06)82006-6

[70] Polimeno P, Magazzu A, Iati MA, Patti F, Saija R, Boschi CD, et al. Optical tweezers and their applications. Journal of Quantitative Spectroscopy and Radiative Transfer. 2018;**218**:131-150. DOI: 10.1016/j.jqsrt.2018.07.013

[71] Ashkin A. Acceleration and trapping of particles by radiation pressure. Physical Review Letters. 1970;**24**:156. DOI: 10.1103/PhysRevLett.24.156

[72] Molloy JE, Padgett MJ. Lights, action: Optical tweezers. Contemporary Physics. 2002;**43**:241-258. DOI: 10.1080/00107510110116051

[73] Curtis JE, Koss BA, Grier DG. Dynamic holographic optical tweezers. Optics Communication. 2002;**207**: 169-175. DOI: 10.1016/S0030-4018(02)01524-9

[74] Balasubramani V, Kuś A, Tu HY, Cheng CJ, Baczewska M, Krauze W, et al. Holographic tomography: Techniques and biomedical applications. Applied Optics. 2021;**60**:B65-B80. DOI: 10.1364/AO.416902

[75] Vijayakumar A, Vinoth B, Minin IV, Rosen J, Minin OV, Cheng CJ. Experimental demonstration of square Fresnel zone plate with chiral side lobes. Applied Optics. 2017;**56**:F128-F133. DOI: 10.1364/AO.56.00F128

[76] Balasubramani V, Tu HY, Lai XJ, Cheng CJ. Adaptive wavefront correction structured illumination holographic tomography. Scientific Reports. 2019;**9**(1):1-7. DOI: 10.1038/s41598-019-46951-w

[77] Smith D, Ng SH, Han M, Katkus T, Anand V, Glazebrook K, et al. Imaging

with diffractive axicons rapidly milled on sapphire by femtosecond laser ablation. Applied Physics B. 2021;**127**: 1-11. DOI: 10.1007/s00340-021-07701-x

[78] Mao T, Ma X, Cuadros AP, Dai X, Wang Z, Zhang X, et al. Static coded aperture in robotic X-ray tomography systems. Optics Express. 2022;**30**: 7677-7693. DOI: 10.1364/OE.449505

[79] Smith D, Gopinath S, Arockiaraj FG, Reddy ANK, Balasubramani V, Kumar R, et al. Nonlinear reconstruction of images from patterns generated by deterministic or random optical masks–concepts and review of research. Journal of Imaging. 2022;**8**(6):174. DOI: 10.3390/jimaging8060174

Chapter 6

Multiplexed Frequency-Selective Incoherent Holography

Wanbin Zhang, Baosheng Li and Jianquan Li

Abstract

We propose a new incoherent optical holographic spectrum stripping reconstruction method, called incoherent multiplexing frequency-selective holography, which compresses two or more on-axis holograms into a single multiplexed on-axis hologram without loss of magnification and resolution. The technique described in this chapter effectively suppresses the background bias term and conjugate term. The acquired spectrum is obtained by stripping in the overlapping confounding correlation terms. The experimental results show the potential of the method in areas such as compressed holography and extended field of view imaging.

Keywords: digital holography, multiplex holography, interferometric imaging, frequency modulation

1. Introduction

Multiplexed holography is the simultaneous capture of multiple complex wavefronts by a single CCD. Each wavefront is encoded into a hologram carrying different information. These complex wavefronts are compressed together and recorded at once. This is a tremendous improvement in acquisition efficiency, and multiplexed holography has the advantage of multiple field of view (FOV). Thus, multiplexed holography has many possible applications. Such as field-of-view multiplexing [1–7], depth-of-field multiplexing [8–12], non-scanning holographic lamination [13–16], and so on. In the development of holography, the laser has good coherence, and it is convenient to form holograms with the interferometer. Multiplexed holography first appeared in the field of coherent optical. In 2014, Frenklach et al. [17] proposed an interferometric method, called interferometry with a triple imaging area, where the camera acquires four holograms at once making the quantitative information collected in the camera exposure triple. In 2016, Rubin et al. [18] compressed six off-axis holograms into one multiplexed off-axis hologram without loss of magnification or resolution. And it can be reconstructed in the form of spatial filtering. The utilization of spatial bandwidth is greatly enhanced. In 2017, Dardikman et al. [19] used the space occupied by the DC term to place two pairs of complex wave fronts possessing conjugates. Eight holograms were reused in a single pass. The DC term needs to be removed by obtaining two phase-shifted holograms.

So far, the experimentally achievable coherent optical multiplexing can be extended to a maximum of eight holograms. The frequency domain obtained from the Fourier transform of the hologram shows that when the cutoff angular frequency required for the multiplexing method is four times the maximum angular frequency of the sample, the spectrum of the eight off-axis wave fronts already fills the two-dimensional frequency domain. No more wavefront spectrum can be accommodated. This is the reason for limiting the number of off-axis coherent optical multiplexes. Our incoherent optical multiplexing scheme can theoretically expand the number of multiplexed wavefronts to an infinite number of dimensions without considering reflectivity and transmittance losses. The frequency selection scheme in this paper needs to be implemented in combination with a single FINCH [20, 21] in a multi-beam setup.

2. Incoherent multiplexed frequency selection scheme

In Section 2, We describe in detail the scheme for stripping and reconstructing object-wave light from multiplexed complex wave fronts, arranged as follows: Section 2.1 presents the dual incoherent frequency selection scheme and experiments. Section 2.2 presents the triple incoherent frequency selection and experiments. Section 2.3 presents the arbitrary multiple incoherent frequency selection scheme. Section 2.4 discusses the advantages and potential developments of incoherent frequency selection over off-axis multiplexing.

2.1 Dual incoherent holographic frequency selection scheme and experiment

The experimental setup is shown in **Figure 1**, where the beam from a incoherent light source (CET-TCX250, 250 W) is passed through a narrow-band filter with a central wavelength of 20 nm. Collimated by a plano-convex lens. The focal length f_1 of the lens L is 150 mm. In the present frequency selection scheme, the polarization direction of the polarizer P is aligned with the long axis of the SLM, and the polarization direction is set to vertical. The symbols O and S in **Figure 1** represent the specimens measured on the two paths. The microscope objective (MO) on both branches serves to amplify the information of the object under test. The phase shifter in the optical path generates the experimentally required π phase shift of the corresponding object wave O(S). The two object beams are projected onto the spatial light modulator (SLM) (FSLM-2 K70-VIS, CAS Microstar) through the converging BS beam splitter on the right side. The mask loaded on the SLM is modulated according to the Fresnel wave band slice principle. The modulation equation can be expressed as $R(x,y) = \frac{1}{2} + \frac{1}{2}Q\left(-\frac{1}{a}\right)e^{i\theta}$. where $Q(b) = \exp\left[i\pi b\lambda^{-1}(x^2+y^2)\right]$ denotes the quadratic phase function. Only about half of the incident light intensity is modulated by the Fresnel mask, while the remaining half is reflected without any modulation, which is the key to FINCH [20, 21]. The monochrome CCD camera (MER-502-79U3M/C) has 2448×2048 pixels, a pixel size of 3.45 μm, and captures at 79 fps.

In this paper, we use the half-wave plate (HWP) as the phase shift module. When the fast axis direction of the HWP is the same as the polarizer polarization direction, the Jones in the optical path is expressed as

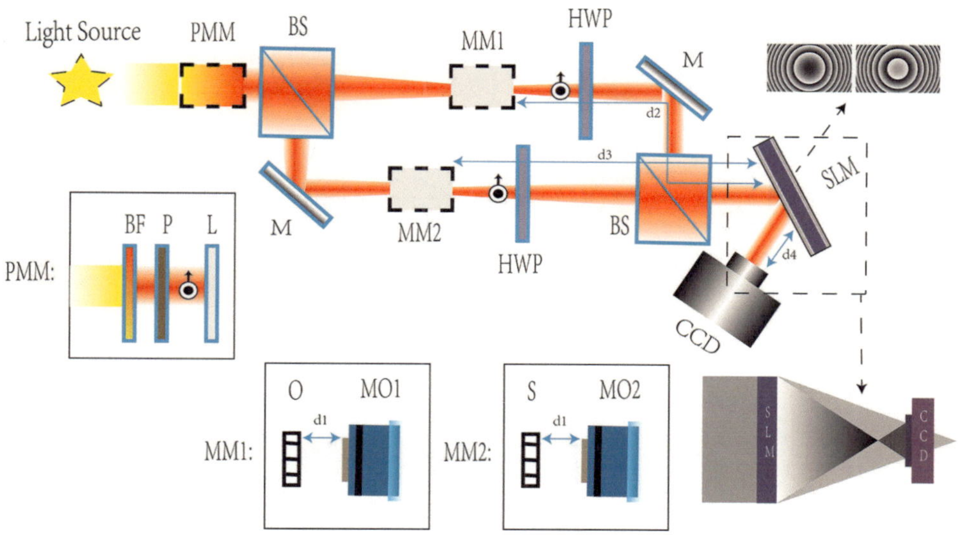

Figure 1.
TPFS-FINCH (two-pack frequency-selective-FINCH) optical multiplexing system. BF, bandpass filter; P, polarizer; PMM: Polarization modulation module; MM1: Microscopic objective module 1; MM2: Microscopic objective module 2; BS, beam splitter; MO, microscope objectives; HWP, half-wave plate; M, mirror; O, object; S, specimen; MO1, ×4, 0.1-NA; MO2, ×10, 0.25-NA; SLM, spatial light modulator; CCD, charge coupled device.

$$E_R = H \cdot E_v = \begin{bmatrix} 1 & 0 \\ 0 & -1 \end{bmatrix} \begin{bmatrix} 0 \\ 1 \end{bmatrix} = \begin{bmatrix} 0 \\ -1 \end{bmatrix} = \begin{bmatrix} 0 \\ e^{i\pi} \end{bmatrix}, \tag{1}$$

where H is the Jones representation of the HWP, and E_v is the Jones representation of the linearly polarized light after passing through the polarizer. According to the Fresnel diffraction theory, the light intensity of the two transmitted samples of light reaching the sensor surface can be expressed as.

$$I_1 = \left| \frac{1}{2} Q\left(\frac{1}{\alpha + d_4}\right) + \frac{1}{2} Q\left(\frac{1}{d_4 - {}^{\alpha a}/_{\alpha - a}}\right) e^{i\theta} \right|^2 \tag{2}$$

$$I_2 = \left| \frac{1}{2} Q\left(\frac{1}{\beta + d_4}\right) + \frac{1}{2} Q\left(\frac{1}{d_4 - {}^{\beta a}/_{\beta - a}}\right) e^{i\theta} \right|^2. \tag{3}$$

Where $\alpha^{-1} = (f_{mo1} - d_1)/[f_{mo1}(d_1 + d_2) - d_1 d_2]$ and $\beta^{-1} = (f_{mo2} - d_1)/[f_{mo2}(d_1 + d_3) - d_1 d_3]$. The distance parameters in the two-pack frequency-selective (TPFS) experiment are a = 150 mm, d1 = 7 mm, d2 = d3 = 235 mm and d4 = 56 mm.

In the frequency selection scheme, if the two holograms are just intensity superimposed, the combined hologram received by the camera should be $I = I_1 + I_2$. If the dual path setup if the conditions of the Mach–Zehnder interferometer are satisfied, the multiple holography recorded by the CCD is an incoherent superposition of all object point intensities by analyzing the interference intensity distribution of the infinite elements, for two three-dimensional objects $g_0(x_0, y_0, z_0)$ and $g_s(x_s, y_s, z_s)$:

$$E(x,y) \cong A\left[C + \iiint g_0(x_0,y_0,z_0)Q\left(\frac{1}{d_4 - \frac{\alpha a}{\alpha - a}} + \frac{1}{\alpha + d_4}\right)e^{i\theta}dx_0dy_0dz_0\right. \tag{4}$$

$$+\iiint g_0(x_0,y_0,z_0)Q\left(\frac{1}{\alpha + d_4} + \frac{1}{d_4 - \frac{\alpha a}{\alpha - a}}\right)e^{-i\theta}dx_0dy_0dz_0$$

$$+\iiint g_s(x_s,y_s,z_s)Q\left(\frac{1}{d_4 - \frac{\beta a}{\beta - a}} - \frac{1}{\beta + d_4}\right)e^{i\theta}dx_sdy_sdz_s$$

$$+\iiint g_s(x_s,y_s,z_s)Q\left(\frac{1}{\beta + d_4} - \frac{1}{d_4 - \frac{\beta a}{\beta - a}}\right)e^{-i\theta}dx_sdy_sdz_s$$

$$+\iiint g_0(x_0,y_0,z_0)Q\left(\frac{1}{d_4 - \frac{\alpha a}{\alpha - a}}\right)e^{i\theta}dx_0dy_0dz_0$$

$$+\iiint g_s(x_s,y_s,z_s)Q\left(\frac{-1}{d_4 - \frac{\beta a}{\beta - a}}\right)e^{-i\theta}dx_sdy_sdz_s$$

$$+\iiint g_0(x_0,y_0,z_0)Q\left(\frac{-1}{d_4 - \frac{\alpha a}{\alpha - a}}\right)e^{-i\theta}dx_0dy_0dz_0$$

$$\left.+\iiint g_s(x_s,y_s,z_s)Q\left(\frac{1}{d_4 - \frac{\beta a}{\beta - a}}\right)e^{i\theta}dx_sdy_sdz_s\right].$$

In order to implement two-pack frequency-selective FINCH (TPFS-FINCH), the corresponding experimental steps are as follows:

1. When a(b) of **Figure 2** is loaded on the SLM and the both phase shift modules do not apply phase shift to the target beam, the intensity distribution recorded by the CCD is E_{10} (E_{20}).

2. Applying a phase shift π exclusively to the object-wave O, i.e. multiplying the object-wave O by the phase factor $e^{i\pi}$. The CCD collected E_{11} and E_{21} intensity distributions after loading a and b of **Figure 2** on the SLM, respectively.

3. On the object wave S, just the phase shift π is applied. The intensity distributions recorded by the CCD are E_{12} and E_{22}, and the phase mask loaded on the SLM is compatible with step (1).

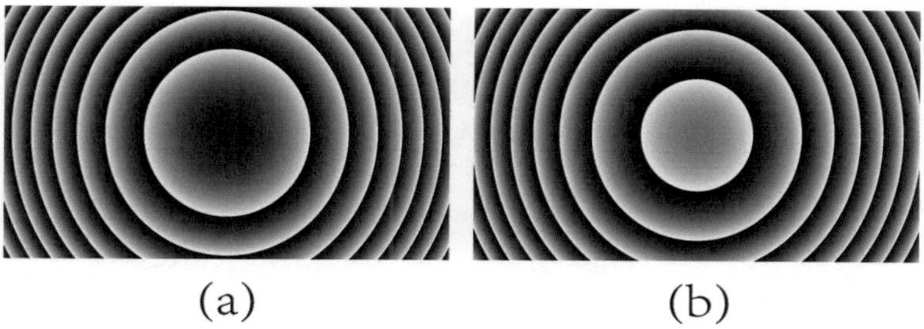

Figure 2.
(a) Fresnel mask template with a gray value of 31. (b) Fresnel mask template with a gray value of 154.

4. The phase shift π is applied to both the object waves O and S at the same time. The phase mask loaded on the SLM is consistent with step (1), and the CCD captured E_{13} and E_{23} intensity distributions.

In TPFS, we take into account the presence of intensity superposition and incoherent optical interference and can remove the spectral information of the object O by the following equation:

$$FT\{O\} = \frac{FT\{\frac{1}{4}(E_{20} - E_{21} + E_{22} - E_{23})\}H_{-z_1} - FT\{\frac{1}{4}(E_{10} - E_{11} + E_{12} - E_{13})\}H_{-z_2}}{2(H_{z_2-z_1} - H_{z_1-z_2})}, \tag{5}$$

$$FT\{S\} = \frac{FT\{\frac{1}{4}(E_{20} + E_{21} - E_{22} - E_{23})\}H_{-z_1} - FT\{\frac{1}{4}(E_{10} + E_{11} - E_{12} - E_{13})\}H_{-z_2}}{2(H_{z_2-z_1} - H_{z_1-z_2})}. \tag{6}$$

At last, when reconstructing the object information of the corresponding respective branches, the transfer function $H_z\left(f_x, f_y, z_i\right) = \exp\left[ikz_i\sqrt{1 - (\lambda f_x)^2 - \left(\lambda f_y\right)^2}\right]$ intercorrelation reconstruction in angular spectral diffraction theory is used. As shown in **Figure 3**.

As can be seen from **Figure 3**, **Figure 3(e)** and **(f)** are the complex wave fronts reconstructed by stripping from the multiplexed hologram **Figure 3(d)**. The intensity distribution of the red solid line also clearly demonstrates that the numbers 4 and 6 have been separated from the compressed hologram. Two-branch multiplexed hologram separation reconstruction has been performed without decreasing the resolution and without increasing the noise.

2.2 Triple incoherent holographic frequency selection scheme and experiment

After successfully implementing the scheme of TPFS-FINCH in the incoherent domain, is it possible to implement the extension of dual frequency selection by adding a dimension to triple frequency selection? This is what needs to be explored in this subsection.

As shown in **Figure 4**, a figure of the experimental setup with triple frequency selection is shown here. The difference from the setup in **Figure 1** is that there is one

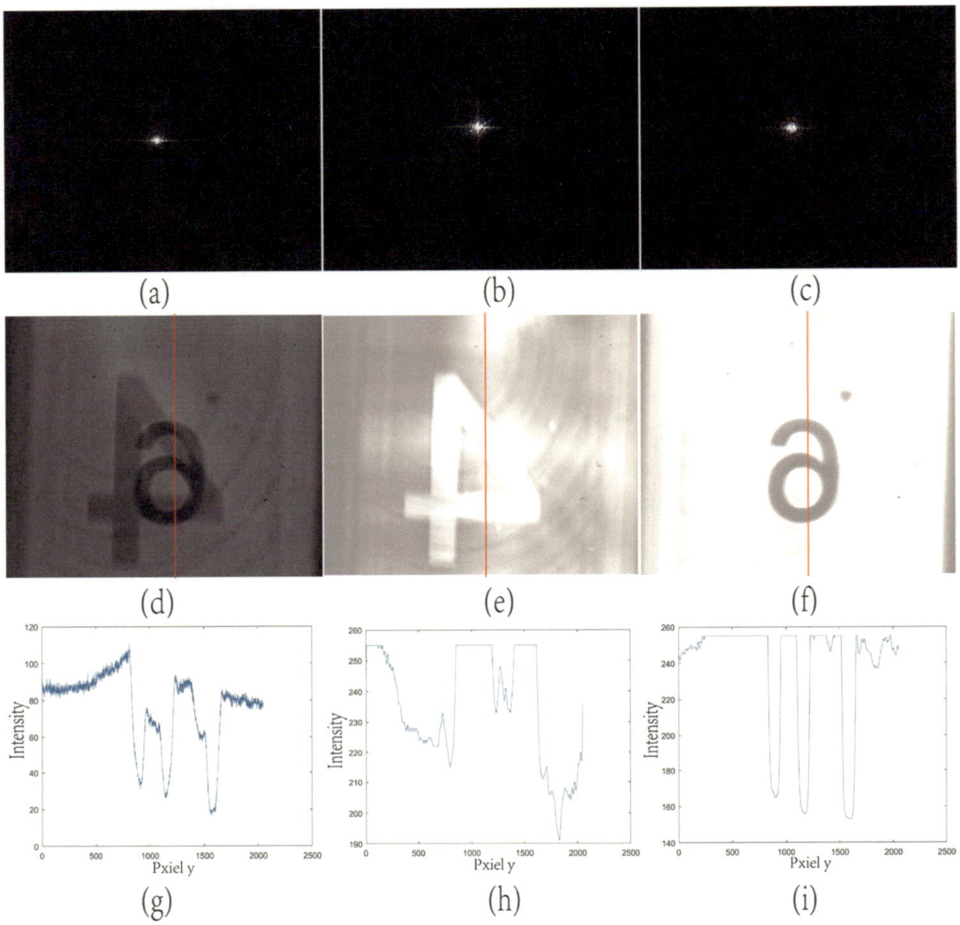

Figure 3.
(a)-(c) the power spectra of (d), (e) and (f), respectively. (e) and (f) the wavefront reconstructed from (d) by means of spectral separation. (g)-(i) the intensity distributions of the red solid lines in the (d)-(f), respectively.

more path carrying information about the object wave light. The phase shifter and phase mask template are the consistent. The experimental steps are as follows.

1. **Figure 2(a)** is loaded onto the SLM with no HWP loaded on all three branches and the intensity distribution captured on the CCD is E_{10}. Placing the phase shifter on the first branch generates a phase shift and the intensity distribution received on the CCD is E_{11}. Placing the phase shifter on the second branch to produce a phase shift, the intensity distribution picked up on the CCD is E_{12}. Placing the phase shifter on the third branch to produce a phase shift, the intensity distribution received on the CCD is E_{13}.

2. **Figure 2(b)** is loaded onto the SLM without HWP on all three branches, and the intensity distribution acquired on the CCD is E_{20}. Placing the phase shifter on the first branch generates a phase shift, and the intensity distribution received on the CCD is E_{21}. Placing the phase shifter on the second branch produces a phase shift, and the intensity distribution picked up on the CCD is E_{22}. Placing the

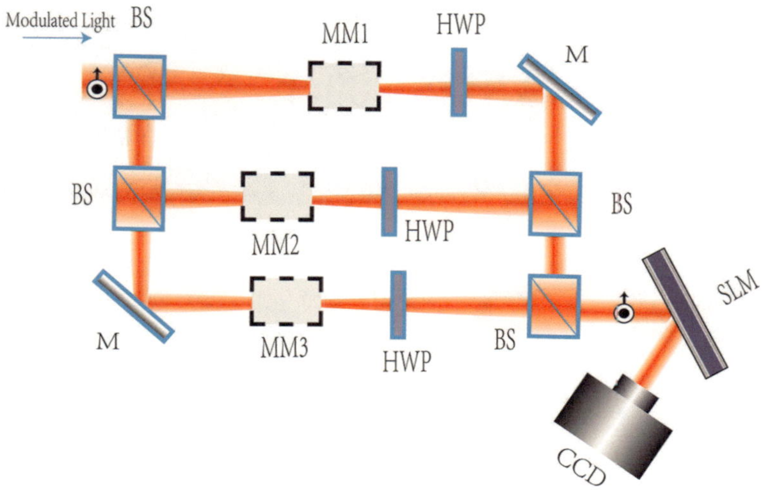

Figure 4.
Triple incoherent holographic frequency selection setup.

phase shifter on the third branch produces a phase shift, and the intensity distribution pictured on the CCD is E_{23}.

Combined with the multiplexed holograms acquired by the camera, the wavefront spectra of objects S_1, S_2 and S_3 can be stripped out by Eq. (7). Then reconstructed with the spatial transfer function $H_z\left(f_x, f_y, z_i\right)$ correlation.

$$FT\{S_n\} = \frac{FT\{E_{20} - E_{2n}\}H_{-z_1} - FT\{E_{10} - E_{1n}\}H_{-z_2}}{2(H_{z_2-z_1} - H_{z_1-z_2})}, n = 1,2,3. \tag{7}$$

The experimental results of three-pack frequency-selective FINCH are shown in **Figure 5**. The numbers 5, 2, and 6 have been separated from the compressed hologram **Figure 5(e)** and the angular spectrum reproduced.

2.3 Arbitrary multiplexed incoherent holographic frequency selection theory scheme

In the field of coherent optical multiplexing, when the idea of multiplexing two channels is further developed to six channels [18], the spectral position in the spatial frequency domain needs to be calculated precisely. And the cross term between the mismatched sample beam and the reference beam should be avoided. These cross terms occupy space in the spatial spectrum required for the additional wavelength channels. Because of the limited spatial frequency domain, the development of coherent optical holographic multiplexing came to an abrupt end at 8PH. The frequency spectrum of the zero-order term and the conjugate term of the incoherent optical in-axis self-interference holography overlap together and can co-occupy the position with the center of the frequency domain. This theoretically eliminates the need to consider how to filter by means of spatial filtering. Using the frequency selection scheme described in this paper the spectrum that is desired to be reconstructed can be stripped out by means of an equation calculation. Based on the experiments of

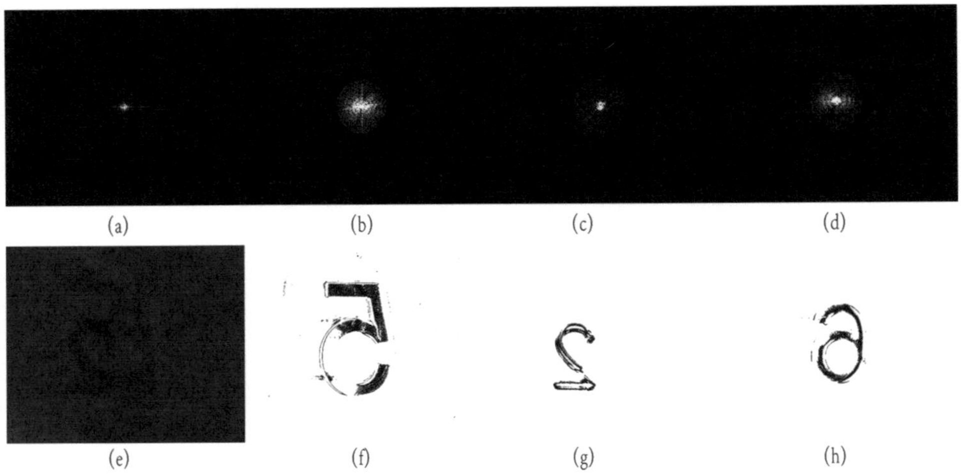

Figure 5.
(a)-(d) the power spectra of (e), (f), (g) and (h), respectively. (f)-(h) is the angular spectrum reproduced by separating the reconstruction from the compressed hologram (e).

multiplexing in Sections 2.1 and 2.2, this subsection describes the theoretical scheme based on the implementation of frequency selection under arbitrary multi-beam settings.

In an arbitrary multiplexed holographic intensity superposition setup, the multiplexed hologram of the n-th branch producing a phase shift when the SLM is loaded with mask a is denoted by E_{an}. Similarly, E_{bn} is used to indicate the multiplexed hologram of the n-th branch generating the phase shift when the SLM is loaded with mask b. E_{a0} and E_{b0} denote multiplexed holograms with all branches not loaded with phase shifts. S_n corresponds to the spectrum of the object wave on the n-th branch.

The equation for the spectrum of the object wave on an arbitrary branch can be expressed as:

$$FT\{S_n\} = \frac{FT\{E_{b0} - E_{bn}\}H_{-z_1} - FT\{E_{a0} - E_{an}\}H_{-z_2}}{2(H_{z_2-z_1} - H_{z_1-z_2})}. \tag{8}$$

At this stage, the arbitrary multiplexed incoherent multiplexing holographic frequency selection scheme has been completed.

2.4 Comparison and discussion

The frequency selection scheme in this paper is compared with studies in the field of coherent light [16–18, 22–27], where a multiplexed holography scheme for coherent light projects two interferometric patterns with different stripe orientations onto the camera, each capturing a different region of the sample or a complex wave front of a different sample. This technique is called interferometry with double imaging area. Since the two FOVs in the coherent light field do not overlap in the spatial frequency domain, they can be completely reconstructed and stitched together without loss of resolution or magnification. The effect of frequency selection is experimentally demonstrated in Sections 2.1 and 2.2 of this paper. The corresponding wavefront spectrum is calculated by Equation. The reconstruction also does not lose resolution and amplification as long as the phase shift is accurate. Stitching different wavefronts together after reconstruction can achieve a multiplexed field of view after reconstructing static or dynamic samples.

According to the calculation of wavefront acquisition efficiency [19], the efficiency of any n-way multiplexed holographic frequency selection is n/8. For in-axis FINCH, a single acquisition of the object wavefront generally requires a three-step phase shift technique to eliminate the conjugate and DC terms. Acquiring n wavefronts requires 3n holograms. Therefore, for the incoherent frequency selection scheme in this paper, the wavefront acquisition improves the efficiency by a factor of 3n/8. Obviously, when n > =3, the advantage of the frequency selection technique of increasing the acquisition efficiency on the basis of ensuring the original FOV of the camera becomes apparent. Frequency selective operation allows us to obtain multiplexed imaging FOVs without adjusting the time interval captured in each sub-hologram to the camera's frame rate. This is particularly effective for situations where the spatial resolution of a static observer is insufficient.

3. Summary

This chapter describes the frequency selection principles and experimental requirements for incoherent multiplexing. The combination of wavefront interference and single-channel FINCH techniques allows a completely new approach to frequency selection. The frequency selection method and the off-axis multiplexing method are compared and analyzed in terms of FOV, efficiency, and applicable scenarios. In general, the incoherent frequency selection principle is expected to be used for a wide range of applications that focus on the advantages of compressed holographic data and incoherent light, including multiplexed field-of-view imaging, fluorescence imaging, and structured light super-resolution imaging.

Acknowledgements

We thank Professor Joseph Rosen for his review comments on this chapter. We sincerely thank Chunlin Shi and Jinjin Song for their assistance with this experiment. Appreciation is extended to Ms. Rongyu Wang and Yimu Zhang for their support to the authors in writing this chapter.

Funding

This work is supported by the Natural Science Foundation of China (NSFC) under Grant Nos. 52177137 and 41371336.

Author details

Wanbin Zhang[1,2]*, Baosheng Li[1,2] and Jianquan Li[1,2]

1 School of Instrument Sciences and Opto-Electronics Engineering, Hefei University of Technology, Hefei, China

2 Anhui Province Key Laboratory of Measuring Theory and Precision Instrument, Hefei University of Technology, Hefei, Anhui, China

*Address all correspondence to: wbzhang0419@163.com

References

[1] Mirsky SK, Shaked NT. First experimental realization of six-pack holography and its application to dynamic synthetic aperture superresolution. Optics Express. 2019;**27**: 26708-26720

[2] Girshovitz P, Frenklach I, Shaked NT. Broadband quantitative phase microscopy with extended field of view using off-axis interferometric multiplexing. Journal of Biomedical Optics. 2015;**20**:111217

[3] Rotman-Nativ N, Turko NA, Shaked NT. Flipping interferometry with doubled imaging area. Optics Letters. 2018;**43**:5543-5546

[4] Girshovitz P, Shaked NT. Doubling the field of view in off-axis low-coherence interferometric imaging. Light Science Applications. 2014;**3**:e151

[5] Roitshtain D, Turko NA, Javidi B, Shaked NT. Flipping interferometry and its application for quantitative phase microscopy in a micro-channel. Optics Letters. 2016;**41**:2354-2357

[6] Tahara T, Awatsuji Y, Nishio K, Ura S, Matoba O, Kubota T. Space-bandwidth capacity-enhanced digital holography. Applied Physics Express. 2013;**6**:22502

[7] Zhang W, Li B, Shi C, Li J. Two-pack frequency-selective incoherent holography by using a dual-beam setup. Optics and Lasers in Engineering. 2022; **156**:107086

[8] Dalgarno PA, Dalgarno HIC, Putoud A, Lambert R, Paterson L, Logan DC, et al. Multiplane imaging and three dimensional nanoscale particle tracking in biological microscopy. Optics Express. 2010;**18**:877-884

[9] Rai MR, Rosen J. Depth-of-field engineering in coded aperture imaging. Optics Express. 2021;**29**:1634-1648

[10] Maurer C, Khan S, Fassl S, Bernet S, Ritsch-Marte M. Depth of field multiplexing in microscopy. Optics Express. 2010;**18**:3023-3034

[11] Ferraro P, Paturzo M, Memmolo P, Finizio A. Controlling depth of focus in 3D image reconstructions by flexible and adaptive deformation of digital holograms. Optics Letters. 2009;**34**: 2787-2789

[12] Kostencka J, Kozacki T, Li˙zewski K. Autofocusing method for tilted image plane detection in digital holographic microscopy. Optics Communication. 2013;**297**:20-26

[13] Schilling BW, Poon T-C, Indebetouw G, Storrie B, Shinoda K, Suzuki Y, et al. Three-dimensional holographic fluorescence microscopy. Optics Letters. 1997;**22**(19):1506-1508

[14] Fang-Yen CM, Choi W, Sung Y, Holbrow CJ, Dasari RR, Feld MS. Video-rate tomographic phase microscopy. Journal of Biomedical Optics. 2011;**16**: 011005

[15] Jin D, Zhou R, Yaqoob Z, So PTC. Dynamic spatial filtering using a digital micromirror device for high-speed optical diffraction tomography. Optics Express. 2018;**26**:428-437

[16] Kim K, Yoon J, Park Y. Simultaneous 3D visualization and position tracking of optically trapped particles using optical diffraction tomography. Optica. 2015;**2**: 343-346

[17] Frenklach I, Girshovitz P, Shaked NT. Off-axis interferometric

phase microscopy with tripled imaging area. Optics Letters. 2014;**39**:1525-1528

[18] Rubin M, Dardikman G, Mirsky SK, Turko NA, Shaked NT. Six-pack off-axis holography. Optics Letters. 2017;**42**: 4611-4614

[19] Dardikman G, Shaked NT. Is multiplexed off-axis holography for quantitative phase imaging more spatial bandwidth-efficient than on-axis holography? Journal of the Optical Society of America. A. 2019;**36**:A1-A11

[20] Rosen J, Brooker G. Digital spatially incoherent fresnel holography. Optics Letters. 2007;**32**:912-914

[21] Rosen J, Brooker G. Fluorescence incoherent color holography. Optics Express. 2007;**15**:2244-2250

[22] Fugal PJ, Shaw RA, Saw EW, Sergeyev AV. Airborne digital holographic system for cloud particle measurements. Applied Optics. 2004;**43**: 5987-5995

[23] Schnell M, Perez-Roldan MJ, Carney PS, Hillenbrand R. Quantitative confocal phase imaging by synthetic optical holography. Optics Express. 2014;**22**:15267-15276

[24] Balasubramani V, Tu H, Lai X, et al. Adaptive wavefront correction structured illumination holographic tomography. Scientific Reports. 2019;**9**: 10489

[25] Rosen J, Vijayakumar A, Kumar M, Rai MR, Kelner R, Kashter Y, et al. Recent advances in self-interference incoherent digital holography. Advances in Optics and Photonics. 2019;**11**:1-66

[26] Wan Y, Liu C, Ma T, Qin Y, Lv S. Incoherent coded aperture correlation holographic imaging with fast adaptive and noise-suppressed reconstruction. Optics Express. 2021;**29**:8064-8075

[27] Vijayakumar A, Rosen J. Spectrum and space resolved 4D imaging by coded aperture correlation holography (COACH) with diffractive objective lens. Optics Letters. 2017;**42**:947-950

Chapter 7

Coded Aperture Correlation Holography (COACH) - A Research Journey from 3D Incoherent Optical Imaging to Quantitative Phase Imaging

Joseph Rosen, Angika Bulbul, Nathaniel Hai and Mani R. Rai

Abstract

Coded aperture correlation holography (COACH) combines incoherent digital holography with coded aperture imaging. COACH is also a method to record incoherent digital holograms of three-dimensional object scenes. Still, COACH can be used for several other incoherent and coherent optical applications. In this chapter, we survey the prime landmarks on the topic of COACH from two major perspectives: architectures and applications of the various systems. We explore the main configurations of hologram recorders in the COACH systems. For each design, we describe some of the recent implementations of these recorders in optical imaging. We conclude the chapter with general ideas on this technology.

Keywords: incoherent holography, digital holography, Fresnel incoherent correlation holography, digital holographic microscopy, phase-shifting interferometry

1. Introduction

Imaging by optical waves has been known in the technology world for centuries [1]. For most of this time, imaging has been direct in the sense that images recorded on the eye retina, photographic film, or electronic sensor have been replicas of the observed scenes. However, the computing revolution of the second half of the twentieth century has opened many possibilities for indirect rather than direct imaging. In indirect imaging, a modified version of the observed scene is transferred from the image sensor to the computer to process and reconstruct the image of the original scene. One of the indirect imaging methods is coded aperture imaging, proposed in the sixties for X-ray imaging [2–4] and later adapted to the visible light using coded phase-masks [5] instead of an array of randomly distributed pinholes used in X-ray imaging [4].

Digital holography [6–8] can also be classified as indirect imaging, although it is special in the sense that the pattern recorded by the image sensor is an interference pattern between two light beams. At least one of the beams originates from the object. However, in the case of incoherent digital holography by self-interference, both interfering beams originate from the object [7]. In 2016, the two different concepts of coded phase-aperture imaging and incoherent digital holography were combined into a new indirect imaging method, dubbed coded aperture correlation holography (COACH) [9]. COACH merges the merits of these two different imaging modalities and enables three-dimensional (3D) imaging with interesting and unexpected features. More specifically, COACH is an electro-optical technique to record digital holograms of two- and three-dimensional scenes, where at least part of the light from the object passes through a coded phase-mask. COACH was initially proposed as an additional method to record incoherent digital holograms without scanning and evolved in several different directions. The COACH concept was inspired by several previous methods and systems [2–5, 10–12] and has already stimulated several studies since then [13–33]; some of them are mentioned in the following. This chapter provides an overview of research activities in the technology of COACH done by several researchers in the field.

About a year after the invention of COACH, a simpler version of it was proposed. This version operates without two-wave interference and can demonstrate some applications, such as 3D imaging. The modified version was called interferenceless COACH (I-COACH) [15]. Usually, the I-COACH is preferred whenever an application can be performed by both COACH and I-COACH with the same quality. This rule of thumb is reasonable because the calibration of a single wave system, such as I-COACH, is simpler, and its noise immunity is higher than that of an incoherent interferometer, such as COACH. However, not all the applications successfully implemented by COACH can be executed by I-COACH, and examples are given in the following.

COACH and I-COACH were initially invented for 3D imaging of incoherently illuminated scenes. Recently, the concept of coherent COACH with and without two-wave interference has been examined [34–36]. A central application of coherent digital holography is quantitative phase imaging (QPI) [37], and hence in the following, we review different ways to implement QPI using COACH [35, 36].

This review consists of six main sections. The development of incoherent COACH and I-COACH architectures, with different modalities and characteristics, are reviewed in the following two sections. We describe the coherently illuminated I-COACH and COACH techniques in the fourth and fifth sections, respectively. The concluding section summarizes the review.

2. Incoherent COACH

Incoherent COACH belongs to the family of self-interference digital holography systems [38]. The general optical configuration of these systems is shown in **Figure 1**. The flow of information starts from the light emitted from each object point in the upper part of **Figure 1**. The light propagates toward a beam splitting unit and is split into two waves. Each wave is modulated differently by a modulation component. The two waves originate from the same object point and hence are mutually coherent, although the light emitted, or reflected, from the object is spatially incoherent. Therefore because of the mutual coherence, the two waves with different wavefronts interfere at the sensor plane. The image sensor accumulates the entire interference patterns of all the input points to an incoherent hologram. A single hologram, or

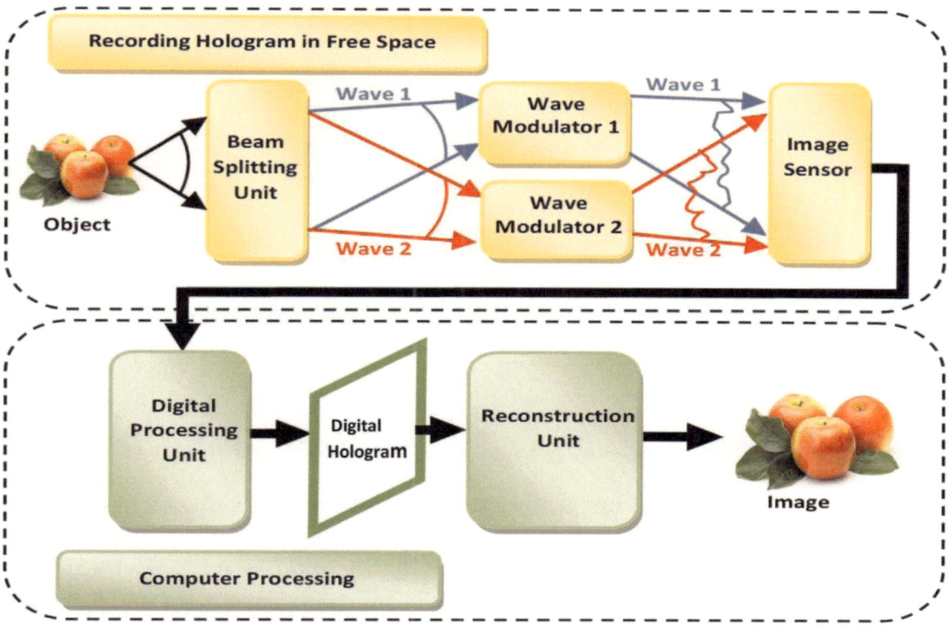

Figure 1.
Recording and reconstruction of holograms in a general self-interference digital holography system.

several acquired holograms, are introduced into a digital computer, where the various operations of the digital processor are schematically shown in the lower part of **Figure 1**. In the case of several holograms, they are superposed into a single digital hologram. Finally, the image of the object is reconstructed from the processed hologram by an appropriate numerical algorithm.

COACH was proposed as a generalized case of Fresnel incoherent correlation holography (FINCH) [10–12], a well-known technique of recording holograms, which also belongs to the self-interference systems. In FINCH, a quadratic phase-mask modulates at least one of the two waves. In COACH, on the other hand, the quadratic phase-mask of FINCH is substituted by a diffractive chaotic phase-aperture. The initial goal of COACH was like FINCH, that is, to acquire a hologram of the 3D observed scene illuminated by quasi-monochromatic spatially incoherent light. COACH's optical scheme is depicted in **Figure 2**. The light from an object is split into two beams, and only one of the object beams is modulated by the chaotic mask termed coded phase-mask (CPM). The modulated beam is coherently interfered with the unmodulated object beam due to their common origin. Because COACH is on-axis system, it needs a phase-shifting procedure and complex hologram synthesis [39]. That means that three holograms are recorded, each of which with the CPM multiplied by a different phase-constant. The three holograms are superposed digitally in the computer such that the result is a complex-valued hologram. This digital hologram is reconstructed into a single image without the twin image and the bias term.

Unlike other well-known incoherent hologram recorders, such as FINCH [10–12, 40] and Michelson-interferometer-based incoherent holographic systems [41–44], COACH does not have a defined image plane where the wavefront can numerically propagate from the hologram to the reconstruction plane. Hence, COACH has different recording and reconstruction procedures. In other words, COACH consists of a two-step recording procedure: a one-time calibration and then imaging. In the

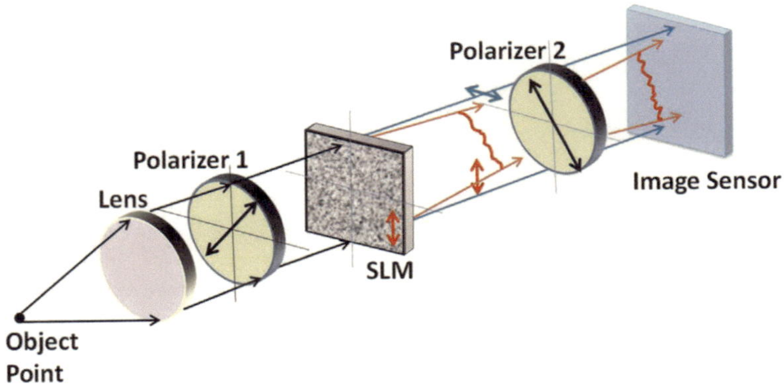

Figure 2.
Schematic diagram of coded aperture correlation holography (COACH). SLM - spatial light modulator.

initial stage of the calibration, one illuminates a moving pinhole along the optical axis, and the image sensor records a point spread hologram (PSH) for every axial location of the pinhole. The set of PSHs is accumulated in a library for later use in the imaging stage. Following the calibration process, an object hologram is recorded under the same restrictions and with the coded apertures as the PSH acquisition. The 3D image of the observed scene is reconstructed by a two-dimensional (2D) cross-correlation between the object hologram and the corresponding elements of the PSH library.

Although FINCH influenced the COACH structure, COACH has different features than FINCH. The image reconstruction has been modified to 2D cross-correlations with guidestar responses instead of the Fresnel back-propagation of FINCH [10–12, 40]. Compared to FINCH, COACH has better axial resolution but worse lateral resolution [9, 45]. However, the main difference is that COACH can do the same holographic 3D imaging without two-wave interference [15]. Nevertheless, several applications can only be performed by a version of the original COACH with two-wave interference. One of such applications is a one-channel-at-a-time incoherent synthetic aperture imager [46], summarized next.

2.1 One-channel-at-time incoherent synthetic aperture

An interesting application for COACH is incoherent imaging with synthetic aperture (SA). SA is a familiar super-resolution method and a conventional technique in astronomy to accomplish image resolution beyond the diffraction limit dictated by the physical aperture [47] of the telescope. Since its invention a century ago [48], incoherent SA imaging was usually realized by at least two optical channels operating simultaneously. The wave interference between two incoming light beams, both originated from the same object, was recorded over time from several viewpoints within the SA region. Then, the interference intensity patterns were processed to produce an image of the object with a resolution equivalent to complete SA [22, 48]. A single-channel SA is possible for cases of imaging systems with coherent light [49], but astronomical imaging is usually done with incoherent light sources. A solution to this double-channel problem of SA incoherent imaging is the lately proposed incoherent single-channel SA technique termed one-channel-at-time incoherent synthetic aperture imager (OCTISAI) [46].

As in many other COACH systems, the CPM of OCTISAI is synthesized using a modified version of the Gerchberg-Saxton algorithm (GSA) [50]. Then, the CPM is divided into N (in the following example $N = 64$) equal parts for the SA implementation. The optical setup of the OCTISAI experiment is shown in **Figure 3** and described next. The system is first calibrated by collimating the light diffracted from a pinhole, where the collimating lens L_1 mimics the far-field imaging condition. A polarizer P_1 polarizes the collimated light to be oriented at 45^0 regarding the active orientation of a spatial light modulator (SLM). The SLM is used as the display on which the CPMs of OCTISAI and all other systems in this chapter are displayed. Only a partial area of the SLM is used at a time, and all other parts are activated in a raster scan mode. Because of the polarization angle, the light is split into two orthogonal linear polarizations beyond the SLM. The CPM modulates one polarized wave, and the other wave passes the SLM without any change. Beyond the polarizer P_2, also oriented at 45^0 to the SLM's active axis, both beams have the same orientation enabling to record a pattern of interference between the two beams. The interference pattern between the modulated and unmodulated beams is captured by the image sensor. Three phase-shifted PSHs for the input point object (pinhole) are recorded for every partial aperture at each position in the SA region. Then, three phase-shifted object holograms are captured for the input object with the same phase-apertures as before in the calibration. Next, using the digital computation capabilities, the entire PSH parts are stitched together into one synthetic PSH. The parts of the object hologram are also processed into one synthetic object hologram by a similar procedure. The final image with the enhanced resolution is obtained by a 2D cross-correlation between the two synthetic holograms.

The complete experiment of OCTISAI is extensively described in [46], and here we briefly describe only the main results. In the experiment, a collection of PSHs was produced using three CPMs, each having a phase-shift $\exp(i\theta_j)$, where $\theta_{1,2,3} = 0°, 120°$, and $240°$. A pinhole of 25 μm diameter was positioned in the input. After the PSH creation, group 3, element 1 of the United States Air Force (USAF) negative resolution chart, replaced the pinhole. We recorded the three object holograms with the same three CPMs used for the PSHs. The synthetic object holograms and PSHs were produced by stitching respective partial holograms and superimposing corresponding

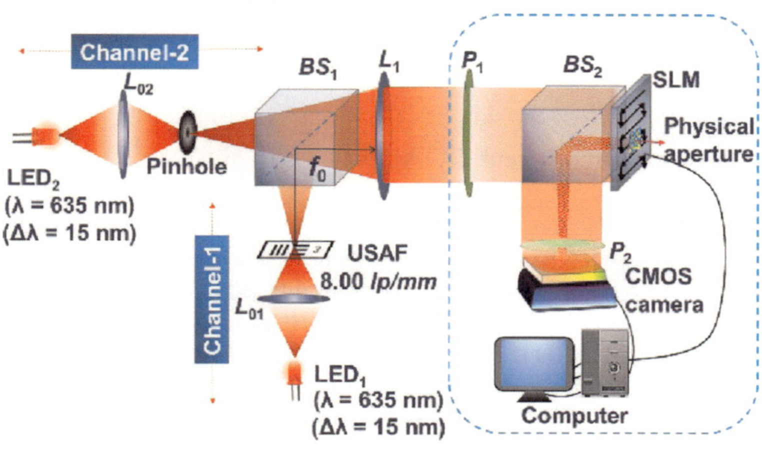

Figure 3.
The tabletop experimental setup for one-channel-at-time incoherent synthetic aperture imager (OCTISAI) inside the blue rectangle, BS_1 *and* BS_2 *– beamsplitters, CMOS camera - Complementary metal-oxide-semiconductor camera,* L_{01}, L_{02}, *and* L_1 *- refractive lenses, LED₁ and LED₂ - identical light-emitting diodes,* P_1 *and* P_2 *– polarizers, SLM - spatial light modulator, and USAF - United States Air Force resolution target. Adapted from [46].*

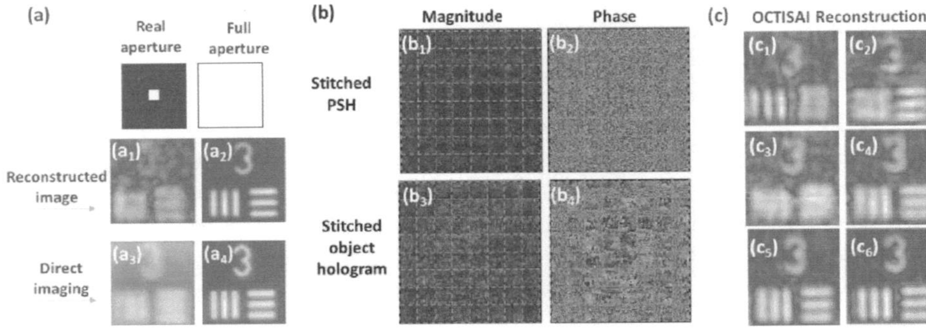

Figure 4.
(a_1, a_2) COACH reconstructed images and (a_3, a_4) direct images of limited aperture (a_1, a_3) and full aperture (a_2, a_4), magnitude and phase of (b_1, b_2) PSH and (b_3, b_4) object holograms of the complete SA, (c_1-c_6) reconstructed images after stitching of (c_1) 8 central horizontal sub-holograms, (c_2) eight central vertical sub-holograms, (c_3) 2 × 2, (c_4) 4 × 4, (c_5) 6 × 6 central sub-holograms, and (c_6) full 64 (8 × 8) sub-holograms. Adapted from [46].

synthetic three-intensity responses. Finally, the object hologram was cross-correlated with the phase-only filtered version of the synthetic PSH. The outcome of this cross-correlation is the final reconstructed image. The COACH images related to the partial and complete apertures are shown in **Figures 4(a_1)** and **(a_2)**, respectively. For comparison, **Figures 4(a_3)** and **(a_4)** show the corresponding images of direct imaging with a setup of a single lens and similar numerical apertures. The stitched holograms after the superposition are shown in **Figure 4(b)**. **Figure 4(c)** presents the reconstructed images for OCTISAI with various area sizes of the SA holograms. **Figures 4(c_1)** and **(c_2)** are produced using the central eight, horizontally [**4(c_1)**] and vertically, [**4(c_2)**] stitched partial holograms, respectively. **Figures 4(c_3)-(c_6)** show the reconstruction results with 2 × 2, 4 × 4, and 6 × 6 central sub-holograms, and the entire 64 sub-holograms. The resolution enhancement by raising the number of stitched partial holograms is demonstrated. Comparing **Figure 4(c_6)** with **Figures 4 (a_1)** and **(a_3)**, one can conclude that OCTISAI's images have higher resolution than the images taken with a limited aperture in both techniques of COACH and direct imaging.

3. Interferenceless incoherent COACH

As mentioned above, interferenceless coded aperture correlation holography (I-COACH) was published in 2017 [15] as a simpler configuration of the earlier proposed COACH [9]. Both systems spatially modify incoherent light by chaotic phase-masks. However, unlike COACH, I-COACH records holograms without two-beam interference. I-COACH is an incoherent 3D imaging method in which the image is digitally obtained by numerical 2D cross-correlation between the hologram of the object and the library of PSHs. The PSHs are recorded once in the calibration mode of the system, before the imaging stage, as shown in **Figure 5**. The same chaotic CPMs modulate the light waves in both the calibration and imaging stages. The modulated light is recorded by a digital camera after propagating in the free space. I-COACH system without two-beam interference can produce similar results as COACH because the intensity point-response of I-COACH on the sensor plane is highly sensitive to the

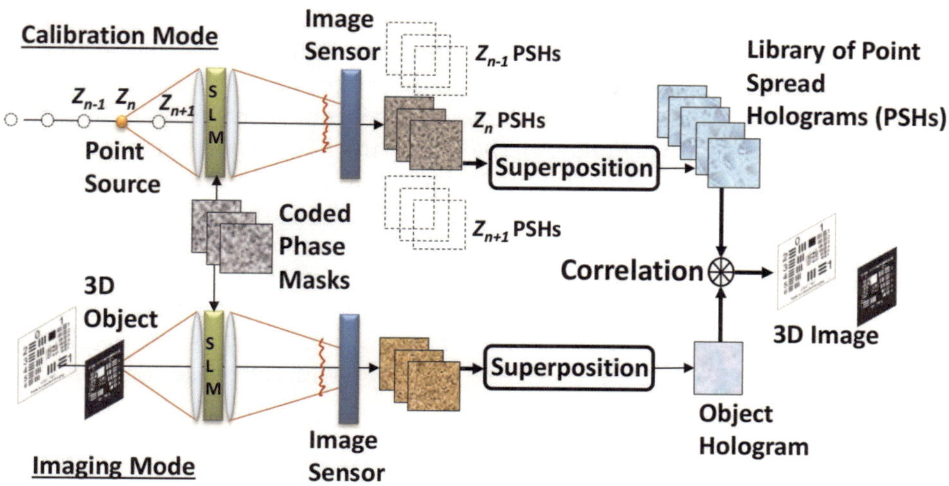

Figure 5.
Schematic diagram of interferenceless coded aperture correlation holography (I-COACH). The upper scheme refers to the calibration mode, whereas the lower describes the imaging mode.

axial location of the input point. Mathematically, the high sensitivity means that the cross-correlation between two intensity responses for two points located at two different axial positions is much smaller than the autocorrelation of each response [15]. Thus, the entire object points can be reconstructed in the 3D image space using 2D cross-correlations between a multi-point object and the library PSHs. The early configuration of I-COACH [15] has been developed into different systems with various architectures and with a variety of algorithms [16–33], each with strengths and weaknesses. The typical design of I-COACH is shown in **Figure 5**, where the same physical setup is schematically depicted in two modes of operation. The upper scheme shows the calibration process, in which the system collects a library of PSHs acquired for an object point positioned at different axial locations. When the library is completed, the same setup works in the imaging mode shown in the lower part of **Figure 5**. An incoherently illuminated 3D object replaces the single point in the system's input. The object intensity response recorded by the sensor is 2D cross-correlated with each PSH of the library. The assembly of cross-correlation results is the desired reconstructed 3D image. This general scheme describes most I-COACH types and has been the basis for developments that have evolved since 2017 [16–31]; one example of an I-COACH application is depth-of-field engineering, briefly described next.

3.1 Depth-of-field engineering

Long depth-of-field (DOF) in imaging systems has been important for many applications [51]. Generally, the DOF is dictated by the numerical aperture of the optical system. Reducing the numerical aperture extends the DOF, but it also unfavorably decreases the lateral image resolution of the system. Several methods have been advanced to extend the DOF of the optical system [51–59] with a minimal resolution decrease. Still, the complicated experimental and computational requirements have stimulated a search for simpler methods. This subsection reviews a new technique proposed first in [60] to engineer the DOF of imaging systems. DOF engineering is done by integrating radial quartic phase-functions (RQPFs) [61, 62] into the incoherent I-COACH shown schematically in **Figure 6**. The phase-mask displayed on the SLM of **Figure 6** is a fusion of three separate phase-masks. The first is the chaotic CPM generated by GSA with the constraints of sparse dots on the camera plane [63] and a constant magnitude on the SLM plane. The second element is a positive diffractive lens used to fulfill the 2D Fourier relations of the GSA between the planes of the SLM and camera. The focal length of the diffractive lens f is determined such that each object is imaged on the camera. In other words, for the distance object-SLM (d_{OS}) and SLM-camera (d_{SC}), the three lengths satisfy the imaging equation $1/f = (1/d_{OS}) + (1/d_{SC})$. The distance object-SLM is chosen as the distance from the center of the object space to the SLM. The third mask is the above-mentioned RQPF implemented to extend the DOF as desired. The RQPF with the phase-function $\exp\left[i2\pi(r/p)^4\right]$ stretches the DOF of the sparse dots created by the CPM, where p is the modulation parameter controlling the length of the DOF, and r is the radial coordinate on the SLM plane. Near the back focal point of the diffractive lens on the camera plane, the RQPF generates sword beams with an almost constant intensity along a controlled propagation distance and a relatively narrow beam-like shape in any transverse plane [61, 62]. The 3D location and the length of the DOF can be determined by changing the parameters of the RQPF and the focal length of the lens. Multiplexing various three-somes of phase-masks (diffractive lens, CPM, and RQPF) with different modulation parameters can create various focusing curves. For instance, an imaging system that can image objects in two non-connected sub-volumes in the object space. In this example, the entire objects inside these sub-volumes remain in focus, while the images outside these sub-volumes are blurred and seen out-of-focus. The unusual DOF enables to image targets in specific sub-volumes simultaneously (or successively), whereas objects in other sub-volumes are blurred. Moreover, the engineered DOF allows to transversely shift an image from one volume relative to another image from another volume. Mutual transverse shifts of sub-volumes can avoid overlap between images when one object is behind or in front of another object.

Back to **Figure 6**, the incoherent light source critically illuminates the observed 3D scene using a lens L_0. In this scheme, an object volume is defined as the volume along z for which the DOF is extended, such that each object inside the volume produces an in-focus image in the output. Off-axis sub-volumes in **Figure 6** indicate images of on-axis objects that are reconstructed out of the z-axis in the output due to an additional linear phase-mask attached to the other three-phase-masks (diffractive lens, CPM,

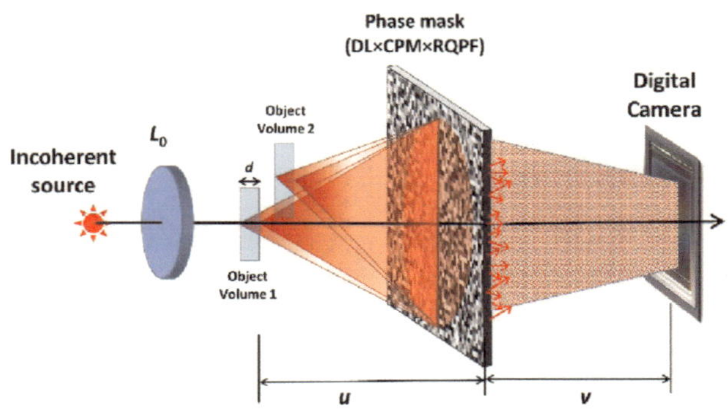

Figure 6.
Optical scheme of the depth-of-field engineering system. DL - diffractive lens, CPM - coded phase-mask, RQPF - radial quartic phase-function. Adapted from [60].

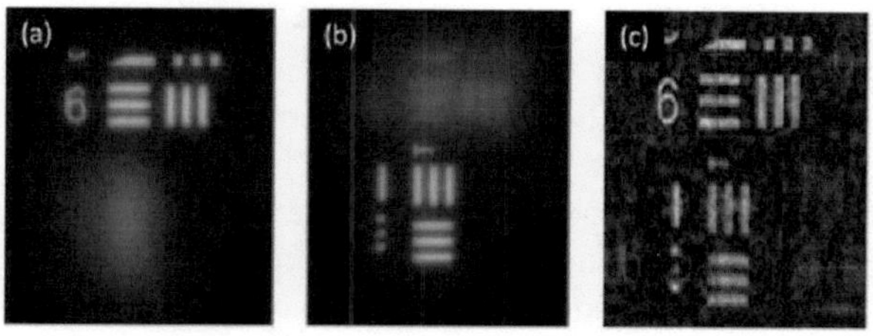

Figure 7.
(a, b) Direct images of the objects with two different lenses and (c) Reconstructed images from a single hologram using depth-of-field engineering. Adapted from [60].

and RQPF). The light emitted from the object scene is modulated by the combination of the phase-masks displayed on the SLM. For any point inside the object volume, the intensity recorded by the camera is distributed in the form of chaotically sparse dots. Like other I-COACH schemes, this dot pattern is used as the PSH, which reconstructs the image of any object by cross-correlation with the object hologram. In previous demonstrations of I-COACH [15, 16], the PSH has been recorded experimentally by illuminating a pinhole positioned at the system input. However, in [60], the PSH is digitally computed based on the known experimental parameters. The object reconstruction is done by a nonlinear cross-correlation [21] between the computed PSH and the object hologram.

Next, we show results of only a single object volume. In other words, the following I-COACH has extended DOF compared to direct imaging with the same numerical aperture. More complicated examples of DOF engineering can be found in Ref. [60]. The proposed technique is verified by an experimental setup like the scheme in **Figure 6**. Unlike typical I-COACH systems, two targets are positioned in two separate channels of the experimental setup such that they are located at each end of the object volume [60]. Two LEDs with refractive lenses separately illuminated the object in each channel. Both objects are from the USAF transmission resolution chart. In one of the channels, the object is element 6 of Group 2, and in the other channel, element 1 of Group 3 is used as the object. The targets are located at distances of 24 and 26 cm from the SLM, respectively. Light from the two targets was combined by a beamsplitter and projected on the SLM. The gap between the SLM and the camera was 22 cm. Direct images of the objects [shown in **Figures 7(a)** and **(b)**] on the camera plane were achieved by displaying only a single diffractive lens on the SLM with the focal length that satisfies the imaging equation, each for a different object in its own depth. For the I-COACH system, the PSH was computed using the optimal CPM that yielded ten randomly distributed dots on the camera plane. The reconstructed images of I-COACH are shown in **Figure 7(c)**. In the case of direct imaging, it is clear from **Figures 7(a)** and **(b)** that the axial gap between the targets was too large to focus both targets at the same time. On the other hand, in the case of the I-COACH with the engineered DOF of 3 cm, the reconstructed images of **Figure 7(c)** show that both targets are in focus without any resolution decrease.

4. Interferenceless coherent COACH

Optical recording of digital holograms with coherent light traditionally involves interference between object and reference waves, complicating the image acquisition

[39]. With the coherent I-COACH, the concept of the coded aperture is adapted from the area of incoherent holography to record digital holograms of three-dimensional coherently illuminated scenes without two-wave interference or any kind of scanning. In addition to the obvious advantages of combining interferenceless holographic systems with coherent light, the proposed method enables relatively rapid image acquisition made possible by its inherent high signal-to-noise ratio (SNR). In [34], the I-COACH method was implemented for generating coherent holograms without interference between reference and object waves. The technique, called interferenceless coherent coded aperture correlation holography (IC-COACH), creates a bi-polar digital hologram of a 3D scene from two camera shots where the scene is illuminated by coherent laser light. The 3D image of the observed scene is reconstructed from the hologram by a deconvolution-like process.

To understand the evolution from incoherent to coherent I-COACH, we briefly summarize the principles of incoherent I-COACH first. Generally, an incoherent I-COACH hologram denotes a 2D function containing an image of a 3D scene, such that the image can be digitally reconstructed from the 2D function. Mathematically, the 2D digital hologram of a 3D object is given by,

$$H_{OBJ}(\bar{r}) = \int I_{OBJ}(\bar{r};z) * p(\bar{r};z)dz, \tag{1}$$

where $*$ is 2D convolution at each z plane, $\bar{r} = (x, y)$ are the transverse coordinates, and $p(\bar{r};z)$ is the PSH of the recording system, which can be a general complex [15] or bi-polar real [16] function. The library of PSHs is a priori acquired in a calibration process with a guidestar, in which each $p(\bar{r};z_j)$ from the PSH library is computed as a response to an object point at z_j. Once the library is ready, and an object hologram is recorded, 2D cross-correlations between the object hologram and each PSH from the library reconstruct each z_j plane of the 3D image. This computation process is based on the linearity of incoherent optical systems with 2D intensity signals expressed by the following familiar convolution,

$$I_{Out}(\bar{r}) = I_{In}(\bar{r}) * |h(\bar{r})|^2, \tag{2}$$

where $h(\bar{r})$ is the coherent point spread function of the optical system. $I_{In}(\bar{r})$ and $I_{Out}(\bar{r})$ are the system input and output intensities, respectively. In contrast to incoherent, coherent optical systems are linear in corresponding to 2D complex amplitudes, and they obey the relation,

$$I_{Out}(\bar{r}) = |A_{In}(\bar{r}) * h(\bar{r})|^2, \tag{3}$$

where $A_{In}(\bar{r})$ is the input 2D complex amplitude fulfilling the equation $I_{In}(\bar{r}) = |A_{In}(\bar{r})|^2$. Because of the nonlinearity of Eq. (3), the implementation of the I-COACH concept in the coherent system is possible only for special cases. Hence, the coherent processor should be adapted in such a way that can satisfy the relation,

$$|A_{In}(\bar{r}) * h(\bar{r})|^2 \approx |A_{In}(\bar{r})|^2 * q(\bar{r}), \tag{4}$$

where $A_{In}(\bar{r})$ represents a broad set of input objects, and we assume that $h(\bar{r})$ and $q(\bar{r})$ are nontrivial functions. Eq. (4) is satisfied if $h(\bar{r})$ is a set of points distributed over the camera plane such that the gap between any two points is wider than the size of $A_{In}(\bar{r})$ [34]. The pattern of the random points on the camera plane is considered the system's PSH. Hence, the CPM replicates the object to a set of the same images chaotically distributed over part of the camera plane. Such CPMs are created by a modified version of GSA [50], in which iterative transformations between the CPM's plane and the spectral plane are done with suitable constraints at each plane. The constraint at the CPM's plane is a constant magnitude distribution because the CPM is displayed on a phase-only SLM. In the spectral plane, which is also the camera plane, the intensity is constrained to be in a shape of randomly distributed dots over all or part of the plane.

The optical configuration of the IC-COACH system of [34] shown in **Figure 8** is based on the classical 4-f spatial filtering system, with the SLM positioned at the Fourier domain and the camera at the image plane. In this setup, the spatial spectrum of the object is modulated by the CPM displayed on the SLM. The CPM was produced by the GSA to duplicate the input object over an ensemble of points randomly distributed at the camera plane. Two different chaotic CPMs are sequentially displayed on the SLM to create two different random sets of replications of the object. These two sets are subtracted from each other to produce a bi-polar object hologram. The ability of IC-COACH to image multi-plane objects is accomplished by multiplexing on the SLM, a few independent CPMs, each of which yields an in-focus different set of dots on a different transverse plane. Each ensemble of out-of-focus dots becomes focused on the camera plane for a point object positioned at the corresponding transverse plane. Lastly, the desired transverse image of the observed 3D scene is reconstructed by cross-correlation between the object hologram and the corresponding PSH. **Figure 9** shows the reconstructed images for two different planes and two different gaps between the object planes, forming two different multi-plane scenes.

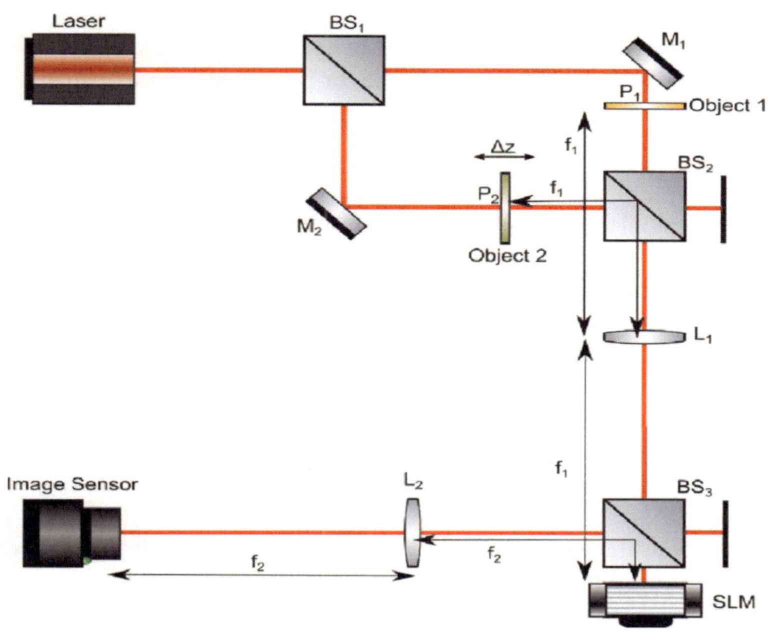

Figure 8.
Experimental setup of interferenceless coherent coded aperture correlation holography (IC-COACH) with two independent illumination channels. $BS_{1,2,3}$: Beamsplitters, $M_{1,2}$: mirrors, and SLM: spatial light modulator. Adapted from [34].

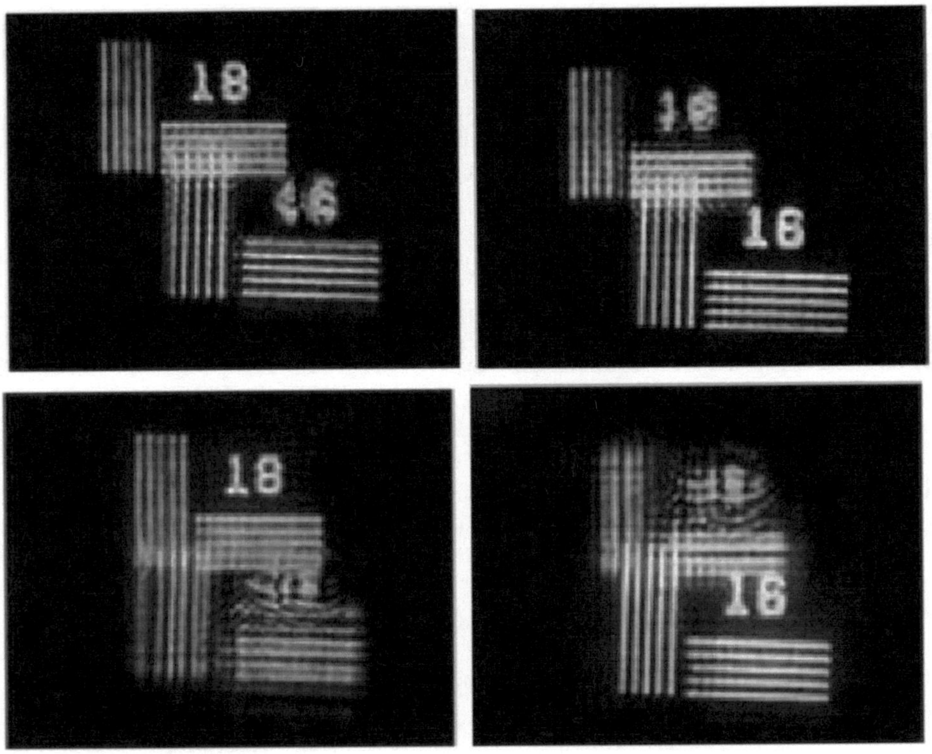

Figure 9.
Image reconstructions of different object planes obtained by a correlation with the corresponding PSH in IC-COACH. Adapted from [34].

Experimental demonstration for imaging diffusely reflective objects also appears in [34], making the IC-COACH system suitable for processing speckle images obtained by coherent illumination.

5. Coherent COACH with two-wave interference

IC-COACH described in the previous section is an adaptation of the incoherent I-COACH to the case of coherent illumination. This system is capable of imaging 3D scenes holographically, but it cannot do phase imaging of any kind. To enable QPI of transparent objects, COACH has been integrated with a Mach-Zehnder interferometer [35, 36]. QPI, in general, is done by capturing the wavefront passing through thin transparent objects and converting it to an optical thickness map of the examined objects. This method is useful for many applications, including label-free biological cell imaging [64, 65] and nondestructive quality tests [66, 67].

Like the previous demonstrations of I-COACH [34, 63], the image of the observed object is projected to randomly and sparsely distributed replications over the camera plane. As before, the replications are obtained by a pseudorandom CPM synthesized by modified GSA [50]. The CPM is displayed on a phase SLM in the configuration of the coherent sparse COACH (CS-COACH) shown in **Figure 10**. The image sensor records the interference pattern between the waves of the image replications and of a reference tilted plane wave as follows:

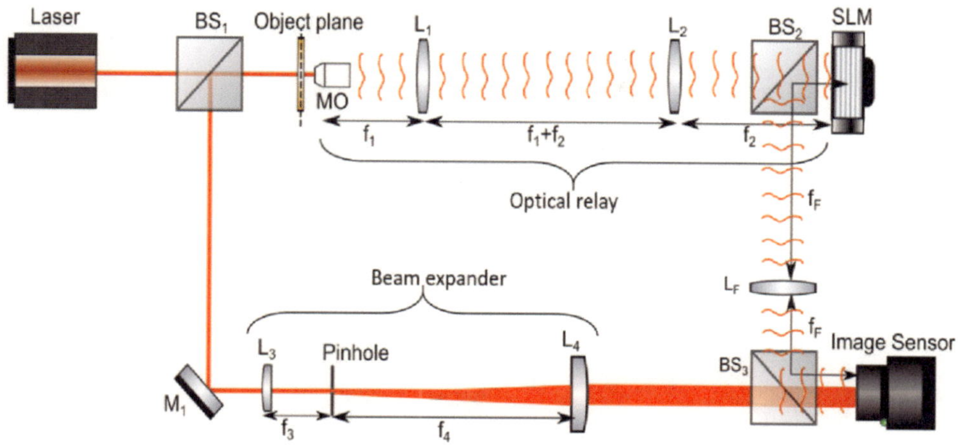

Figure 10.
Optical configuration of coherent sparse COACH. MO: Microscope objective, $BS_{1,2,3}$: Beamsplitters, M_1: Mirror, and SLM: Spatial light modulator. Adapted from [36].

$$I(x,y) = \left| O(x,y)\exp[j\phi(x,y)] * \sum_{i=1}^{N}\delta(x-x_i, y-y_i) + R\cdot \exp\left[j\frac{2\pi}{\lambda}(x\sin\theta_x + y\sin\theta_y)\right]\right|^2, \tag{5}$$

where $O(x, y)$ is the object amplitude, and $\phi(x, y)$ is its phase, R is the reference wave amplitude, λ is the illumination wavelength, N is the number of image replications, (x_i, y_i) are the displacement values of the i-th replica from the camera origin, and (θ_x, θ_y) are the angles between the object and reference waves in the x-z and y-z planes, respectively. It should be noted that off-axis holography is used to acquire holograms by a single camera shot. A digital filtering process in the spatial frequency domain eliminates the bias term and the twin image from the recorded intensity pattern. The processed object hologram is:

$$H_{OBJ}(x,y) = R^* \cdot O(x,y)\exp[j\phi(x,y)] * \sum_{i=1}^{N}\delta(x-x_i, y-y_i). \tag{6}$$

This hologram includes several randomly distributed replications of the object over the image plane. Like the procedure explained in [34, 63], the reconstruction of the object's complex amplitude is performed by 2D cross-correlation between the object hologram H_{OBJ} and the PSHs. **Figures 11(a)** and **(b)** show the phase-image of polystyrene microspheres (FocalCheck, 6 μm diameter) with the proposed CS-COACH method. For comparison purposes, the phase-images extracted from a regular Mach-Zehnder interferometer using conventional off-axis holography are shown in **Figures 11(c)** and **(d)**. It is apparent that the image of CS-COACH has higher SNR than the conventional technique. This advantage is attributed to the averaging procedure over several replications accompanied by the reconstruction using cross-correlation. Noise reduction is one of the several advantages of CS-COACH in comparison to open-aperture equivalent systems. Another advantage presented in Ref. [36] is extending the field-of-view (FOV) of the imaging system. Extended FOV realized with the same focal length of the microscope objective and without sacrificing the image resolution is an important advantage in microscopy.

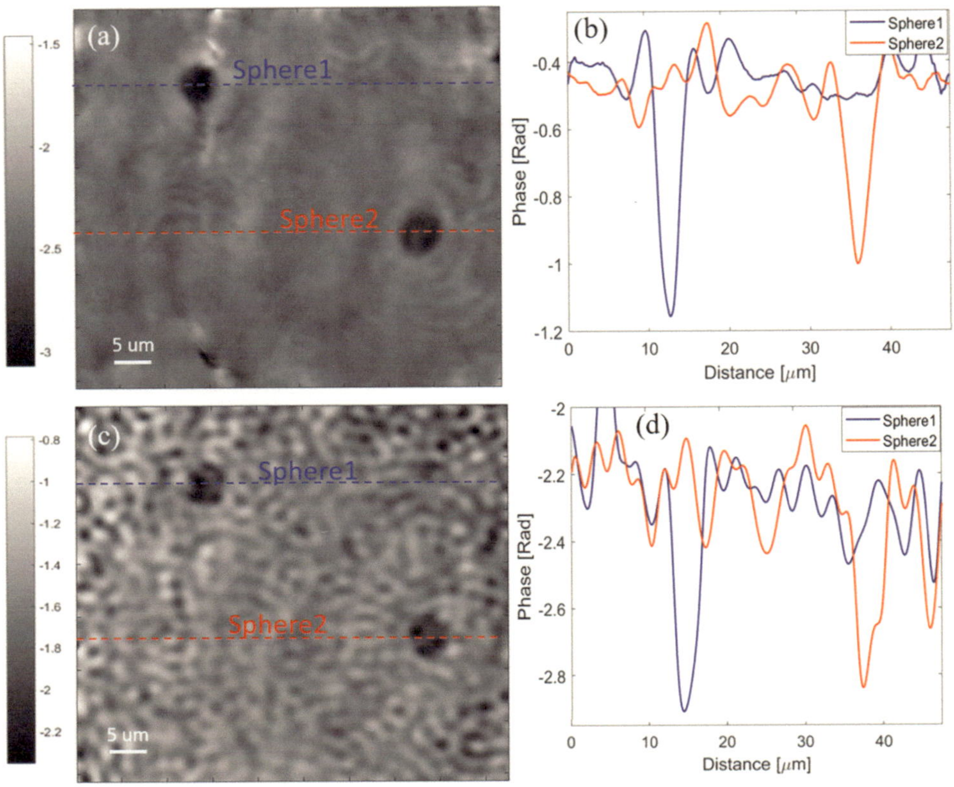

Figure 11.
Reconstructed phase images of polystyrene microspheres and the phase cross-sections using (a)-(b) CS-COACH and (c)-(d) conventional off-axis holography. Units of the left panel color bars are radian. Adapted from [36].

6. Discussion and summary

For all its forms, COACH is a rapidly evolving technology because of the desire to enhance the resulting images and due to the new applications supported by the method. Any technology of imaging is expected to be as quickly as possible with the least camera shots. While the early version of I-COACH [15] operated with three camera shots taken under three independent CPMs, the number of shots and CPMs was decreased to two in [16]. By multiplexing two CPMs in space instead of time as before [15, 16], a single-camera shot was applied in Ref. [18]. I-COACH [19] and CS-COACH [36] with extended FOV were demonstrated by calibrating the systems with extended PSHs beyond the conventional FOV. The numerical reconstruction procedure was changed in [21] by substituting the ordinary linear cross-correlation with new nonlinear cross-correlation optimized to yield a correlation distribution with the lowest entropy. A different nonlinear cross-correlation with other cost-function in the optimization process was employed in [26, 30]. Some of the noise on the resulting images in the early versions [15, 16] appeared because of the low-intensity level per pixel of the PSH on the sensor plane. This difficulty was treated in [63] by imposing a PSH with the structure of sparse dots of light intensity distributed chaotically inside a limited region. The same problem was differently solved in [30] with PSHs of a ring shape. The electro-optical calibration in the upper part of **Figure 5** was changed by a pure digital technique of synthesizing the library of PSHs in the computer [68]. Lateral resolution can be considered one of the holy grails of optical imaging. Improving the lateral resolution by I-COACH has been treated in [23, 45, 69, 70] by different approaches. Usually, I-COACH's lateral and axial resolutions are the same as those of lens-based imaging systems with the same numerical aperture. The methods of [23, 45, 69, 70] improve the lateral resolution beyond the diffraction limit enforced by the finite numerical aperture of optical systems. In [23], resolution-enhanced images of the observed objects are reconstructed by a nonlinear cross-correlation between object holograms and PSHs. In [69, 70], a CPM displayed on the SLM was introduced between the object and the input aperture of a regular lens-based imager. Thus, the effective numerical aperture was increased beyond the characteristic numerical aperture of the imaging system. The effective numerical aperture and the improved resolution limits can be tuned by altering the scattering degree of CPMs [69, 70]. Other applications of COACH and I-COACH and their context in a frame of systems with dynamic diffractive phase-apertures are reviewed in [17, 71, 72].

To conclude this review, we note that COACH for all its modes is based on the extension of the resources available for imaging in a few ways. First, the real-valued aperture function of ordinary direct imaging is replaced with the complex-valued aperture function of COACH. Second, the COACH aperture is modified over time in the multiple-shot versions. Finally, an additional stage of digital processing is integrated with the optical system. These additional resources add to the COACH system new capabilities and unique features. Even though I-COACH is a simpler form of COACH and thus is preferred for many 3D imaging projects, there are some unusual applications in which COACH with two-beam interference is required. Incoherent synthetic aperture imagers [20, 22], the hybrid FINCH-COACH system [45], and quantitative phase-imagers [35, 36] are characteristic examples of systems that two-beam interference is necessary for their operations. However, other applications can be implemented successfully on I-COACH; some are presented herein others might be proposed in the future.

Author details

Joseph Rosen*, Angika Bulbul, Nathaniel Hai and Mani R. Rai
School of Electrical and Computer Engineering, Ben-Gurion University of the Negev, Beer-Sheva, Israel

*Address all correspondence to: rosenj@bgu.ac.il

References

[1] Mait JN, Euliss GW, Athale RA. Computational imaging. Advances in Optics and Photonics. 2018;**10**:409-483

[2] Ables JG. Fourier transform photography: A new method for X-ray astronomy. Proceedings of the Astronomical Society of Australia. 1968; **1**:172-173

[3] Dicke RH. Scatter-hole cameras for X-rays and gamma rays. Astrophysics Journal. 1968;**153**:L101

[4] Fenimore EE, Cannon TM. Coded aperture imaging with uniformly redundant arrays. Applied Optics. 1978; **17**:337-347

[5] Chi W, George N. Optical imaging with phase-coded aperture. Optics Express. 2011;**19**:4294-4300

[6] Goodman JW, Lawrence RW. Digital image formation from electronically detected holograms. Applied Physics Letter. 1967;**11**:77-79

[7] Liu J-P, Tahara T, Hayasaki Y, Poon T-C. Incoherent digital holography: A review. Applied Sciences. 2018;**8**:143

[8] Javidi B, Carnicer A, et al. Roadmap on digital holography. Optics Express. 2021;**29**:35078-35118

[9] Vijayakumar A, Kashter Y, Kelner R, Rosen J. Coded aperture correlation holography—a new type of incoherent digital holograms. Optics Express. 2016; **24**:12430-12441

[10] Rosen J, Brooker G. Digital spatially incoherent Fresnel holography. Optics Letters. 2007;**32**:912-914

[11] Brooker G, Siegel N, Wang V, Rosen J. Optimal resolution in Fresnel incoherent correlation holographic fluorescence microscopy. Optics Express. 2011;**19**:5047-5062

[12] Bouchal P, Kapitán J, Chmelík R, Bouchal Z. Point spread function and two-point resolution in Fresnel incoherent correlation holography. Optics Express. 2011;**19**:15603-15620

[13] Vijayakumar A, Rosen J. Spectrum and space resolved 4D imaging by coded aperture correlation holography (COACH) with diffractive objective lens. Optics Letters. 2017;**42**: 947-950

[14] Vijayakumar A, Kashter Y, Kelner R, Rosen J. Coded aperture correlation holography system with improved performance. Applied Optics. 2017;**56**: F67-F77

[15] Vijayakumar A, Rosen J. Interferenceless coded aperture correlation holography—a new technique for recording incoherent digital holograms without two-wave interference. Optics Express. 2017;**25**: 13883-13896

[16] Kumar M, Vijayakumar A, Rosen J. Incoherent digital holograms acquired by interferenceless coded aperture correlation holography system without refractive lenses. Scientific Reports. 2017;**7**:11555

[17] Rosen J, Anand V, Rai MR, Mukherjee S, Bulbul A. Review of 3D imaging by coded aperture correlation holography (COACH). Applied Sciences. 2019;**9**:605

[18] Rai MR, Vijayakumar A, Rosen J. Single camera shot interferenceless coded aperture correlation holography. Optics Letters. 2017;**42**:3992-3995

[19] Rai MR, Vijayakumar A, Rosen J. Extending the field of view by a scattering window in I-COACH system. Optics Letters. 2018;**43**:1043-1046

[20] Bulbul A, Vijayakumar A, Rosen J. Partial aperture imaging by systems with annular phase coded masks. Optics Express. 2017;**25**:33315-33329

[21] Rai MR, Vijayakumar A, Rosen J. Nonlinear adaptive three-dimensional imaging with interferenceless coded aperture correlation holography (I-COACH). Optics Express. 2018;**26**: 18143-18154

[22] Bulbul A, Vijayakumar A, Rosen J. Superresolution far-field imaging by coded phase reflectors distributed only along the boundary of synthetic apertures. Optica. 2018;**5**:1607-1616

[23] Rai MR, Vijayakumar A, Ogura Y, Rosen J. Resolution enhancement in nonlinear interferenceless COACH with a point response of subdiffraction limit patterns. Optics Express. 2019;**27**: 391-403

[24] Ji T, Zhang L, Li W, Sun X, Wang J, Liu J, et al. Research progress of incoherent coded aperture correlation holography. Laser and Optoelectronics Progress. 2019;**56**:080005 (in Chinese)

[25] Anand V, Ng SH, Maksimovic J, et al. Single shot multispectral multidimensional imaging using chaotic waves. Scientific Reports. 2020;**10**:13902

[26] Liu C, Man T, Wan Y. Optimized reconstruction with noise suppression for interferenceless coded aperture correlation holography. Applied Optics. 2020;**59**:1769-1774

[27] Jiang Z, Yang S, Huang H, He X, Kong Y, Gao A, et al. Programmable liquid crystal display based noise reduced dynamic synthetic coded aperture imaging camera (NoRDS-CAIC). Optics Express. 2020;**28**: 5221-5238

[28] Dubey N, Rosen J, Gannot I. High-resolution imaging with an annular aperture of coded phase masks for endoscopic applications. Optics Express. 2020;**28**:15122-15137

[29] Anand V, Ng SH, Katkus T, Juodkazis S. Spatio-spectral-temporal imaging of fast transient phenomena using a random array of pinholes. Advanced Photonics Research. 2021;**2**: 2000032

[30] Wan Y, Liu C, Ma T, Qin Y, lv S. Incoherent coded aperture correlation holographic imaging with fast adaptive and noise-suppressed reconstruction. Optics Express. 2021;**29**: 8064-8075

[31] Anand V, Ng SH, Katkus T, Juodkazis S. White light three-dimensional imaging using a quasi-random lens. Optics Express. 2021;**29**: 15551-15563

[32] Dubey N, Kumar R, Rosen J. COACH-based Shack-Hartmann wavefront sensor with an array of phase coded masks. Optics Express. 2021;**29**: 31859-31874

[33] Yu X, Wang K, Xiao J, Li X, Sun Y, Chen H. Recording point spread functions by wavefront modulation for interferenceless coded aperture correlation holography. Optics Letters. 2022;**47**:409-412

[34] Hai N, Rosen J. Interferenceless and motionless method for recording digital holograms of coherently illuminated 3-D objects by coded aperture correlation holography system. Optics Express. 2019;**27**:24324-24339

[35] Hai N, Rosen J. Doubling the acquisition rate by spatial multiplexing of holograms in coherent sparse coded aperture correlation holography. Optics Letters. 2020;**45**:3439-3442

[36] Hai N, Rosen J. Coded aperture correlation holographic microscope for single-shot quantitative phase and amplitude imaging with extended field of view. Optics Express. 2020;**28**:27372-27386

[37] Balasubramani V et al. Roadmap on digital holography-based quantitative phase imaging. Journal of Imaging. 2021; **7**:252

[38] Rosen J, Vijayakumar A, Kumar M, Rai MR, Kelner R, Kashter Y, et al. Recent advances in self-interference incoherent digital holography. Advances in Optics and Photonics. 2019;**11**:1-66

[39] Yamaguchi I, Zhang T. Phase-shifting digital holography. Optics Letters. 1997;**22**:1268-1270

[40] Rosen J et al. Roadmap on recent progress in FINCH technology. Journal of Imaging. 2021;**7**:197

[41] Kim MK. Adaptive optics by incoherent digital holography. Optics Letters. 2012;**37**:2694-2696

[42] Watanabe K, Nomura T. Recording spatially incoherent Fourier hologram using dual channel rotational shearing interferometer. Applied Optics. 2015;**54**: A18-A22

[43] Nobukawa T, Muroi T, Katano Y, Kinoshita N, Ishii N. Single-shot phase-shifting incoherent digital holography with multiplexed checkerboard phase gratings. Optics Letters. 2018;**43**: 1698-1701

[44] Nobukawa T, Katano Y, Goto M, Muroi T, Kinoshita N, Iguchi Y, et al. Incoherent digital holography simulation based on scalar diffraction theory. Journal of Optical Society of America A. 2021;**38**:924-932

[45] Bulbul A, Rosen J. Coded aperture correlation holography (COACH) with a superior lateral resolution of FINCH and axial resolution of conventional direct imaging systems. Optics Express. 2021; **29**:42106-42118

[46] Bulbul A, Rosen J. Super-resolution imaging by optical incoherent synthetic aperture with one channel at a time. Photonics Research. 2021;**9**:1172-1181

[47] Merkle F. Synthetic-aperture imaging with the European very large telescope. Journal of Optical Society of America A. 1988;**5**:904-913

[48] Michelson AA, Pease FG. Measurement of the diameter of α-Orionis by the interferometer. Astrophysics Journal. 1921;**53**:249-259

[49] Ilovitsh A, Zach S, Zalevsky Z. Optical synthetic aperture radar. Journal of Modern Optics. 2013;**60**:803-807

[50] Gerchberg RW, Saxton WO. A practical algorithm for the determination of phase from image and diffraction plane pictures. Optik. 1972;**35**:227-246

[51] Narayanswamy R, Johnson GE, Silveira PE, Wach HB. Extending the imaging volume for biometric iris recognition. Applied Optics. 2005;**44**: 701-712

[52] Pieper RJ, Korpel A. Image processing for extended DOF. Applied Optics. 1983;**22**:1449-1453

[53] Li S, Kwok JT, Wang Y. Multifocus image fusion using artificial neural networks. Pattern Recognition Letters. 2002;**23**:985-997

[54] Dowski ER, Cathey WT. Extended depth of field through wave-front coding. Applied Optics. 1995;**34**: 1859-1866

[55] Tucker S, Cathey WT, Dowski E Jr. Extended DOF and aberration control for inexpensive digital microscope systems. Optics Express. 1999;**4**:467-474

[56] Le VN, Chen S, Fan Z. Optimized asymmetrical tangent phase mask to obtain defocus invariant modulation transfer function in incoherent imaging systems. Optics Letters. 2014;**39**: 2171-2174

[57] Liao M, Lu D, Pedrini G, Osten W, Situ G, He W, et al. Extending the depth-of-field of imaging systems with a scattering diffuser. Scientific Reports. 2019;**9**:7165

[58] Mikula G, Kolodziejczyk A, Makowski M, Prokopowicz C, Sypek M. Diffractive elements for imaging with extended depth of focus. Optical Engineering. 2005;**44**:058001

[59] Zhai Z, Ding S, Lv Q, Wang X, Zhong Y. Extended depth of field through an axicon. Journal of Modern Optics. 2009;**56**:1304-1308

[60] Rai MR, Rosen J. Depth-of-field engineering in coded aperture imaging. Optics Express. 2021;**29**:1634-1648

[61] Rosen J, Salik B, Yariv A. Pseudo-nondiffracting beams generated by radial harmonic functions. Journal of Optical Society of America. A. 1995;**12**: 2446-2457

[62] Rosen J, Salik B, Yariv A. Pseudo-nondiffracting beams generated by radial harmonic functions: Erratum. Journal of Optical Society of America A. 1996;**13**:387

[63] Rai MR, Rosen J. Noise suppression by controlling the sparsity of the point spread function in interferenceless coded aperture correlation holography (I-COACH). Optics Express. 2019;**27**: 24311-24323

[64] Habaza M, Kirschbaum M, Guernth-Marschner C, Dardikman G, Barnea I, Korenstein R, et al. Rapid 3D refractive-index imaging of live cells in suspension without labeling using dielectrophoretic cell rotation. Advanced Science. 2017;**4**: 1600205

[65] Rivenson Y, Liu T, Wei Z, Zhang Y, de Haan K, Ozcan A. Phase Stain: The digital staining of label-free quantitative phase microscopy images using deep learning. Vol. 8. Light: Science and Application; 2019. pp. 1-11

[66] Charrière F, Kühn J, Colomb T, Montfort F, Cuche E, Emery Y, et al. Characterization of microlenses by digital holographic microscopy. Applied Optics. 2006;**45**:829-835

[67] Niu M, Luo G, Shu X, Qu F, Zhou S, Ho YP, et al. Portable quantitative phase microscope for material metrology and biological imaging. Photonics Research. 2020;**8**:1253-1259

[68] Kumar M, Vijayakumar A, Rosen J, Matoba O. Interferenceless coded aperture correlation holography with synthetic point spread holograms. Applied Optics. 2020;**59**:7321-7329

[69] Rai MR, Vijayakumar A, Rosen J. Superresolution beyond the diffraction limit using phase spatial light modulator between incoherently illuminated objects and the entrance of an imaging system. Optics Letters. 2019;**44**: 1572-1575

[70] Rai MR, Rosen J. Resolution-enhanced imaging using interferenceless

coded aperture correlation holography with sparse point response. Scientific Reports. 2020;**10**:5033

[71] Rosen J, Hai N, Rai MR. Recent progress in digital holography with dynamic diffractive phase apertures [Invited]. Applied Optics. 2022;**61**:B171-B180

[72] Anand V, Rosen J, Juodkazis S. Review of engineering techniques in chaotic coded aperture imagers. Light. Advanced Manufacturing. 2022;**3**:24

Chapter 8

A Mapping Relationship-Based near-Field Acoustic Holography

Haijun Wu and Weikang Jiang

Abstract

A mapping relationship-based near-field acoustic holography (MRS-based NAH) is a kind of innovative NAH by exploring the mapping relationship between modes on surfaces of the boundary and hologram. Thus, reconstruction is converted to obtain the coefficients of participant modes on holograms. The MRS-based NAH supplies an analytical method to determine the number of adopted fundamental solution (FS) as well as a technique to approximate a specific degree of mode on patches by a set of locally orthogonal patterns explored for three widely used holograms, such as planar, cylindrical, and spherical holograms. The NAH framework provides a new insight to the reconstruction procedure based on the FS in spherical coordinates. Reconstruction accuracy based on two types of errors, the truncation errors due to the limited number of participant modes and the inevitable measurement errors caused by uncertainties in the experiment, are available in the NAH. An approach is developed to estimate the lower and upper bounds of the relative error. It supplies a tool to predict the error for a reconstruction under the condition that the truncation error ratio and the signal-to-noise ratio are given. The condition number of the inverse operator is investigated to measure the sensitivity of the reconstruction to the input errors.

Keywords: near-field acoustic holography, mapping relationship, integral identity, acoustic measurement, spherical fundamental solutions

1. Introduction

To locate the position and target the strength of noise for a vibrating structure, near-field acoustic holography (NAH) had been widely adopted as an effective tool. It has a significant influence on the noise diagnostics, which gives a permission to get all desired acoustic quantities, such as pressure, particle velocity, and sound power, from a number of discrete field measurement.

It was originally developed by Willams, Manynard, etc., to reconstruct surface velocity of a rectangular plane with Fourier transform technique [1–3]. Initially, the Fourier-based NAH decomposes the field pressure into k-space (wave number space) for baffled problems. In other words, the field pressure is expanded into plane waves, and the reconstruction procedure is to obtain coefficients of the plane waves based on measured pressure. Although different from the k-space decomposition, concept of

Fourier transformation was inherent to the 3D cylindrical and spherical NAH problems as the in-depth discussions in Ref. [4].

Since it was proposed [1], varieties of approaches had been proposed and their superiorities had been proven in various applications, which resulted in several categories according to their underlying theories. Statistical optimal NAH [5–7] uses the elemental waves to approach the acoustic field, in which the surface-to-surface projection of the sound field is performed by using a transfer matrix defined in such a way that all propagating waves and a weighted set of evanescent waves are projected with optimal average accuracy [6]. Boundary element method (BEM)-based NAH [8–13] is appropriate for arbitrarily shaped model in which a general transformer matrix between the surfaces of structure and hologram is derived from the integral equation. Among the BEM-based NAH, two types of integration equation are adopted: the directive formulation (Helmholtz integral equation) and indirect formulation (single- or double-layer integral equation). The quantities reconstructed by the NAH derived from directive formulation have clear physical meaning [8–10], while the ones obtained by NAH derived from the indirect formulation are not the real physical quantities [11–13]. The equivalent source method (ESM) [14–19], also named as wave superposition algorithm (WSA) [16, 20, 21], was proposed by Koopman [22] for solving acoustic radiation problems of closed sources. ESM assumes that the field is generated by a series of simple sources such as monopoles and dipoles, and numerical integration is not needed in determining the source strength for a set of prescribed positions. Despite versatility of the ESM and various successful applications, "retreat distance" between the actual source surface and the virtual source cannot be well defined and deserves more attention in the application [23]. The Helmholtz equation least square method (HELS) [24–26] adopted the spherical wave expansion theory to reconstruct acoustic pressure field from a vibrating structure. Coefficients of the spherical wave function, the fundamental solution (FS) for the Helmholtz equation, are determined by requiring the assumed form of solution to satisfy the pressure boundary condition at the measurement points. Since the spherical wave functions solve the Helmholtz equation directly, it is immune to the nonuniqueness difficulty inherent in BEM-based NAH [27]. However, HELS works better for spherical or chunky model than elongated model due to the specific basis function [25].

Essentially speaking, NAH is to achieve the desired acoustic quantities by the measured physical quantities such as sound pressure in the field. Most of the methods explicitly require the transfer operator $\mathrm{T}(\mathbf{y}, \mathbf{x})$ between desired acoustic quantities $f(\mathbf{y})$and measured physical quantities $\overline{p}(\mathbf{x})$. They built a linear system of$f(\mathbf{y}) = inv(\mathrm{T}(\mathbf{y}, \mathbf{x}))\overline{p}(\mathbf{x})$ in which $inv(\star)$ represents an inverse operator, by either a general numerical method (BEM-based NAH) [8–13], or specific basis spaces such as a general Fourier basis (Fourier-based NAH) [1–4], simplified monopoles, dipoles (ESM and WSA) [14–16, 18, 20, 21], and fundamental solutions (HELS) [24–27]. The reconstruction procedure is therefore to solve the linear system to obtain the physical quantities on the boundary, such as pressure or normal velocity in BEM-based NAH, the source strength of equivalent source in ESM, coefficients of basis functions in Fourier-based NAH and HELS, and following by an extrapolation process to achieve desired acoustic quantities.

Unfortunately, all the proposed methods are very sensitive to errors which may cause reconstruction to fail. It is primarily due to abundant adoption of basis functions in the transfer operator which amplifies the errors in the inverse process. That is the reason why there have been numerous studies focusing on the development of regularization methods to stabilize this inverse problem, such as truncated singular value decomposition [28] and the Tikhonov regularization [29]. Thus, construction of transfer operator is not a trivial process but is crucial to the feasibility and accuracy of the NAH. Concerning the theory development and practical measurement, it naturally arises a question whether there exists a guideline to determine the number and location of generalized basis function as well as measurement to obtain their coefficients for a given shape of source surface and prescribed tolerance.

The number of FS as well as number and position of the microphones array in the measurement are not well studied for the category of NAH based on the FS. Thus, one advantage of the mapping relationship-based NAH (MRS-based NAH) is the available guideline to the determination of the number of FS and measurement configuration in the FS-based NAH by exploring the mapping relationship between the modes in FS between surface and hologram, and investigating approximation of the modes with a set of locally orthogonal patterns.

As errors are inevitable in the practical measurement, it is curious to know how the errors go through the inverse operation and what influence imposed on the accuracy of the reconstruction results. To the best knowledge of authors, few works are devoted to the errors analysis of the NAH by comparing with that for the regularization methods. It is because that the NAH was usually viewed as a very ill-posed inverse problem for which regularized solution is the primary task. Thus, it is difficult to predict or estimate the reconstruction accuracy. Instead of a predictable way, numerical simulation and experimental validation are two frequently adopted methods to investigate the performance of NAH for different parameters [19, 30, 31]. For practical problems, it is hard to estimate the accuracy of the reconstructed results. Thus, one merit of our approach is the availability for predicting the reconstructed accuracy for a specific setup of the MRS-based NAH.

2. The mapping relationship-based NAH

2.1 Theorem development

As shown in **Figure 1**, assume that the fluid is homogenous, inviscid, and compressible and only undergoes small translation movement. The time harmonic sound pressure radiated from a vibrating structure into an infinite domain Ω is described by the well-known Helmholtz equation:

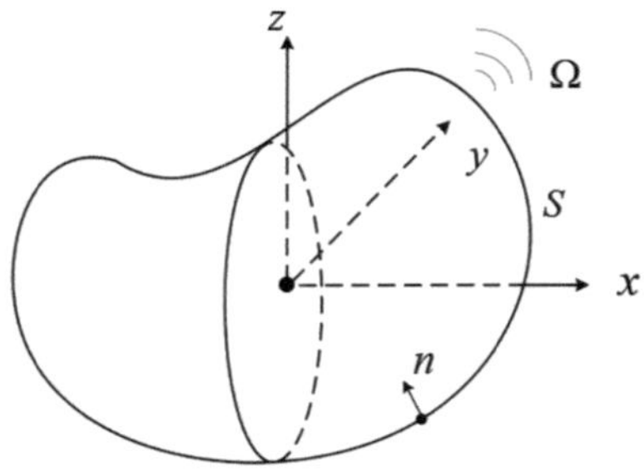

Figure 1.
Exterior acoustic problem of a vibrating structure in free space.

$$\nabla^2 p(\mathbf{x}) + k^2 p(\mathbf{x}) = 0 \text{ for } \mathbf{x} \in \Omega \tag{1}$$

where k is the wave number, relating to the acoustic speed c and angular frequency ω by $k = \omega/c$, and $\mathbf{x}$ is a point in the domain. The time component is assumed to be $e^{-i\omega t}$.

The fundamental solution of the governing formulation Eq. (1) in the spherical coordinates is

$$S_n^m(k, \mathbf{x}) = h_n(k\|\mathbf{x}\|)Y_n^m(\theta, \phi), |m| \leq n \mathrm{n} = 0,1,2, \dots ,N \tag{2}$$

where variables θ and ϕ are the polar angles of a point in the spherical coordinates and N is the truncated degree in a series expansion. j_n and h_n are the nth spherical Bessel function and spherical Hankel of the first kind, respectively. Y_n^m is the normalized spherical harmonic function:

$$Y_n^m(\theta, \phi) = \frac{1}{\sqrt{2\pi}} \bar{P}_n^m(\cos\theta)e^{im\phi} \tag{3}$$

where $\bar{P}_n^m$ is the normalized associated Legendre function [32].

Normal gradient in the direction $n(\mathbf{x})$ for the fundament solution Eq. (3) is as follows:

$$q_n^m(\mathbf{x}) = \frac{\partial S_n^m(\mathbf{x})}{\partial n(\mathbf{x})} = \frac{\partial h_n(k\|\mathbf{x}\|)Y_n^m(\theta, \phi)}{\partial n(\mathbf{x})} \tag{4}$$

which is related to the normal velocity v_n by the Euler formulation:

$$q(\mathbf{x}) = ik\rho c v_n(\mathbf{x}) \tag{5}$$

It should be noted that Eqs. (2) and (5) are related as a solution pair for exterior acoustic problems, which means giving one as the boundary condition, the other will be the solution. They form a set of pressure/velocity modes on the boundary of a vibrating structure, which are generally independent on nonspherical surfaces and orthogonal on spherical surfaces. To facilitate derivations, we refer the velocity modes as the normal gradient q instead of the normal velocity v_n. Assume that a structure is vibrating in one of its velocity modes Eq. (5), and the radiated pressure must be in the form of Eq. (2), which can be derived by making use of the equivalent source method (ESM) and boundary integral equation (BIE) [33].

Based on the model decomposition theorem and the mapping relationship, the boundary velocity $v(\boldsymbol{y})$ and the radiated sound pressure $p(\boldsymbol{x})$ on the hologram can be expressed for a given set of participant coefficients $a_{n'}$ as:

$$\begin{cases} v(\boldsymbol{y}) = \sum\limits_{n'=0}^{N'} a_{n'} v_{n'}(\boldsymbol{y}), \boldsymbol{y} \in S \\ p(\boldsymbol{x}) = \sum\limits_{n'=0}^{N'} a_{n'} p_{n'}(\boldsymbol{x}), \boldsymbol{x} \in \Omega \end{cases} \tag{6}$$

where $v_{n'=n^2+n+m+1} = q_n^m$ and $p_{n^2+m+1} = S_n^m$.

Eq. (6) is the basement of the MRS-based NAH but must be properly truncated. The subscription convention in Eq. (6) is convenient for the discretized linear operation. Obviously, the truncation number N' in Eq. (6) and N in Eq. (2) are related by

$N' = (N+1)^2$. The truncation number is crucial to the pressure evaluation as well as the number and positions of microphone array required in the NAH.

Determination of the number of most efficient modes is converted to seeking the truncation number of radiation efficiency σ_n making the upper bound of the radiated sound power converge to a given tolerance. A relative error of the upper bounded radiated sound power caused by new added modes for degree N in Eq. (2) with respective to the one produced by existing modes for degree less than N is defined as:

$$\varepsilon_N = \frac{(2N+1)\sigma_N(k\tilde{r})}{\sum_{n=0}^{N-1}(2n+1)\sigma_n(k\tilde{r})} \tag{7}$$

where $\tilde{r}$ is the equivalent radius of the vibrating structure. The equivalent radius is the description of the spherical source which has the same average radiated sound power per unit area as that of the vibrating structure or holograms. Since the radiated sound power from the equivalent source and the vibrating structure as well as the holograms should be same, the requirement of same average radiated sound power per unit area makes the equivalent radius satisfy

$$\tilde{r} = \sqrt{S/4\pi} \tag{8}$$

where S represents area of structure in the determination of equivalent radius $\tilde{r}$.

Fortunately, for a specific dimensionless value $k\tilde{r}$, the radiation efficiency $\sigma_{N>N_c}$ is a strictly decreasing function with respective to the degree N after a certain degree N_c which can be obtained by its closed-form expression. Radiation efficiency σ_N and the relative error ε_N for the dimensionless size $0.1 \le k\tilde{r} \le 10$ and the varying degree from 0 to 7 are presented in **Figure 2a** and **b**, respectively. It shows that σ_N is a monotonously decreasing function and clearly distinguishes from each degree for $k\tilde{r} < 2.0$. There is a plateau on which σ_N starts to overlap for $k\tilde{r} > 2.0$. It means that those degrees of modes contribute to the radiated sound power almost equally and no one can be neglected, which is verified by the relative errors in **Figure 2b**. Therefore, more degrees of modes are needed to make the radiated sound power converge for larger dimensionless value $k\tilde{r}$. The relative error ε_N presented in **Figure 2b** can be used as a reference to determine the degree of the most efficient modes for $0.1 \le k\tilde{r} \le 10$, or in other words the

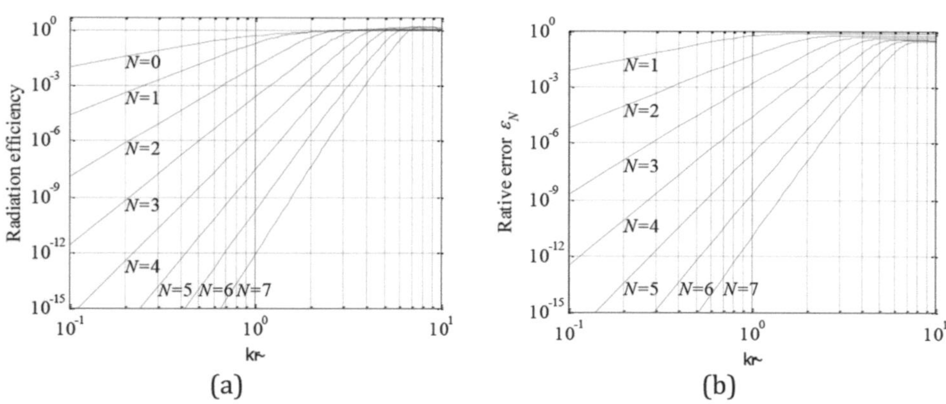

Figure 2.
(a) Radiation efficiency of a sphere, and (b) relative errors of upper bounded radiated sound power, for varying degrees and $k\tilde{r}$.

truncation number N for the FS-based NAH. However, Eq. (7) only relates to the size of the structure and has nothing to do with the field point or size of hologram.

2.2 The NAH procedure

Suppose a set of at least independent velocity modes on the boundary, denoted as $v_i(\mathbf{x} \in S), (i = 1, 2, \ldots)$, can produce a set of independent pressure modes $p_i(\mathbf{x} \in \Gamma), (i = 1, 2, \ldots)$ on the measurement surface Γ, correspondingly, and they form a pair of bijective mapping relationship.

Generally, the pressure patterns $p_i(\mathbf{x} \in \Gamma)$ are non-orthogonal on the hologram, and an orthogonalization process is required, which can be done by the Gram-Schmidt approach as:

$$u_i(\mathbf{x}) = p_i(\mathbf{x}) - \sum_{j=1}^{i-1} \frac{\langle p_i(\mathbf{x}), u_j(\mathbf{x}) \rangle}{\|u_j(\mathbf{x})\|} u_j(\mathbf{x}) \tag{9}$$

where the inner product $\langle p, u \rangle = \int_\Gamma p(\mathbf{x}) u^*(\mathbf{x}) d\Gamma(\mathbf{x})$ and the norm $\|u\| = \sqrt{\langle u, u \rangle}$. After some algebraic operations, the normalized orthogonal modes $e_i(\mathbf{x}) = u_i(\mathbf{x}) / \|u(\mathbf{x})\|$ can be expressed in the following form:

$$\mathbf{E} = \mathbf{pR} \tag{10}$$

where $\mathbf{E} = [e_1(\mathbf{x}), e_2(\mathbf{x}), \cdots], \mathbf{p} = [p_1(\mathbf{x}), p_2(\mathbf{x}), \cdots]$ and $\mathbf{R}$ is a real upper triangular square matrix. As indicated in Eq. (10), the normalized orthogonal modes are actually a linear combination of the independent pressure modes p_i. It is remarkable that evaluation of the inner produce is performed on holograms which are generally in smooth shapes, such as the three typical holograms in Section 0. Thus, they can be computed on exact geometries which are a void of discretization errors. Furthermore, orthogonalization of the modes on simple shaped holograms will yield a translation matrix $\mathbf{R}$ with good numerical characteristics, such as the condition number.

Assume the radiated pressure $p(\mathbf{x})$ is decomposed into the normalized orthogonal modes $e_i(\mathbf{x})$ on the hologram as:

$$p(\mathbf{x}) = \sum_{i=1}^{\infty} \alpha_i e_i(\mathbf{x}) \tag{11}$$

where

$$\alpha_i = \langle p(\mathbf{x}), e_i(\mathbf{x}) \rangle_\Gamma \tag{12}$$

Once those participant coefficients are obtained by the measured pressure, substituting Eq. (10) into Eq. (12) yields

$$p(\mathbf{x}) = \sum_{i=1}^{\infty} \lambda_i p_i(\mathbf{x}) \tag{13}$$

where the coefficients are

$$\lambda_i = \sum_{j=i}^{\infty} R_{ij} \alpha_j \tag{14}$$

Due to the unique mapping relationship between surfaces of vibrating structure and hologram, reconstruction for the boundary quantities can be performed by multiplying the corresponding modes with the same set of participant coefficients λ_i on the surface of the vibrating structure.

Thus, acoustic holography is converted to seek explicit descriptions of the mapping relationship between the modes on the boundary and modes on the field and design a proper experimental setup for obtaining the participant coefficients of the modes on the measurement surface. The modes on the boundary are free of restrictions for their form of expression, which could be in any well-studied analytical functions or in generally numerical representations. However, it should be expected to have a capacity of fast convergent ratio in the decomposing of boundary quantities and generate a radiated pressure on the hologram which is easy to be determined by the experiment. In the current work, the FS in spherical coordinates for the Helmholtz equation Eq. (2) and its normal gradient Eq. (5) are chosen as the pressure and velocity modes. Merits of choosing those forms of modes are twofold. First, the radiated modes on the field are also the in the same form; and second, the most effective modes contributing to the field pressure are easy to be determined. Henceforth, the radiated pressure modes in Eq. (13) are chosen as $p_i = p_{n^2+m+n+1} = S_n^m$ in our analysis to facilitate derivations. So do the velocity modes q_i.

2.3 Setup of the microphone array

Since the modes are distributed on an enclosing surface, the holograms should form an enveloping surface enclosing the vibrating structure. Otherwise, partially measured pressure cannot represent the modes completely and consequently cannot be applied to reconstruct the boundary information based on the mapping relationship.

The distribution of a specific mode varies on different holograms. Generally, the measurements are subject to the experimental resource such as microphones and permissible space. How to accurately recognize the field pressure modes is one of the crucial factors to NAH. In practice, microphones are preferred to be placed on planar, cylindrical, or spherical surfaces which are easy to be set up but generally not conformal to the vibrating structure, as shown in **Figure 3**.

For the enclosing planar holograms, as shown in **Figure 3a**, each pressure mode is divided and projected onto six patches. On each patch, the measured pressure should be able to accurately represent the projected pressure modes. However, once the pressure is discretely sampled, the spectrum or the number of participant modes on

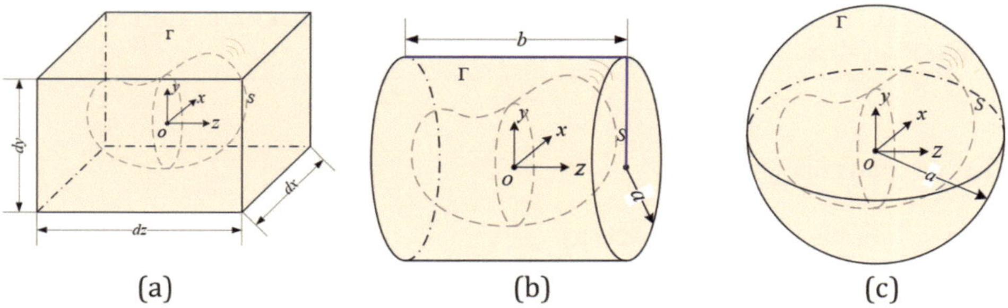

Figure 3.
Three typical holograms: (a) planar holograms, (b) cylindrical holograms, and (c) spherical holograms.

that patch is truncated. Therefore, the primary task in the measurement is to set up the microphone arrays properly with an aim to approximate all the projected pressure modes on each patch actually. On each planar surface, the pressure modes can be expressed by two sets of locally orthogonal polynomials such as polynomial $f_n(x)$ and $g_m(y)$ of degree n and m in each direction, respectively. The most significant degrees in each direction can be numerically obtained directly by approximating the analytical pressure mode, Eq. (2), with the polynomial expansions $f_n g_m$, as

$$S_n^m(k, \mathbf{x}) = \sum_{n'=0}^{N'} \sum_{m'=0}^{M'} a_{n'}^{m'} f_{n'}(x(\mathbf{x})) g_{m'}(y(\mathbf{x})) \tag{15}$$

where $a_{n'}^{m'}$ is the coefficients. Once the polynomial degrees N' and M' on each planar patch are determined, the microphones are placed at the abscissas of Gaussian quadrature points on the patch, which results in $(N' + 1) \times (M' + 1)$ microphone positions [33].

A closed cylindrical measurement surface, as shown in **Figure 3b**, has three patches, one left circular planar patch $\Gamma_l(\mathbf{x}| - b/2 = \|\mathbf{x}\| \cos\theta)$, one right $\Gamma_r(\mathbf{x}|b/2 = \|\mathbf{x}\| \cos\theta)$ circular planar patch, and one cylindrical surface $\Gamma_c(\mathbf{x}|a = \|\mathbf{x}\| \sin\theta)$ in the central portion. Similar to the planar hologram, it needs to select a set of locally orthogonal patterns to approximate the pressure modes on each patch. In this case, arguments θ and ϕ in cylindrical coordinates are the two independent variables to define patches. All three patches possess a complete description for the variable ϕ in the range $[0, 2\pi]$. In light of the expression for pressure modes, Eq. (2), normalized basis $g_m(\phi(\mathbf{x})) = e^{im\phi(\mathbf{x})}/\sqrt{2\pi}$ is selected as one set of the orthogonal patterns in the ϕ direction. Thus, the determination of the truncation number for another set of local basis function $f_{n'}$ is simplified to approximate the following function $\mathcal{F}_n(k, \mathbf{x}) = h_n(k\|\mathbf{x}\|)\bar{P}_n^m(\cos\theta)$.

A spherical measurement surface is shown in **Figure 3c**, which is a conformal patch to the spherical coordinates upon which the FS is obtained. Field modes on the spherical surface are orthogonal. Determination of the field modes on the spherical hologram is actually to identify the spherical harmonic functions based on the measured pressure. Due to conformality of the hologram to the coordinate system of the spherical FS, an analytical way is available to determine the number and position of the measurement. The quadrature technique on a sphere is well studied and widely used in the computational acoustics [34]. Therefore, the participant coefficients in Eq. (12) can be accurately evaluated by $(N + 1)$ point Gaussian quadrature and $(2N + 1)$ point square quadrature for variables of θ and ϕ on the spherical surface for the pressure mode $S_N^m(|m| \leq N)$. Thus, it is free of numerical searching in the determination of the truncation number. In addition, the total number of measurements is the smallest than that requested by the other two holograms.

3. Error analysis

3.1 Error bounds on pressure energy

The NAH is an inverse problem and thus poses significant challenges to the stable and accurate solution. However, a practical measurement is prone to errors and always incorporates uncertainties, such as random fluctuations, effect of rapid decay of the evanescent waves. Generally, the great affection to the reconstruction by the inevitable measurement errors is largely due to over-selected number of the basis (either in numerical or analytical form) which results in an ill-posed inverse operator. Fortunately, the number of basis or modes can be well estimated by an analytical way as introduced in Section 2. Thus, a pre-regularization process is embedded in the MRS-based NAH.

On the holograms, the error included pressure is simply modeled as:

$$\mathrm{p}(\mathbf{x}) = \mathrm{p}_0(\mathbf{x}) + \mathrm{n}(\mathbf{x}), \text{or } \mathbf{P} = \mathbf{P}_0 + \mathbf{n} \tag{16}$$

where $\mathbf{P_0}$ represents the source pressure and $\mathbf{n}$ is the noise terms. Denote the signal-to-noise ratio (SNR) on the hologram as:

$$\mathrm{SNR} = 10\log_{10}\frac{W_{P_0}}{W_{Noise}} = 10\log_{10}\frac{\int_{S_h}|\mathrm{P}_0(\mathbf{x})|^2 dS(\mathbf{x})}{\int_{S_h}|\mathrm{n}(\mathbf{x})|^2 dS(\mathbf{x})} \tag{17}$$

where W_{P_0} and W_{Noise} represent energies of source pressure and noise pressure, respectively. Under the condition that the source pressure and noise pressure can be completely decomposed by a set of modes, the SNR can be reformatted by the Parseval law as:

$$\mathrm{SNR} = 10\log_{10}\frac{\|\boldsymbol{\alpha}_0\|_2^2}{\|\boldsymbol{\alpha}_{Noise}\|_2^2} \tag{18}$$

where $\boldsymbol{\alpha}_0$ and $\boldsymbol{\alpha}_{Noise}$ are coefficients of the locally normalized orthogonal patterns on the holograms for the source pressure and noise term, and correspondingly the coefficients for pressure $\mathbf{P}$ is $\boldsymbol{\alpha} = \boldsymbol{\alpha}_0 + \boldsymbol{\alpha}_{Noise}$.

By taking advantages of the mapping relationships, Eq. (11) is used to evaluate the reconstructed pressure on the surface of vibrating structure after the coefficients $\boldsymbol{\lambda}$ of FS are obtained on the holograms. Decompose the FS in Eq. (11) on the surface of vibrating structure as same as that in Eq. (10) but with symbol $\boldsymbol{S}$ substituting for pressure mode $\boldsymbol{p}$ as:

$$\boldsymbol{S}_\Gamma = \mathbf{E}_\Gamma \mathbf{R}_\Gamma^- \tag{19}$$

where $\mathbf{E}_\Gamma$ is the column normalized modes on the surface of vibrating structure and $\mathbf{R}_\Gamma$ is a translation operator which is an upper triangular square matrix. In light of Eq. (14) and Eq. (19), reconstructed sound pressure on surface of the vibrating structure, $\mathbf{p}_\Gamma$ in the Eq. (13) can be expressed as:

$$\mathbf{p}_\Gamma = \mathbf{E}_\Gamma \mathbf{R}_\Gamma^- \mathbf{R}\boldsymbol{\alpha} \tag{20}$$

Therefore, it could be observed that the reconstruction process is to translate the local coefficients $\boldsymbol{\alpha}$ obtained on the holograms to that on the surface of vibrating structure by the translator $\mathbf{R}_\Gamma^- \mathbf{R}$, and then the reconstructed pressure is evaluated by the modal decomposition method. Stability of the translation is largely dependent on the product of the two translators $\mathbf{R}_\Gamma^-$ and $\mathbf{R}$ which are closely related to the geometric information of the structure and holograms, respectively.

According to the Parseval law, the reconstructed pressure energy on the surface of vibrating structure is

$$W_{P_\Gamma} = \left\|\mathbf{R}_\Gamma^- \mathbf{R}\boldsymbol{\alpha}\right\|_2^2 = \boldsymbol{\alpha}^* \mathbf{T}_\mathbf{R} \boldsymbol{\alpha} \tag{21}$$

where $\mathbf{T_R} = \mathbf{R}^* \mathbf{R}_\Gamma^{-*} \mathbf{R}_\Gamma^- \mathbf{R}$ is a Hermitian matrix. Therefore, there is an eigen dec omposition of $\mathbf{T_R} = \mathbf{Q}^* \boldsymbol{\Lambda} \mathbf{Q}$ in which $\mathbf{Q}$ is a unitary complex matrix whose columns comprise an orthonormal basis of eigenvectors of $\mathbf{T_R}$, and $\boldsymbol{\Lambda}$ is a real diagonal matrix whose main diagonal entries are the corresponding eigenvalues. Assume the eigenvalues are sorted in a descending order, such as $\Lambda_i \geq \Lambda_j$ for $j > i$. The lower and upper bounds of the reconstructed pressure energy are easy to be obtained

$$\Lambda_d \|\boldsymbol{\alpha}\|_2^2 \leq W_{P_\Gamma} \leq \Lambda_1 \|\boldsymbol{\alpha}\|_2^2 \tag{22}$$

where d is the dimension of the matrix. In practice, the relative error of reconstructed pressure energy on the surface of the vibrating structure is more concerned. Obviously, the bounds of the exact pressure energies $W_{P_{0\Gamma}}$ and noise generated pressure energies $W_{P_{\text{noise}\Gamma}}$ can be obtained with coefficients $\boldsymbol{\alpha}$ replaced with $\boldsymbol{\alpha}_0$ and $\boldsymbol{\alpha}_{Noise}$ in Eq. (22), respectively. Thus, bounds for the relative errors $\varepsilon_{W_\Gamma} = W_{P_{\text{noise}\Gamma}} / W_{P_{0\Gamma}}$ are

$$cond(\mathbf{T_R})^{-} 10^{-SNR/10} \leq \varepsilon_{W_\Gamma} \leq cond(\mathbf{T_R}) 10^{-SNR/10} \tag{23}$$

where $10^{-SNR/10} = \|\boldsymbol{\alpha}_{Noise}\|_2^2 / \|\boldsymbol{\alpha}_0\|_2^2$, and $cond(\mathbf{T_R}) = \boldsymbol{\Lambda}_1 / \boldsymbol{\Lambda}_d$ is the condition number of the translator matrix $\mathbf{T_R}$. Eq. (23) can be reexpressed as:

$$SNR - \log_{10} cond(\mathbf{T_R}) \leq SNR_\Gamma \leq SNR + \log_{10} cond(\mathbf{T_R}) \tag{24}$$

in which $SNR_\Gamma = -\log_{10} \varepsilon_{W_\Gamma}$ is the signal-to-noise ratio of the reconstructed pressure on the model's surface.

3.2 The modified error bounds

Above analysis is based on an assumption that the pressure can be completely decomposed by a set of modes. Otherwise, the Parseval law cannot be applied equivalently in evaluating the pressure energy. However, the complete set of modes is hardly to be satisfied in decomposing the radiated pressure of a realistic radiator, but an incomplete set is applied to approximately decompose the radiated pressure within a given tolerance. Therefore, a compromise on accuracy and robustness is made by truncating the series expansion $\mathrm{N}' = (N+1)^2$, Eq. (6), with a proper number N such as that given by Eq. (7). Due to the truncation, the measured pressure energy evaluated by the modal decomposition method on holograms is not equal to the true quantities, and so is the noise energy. It is also reasonable to define SNR on the hologram by Eq. (18), because once the radiated pressure is spatially measured by limited number of microphones, such as the way introduced in Section 2.2, the higher frequency components are filtered out which will not go through the inverse system.

Suppose the exact pressure on the surface of the vibrating structure is $p_{0,\Gamma}$ and the corresponding reconstructed pressure is p_Γ. Therefore, the relative error of the reconstructed pressure energy in discretized form is

$$\varepsilon_{W_\Gamma} = \frac{\left\| \mathbf{p}_{0,\Gamma} - \mathbf{p}_\Gamma \right\|_2^2}{\left\| \mathbf{p}_{0,\Gamma} \right\|_2^2} \tag{25}$$

According to the derivation in the appendix in the ref. [35], the relative errorε_{W_Γ} can be expressed as:

$$\varepsilon_{W_\Gamma}(c_1, c_2, c_3, c_4) = \frac{c_4 + c_3 10^{-SNR/10} + c_1\sqrt{c_4}\sqrt{c_3}10^{-SNR/20}}{1 + c_4 + c_2\sqrt{c_4}} \tag{26}$$

It reaches the lower bound at $c_1 = -2$ and $c_2 = 2$, and the upper bound at $c_1 = 2$ and $c_2 = -2$. However, Eq. (26) is a nonlinear function for variables c_3, and constrained nonlinear optimization algorithm is sought to find the lower and upper bounds, as the minimum of a problem specified by:

$$\min_{(c_3)} f(c_3) \textit{such that } cond(\mathbf{T_R})^- \leq c_3 \leq cond(\mathbf{T_R}) \tag{27}$$

where the objective function is

$$f(c_3) = \begin{cases} \varepsilon_{W_\Gamma}(-2,\ 2,\ c_3,\ c_4), \textit{for lower bound} \\ -\varepsilon_{W_\Gamma}(2,\ -2,\ c_3,\ c_4), \textit{for upper bound} \end{cases} \tag{28}$$

In the above analysis, the variables SNR and c_4 are supposed to be given. The SNR of the environment can be estimated by measurement. For the ideal cases in which there is no noise included, equivalent to SNR $= \infty$, the lower bound of the relative error is easy to be obtained as:

$$\varepsilon_{W_\Gamma} \geq \frac{c_4}{1 + c_4 + 2\sqrt{c_4}} \tag{29}$$

which is only related to the c_4 and in turn related to the number of adopted participant modes. The actual reconstructed error of a realistic problem or the case with small SNR are not expected to have a lower bound less than that estimated for no noise included case. Thus, the lower bound of the reconstructed pressure energy can be estimated by Eq. (29).

The variables c_4 is a crucial parameter for the error bounds estimation, which can either be evaluated by numerical simulation for specific problems or estimated by an analysis method. However, numerical simulation is hard to be realized for practical problems, since the source is not clear and it is the reason why the NAH is needed. Therefore, it is demand for developing a general analysis approach to properly estimate the variable c_4. Whereas, it is out of our mathematical ability as well as the range of the current work. In the following numerical examples, variable c_4 is estimated by the numerical method, a combination of finite element method (FEM) and BEM.

3.3 Characteristics of the translator

To investigate how much the output value of a function, such as the reconstructed quantities, can change for a small variation, such as the errors introduced in the experiment, in the input arguments, condition number of the function is one of the frequently used measure. Therefore, investigation of the condition number of translators in the NAH can somehow describe the stability of the reconstruction. Generally, numerical approach is applied to compute the condition number. However, if both shapes of the structure and holograms are conformal to sphere, a simple asymptotic expression of the condition number is available. The radii of spherical structure and holograms are denoted as r_Γ and r. Translators $\mathbf{R}_\Gamma$ and $\mathbf{R}$ are all diagonal matrices, taking the $\mathbf{R}$ as an example:

$$\mathbf{R} = \text{diag}\left\{ \bigcup_{n=0}^{N} \left[\overbrace{|rh_n(kr)|^{-} \quad \cdots \quad |rh_n(kr)|^{-}}^{2n+1} \right] \right\} \tag{30}$$

where diag$\{\mathbf{v}\}$ means a square diagonal matrix with the elements of vector $\mathbf{v}$ on the main diagonal, and $\bigcup_{n=0}^{N}\mathbf{v}_n$ returns a vector combing all the subvector $\mathbf{v}_n$. Thus,

$$\mathbf{T}_\mathbf{R} = \text{diag}\left\{ \bigcup_{n=0}^{N} \left[\overbrace{\frac{|r_\Gamma h_n(kr_\Gamma)|^2}{|rh_n(kr)|^2} \quad \cdots \quad \frac{|r_\Gamma h_n(kr_\Gamma)|^2}{|rh_n(kr)|^2}}^{2n+1} \right] \right\} \tag{31}$$

According to the analysis in Ref. [36], the asymptotic expression of $|xh_n(x)|$ for $n \gg x$ is

$$|xh_n(x)| \sim \sqrt{\frac{2}{e}} \left(\frac{2l+1}{e} \right)^l x^{-n} e^{x^2/4n} \tag{32}$$

which is actually the absolute value of the imagine part of the spherical Hankel function, since the real part goes rapidly to zero for $n \gg x$; therefore,

$$cond(\mathbf{T}_\mathbf{R}) = \frac{|r_\Gamma h_N(kr_\Gamma)|^2}{|rh_N(kr)|^2} \sim \left(\frac{r}{r_\Gamma} \right)^{2N} e^{k^2(r_\Gamma^2 - r^2)/2N} \sim \left(\frac{r}{r_\Gamma} \right)^{2N} \tag{33}$$

due to that $e^{k^2(r_\Gamma^2 - r^2)/2N}$ approximate rapidly to one for $N \gg kr_\Gamma$ and $N \gg kr$.

To investigate how much the reconstructed coefficients and in turn the pressure can change for a small variation in the local coefficients $\boldsymbol{\alpha}$, condition number of the translator $\mathbf{R}_\Gamma^- \mathbf{R}$ is the quantity needs to be studied. In light of the previous analysis, the condition number of the translation operator $\mathbf{R}_\Gamma^- \mathbf{R}$ for spherical structure and holograms satisfies

$$\text{cond}\left(\mathbf{R}_\Gamma^- \mathbf{R}\right) = \frac{|r_\Gamma h_N(kr_\Gamma)|}{|rh_N(kr)|} \sim \left(\frac{r}{r_\Gamma} \right)^N \tag{34}$$

Actually, the condition number of translator $\mathbf{T}_\mathbf{R}$ is a square power of that for the translator $\mathbf{R}_\Gamma^- \mathbf{R}$, i.e. $cond(\mathbf{T}_\mathbf{R}) = \text{cond}\left(\mathbf{R}_\Gamma^- \mathbf{R}\right)^2$. Eq. (34) indicates that the condition number has a geometric growth with N under the condition $N \gg kr_\Gamma$ and $N \gg kr$. Correspondingly, the larger condition number of the translator will obviously increase the sensibility to the inevitable errors introduced in the experiment. In addition, the large ratio of r/r_{Γ} could also increase the condition number for a fixed degree N. It implies that the same distance of the measurement to a smaller size surface will result in a reconstruction which is more prone to be contaminated. The condition numbers of the translator $\mathbf{R}_{\Gamma}^{-}\mathbf{R}$ for a spherical model with radius being 0.1 m and spherical holograms at frequency 601 Hz are given in **Figure 4**, which clearly validates the asymptotic form Eq. (34) to the exact one, for $N \gg kr_{\Gamma}$ and $N \gg kr$.

The asymptotic expression of the condition number of the translator $\mathbf{R}_{\Gamma}^{-}\mathbf{R}$ for general models and holograms are hardly to be obtained. Numerical method is resorted to get the condition number once the frequency, geometrical information of the model as well as holograms are supplied. To explore the influence of shapes of the model and holograms to the condition number, case studies of a cubic model with planar holograms and spherical holograms, which are representatives as conformal measurement and non-conformal measurement, are performed. The cubic model is same as that used in the numerical examples in Section 4. The planar holograms and spherical hologram also have the same configuration as that in the numerical examples. Two frequencies 601 Hz and 1333 Hz are analyzed. The condition numbers obtained by numerical method and asymptotic are presented in **Figure 5**. It is noticeable that the asymptotic condition number of the cubic model with planar holograms are evaluated with their equivalent radii. It can be observed that the asymptotic expression for planar holograms, which is a conformal hologram to the model, approximates to that obtained numerically very well for the frequency 601 Hz. However, the approximation does not show up for the frequency 1333 Hz up to the degree $N = 10$. It is because that the larger dimensionless value $k\tilde{r}$ needs larger degree to satisfy the condition of the asymptotic expression. The asymptotic estimation of the condition number does not work well for the spherical holograms due to the disparity in shapes with the cubic model. Even the ratio $r/\tilde{r}_{\Gamma}$ for the spherical hologram is smaller than that of the planar hologram, the condition number for the spherical hologram is inclined to be larger than that for the planar holograms as illustrated in **Figure 5**. It implicates that a cubic model with spherical hologram, a representative case of non-conformal measurement, may be more sensitive to measurement errors than that with planar hologram.

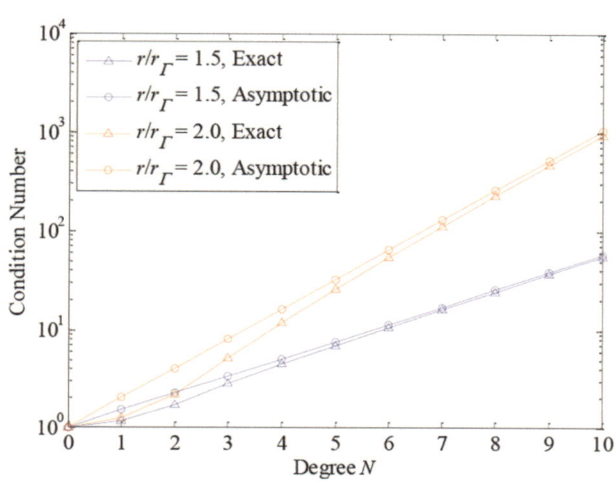

Figure 4.
The exact and asymptotic condition numbers of the translator $\mathbf{R}_{\Gamma}^{-}\mathbf{R}$ for different ratio r/r_{Γ} for spherical model and holograms.

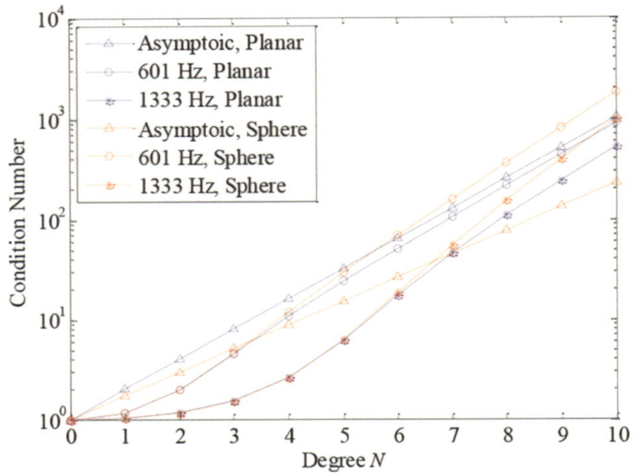

Figure 5.
The asymptotic and numerical condition numbers of the translator $\mathbf{R}_{\Gamma}^{-}\mathbf{R}$ for a cubic model with planar holograms and spherical holograms at two frequencies.

4. Experiment study

4.1 Numerical simulation

The necessary number of participant modes is hard to be obtained exactly for a realistic problem, and truncation error is introduced. In this case, the radiated source pressure on holograms is generated from a vibrating cubic model driven by a harmonic excitation. As shown in **Figure 6**, the cubic model is of size 0.2m × 0.2m × 0.2m and is excited by a harmonic force along z-axial at a specified position $(0.4a, 0.4a, -0.5a)$, and the four corners at the bottom are constrained. Thicknesses of the six walls are set as 0.004 m, and the steel material is assigned to the model. The harmonic response is obtained by a commercial finite element software at frequencies 601 Hz, which is chosen closely to the one modal frequency with an aim to obtain a uniformly distributed velocity on the surface. Once the boundary velocities are obtained, the radiated sound pressure at the measurement positions are computed by the boundary element method [37] as the inputs for the reconstruction. It is to simulate the realistic radiator whose exact number of efficient modes is hardly to be obtained but estimated by a reasonable guideline. Both of the cases include a specific amount of noise with prescribed SNR on the hologram. The noise is generated with

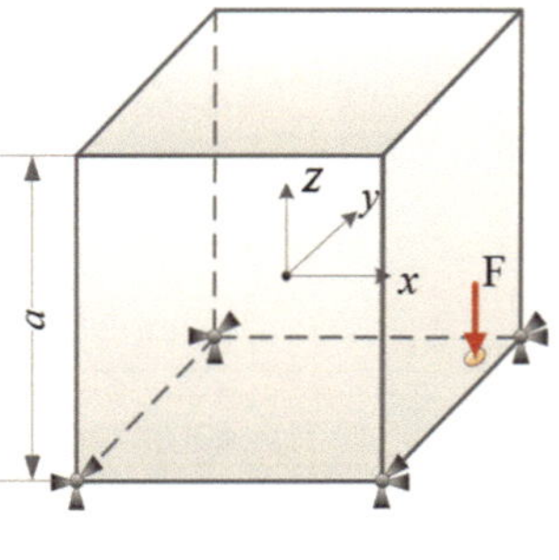

Figure 6.
Configuration of the cubic model excited by harmonic force F (denoted by the red arrow) along the z-axis.

$$\mathbf{n} = \gamma \mathbf{S} \boldsymbol{\lambda}_{Noise} \tag{35}$$

where $\boldsymbol{\lambda}_{Noise}$ is a Gaussian random vector, and $\gamma = 10^{-SNR/20} \left\| \mathbf{R}^{-1} \boldsymbol{\lambda}_{Noise} \right\|_2^{-} \sqrt{W_{P_0}}$ is an energy-related variable to make sure the generated noise **n** can form a specified SNR.

In this section, investigations are devoted to the reconstructions of the NAH with spherical holograms. To use the MRS-based NAH, the primary task is to determine the number of necessary modes. According to the method introduced in Section 2.1, the equivalent radius of the cubic model is $\tilde{r} = \sqrt{3/2\pi a}$ which results in dimensionless variables $k\tilde{r}$ equal to 1.53. The energy criteria Eq. (7) are adopted to determine the necessary number of FS due to its reasonable compromise on the accuracy and efficiency in the reconstruction. The tolerance is set as ε_T =1E-1, and it yields the necessary number of FS being 3 by referring to the diagram in **Figure 2**. The number and position of microphones are consequently determined by the approach in Section 2.3. To investigate the influence of the necessary number to the final results, reconstructions are performed for the necessary number ranging from 3 to 6 for each case.

There are errors due to the truncation of the participant modes. Therefore, reconstructions are firstly performed for the pressure $\boldsymbol{p}_{num}$ obtained numerically by the FEM and BEM which are also treated as the exact source pressures. The reconstructions are considered as the references of no-noise included measurement. Contour plots of the reconstructed pressure for reference cases are given in **Figure 7**. In spite of a slight disparity in the quantity, it can be observed that the reference reconstructions are very satisfied with the simulated pressure, because the significant pressure distributions are well reconstructed. Later on, different SNRs ranging from 4 dB to 28 dB with the increment being 4 dB are added to the simulated pressure to validate the robustness of the NAH to the noise which is unavoidable in the realistic experiment. The noise is obtained by Eq. (35).

Relative errors of reconstructed pressure energy on the model's surface is defined as:

$$\varepsilon_{W_r} = \frac{\left\| \boldsymbol{p}_{recon} - \boldsymbol{p}_{num} \right\|_2^2}{\left\| \boldsymbol{p}_{num} \right\|_2^2} \tag{36}$$

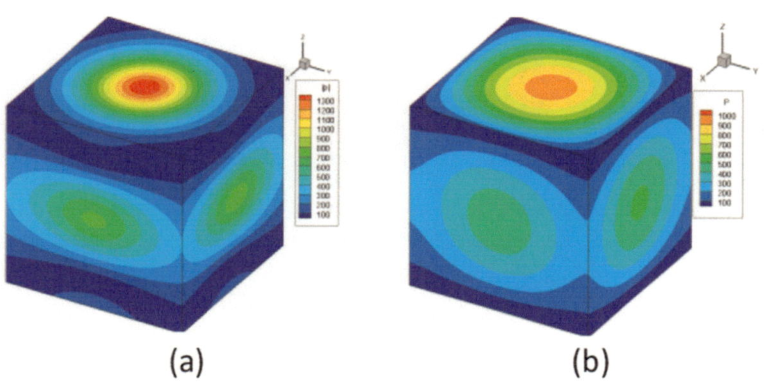

(a) (b)

Figure 7.
The reconstructed pressure on the boundary based on the no-noised included measurement on (a) the spherical holograms with the necessary number of modes N = 3 for (b) the simulated results at frequency 601 Hz.

where $\boldsymbol{p}_{recon}$ is the reconstructed pressure based on different SNRs. To estimate the error, variable c_4 is required to be supplied by:

$$c_4 = \frac{\|\boldsymbol{p}_{recon}\|_2^2}{\|\boldsymbol{p}_{recon} - \boldsymbol{p}_{num}\|_2^2} \tag{37}$$

based on the simulated results. According to bounds Eq. (26), normalized relative errors of pressure energy are defined as:

$$\tilde{\varepsilon}_{W_r} = \frac{\left(10^{SNR/10}\varepsilon_{W_r} - cond(\mathbf{T_R})^{-}\right)}{cond(\mathbf{T_R}) - cond(\mathbf{T_R})^{-}} \tag{38}$$

for different SNRs and holograms, which should satisfy $\tilde{\varepsilon}_{W_r} \in [0, 1]$.

The errors are plotted in **Figure 8**. **Figure 8a** depict that the most necessary number of modes for the boundary pressure reconstruction is $N = 3$. However, the reconstruction with $N = 4$ is superior to that of $N = 5$ in the sense of stability. Because the condition numbers of translators for $N = 5$ are larger which results in an inverse operation sensitive to errors. Actually, condition number for $N = 5$ is more than six times of that for $N = 4$ which reduces the reconstruction accuracy by the amplified errors. It is also the reason why the results obtained with $N = 5$ is worse than $N = 4$ for SNRs smaller than 20 dB even that it has a smaller radiation efficiency error in **Figure 2**. The decreased results for the reconstructed pressure with increased number of necessary modes illustrate that over-selected number of modes may result in accuracy loss, especially for models with irregular shapes as well as small SNRs. It is because that over-selected number of modes will yield translators with larger condition numbers which are likely to amplify the errors in the reconstruction. It is also observed from **Figure 8** that the two types of holograms can deliver the same level of accuracy with proper selected number of modes. However, the spherical hologram is more preferable since it requires the smallest number of microphones as compared with others.

The normalized errors are presented in of **Figure 8b**. The normalized errors are all less than 1 and increase along with the SNRs. The small errors for small SNRs and participant number of modes are due to the fact their lower bounds are

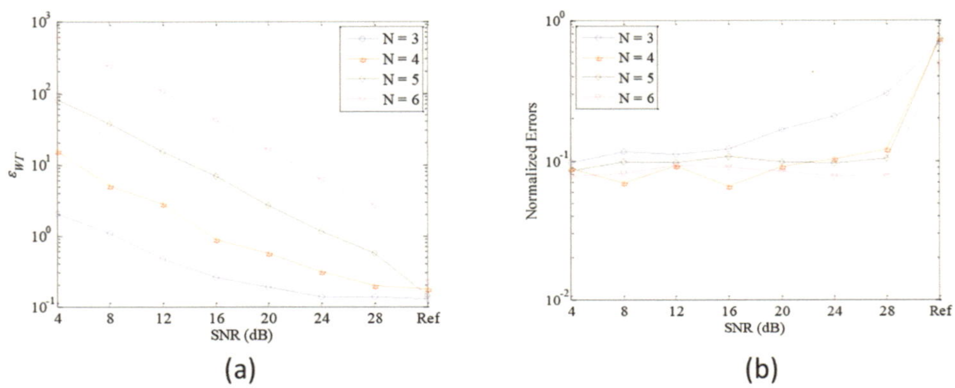

Figure 8.
Errors of the reconstructed pressure energy with spherical holograms at 601 Hz for (a) the relative error ε_{W_T} *and (b) the normalized error* $\tilde{\varepsilon}_{W_T}$.

underestimated, while the upper bounds are overestimated by the approach in Section 3.3. The extreme small normalized errors for case $N = 3$ are due to the dominated truncation errors for small number of adopted participant modes. As indicated for the reference cases, also the no-noise included cases, the reconstructed errors are closer to 1, or in other words more approximated to the estimated upper bounds. More important than the cases with small SNRs for which the estimated lower bounds are almost 0, the no-noise included cases (with infinity large SNR) can supply more reasonable lower bounds by the approach in Section 3.3. The numerical examples clearly demonstrated the validity of the proposed bounds estimation.

4.2 A practical experiment

An experiment is set up to explore the performance of the MRS-based NAH in this section. The source is in the same size and possesses the same material property as the one in the Section 4.1. Reconstruction is only preformed on the spherical hologram, since it requires the minimum number of microphones by comparing with the other two types of holograms.

An equipment is designed to facilitate the measurement. As shown in **Figure 9a**, the measurement on a spherical hologram is realized by rotating a half circular album arm, on which the microphones are mounted, around an axial which is the z-axial. The cubic model is placed at the center of the spherical hologram by hanging in a portal frame with a rigid hollow rod. A single-point drive is applied to the model by a small exciter on the top surface, as shown in **Figure 9b**. To make sure a uniform velocity distribution is generated on the surface, the analyzing frequency is selected closely to one of modal frequencies, which is 634 Hz. The model has an equivalent radius $\tilde{r} = 0.138$m, and the measurements are performed on a spherical hologram apart from the equivalent sphere by $\Delta = 0.1$ m. In light of Eq. (13), the necessary degree of the FS is adopted as $N = 3$. Correspondingly, the minimum required microphones along the θ and ϕ direction are 4 and 7, respectively. Here, we made an oversampling by placing five microphones along the θ direction and taking nine sequential measurements along ϕ direction.

To validate the reconstructed results, the same 5 by 9 measurements are performed on a spherical validation surface Ω with radius being 0.18 m. In light of Eq. (13), the

(a) (b)

Figure 9.
The experimental setup: (a) overview of the configuration and (b) details of the model.

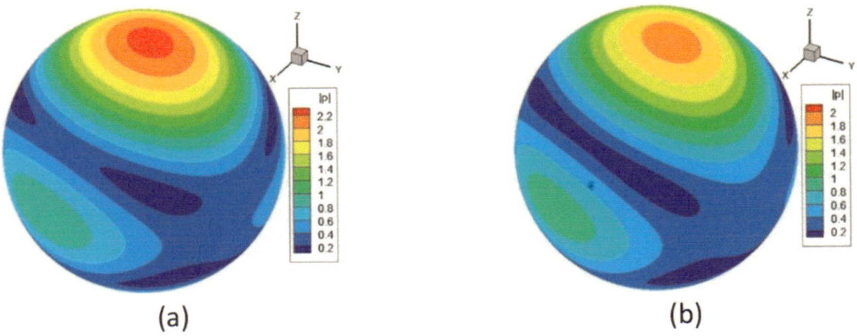

Figure 10.
The radiated pressure distribution on a spherical validation surface for (a) reconstructed from the hologram and (b) obtained from the direct measurement.

relative difference of reconstructed pressure with respective to the measured one on the validation surface is denoted by:

$$\varepsilon_{W_\Omega} = \frac{\left\| \sum_{n=0}^{(N+1)^2} (\lambda_n - \lambda_{n,\Omega}) p_n(k,\ \mathbf{x}) \right\|_2}{\left\| \sum_{n=0}^{(N+1)^2} \lambda_{n,\Omega} p_n(k,\ \mathbf{x}) \right\|_2} \tag{39}$$

where λ_n and $\lambda_{n,\Omega}$ are coefficients obtained on the hologram and validation surfaces, respectively. **Figure 10** depicts a comparison of the radiated pressure distribution on a spherical surface between the one reconstructed from the hologram and the one measured directly on the validation surface. It is clearly observed that the reconstructed pressure has a very satisfactory distribution agreement with the measured one for which the ε_{W_Ω} is 4.5% and the relative error of the maximum pressure is 21.8%.

The experiment is done in a semi-anechoic chamber which can reduce the influences of the environmental noise. Even that, positional errors and some other uncertainties are inevitable to be included in the measured signals which in turn affects the reconstruction results. The error analysis in Section 3 ascribes them to the SNR. How those relate to the SNR is crucial to the error estimation, which needs more investigation.

5. Conclusions

A NAH based on the mapping relationship between modes on surfaces of structure and hologram is introduced. The modes adopted in the NAH are FS of the Helmholtz equation in spherical coordinates which are generally independent and not orthogonal except on the spherical surface. The NAH framework provides a new insight to the reconstruction procedure based on the FS in spherical coordinates. The modes on the surface of structure and hologram form a bijective mapping. Number of modes prescribed in the MRS-based NAH is crucial to total number of measurement as well as the final reconstruction accuracy. An approach is proposed to estimate the necessary degree of effective modes. It is built on the energy criteria by exploring the radiation efficiency of the modes on the equivalent spherical source. An upper bounded error is derived for the radiated sound power of a vibrating structure with degree of modes up to a specific value. A relatively small value of degree is given by this approach.

Once the necessary degree of modes is determined, the number and position of microphones, which are also very crucial to the NAH, are investigated. Techniques to approximate the modes on three types of holograms by a set of locally orthogonal patterns are developed. A numerical algorithm is needed to determine the tight bounds for two locally orthogonal patterns on the planar patch. Due to the completeness of polar angles on cylindrical holograms, the algorithm is reduced by one dimension and the number of degree and positions are analytically determined for the local patterns along the polar direction. The number and position of measurement on the spherical hologram are determined by a purely analytical method because of its conformity to the coordinates of modes.

Errors are inevitable to be encountered in the NAH experiment. It is found that the reconstruction accuracy is subjected to two kinds of errors, one is the SNR and another one is the truncation error due to the limited number of participant modes adopted in the MRS-based NAH. An error model is built, and the relative error of the reconstructed pressure energy on the surface of the vibrating structure is derived. The lower and upper bounds of the relative error can be achieved numerically by a constrained nonlinear optimization algorithm. However, the approach generally yields underestimation of the lower bound and the overestimation of the upper bound, especially for MRS-NAH with large condition numbers. Alternatively, a reasonable lower bound is obtained by considering the case without noise or equivalently with positive infinite SNR. It eliminates the influence of the condition number of the inverse translator and is only related to the truncation errors. Thus, it is feasible to predicate the lower error of a reconstruction with the MRS-based NAH once the truncation error is given, which is validated by numerical examples. Proper estimation of the truncation errors is highly related to the reasonable estimation of the lower bound, which deserves more investigation.

Numerical examples are set up to validate the error analysis of the MRS-based NAH. It clearly demonstrates that the reconstructed results agree well with the simulated results. Physical experiment is designed to further demonstrate the feasibility and performance of the MRS-based NAH. The reconstructed results demonstrate a very satisfactory agreement with the direct measured one with respective to the quantities as well as the distribution on the validated surface. However, to estimate the performance with respective to the actual quantities by the proposed approach, it is desirable to investigate the influences of positional errors and other uncertainties on the SNR.

Acknowledgements

The work is supported by the National Natural Science Foundation of China (Grant No. 11404208).

Notes/thanks/other declarations

Place any other declarations, such as "Notes," "Thanks," etc. in before the References section. Assign the appropriate heading. Do NOT put your short biography in this section. It will be removed.

Author details

Haijun Wu[1,2*] and Weikang Jiang[1,2]

1 State Key Laboratory of Mechanical System and Vibration, Shanghai Jiao Tong University, Shanghai, China

2 Institute of Vibration, Shock and Noise, Collaborative Innovation Center for Advanced Ship and Deep-Sea Exploration, Shanghai Jiao Tong University, Shanghai, China

*Address all correspondence to: haijun.wu@sjtu.edu.cn

References

[1] Williams EG, Maynard JD, Skudrzyk E. Sound source reconstructions using a microphone array. Journal of the Acoustical Society of America. 1980;**68**(1):340-344

[2] Maynard JD, Williams EG, Lee Y. Nearfield acoustic holography: I. theory of generalized holography and the development of NAH. Journal of the Acoustical Society of America. 1985; **78**(4):1395-1413

[3] Veronesi WA, Maynard JD. Nearfield acoustic holography (NAH) II. Holographic reconstruction algorithms and computer implementation. Journal of the Acoustical Society of America. 1987;**81**(5):1307-1322

[4] Williams EG. Fourier Acoustics Sound Radiation and Nearfield Acoustical Holography. San Diego, Calif: Academic Press; 1998

[5] Steiner R, Hald J. Near-field acoustical holography without the errors and limitations caused by the use of spatial DFT. International Journal of Acoustics and Vibration. 2001;**6**(2): 83-8989

[6] Cho YT, Bolton JS, Hald J. Source visualization by using statistically optimized near-field acoustical holography in cylindrical coordinates. Journal of the Acoustical Society of America. 2005;**118**(4):2355-2364

[7] Hald J. Basic theory and properties of statistically optimized near-field acoustical holography. Journal of the Acoustical Society of America. 2009; **125**(4):2105-2120

[8] Kim GT, Lee BH. 3-D sound source reconstruction and field reprediction using the Helmholtz integral equation. Journal of Sound and Vibration. 1990; **136**(2):245-261

[9] Bai MR. Application of BEM (boundary element method)-based acoustic holography to radiation analysis of sound sources with arbitrarily shaped geometries. Journal of the Acoustical Society of America. 1992;**92**(1):533-549

[10] Veronesi WA, Maynard JD. Digital holographic reconstruction of sources with arbitrarily shaped surfaces. Journal of the Acoustical Society of America. 1989;**85**(2):588-598

[11] Zhang Z et al. A computational acoustic field reconstruction process based on an indirect boundary element formulation. Journal of the Acoustical Society of America. 2000;**108**(5 I): 2167-2178

[12] Zhang Z et al. Source reconstruction process based on an indirect variational boundary element formulation. Engineering Analysis with Boundary Elements. 2001;**25**(2):93-114

[13] Schuhmacher A et al. Sound source reconstruction using inverse boundary element calculations. Journal of the Acoustical Society of America. 2003; **113**(1):114-127

[14] Johnson ME et al. An equivalent source technique for calculating the sound field inside an enclosure containing scattering objects. Journal of the Acoustical Society of America. 1998; **104**(3 I):1221-1231

[15] Jeon IY, Ih JG. On the holographic reconstruction of vibroacoustic fields using equivalent sources and inverse boundary element method. Journal of the Acoustical Society of America. 2005; **118**(6):3473-3482

[16] Sarkissian A. Method of superposition applied to patch near-field acoustic holography. Journal of the Acoustical Society of America. 2005; **118**(2):671-678

[17] Bi CX et al. Nearfield acoustic holography based on the equivalent source method. Science in China, Series E: Technological Sciences. 2005;**48**(3): 338-353

[18] Bi CX, Chen XZ, Chen J. Sound field separation technique based on equivalent source method and its application in nearfield acoustic holography. Journal of the Acoustical Society of America. 2008;**123**(3): 1472-1478

[19] Bi CX, Bolton JS. An equivalent source technique for recovering the free sound field in a noisy environment. Journal of the Acoustical Society of America. 2012;**131**(2):1260-1270

[20] Song L, Koopmann GH, Fahnline JB. Numerical errors associated with the method of superposition for computing acoustic fields. Journal of the Acoustical Society of America. 1991;**89**(6): 2625-2633

[21] Fahnline JB, Koopmann GH. A numerical solution for the general radiation problem based on the combined methods of superposition and singular-value decomposition. Journal of the Acoustical Society of America. 1991; **90**(5):2808-2819

[22] Koopmann GH, Song L, Fahnline JB. A method for computing acoustic fields based on the principle of wave superposition. Journal of the Acoustical Society of America. 1989;**86**(6): 2433-2438

[23] Bai MR, Chen CC, Lin JH. On optimal retreat distance for the equivalent source method-based nearfield acoustical holography. Journal of the Acoustical Society of America. 2011;**129**(3):1407-1416

[24] Wu SF. On reconstruction of acoustic pressure fields using the Helmholtz equation least squares method. Journal of the Acoustical Society of America. 2000;**107**(5 I): 2511-2522

[25] Wang Z, Wu SF. Helmholtz equation-least-squares method for reconstructing the acoustic pressure field. Journal of the Acoustical Society of America. 1997;**102**(4):2020-2032

[26] Wu SF, Yu JY. Reconstructing interior acoustic pressure fields via Helmholtz equation least-squares method. Journal of the Acoustical Society of America. 1998;**104**(4):2054-2060

[27] Wu SF. Methods for reconstructing acoustic quantities based on acoustic pressure measurements. Journal of the Acoustical Society of America. 2008; **124**(5):2680-2697

[28] Thite AN, Thompson DJ. The quantification of structure-borne transmission paths by inverse methods. Part 1: Improved singular value rejection methods. Journal of Sound and Vibration. 2003;**264**(2):411-431

[29] Williams EG. Regularization methods for near-field acoustical holography. Journal of the Acoustical Society of America. 2001;**110**(4): 1976-1988

[30] Lu HC, Wu SF. Reconstruction of vibroacoustic responses of a highly nonspherical structure using Helmholtz equation least-squares method. Journal of the Acoustical Society of America. 2009;**125**(3):1538-1548

[31] Bi CX et al. Reconstruction of the free-field radiation from a vibrating structure based on measurements in a noisy environment. Journal of the Acoustical Society of America. 2013; **134**(4):2823-2832

[32] Abramowitz M, Stegun IA. Handbook of Mathematical Functions with Formulas, Graphs, and Mathematical Tables. Washington: U.S. Govt. Print. Off; 1964

[33] Wu HJ, Jiang WK, Zhang HB. A mapping relationship based near-field acoustic holography with spherical fundamental solutions for Helmholtz equation. Journal of Sound and Vibration. 2016;**373**(7):66-88

[34] Wu HJ, Liu YL, Jiang WK. A fast multipole boundary element method for 3D multi-domain acoustic scattering problems based on the Burton-miller formulation. Engineering Analysis with Boundary Elements. 2012;**36**(5):779-788

[35] Wu HJ, Jiang WK. Experimental study of the mapping relationship based near-field acoustic holography with spherical fundamental solutions. Journal of Sound and Vibration. 2017;**394**: 185-202

[36] Rahola J. Diagonal forms of the translation operators in the fast multipole algorithm for scattering problems. BIT Numerical Mathematics. 1996;**36**(2):333-358

[37] Wu HJ, Liu YJ, Jiang WK. A low-frequency fast multipole boundary element method based on analytical integration of the hypersingular integral for 3D acoustic problems. Engineering Analysis with Boundary Elements. 2013; **37**(2):309-318

Section 2

Recent Developments and Applications

Chapter 9

Synthesis of Nano-Optical Elements for Forming 3D Images at Zero Diffraction Order

Anton Goncharsky and Svyatoslav Durlevich

Abstract

A method is proposed to compute and synthesize a nano-optical element to produce a new visual effect: a 3D image formed in the vicinity of zero diffraction order. Usual relief rainbow holograms or OVDs can form 3D effect, but at +1 or − 1 diffraction order only and they provide 3D parallax in left/right direction only, and after rotation/inclination of an element, a 3D image changes its color and further disappears completely. The new visual effect provides with full 3D parallax. Moreover, a 3D zero-order image is well visible when an optical element is rotated through 360 degrees; the color of 3D image does not depend on the viewing angle. A synthesis technology is developed incorporating the computation of scattering patterns in elementary areas, computation of the phase function of the entire optical element, and the formation of its microrelief by using e-beam lithography. The microrelief consists of multilevel kinoforms with an accuracy of 10 nm in terms of depth. It was demonstrated by experimental results that the new visual effect is easy for visual perception under white light illumination. A sample of nano-optical element is manufactured, which when illuminated by white light, forms a 3D image in the vicinity of zero-order of diffraction (video available at: https://bit.ly/3QtzxbI).

Keywords: DOE, CGH, 3D imaging, e-beam lithography, kinoform, nano-optical elements, ill-posed problems, surface relief holograms

1. Introduction

The development of methods for synthesizing optical elements to form 3D images began in the 1970s after Denis Gabor was awarded the Nobel Prize for the development of holographic recording principle [1]. Gabor's follower [2] developed method for recording 3D holograms, which formed 3D images when illuminated by a white point light source. It was the invention of the surface relief holograms, which became known as rainbow holograms. These holograms formed visual 3D parallax in the left/right direction only, and when tilted up/down, the 3D image changes its color, that's why they were called rainbow holograms. The 3D effect was formed in 1st diffraction order. In 1976, it was suggested the approach for forming 3D image at 1st diffraction order using binary computer-generated holograms [3]. Later, it was invented efficient technology of rainbow holograms microrelief replication (embossing technology) and rainbow holograms became widely used for protection against counterfeit [4]. The first optical security element used on Visa credit cards was a rainbow hologram with the original recorded on the optical table using an analog laser recording technique. At the same time, such optical elements appeared on bank notes and IDs.

The microrelief can be formed using both electron-beam lithography and optical recording methods. Modern approaches for synthesizing optical elements using laser microrelief recording have a resolution of 0.5 microns at best [5, 6], which is insufficient for the recording of asymmetric microrelief of optical elements that produce 3D full-parallax images. In this chapter, we form asymmetric microrelief of nano-optical elements using e-beam lithography with a resolution of 0.1 microns [7, 8]. This makes it possible to synthesize nano-optical elements that cannot be reproduced using standard widespread methods of microrelief recording based on optical origination techniques.

E-beam lithography has already been used to synthesize nano-optical elements that form 3D images [9, 10]. In these studies, a 3D image was formed at the first order of diffraction with the microrelief of the optical element shaped using symmetrical structures. The resulting 3D images can be observed only within a limited range of viewing angles near the first order of diffraction when the element is tilted left/right and up/down; however, the 3D image formed disappears when the element is rotated.

We discuss the possibilities of synthesizing nano-optical elements to form 3D images at the zero-order of diffraction. This is the first time that methods of synthesizing nano-optical elements to form 3D images in the zero-order of diffraction have been developed. For such elements, the 3D effect can be observed near the zero-order of diffraction both by tilting the element and rotating it by 360 degrees.

2. Formulation of the problem of the synthesis of nano-optical elements for the formation of 3D images at zero diffraction order

The standard scheme for observing a 3D image formed by a rainbow hologram [2] is shown in **Figure 1a**, where the area of observation (the area in which the observer's eyes can be located) is indicated by the yellow band. The 3D image is formed in the first order of diffraction, and the observation area is a limited narrow band. An observer can see 3D parallax in the left/right direction only. When the optical element is slightly tilted or rotated, observer's eyes leave the area of observation, and the 3D image disappears for an observer.

Figure 1b shows the observation scheme of a 3D image with full parallax at the 1st diffraction order [11]. The nano-optical element consists of calculated binary kinoforms recorded by e-beam lithography. Here the area of observation is a thick rectangle. An observer can see 3D parallax in both left/right and up/down directions. However, the viewing angle range is also limited, and after rotation of the optical element observer's eyes leave the area of observation and the 3D image disappears.

Figure 1c shows the new proposed authors' observation scheme of a 3D image that is formed in the vicinity of the zero order of diffraction. The observation area is a large square centered on the zero order. As long as observer's eyes are within this region, an observer sees a full parallax 3D image, and the 3D image is observed over a wide range of tilt angles and even when the optical element is rotated through a full range of 360 degrees.

DOI: http://dx.doi.org/10.5772/intechopen.106145

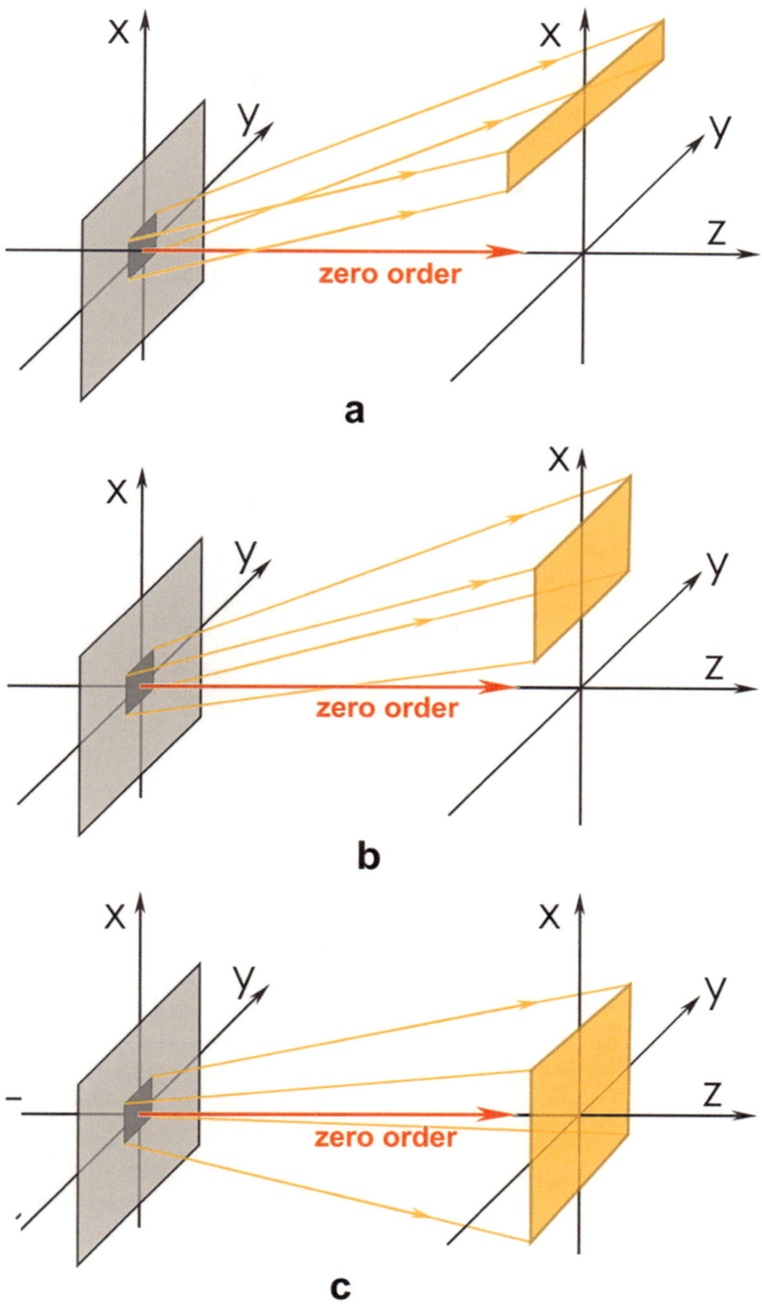

Figure 1.
Scheme for observing 3D image: a) formed by rainbow hologram at 1st diffraction order, b) formed by CGH with full 3D parallax at 1st diffraction order, and c) in the vicinity of zero order.

Figure 2 schematically shows the formation of 3D images by a flat reflecting optical phase element at diffraction angles within plus or minus 30° of the zero order.

The optical element is located in the $z = 0$ plane. **Figure 2** shows a fragment of observation points (five points along the horizontal direction and three points along the vertical direction). The centers of observation points are marked with the letter

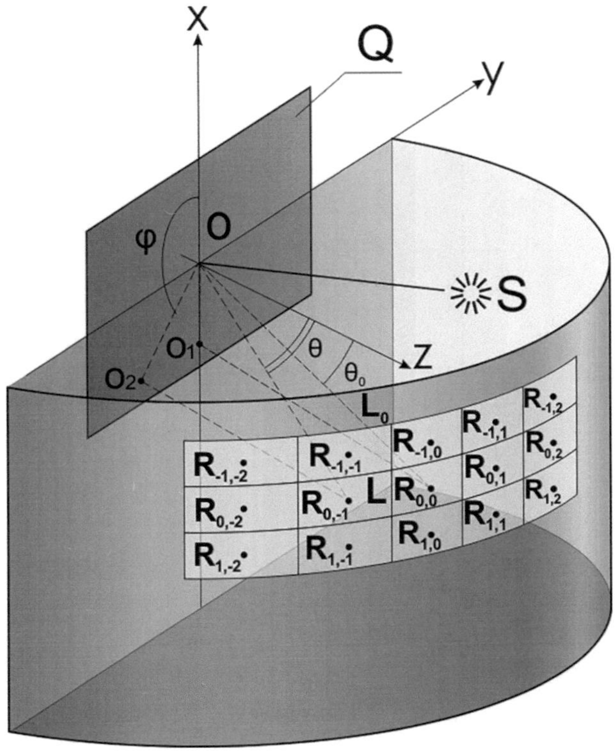

Figure 2.
Schematic diagram of the formation of 3D images by a flat optical element.

"R". The number of frames is several hundred for real optical elements that form a zero-order 3D image. The light source "S" is located in the Oxz plane of the Cartesian coordinate system. The source is located at an angle of θ_0 to the Oz axis. The direction to the zero order is denoted as L_0. At different angles φ,θ, the observer sees different 2D frames K_n, n = 1 ... N of a 3D image. Here, φ and θ are angles in the spherical coordinate system. The angle θ is counted from the Oz axis, and φ is the azimuthal angle. The ray L in **Figure 2** points towards one of the observation points and has angular coordinates φ,θ. Let us assume that the angles (φ_n, θ_n) set the directions towards the observation point of frame K_n, n = 1 ... N.

Figure 3 shows the observation scheme in the O*xz* plane at small diffraction angles. The diffraction angle is defined as the angle between the zero order of diffraction and the direction towards the observation point. Let us denote the diffraction angle as β. For small diffraction angles, the angle is defined by the formula $\beta = \theta - \theta_0$. A 3D image is observed at diffraction angles within 30° of the zero order of diffraction. The angle θ_0 between the light source S and the normal to the plane of the optical element coinciding with the Oz axis in the diagram determines the zero-order diffraction by beam L_0.

Synthesis of a nano-optical element to form zero-order 3D images is quite a complex and challenging task. If we use a grid with a step of 0.1 x 0.1 micron for an optical element of, for example, 28×33 mm^2 size, then the number of points at which it is necessary to calculate the phase function of the optical element is about 10^{11}. However, the proposed method for calculating the phase function of an optical element efficiently solves this problem.

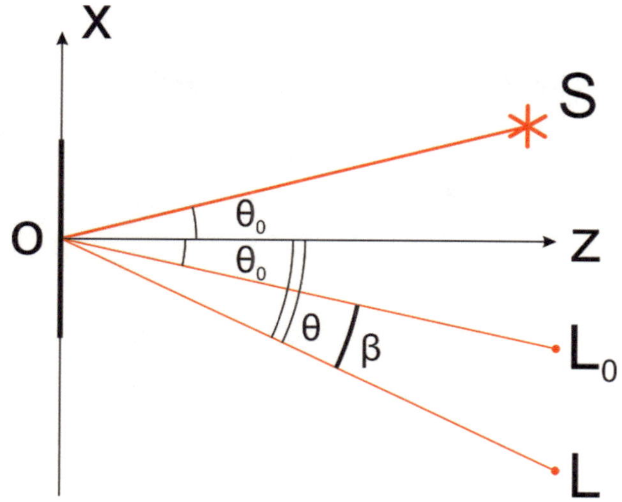

Figure 3.
Scheme of observation at small diffraction angles.

The method that we propose for the first time in this chapter allows the use of different 3D models to form 3D greyscale images. To demonstrate the method for calculating the phase function of the diffractive optical element (DOE) a 3D object was chosen. **Figure 4** shows a computer 3D model of the object.

Figure 5 shows a fragmented set of the 2D frames of the 3D computer model, and **Figure 6** presents a scheme of partitioning the flat optical element into the elementary regions (G_{ij}). The size of the elementary region does not exceed 80 microns, which is beyond the resolution of the human eye.

Figure 4.
3D model of the object.

Figure 5.
Fragment of the 2D-frames of the 3D computer model.

Figure 7 schematically shows the formation of the angular pattern in the elementary area G_{ij} i = 1 ... L, j = 1 ... M. The formation involves all rays from the center of the elementary area to all observation points (R). The ray L_n directed towards the center of the observation point K_n is defined by the angles φ_n, θ_n. The number of rays coincides with the number of 2D frames of the 3D image and is equal to several hundred. The intensity of beam L_n in the direction (φ_n,θ_n) for each n, n = 1 ... N, is determined as follows. The brightness of point (x_i,y_j) in frame K_n determines the intensity of beam L_n.

The angular pattern of the light scattered from each elementary region G_{ij} is formed at all observation angles (φ_n, θ_n) of the 3D image. Here, n = 1 ... N. The angular pattern of the region G_{ij} is a set of N rays, and each ray L_n has a given intensity. **Figure 8** shows the angular patterns computed for twelve elementary G_{ij} regions.

Thus, in the approximation of geometrical optics, radiation angular patterns are determined for all elementary regions G_{ij}. In the next step, we use the determined angular patterns to compute the phase function of the optical element for each elementary region G_{ij}.

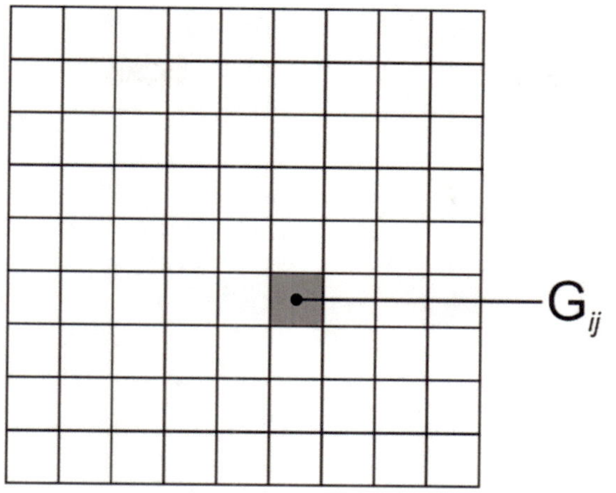

Figure 6.
Schematic diagram of the partitioning of an optical element into elementary regions.

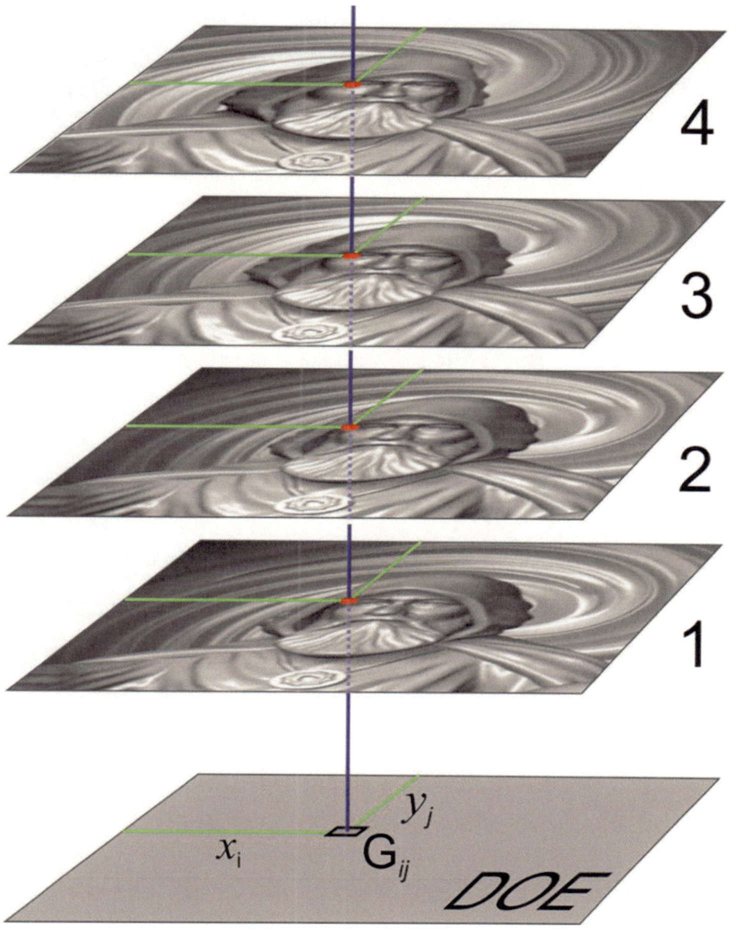

Figure 7.
Schematic diagram of the formation of the angular pattern of the area G_{ij}.

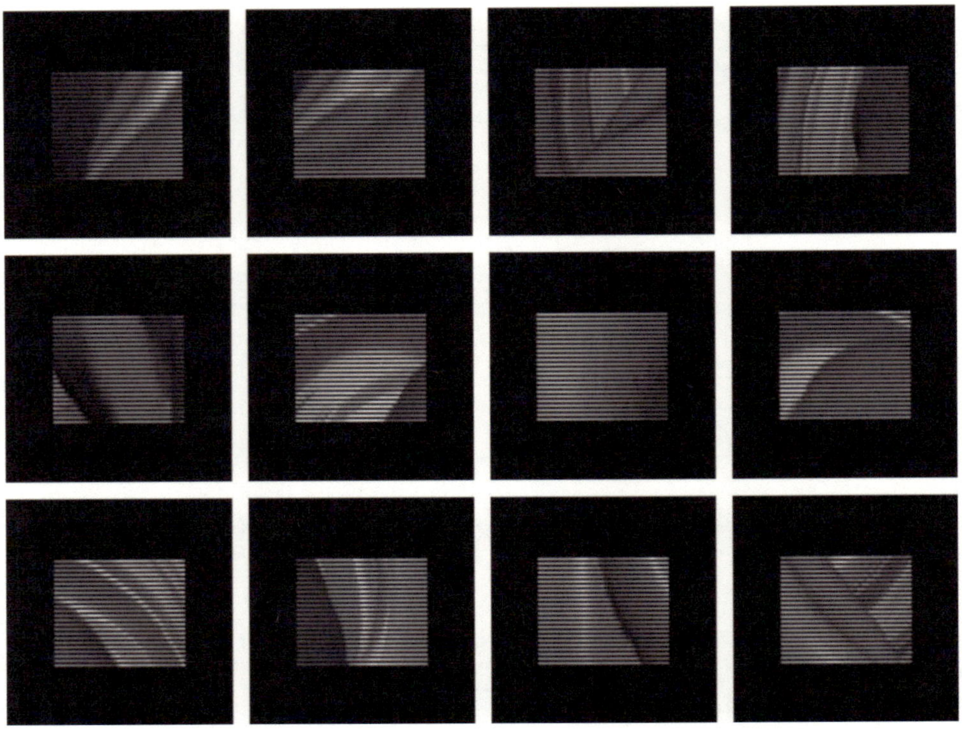

Figure 8.
Angular patterns of twelve different areas G_{ij}.

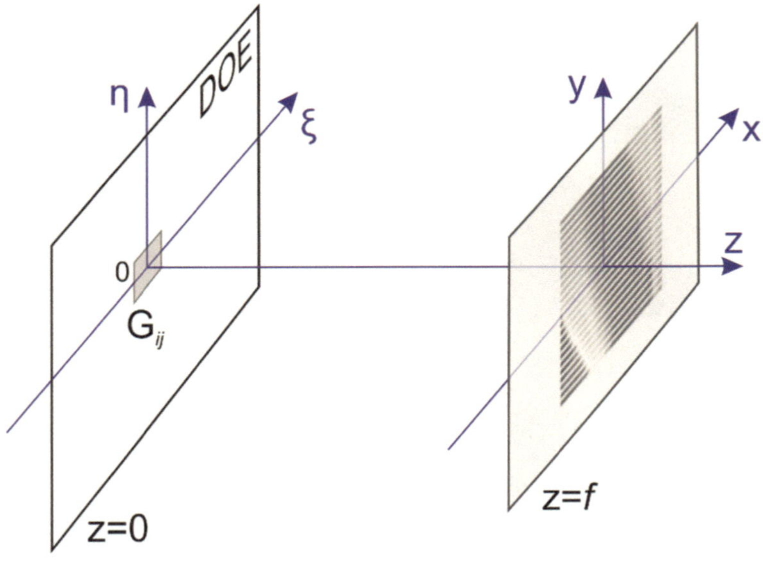

Figure 9.
Optical scheme of the formation of the image in elementary region G_{ij}.

3. Method for computing the phase functions in elementary regions

We use the scalar Fresnel wave model to compute the phase functions in the all elementary G_{ij} regions. In this Fresnel model, the scalar wave field $u(x,y,f)$ in the $z = f$ plane is related to the scalar wave field $u(\xi,\eta,0)$ by the following formula:

$$\gamma \iint_{G_{ij}} u(\xi,\eta,0) \exp\left\{ ik \frac{(x-\xi)^2 + (y-\eta)^2}{2f} \right\} d\xi d\eta = u(x,y,f). \tag{1}$$

Here $k = 2\pi/\lambda$ and $\gamma = \exp.(ikf)/i\lambda f$ is a given constant where λ is a wavelength. **Figure 9** shows a scheme of the formation of the 2D image formed by the angular pattern of the elementary region G_{ij}. The plane wave falls onto the reflecting flat phase optical element whose microrelief forms an image in the $z = f$ plane.

The peculiarity of the inverse problem of forming a 2D image is that the right-hand side of Eq. (1) does not contain the wave function $u(x,y,f)$ but only its absolute value $F(x,y) = |u(x,y,f)|$.

Let us represent the wave function on the plane z = 0 in the form $u(\xi,\eta,0) = \overline{u}(\xi,\eta) \exp(ik\varphi(\xi,\eta))$. Here $\overline{u}(\xi,\eta)$ is the amplitude and $\varphi(\xi,\eta)$ is the phase function of the planar optical element in the elementary region G_{ij}, We, thus, have the following operator equation:

$$A\varphi(\xi,\eta) = F(x,y). \tag{2}$$

In the Fredholm operator equation of the first kind (2) $F(x,y)$ is a given function. The operator A is defined by the following relation:

$$A\varphi = \left| \gamma \iint_{G_{ij}} \bar{u}(\xi,\eta) \exp(ik\varphi(\xi,\eta)) \exp\left\{ ik \frac{(x-\xi)^2 + (y-\eta)^2}{2f} \right\} d\xi d\eta \right|. \tag{3}$$

Eq. (3) is a nonlinear operator equation with respect to the desired function $\varphi(\xi,\eta)$ and describes an ill-posed problem. Efficient numerical algorithms were developed to solve ill-posed nonlinear problems [12, 13]. However, one of the most efficient methods for the approximate solution of Eq. (2) is the method proposed by Lesem and his colleagues in 1969 [14]. This method later came to be called the Gerchberg-Saxton algorithm [15]. Many studies were dedicated to investigating this algorithm [16–18]. All these methods have the same property. The value of the functional decreases monotonically quite rapidly during the first 10 iterations, and then the decrease rate falls off rapidly.

We follow Lesem [14] to use an algorithm for the approximate solution of nonlinear Eq. (2). Let us introduce the following notation:

$$\Phi\{\nu\}(x,y) = \gamma \iint_{G_{ij}} \nu(\xi,\eta) \cdot exp\left(ik \frac{(x-\xi)^2 + (y-\eta)^2}{2f} \right) d\xi d\eta. \tag{4}$$

Here $\Phi\{\nu\}(x,y)$ is the Fresnel transform of function v. We construct the iterative process of building the phase function that is an approximate solution of inverse problem (3) as follows. Four steps have to be taken to perform one iteration in the iterative algorithm for solving problem (3). Let $v^{(m)}(x,y)$ be given at the m-th

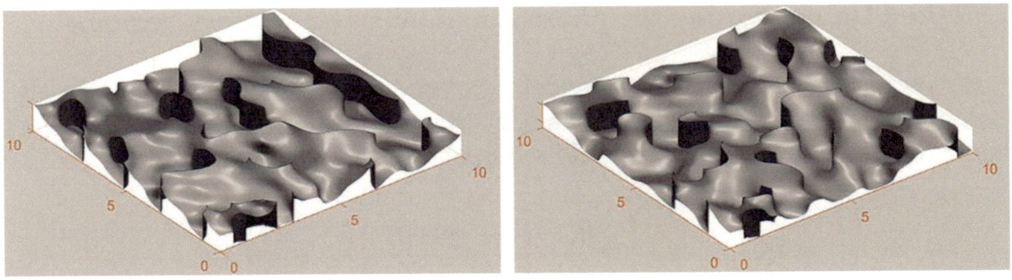

Figure 10.
Calculated fragments of the microrelief in the two elementary G_{ij} regions.

iteration. We write function $v^{(m)}(x,y)$ in the form $v^{(m)}(x,y) = A_0\exp(ik\varphi_0{}^{(m)}(x,y))$ and function $w^{(m)}(x,y)$ in the form $w^{(m)}(x,y) = A_1\exp(ik\varphi_1{}^{(m)}(x,y))$. Both A_0 and A_1 are real functions. Let $A_0(x,y)$ be the given intensity distribution of incident light in the $z = 0$ plane. As we are considering phase only DOE, then the amplitude $A_0(x,y)$ will be equal to one in the elementary region G_{ij}. Let $A_1(x,y) = F(x,y)$ be the given intensity distribution in the focal plane $z = f$. The algorithm for solving the inverse problem consists of the following four steps performed in sequence:

$$\begin{aligned}
&1)\ \varphi_1^{(m)}(x,y) = arg\left(\Phi\{v^{(\mathrm{m})}\}(x,y)\right)\\
&2)\ w^{(m)}(x,y) = A_1(x,y)\,exp\left(ik\varphi_1^{(m)}(x,y)\right)\\
&3)\ \varphi_0^{(m+1)}(x,y) = arg\left(\Phi^{-1}\{w^{(m)}\}(x,y)\right)\\
&4)\ v^{(m+1)}(x,y) = A_0(x,y)\,exp\left(ik\varphi_0^{(m+1)}(x,y)\right)
\end{aligned} \tag{5}$$

The function $\varphi_0^{(m+1)}$ is an approximate solution of Eq. (2). A phase distribution equal to a constant can be used as an initial approximation. The phase function $\varphi(\xi, \eta)$ computed by the iterative process (5) uniquely determines the microrelief in the region G_{ij}. For example, for a normal wave incident on an optical element, the depth of the microrelief in the region G_{ij} is equal to 0.5 $\varphi(\xi, \eta)$ for any point (ξ, η) in this region.

Figure 10 shows two calculated fragments of the microrelief of the multilevel kinoform in the two elementary G_{ij} regions. The fragments size is $10 \times 10\ \mu m^2$. The depth of the microrelief does not exceed 0.5λ and is equal to approximately 0.3 μm.

Thus, the solution of the inverse problem for each elementary region, G_{ij}, i = 1... L, j = 1... M, yields the microrelief on the entire area of the nano-optical element. The above algorithm for computing the phase function can be applied to the 3D model of any 3D object.

4. Example of the synthesis of multilevel nano-optical element for forming 3D images at zero diffraction order

To demonstrate the efficiency of the proposed method, we made a $28 \times 33\ mm^2$ nano-optical element to form a zero-order 3D image. A $28 \times 33\ mm^2$ flat optical element was partitioned into 369,600 $50 \times 50\ \mu m^2$ elementary G_{ij} regions, i = 1... L, j = 1... M, as shown in **Figure 6**. The number of frames N was 1440 (60 frames horizontally, 24 frames vertically). We compute the microrelief of the flat optical

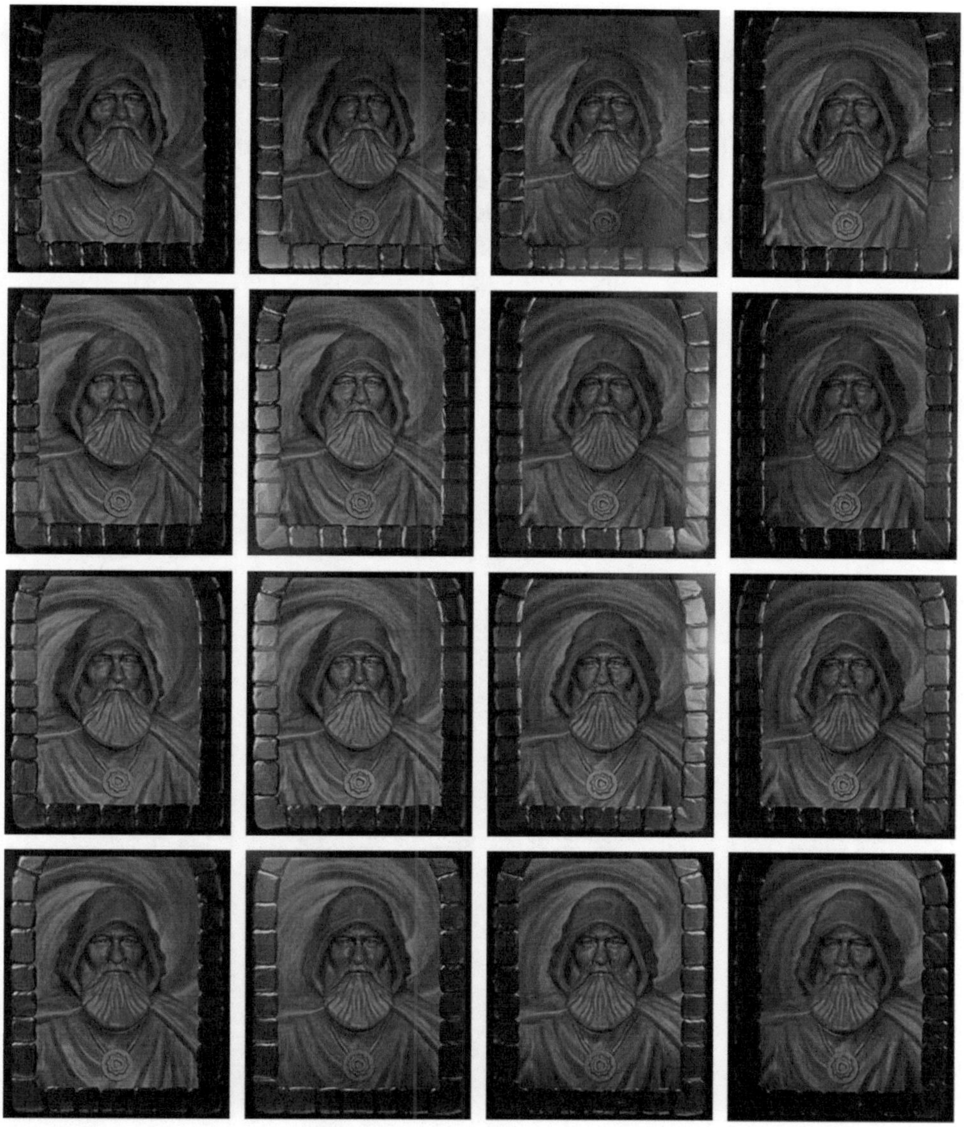

Figure 11.
3D images formed by the nano-optical element at different angles under green light source (see video V1 and video V2) (video available at: https://bit.ly/3QtzxbI).

element at the fixed green wavelength λ = 547 nm for each elementary region G_{ij}. To compute the phase function in the area G_{ij}, it was used a 500 $\times$ 500 grid to solve the inverse problem (2) of computing the phase functions in the elementary regions, and it takes more than 4 hours to compute the phase function for the entire optical element on a PC (AMD Phenom II X6 3.2 GHz CPU and 16 Gb DDR3 memory).

We used a shaped beam e-beam lithography system with a minimum beam size of 0.1 μm x 0.1 μm to record the microrelief of the nano-optical element and used a positive PMMA electron beam resist to form the nano-structures. The maximum microrelief depth was 0.3 μm, and the depth accuracy of microrelief formation was 10 nanometers in terms of depth. The nickel master shim of the diffractive optical element was produced using a standard electroforming procedure.

Figure 11 shows photographs of the nano-optical element taken from different viewing angles at diffraction angles of plus or minus 30° relative to the zero order of diffraction. A cell phone flash with green color filter was used as the light source.

We computed the microrelief at the fixed wavelength of λ = 547 nm, which corresponds to green light, and the quality of the images formed are good when the nano-optical element is under green light source illumination. However, of course, the main interest is how the element will be seen when illuminated with a white light source. **Figure 12** shows photographs of the same nano-optical element taken from different viewing angles by using a white light source, cell phone flash without any filter.

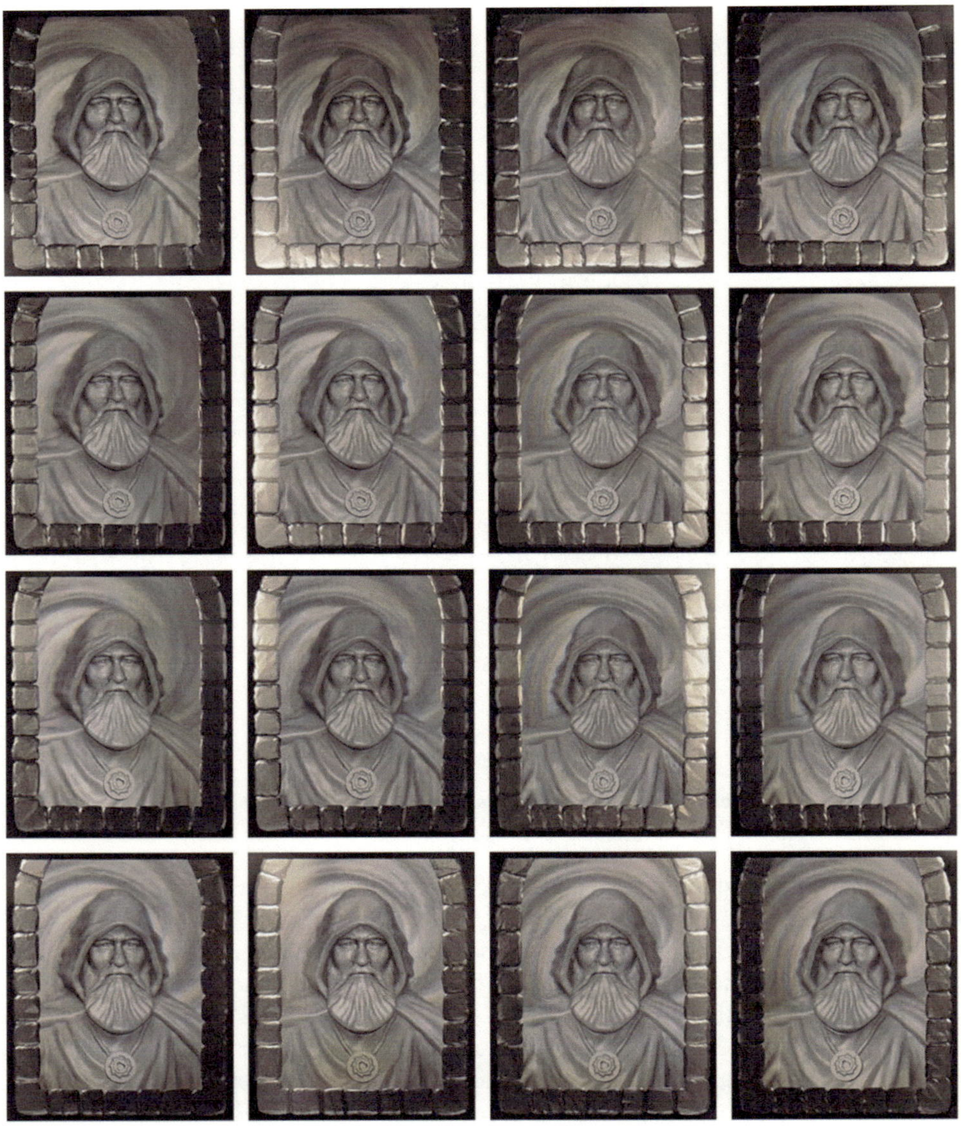

Figure 12.
3D images formed by the nano-optical element at different angles under white light source (see video W1 and video W2) (video available at: https://bit.ly/3QtzxbI).

As can be seen from the **Figure 12**, the formed 3D image remains clear and contrasting despite the illumination with a white light source. The resulting 3D image can be observed well when illuminated by white light, and the observer sees the 3D image with full parallax both when tilting the optical element and when rotating it by 360 degrees. In addition, unlike rainbow holograms, the color of the formed 3D image does not depend on the viewing angle. That is, it turns out that the formed 3D image behaves like a real 3D object in a full range of viewing angles.

5. Discussion and conclusion

In this chapter, we develop methods for synthesizing nano-optical elements to form 3D images at the zero-order diffraction for the first time. The synthesis methods include both the computation of the phase function of the nano-optical element and the formation of its microrelief by using of e-beam lithography. From a mathematical point of view, the computation of the phase function is a typical inverse problem, which we solve in two steps. In the first step, we use all the image frames that define a 3D object to generate the angular patterns in each elementary region. In the second stage, we compute the phase functions of the nano-optical element in each elementary region. The latter problem reduces to solving a nonlinear integral equation. Despite the large number of elementary regions (~370,000), a personal computer is definitely sufficient to compute the phase function of the entire nano-optical element.

We used e-beam lithography for the formation of the microrelief. The accuracy of microrelief formation is 10 nanometers in terms of depth. We produced a sample nano-optical element that forms a 3D image in the zero order of diffraction. The resulting 3D image can be observed when illuminated by white light. A 3D image can also be formed in the first order of diffraction, as we did, for example, in our earlier study [11], using a binary microrelief. In this case, the diffraction efficiency of the optical element does not exceed 40%. The use of multilevel microrelief makes it possible not only to increase the diffraction efficiency but also to significantly widen the viewing angles of the 3D image. The observer sees a 3D image with full parallax both when tilting the optical element and even when rotating it by 360 degrees. The 3D image is stable and behaves like a real 3D object. The authors also believe that it is possible to use 2D frames captured from 3D computer models with some animation, and thus the formation of 3D zero order images with the effect of animation is possible; however, this is the subject of future numerical and real experiments on the formation of 3D animated images in the zero order of diffraction.

The structure of an optical element forming a 3D image in the zero order of diffraction can be modified to make the kinoforms fill the G_{ij} regions partially rather than completely [19, 20]. The remaining parts of the elementary G_{ij} regions can be filled with diffraction gratings with periods less than 0.6 μm. These diffraction gratings can form an additional 2D image visible over the entire area of the DOE at diffraction angles greater than 60 degrees.

The nano-optical element can be replicated using standard embossing equipment for the production of surface relief holograms. The synthesis methods developed are designed first of all to protect bank notes, IDs and brands against counterfeiting. The technology of the synthesis of nano-optical elements is expensive, knowledge intensive and not widespread, thereby ensuring high protection level of the developed DOEs against counterfeiting.

Methods for calculating the phase functions of nano-optical elements can be used in advanced 3D displays and 3D projectors. At present, supercomputer technologies are widely used to speed up calculations. The phase function in each elementary region is calculated independently, which makes it easy to parallelize the numerical algorithm. The use of a graphics processing unit (GPU) cluster can speed up the calculation of the phase function of a nano-optical element by hundreds or even thousands of times. At present, processors with hundreds thousand cores have been developed and are available [21]. The use of such technologies can make it possible to compute the phase function of the entire optical element in a fraction of a second, thereby opening up opportunities for the synthesis of animated 3D images in prospective 3D design systems and 3D displays [22, 23].

Author details

Anton Goncharsky* and Svyatoslav Durlevich
Research Computer Center, M.V. Lomonosov Moscow State University, Moscow, Russia

*Address all correspondence to: gonchar@srcc.msu.ru

References

[1] Gabor D. A new microscopic principle. Nature. 1948;**161**:777-778

[2] Benton SA. Hologram reconstructions with extended incoherent sources. Journal of the Optical Society of America. 1969;**59**:1545-1546

[3] Yatagai T. Stereoscopic approach to 3-D display using computer-generated holograms. Applied Optics. 1976;**15**: 2722-2729

[4] Van Renesse RL. Optical Document Security. Artech House, Boston, USA: Artech House Optoelectronics Library; 2005

[5] Van Renesse RL. Security aspects of commercially available dot matrix and image matrix origination systems. In: Proceeding to SPIE International Conference on Optical Holography and Its Applications. 2004

[6] Blanche PA. Optical Holography: Materials, Theory and Application. Elsevier; 2020

[7] Manfrinato VR, Zhang L, Su D, Duan H, Hobbs RG, Stach EA, et al. Resolution limits of electron-beam lithography toward the atomic scale. Nano Letters. 2013;**13**:1555-1558

[8] Rai-Choudhury P. Handbook of Microlithography, Micromachining, and Microfabrication: Microlithography. Bellingham, Washington, USA: SPIE Optical Engineering Press; 1997

[9] Goncharsky AV, Goncharsky AA, Durlevich S. Diffractive optical element for creating visual 3D images. Optical Express. 2016;**24**:9140-9148

[10] Goncharsky AA, Durlevich S. Cylindrical computer-generated hologram for displaying 3D images. Optical Express. 2018;**26**: 22160-22167

[11] Goncharsky A, Goncharsky A, Durlevich S. High-resolution full-parallax computer-generated holographic stereogram created by e-beam technology. Optical Engineering. 2017;**56**:063105

[12] Tikhonov AN, Goncharsky AV, Stepanov VV, Yagola AG. Numerical Methods for the Solution of Ill-Posed Problems. Dordrecht, Netherlands: Kluwer; 1995

[13] Groetsch CW, Neubauer A. Regularization of ill-posed problems: Optimal parameter choice in finite dimensions. Journal of Approximation Theory. 1989;**58**:184

[14] Lesem LB, Hirsch PM, Jordan JA. The Kinoform: A new wavefront reconstruction device. IBM Journal of Research and Development. 1969;**13**: 150-155

[15] Gerchberg RW, Saxton WO. A practical algorithm for the determination of the phase from image and diffraction plane pictures. Optik. 1972;**35**:237-246

[16] Fienup JR. Phase retrieval algorithms: A comparison. Applied Optics. 1982;**21**:2758

[17] Škeren M, Richter I, Fiala P. Iterative Fourier transform algorithm: Comparison of various approaches. Journal of Modern Optics. 2002;**49**: 1851-1870

[18] Wyrowski F. Diffractive optical elements: Iterative calculation of quantized, blazed phase structures.

Journal of the Optical Society of America. A. 1990;7:961

[19] Goncharsky AA. Patent EP3842252B1 "Microoptical System for the Formation of the 3D Image in the Zero Diffraction Order" (granted on 25.05.2022)

[20] Goncharsky AA, Goncharsky AV, Durlevich SR, et al. Synthesis of nano-optical elements for zero-order diffraction 3D imaging. Scientific Reports. 2022;**12**:8639

[21] Moore S. AI System Beats Supercomputer in Combustion Simulation. IEEE.org. IEEE Spectrum Magazine; 2020. Availbale form: https://spectrum.ieee.org/ai-system-beats-supercomputer-at-key-scientific-simulation

[22] Park J-H. Recent progress in computer-generated holography for three-dimensional scenes. Journal of Information Display. 2017;**18**:1-12

[23] Wakunami K, Hsieh P-Y, Oi R, Senoh T, Sasaki H, Ichihashi Y, et al. Projection-type see-through holographic three-dimensional display. Nature Communications. 2016;7:12954

Chapter 10

Volume Holographic Structuring of Special Hydrogel Films by Photochemical Crosslinking

Tina Sabel-Grau, Arina Tyushina, Rahima Rahman, Cigdem Babalik, Zhenfang Zhang and Marga C. Lensen

Abstract

Multifunctional biomedical materials capable of integrating optical functions open up promising new possibilities for the application of photosensitive materials. For example, they are highly desirable for advanced intraocular lens (IOL) implants. For this purpose, we propose hydrogels, based on poly(ethylene glycol) (PEG) prepolymers, which are photochemically crosslinkable and thereby patternable. Various photoinitiators are used and investigated spectroscopically; those with high sensitivity in the optical region of the spectrum are advantageous. Hydrogel films have been obtained, which are applicable for light-based patterning and, hence, for functionalization of both surface and volume: It is shown that a local change in optical properties can be induced in special hydrogel films by photochemical crosslinking. Such a local light-induced material response forms the basis for volume holographic patterning. Cytocompatibility of hydrogels and compositions is evaluated via cytotoxicity tests. Exploiting the interrelationship between structure and function is highly relevant for biomedical materials with multifunctionality.

Keywords: photosensitive materials, hydrogels, photopolymers, volume holographic materials, photocuring, multifunctional biomedical biomaterials, light-responsive materials

1. Introduction

Multifunctional biomedical materials capable of integrating optical functions are highly desirable for many applications, such as advanced intra-ocular lens (IOL) implants [1].

IOLs are needed in the case of cataracts; the irreversible turbidity of the natural lens of the eye, which is one of the most common causes of global blindness and which can only be treated by replacing the clouded lens with an artificial IOL implant (see **Figure 1**).

Intraocular lens implantations have been performed since 1949. The first artificial lenses were made of polymethyl methacrylate (PMMA). Since the beginning of the 1980s, soft lenses made of silicone have been implanted in a folded state. An as-yet

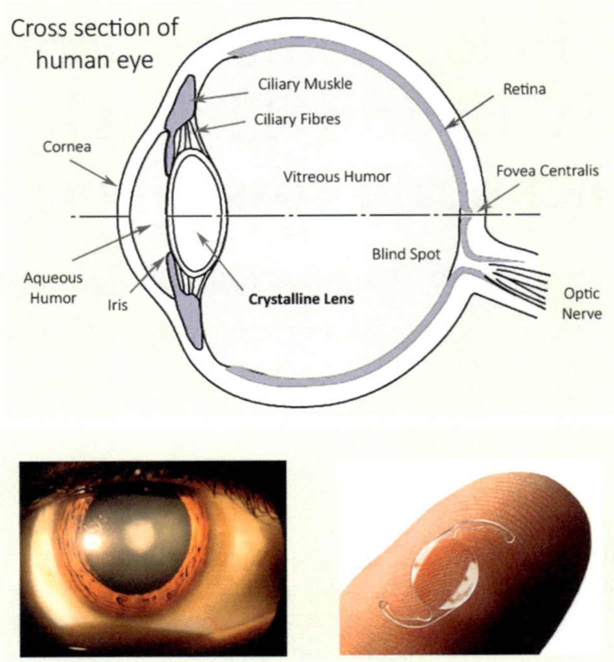

Figure 1.
Schematic cross section of the human eye (top) and exemplary images of the eye with cataract, i.e., pathological turbidity of the crystalline lens (bottom left) and intraocular lens implant (bottom right) [2]. Photos via Wikimedia Commons.

unsolved problem is the preservation or restoration of the accommodation ability, i.e., the dynamic adjustment of the refractive power of the eye by tension and relaxation of the ciliary muscle and zonular fibers. The lens contributes only one-third of the total refractive power of the eye. Most of the refractive power (about 40 of 60 diopters) is generated by the cornea alone. However, accommodation of the lens can change the refractive power by about 10 diopters, or up to 20%.

To counteract the loss of accommodative power due to IOLs, bifocal lenses were introduced in the early 1990s. These simultaneously form a distance image and a near image on the fovea (the area of the sharpest vision of the retina). Disadvantages that have to be accepted for this are photic phenomena such as significantly poorer contrast vision, increased glare sensitivity, poorer imaging quality, as well as an increased stray light component (halo) around the respective primary focal point (see also **Figure 2**, below) [5]. Intraocular lenses of the latest generation—ultraflat, bifocal IOLs with diffractive-refractive optics—have reduced the manifestation of photic phenomena.

Among the state-of-the-art IOLs are modern foldable hydrogel lenses [6]. Persistent problems with IOLs include postoperative calcification [7] and secondary cataract [8]. Such postoperative opacification occurs in vivo through interaction with the biological environment. The underlying processes are still not well understood.

This is where the idea of volume holographic structuring comes into play.

Volume holography is a very promising field of application for photo-responsive polymers: diffractive structures are induced by a spatially modulated holographic

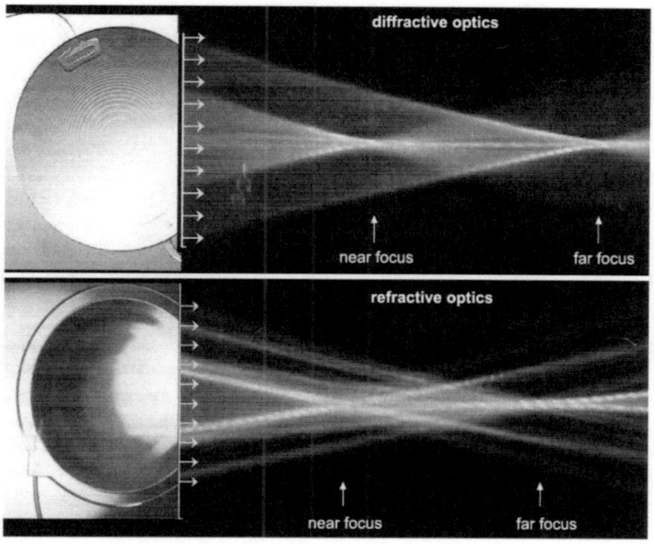

Figure 2.
Beam paths of a diffractive and a refractive multifocal intraocular lens (modified from [3], reprinted with permission from [4], Copyright 2008 SLACK Incorporated).

exposure [9]. Holographic elements such as diffractive structures can accommodate classical optical functions, while at the same time being extremely flat in shape and low in weight. This gives rise to a great potential for replacing classical refractive optical systems or extending them with new functionalities. In any case, the prerequisite is the availability of suitable, photo-patternable materials that can exhibit function through structure. An example of such diffractive structures with classical optical function includes holographic lenses (HL), which owe their optical functionality, i.e. focusing the light onto the retina, to their micro-/nano-structure (see **Figure 3**).

Prospective IOLs, based on multifunctional biomedical material with integrated optical functionality, could fulfill their function—i.e., to focus the light onto the retina (see **Figures 1** and **2**)—with an optically structured volume [2]. As a result, the shape and surface of the IOL would remain free and available for other purposes. This leaves subsequent surface modifications optional to achieve specific interactions with the biological environment. To realize this, we propose a strategy that integrates optical structuring of the volume, combined with specific modification of the surface. Therefore, volume holographic structuring can be applied for the integration of three-dimensional optical structures with specific functionality in terms of diffractive properties. The optical functionality of an IOL with integrated holographic lens as a diffractive element consists in focusing the light onto the retina. Multiple holographic elements may be combined in stacks, so that the functionality of the individual elements overlap. The selectivity of a stack then results in a superposition of Bragg selectivity of the individual elements [11]. Thus, for example, multiple foci can be combined by multiplexing. Bifocal holographic lenses can also be created by combining different phenomena [12]. Azo compounds [13, 14] are suitable for this purpose, for example, and the formation mechanisms here are due to light-induced anisotropy and surface relief gratings.

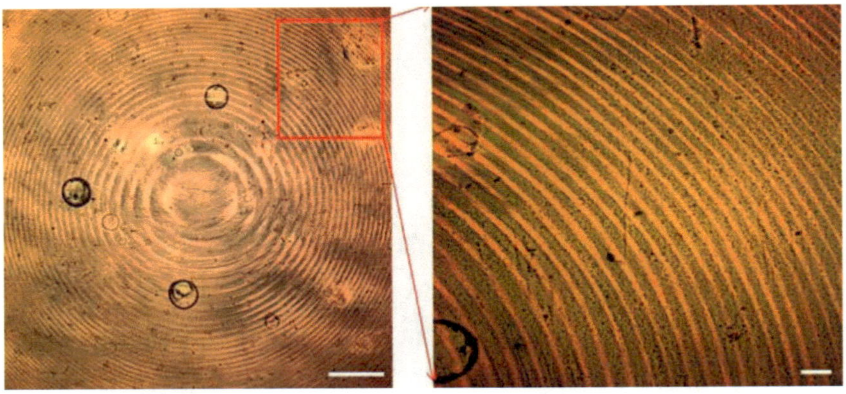

Figure 3.
Holographic lens (HL) on gold-coated substrate. Scale bars: 200 μm (left) and 20 μm (right). Reprinted with permission from [10]. Copyright 2015 American Chemical Society.

In the case of modern intraocular lenses, diffractive structures are already being used to improve the optical functionality of the implant [15]. **Figure 2** shows the advantage of diffractive optics over purely refractive multifocal IOLs.

However, at the current stage of development, only surface structuring has been used. In optically structuring the volume of the IOL, whether in addition or instead of surface patterning, lies a hitherto untapped opportunity with great potential. But new light-sensitive biomaterials are needed to exploit this potential.

Advanced IOL implants are just an example of why new materials for volume holography are the subject of active research [16, 17].

The versatile applications of volume holographic materials are numerous and range from recording media, holographic data storage, self-written waveguides and wavelength-selective devices, to solar energy concentrators and diffractive elements for biomedical applications [1, 9, 18, 19]. The long list of potential applications is driving the need for suitable materials for volume holography.

Most photosensitive materials are designed either for volume or for surface patterning. However, some systems have been described for hybrid volume/surface gratings, where a periodic modulation of the surface is observed in addition to a volume phase grating [20–22]. Such dual grating structures are of great interest for many applications and offer special opportunities for a deeper understanding of the underlying grating formation mechanisms [23].

As a volume holographic material, photosensitive polymers represent a particularly interesting group among stimuli-responsive polymeric materials, characterized by their ability to be used in a noninvasive and easily controllable manner [24]. Light as a stimulus enables optical patterning by applying volume holography as a single-step method to fabricate diffractive 3D micro- and nanostructures [25].

The formation of holographic gratings in polymers is determined by the interplay of polymerization and diffusion, where the chemical gradient is always the driving force: If a component is consumed (e.g. by bonding and crosslinking) or changed (e.g. by isomerization), a concentration gradient is created. This results in diffusion of the consumed or changed components and is accompanied by a material transport, which creates the permanent grating [9].

While light-induced polymerization-diffusion processes enable the structuring of any optically transparent, photosensitive material that exhibits sufficient mobility in the sense of diffusion, permanent structures can only be achieved if the diffusing species are incorporated into a network, e.g., a chemically crosslinked gel.

Employing photosensitive hydrogels combines the advantages of hydrogels and light; the hydrated structure of the hydrogel makes them favorable as a bio-compatible material, and the use of light as a patterning tool allows dynamic control of their properties. The combination of these advantages makes these systems attractive for biomedical applications [26].

Hydrogels, e.g., based on poly(ethylene glycol) (PEG), are generally highly interesting for application as scaffolds for tissue engineering. They are cytocompatible and intrinsically resistant to protein and cell adhesion, and they can be controlled and functionalized on the molecular level [27]. Numerous applications open up based on explicit control over molecular structure and mechanical properties, such as elasticity and degree of crosslinking, as well as surface morphologies [28].

As biomaterials, hydrogels offer versatile possibilities for (photo)chemical crosslinking and are suitable for the generation of topographic surface structures as well as for three-dimensional, macroporous scaffolds. Acrylate-terminated. PEG macromers undergo chemical crosslinking in the presence of photoinitiators that generate radicals when exposed to light [27]. This makes PEG-DA hydrogels interesting as scaffolds into which desired bioactivity can be tailored via light-induced patterning. In this context, it has already been shown that hydrogels can also be structured photolithographically by exploiting diffusion processes [29]. The latter phenomenon forms the basis for volume holography. However, in addition to the possibilities and advantages, holographic gratings in hydrogels also bring new challenges, for example, with respect to temporal stability—this concerns the storage of time-stable holographic gratings in hydrogel matrices [30].

When a new photosensitive material is designed, the type of photoinitiator (PI) used is key for the specific photo-response of a certain material. The properties of the PI have strong influence on holographic grating formation in the respective material [31]. It also influences how well certain conditions are met, such as resistance to humidity [32]. Eosin-Y (EY) and Erythrosin B (EB) are used as PIs for holographic grating formation in an AA/PVA photopolymer [33]. EY features excellent spectroscopic properties, which makes it suitable for use with light sources in the visible range and safe for living organisms [34]. EB is only suitable for free radical polymerization [33].

Photochromic hydrogels, capable of photoinduced reversible color changes, exhibit excellent reversible conversion behavior, which can also be used for repeatable writing of optical information [35]. Photomechanical hydrogels based on typical molecular photoswitches, such as azobenzene, function as smart materials that respond to light [36].

The azobenzene moieties switch from trans to cis conformation upon exposure to UV light and reverse their conformation upon exposure to the visible light. Also, these gels proved to be thermally responsive as well: thermal cis → trans isomerization occurs spontaneously due to the thermodynamic stability of the trans isomer [37]. Thereby, in case of azobenzene-based light-driven real-time information-transmitting systems, the information is expected to be transmitted at the molecular scale with response times ultimately within the nanosecond or picosecond range [38]. In this process, azo-dye molecules can also be used as diffusing dopants to create holographic gratings [39]. Finally, the use of mutually reactive acrylate functionalities on both the azobenzene derivatives and the prepolymer macromolecules ensures the fixation of the diffraction gratings.

2. Material composition and processing

2.1 PEG-DA hydrogels

2.1.1 Chemicals

Poly(ethylene glycol) diacrylate (PEG, Mn 575) and 2-hydroxy-40-(2-hydroxyethoxy)-2-methylpropiophenone (photoinitiator (PI)—Irgacure 2959), Erythrosin B and Eosin Y were from Sigma-Aldrich Chemie GmbH (Steinheim, Germany). The chemical structures of the hydrogel components are shown in **Figure 4**.

2.1.2 Synthesis of PEG-Based Photopolymer with PIs

PEG-DA was used as a precursor. It was mixed with PI (Irgacure, EB and EY, respectively).

PI concentration was kept between 1% and 6 ppm. The chemical structures of different PIs are shown in **Figure 4**. For good mixing of both substances, the mixture was sonicated for around 30 min. At first, the mixture was converted in a cuvette and measured with a UV–Vis spectrometer to obtain spectra before cross-linking. Then, the mixture was dispensed on a glass slide and covered with a thin glass cover slip to achieve a flat and thin hydrogel sample. The glass-sandwich was placed under a UV-light source (366 nm) for 60 min and the glass coverslip was peeled off. A flat, thin standalone hydrogel film was received and also prepared for UV–Vis measurement. For this purpose, the hydrogel was placed on a thin glass coverslip and measured with a UV–Vis spectrometer to obtain spectra after crosslinking.

2.2 Functionalized 8-arm PEG-OA

A composition based on an 8-arm star-shaped PEG with acrylate groups (8PEG-OA) (octa-acrylate) was chosen for the preparation of the hydrogel film,

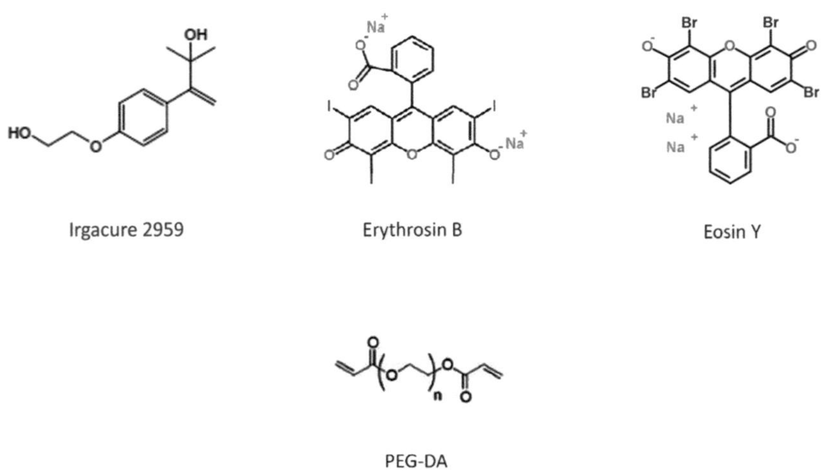

Figure 4.
Irgacure 2959, Erythrosin B and Eosin Y are used as photoinitiators (PI); PEG-DA as a precursor for PEG hydrogel [1].

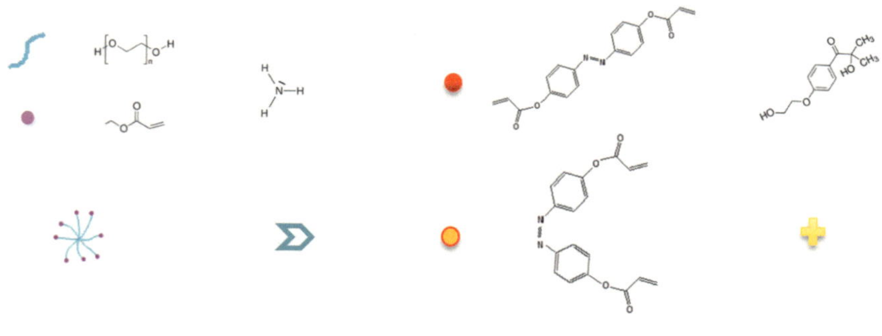

Figure 5.
Components of functionalized 8-arm PEG.

and an azo compound (azobenzene-functionalized acrylates) was added [37]. The components and their chemical structures can be found in **Figure 5**.

This mixture was placed between two glass slides with the addition of photoinitiator (2%), ammonia (30%), and water (100%). Ammonia acts as a crosslinking agent and induces the gel formation [40]. Ammonia as a crosslinker for gel formation is a great tool to make gels that do not disintegrate and which can be finally fixed by photochemical crosslinking of residual acrylate groups (i.e., those who have not been consumed by reaction with ammonia). Subsequently, the sample was further photochemically crosslinked under ultraviolet light (UV).

3. Results and discussion

3.1 Gel formation

PEG-DA was mixed with PI (Irgacure, EB, and EY, respectively). Films were prepared by photocrosslinking with 366 nm for 1 h. In terms of optical transparency, mechanical integrity, flexibility, and stability, the new gels compare well with other gels based on PEG-DA [41–43].

For PEG-DA with Irgacure, EB and EY as PI, a certain minimum concentration of PI was needed to make the gel. The optimal concentration for the different PIs is shown in **Table 1**. We found a suitable PI concentration for producing the new hydrogels with PEG-DA and different PIs to be 0.025% for Irgacure, 0.1% for EB and 0.5% for EY, respectively.

3.2 UV: VIS spectra

Figure 6 shows the UV–Vis spectra before and after crosslinking for the novel PIs. In general, for all new PEG-DA hydrogels, we find the absorption band of the dye in the polymer matrix shifted somewhat toward higher wavelengths compared to the pure PIs [44, 45], while crosslinking tends to cause a small shift toward lower wavelengths again, as already substantiated in the literature [46].

While Irgacure, which is an often employed and suitable photoinitiator for biomaterials research, has an absorption maximum in the UV/Vis-spectrum around 300 nm [44], the novel dyes under investigation display a strong absorption of visible light with wavelengths up to 550 nm (see **Figure 6**).

Photoinitiator (PI)	PI Concentration to Make Gel with PEG-DA
Irgacure 2959	0.025%
EB	0.1%
EY	0.5%

Table 1.
PI concentration to make gel for the different PIs [1].

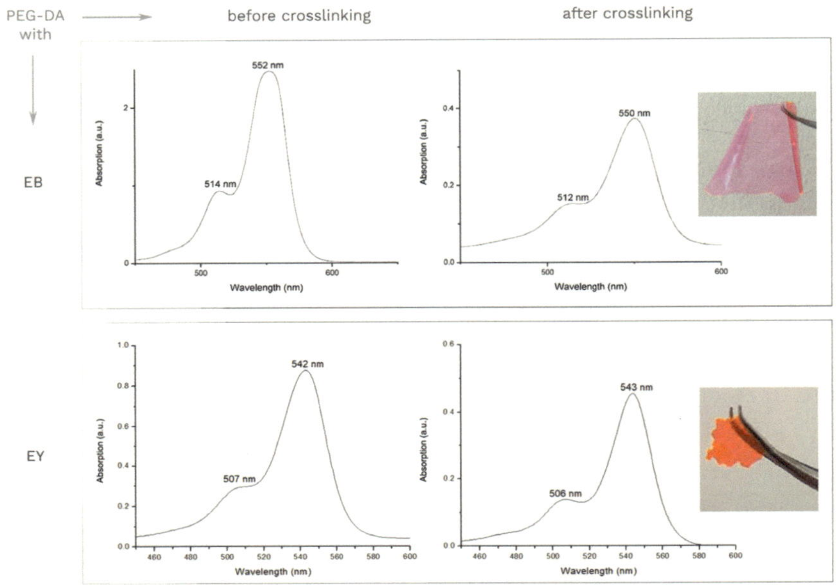

Figure 6.
UV–Vis spectra for the novel PIs (EB and EY) with PEG-DA before and after crosslinking, respectively. The hydrogels with new PIs (EB and EY) exhibit good spectroscopic properties, so they can be used with light sources in the visible range [1].

The major advantage of the new PIs (EB and EY) over the standard Irgacure is their high sensitivity in the optical range of the spectrum, which enables optical patterning—e.g., by volume holography.

For further optimization, the respective optical response must be determined depending on the composition. In some cases (such as with an organic cationic ring-opening polymerization system), competing effects can be observed with respect to the contribution to the formation of the optical grating [47]. It is also known that optical shrinkage can have a significant effect on the formation of gratings [48] and that the amount of PEG in composition affects film shrinkage, as well as its optical properties [49]. In addition, we have found that photoinitiators can also contribute to light-induced changes in optical properties and subsequent pattern formation [50].

3.3 Cytotoxicity tests

A live/dead staining assay has been used to study the cell viability after incubation with PI EB and EY and also EB 0.1% with PEG-DA before and after crosslinking for

24 h. In the live/dead staining assay, dead cells turn up red, while living cells turn up green when observed with a fluorescence microscope.

Figure 7 shows fluorescent images of cell tests by live/dead cell assay after being incubated for 24 h with cell line L929. Cell tests shown in **Figure 7** demonstrate cytocompatibility of PEG-DA hydrogels and PIs, confirming its suitability for biomedical applications.

3.4 Local change in optical properties

The refractive index of the functionalized hydrogel film based on 8-arm PEG was determined by ellipsometry. Results are shown in **Table 2**.

Comparison of exposed and unexposed areas shows a decrease in refractive index due to UV exposure of the order of 10^{-3} in case of functionalized hydrogel. In addition, a dependence of the contrast on the azo content was found: An azo content of 0.3% results in a contrast of 0.005. A higher azo content of 0.6% shows a higher contrast of 0.007.

The change in optical properties found can be attributed to transformation processes in the molecular structure. UV exposure causes the material to crosslink, whereby double bonds are converted to single bonds. Single bonds have a lower polarizability compared to double bonds. The transformation should therefore be accompanied by a reduction of the refractive index. On the other hand, the crosslinking arguably causes an increase in the density of the material, which is linked to an increase in the refractive index. Experimentally, a decrease in refractive index was

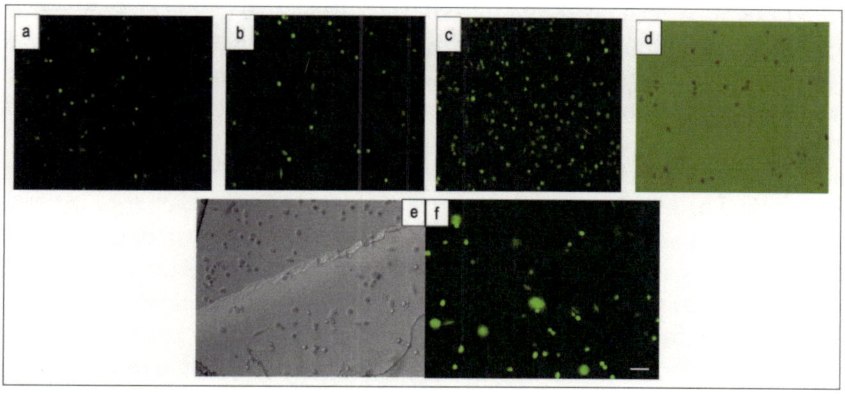

Figure 7.
Fluorescence microscopy images of cells test by live/dead cell assay with L929 after incubated for 24 h. (a) Control cells; (b) EB; (c) EY; (d) EB 0.1% with PEG-DA; (e) optical micrograph of cross-linked EB/PEG-DA hydrogel; (f) fluorescence image of cross-linked EB/PEG-DA hydrogel. Scale bar depicts 50 μm [1].

Refractive Index	Sample A (without Azo)	Sample B (0.3% Azo)	Sample C (0.6% Azo)
Unexposed	1.417 ± 0.006	1.423 ± 0.002	1.429 ± 0.002
UV exposed	1.423 ± 0.008	1.418 ± 0.002	1.422 ± 0.002
Contrast		−0.005	−0.007

Table 2.
Elypsometrically determined local refractive index n of the hydrogel layers [51].

observed in the course of UV exposure-induced crosslinking. Thus, the increase in polarizability seems to outweigh the increase in density of the material.

Altogether, the decrease in refractive index found over the course of UV irradiation-induced crosslinking can be attributed to the increase in polarizability and seems to outweigh the increase in density of the material.

As we know, an inducible refractive index contrast of the order of 10^{-3} allows a diffraction efficiency close to 100% for 200 μm thick layers [9]. It should be noted that the experiments performed here to determine the refractive index contrast do not even involve diffusion processes. Thus, the contrast is not due to material transport, but only due to changes in molecular structure. Since azo components are involved here, trans→cis isomerization is also a possibility for the contrast-effective change in molecular structure [52]. However, by integrating polymerization-diffusion processes in the form of holographic patterning, a much higher contrast can therefore be expected.

4. Targeted volume holographic structuring

4.1 Grating formation mechanisms

A fundamental prerequisite for the development of a multifunctional optical material is a better understanding of the processes underlying optical patterning, such as the interplay of polymerization and diffusion in holographic gratings.

The general approach is to consider the holographic grating formation as a consequence of photopolymerization and mass transport processes: local polymerization is induced by a light pattern projected into the photosensitive medium. Polymerization proportional to the light intensity results in the induction of a chemical gradient, followed by monomer diffusion and subsequent polymerization. The resultant grating is achieved as a periodic modulation of optical properties, corresponding to the recording light pattern [9].

Among the various effects that can contribute to grating formation via a local change in optical properties are: basically, the exposure-induced change in refractive index due to crosslinking of acrylate groups and the change in density due to crosslinking. In the case of azo components, the cis/trans isomerization causes local changes in molecular structures or configurations, in connection with changes in polarizability with direct effects on the refractive index [52]. In addition, cis-trans isomerization can also contribute to a change in density [53]. Then the important aspect of photo-triggered mass transport in the cause of component diffusion is to be considered: both with respect to diffusion of hydrogel components PEG [29] and with respect to diffusion of the azo-dye [39].

4.2 Generation of one-dimensional diffractive surface gratings

The functionalized hydrogel layers were UV-crosslinked under a photomask (1D modulated; lattice constant 20 μm). In this process, diffractive surface gratings were formed. **Figure 8** shows a layer with optical functionality, as well as the corresponding surface characteristics (topography profile from atomic force microscopy studies) of the patterned hydrogel.

It can be assumed that the diffusion of material plays a substantial role in the surface relief formation: the simultaneous formation of surface relief gratings and volume holographic phase gratings by holographic interference exposure in epoxy-based material can be explained based on photo-triggered mass transport, resulting from the light-induced polymerization–diffusion process in the course of volume hologram formation [23].

4.3 Anticipated grating formation in functionalized hydrogel

The anticipated grating formation in the functionalized hydrogel is schematically shown in **Figure 9**.

In the first step, the mixture forms a gel by reaction of 8PEG and crosslinker (ammonia), leaving uncrosslinked functional groups (acrylate groups). In the second step, the structured exposure is performed by a photomask (or, if volume holographic structuring is used, by an interference pattern). The photoinitiator is consumed, with photoinduced attachment of the dopant (here, azobenzene-functionalized acrylates). In addition, trans→cis isomerization occurs under UV irradiation. The third step, diffusion of uncrosslinked dopant (and possibly - depending on the exposure dose - also crosslinking and isomerization of the diffused components), forms the permanent optical grating.

5. Conclusions

It was shown that a local change in optical properties can be induced in special functionalized hydrogel films by photochemical crosslinking, while we found an overall decrease in refractive index over the course of UV irradiation-induced crosslinking.

The strategy of combining optical structuring of the volume and specific modification of the surface is of particular interest for the design of advanced intraocular lens (IOL) implants: based on a multifunctional biomedical material with integrated optical functionality and following the principle of *function by structure*, such a new type of IOL should achieve improved functionality [2].

In addition to enhanced functionality, moving optical functionality from the surface into the volume of the IOL implant brings other advantages such as the free

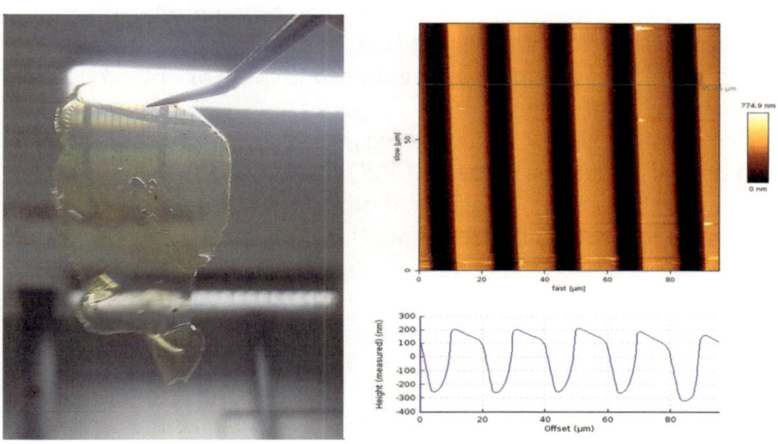

Figure 8.
Diffractive structures in hydrogel: hydrogel film with optical functionality (left) and corresponding surface topography (right) [51].

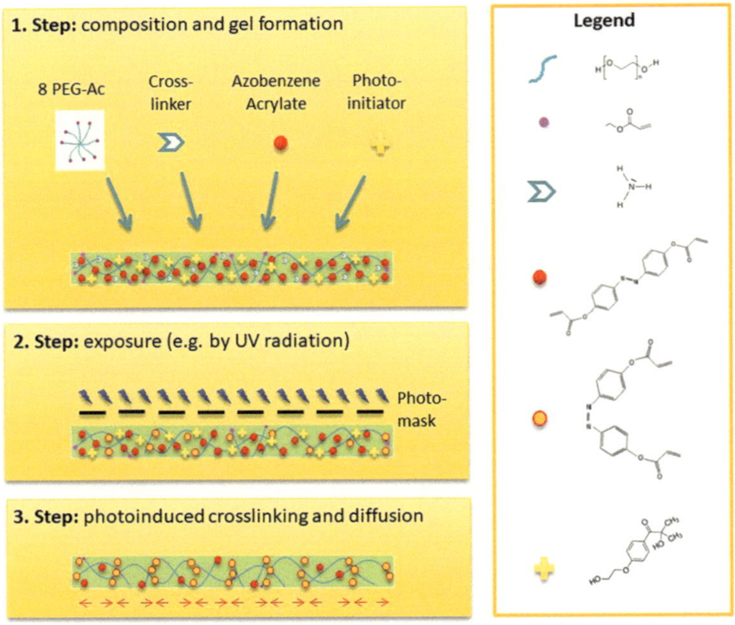

Figure 9.
Anticipated grating formation in functionalized hydrogel [51].

interface for specific interaction with the biological environment. Since the existing problems with conventional IOLs, such as postoperative clouding, emerge in vivo from interactions with the biological environment, they could be better solved with free surface IOLs.

The next step toward such a multifunctional optical material is to better understand the processes underlaying optical structuring, e.g., the interplay of polymerization and diffusion in the case of holographic gratings. Here, the general approach is to understand holographic grating formation as a consequence of photopolymerization and mass transport processes: local polymerization is induced by a light pattern projected into the photosensitive medium. Polymerization proportional to the light intensity creates a chemical gradient, resulting in monomer diffusion and subsequent polymerization. The final grating is a periodic modulation of optical properties, following the recording light pattern [9].

An exciting question is now, what role the individual components play in the formation of optical structures in cases of PEG-DA hydrogels with EB and EY. And furthermore, it would have to be clarified if other additives—such as crosslinker or dopant, e.g., in the form of azobenzene-functionalized acrylates or gold nanoparticles— may show a positive effect on the formation of optical patterns.

We have created one-dimensional diffractive surface gratings, using a photomask. Such structured hydrogel layers show optical functionality, and the related surface topography was visualized in the atomic force microscope. The observed optical functionality may be caused by the measured surface grating, or additionally by a possibly generated correlated volume phase grating. Which contribution each of the two have, will have to be clarified by further investigations, preferably by the generation of holographic gratings in hydrogel layers and exact analysis of their corresponding diffraction properties. Then the material response can be determined (and optimized in connection with adapted material composition) from the analysis of the refractive index contrast as a function of the exposure conditions.

Acknowledgements

This research was funded by Deutsche Forschungsgemeinschaft (DFG, German Research Foundation), grant number SA 2990/1-1. The APC was funded by DFG.

Conflict of interest

The authors declare no conflict of interest.

Author details

Tina Sabel-Grau*, Arina Tyushina, Rahima Rahman, Cigdem Babalik, Zhenfang Zhang and Marga C. Lensen*
Nanopatterned Biomaterials (Secr. C 1), Department of Chemistry, Technische Universität Berlin, Berlin, Germany

*Address all correspondence to: tina@physik.tu-berlin.de and marga@lensenlab.de

References

[1] Sabel-Grau T, Tyushina A, Babalik C, Lensen MC. UV-VIS curable PEG hydrogels for biomedical applications with multifunctionality. Gels. 2022;**8**(164):8

[2] Sabel T, Lensen MC. Function and structure – Combined optical functionality and specific bio-interaction for multifunctional biomedical materials. Journal of Engineering Materials and Technology. 2017;**1**(2):10-12

[3] Bahner KM. Kombination von diffraktiver und refraktiver multifokaler Intraokularlinse nach Kataraktoperation zur optimalen Korrektion der Refraktion. Gießen, Germany; Justus-Liebig-Universität Gießen; 2010

[4] Terwee T, Weeber H, Van Der Mooren M, Piers P. Visualization of the retinal image in an eye model with spherical and aspheric, diffractive, and refractive multifocal intraocular lenses. Journal of Refractive Surgery. 2008;**24**:223-232

[5] Dick B, Eisenmann D, Fabian E, Schwenn O. Refraktive Kataraktchirurgie mit multifokalen Intraokularlinsen. Berlin Heidelberg: Springer; 1999

[6] Izak AM, Werner L, Pandey SK, Apple DJ. Calcification of modern foldable hydrogel intraocular lens designs. Eye. 2003;**17**(3):393-406

[7] Kanclerz P, Yildirim T, Khoramnia R. Microscopic characteristics of late intraocular lens opacifications. Archives of Pathology & Laboratory Medicine. 2020;**145**:1-9

[8] Lundgren B, Jonsson E, Rolfsen W. Secondary cataract. Acta Ophthalmologica. 2009;**70**:25-28

[9] Sabel T, Lensen MC. Volume holography: Novel materials, methods and applications. In: Naydenova I, Babeva T, Nazarova D, editors. Holographic Materials and Optical Systems. Rijeka, Croatia: InTech; 2017

[10] Zhao Q, Yetisen AK, Sabouri A, Yun SH, Butt H. Printable nanophotonic devices via holographic laser ablation. ACS Nano. 2015;**9**(9):9062-9069

[11] Akbari H, Naydenova I, Persechini L, Garner SM, Cimo P, Martin S. Diffractive optical elements with a large angle of operation recorded in acrylamide based photopolymer on flexible substrates. International Journal of Polymer Science. 2014;**2014**:918285

[12] Martinez-Ponce G, Petrova T, Tomova N, Dragostinova V, Todorov T, Nikolova L. Bifocal-polarization holographic lens. Optics Letters. 2004;**29**:1001-1003

[13] Sabat RG. Superimposed surface-relief diffraction grating holographic lenses on azo-polymer films. Optics Express. 2013;**21**(7):8711

[14] Kulikovska O, Kulikovsky L, Goldenberg LM, Stumpe J. Generation of microstructures in novel supramolecular ionic materials based on azobenzene. Organic Optoelectronics and Photonics III. 2008;**6999**:69990I

[15] Xu X, Zhu M-M, Zou H-D. Refractive versus diffractive multifocal intraocular lenses in cataract surgery: A meta-analysis of randomized controlled trials. Journal of Refractive Surgery. 2014;**30**(9):634-644

[16] Guo J, Gleeson MR, Sheridan JT. A review of the optimisation of

photopolymer materials for holographic data storage. Physics Research International. 2012, no. Article ID 803439:16

[17] Barachevsky VA. The current status of the development of light-sensitive media for holography (a review). Optics and Spectroscopy. 2018;**124**:373-407

[18] Gallego S et al. Two diffusion photopolymer for sharp diffractive optical elements recording. Optics Letters. 2015;**40**(14):3221-3224

[19] Naydenova I et al. Photopolymer holographic optical elements for application in solar energy concentrators. In: Mihaylova E, editor. Holography - Basic Principles and Contemporary Applications. London: InTech; 2013. pp. 129-145

[20] Vorzobova ND, Sokolov PP, Veselov VO, Schelkanova IJ. Holographic formation and diffractive properties of hybrid periodic structures. Applied Optics. 2018;**57**(12):3323-3328

[21] Naydenova I, Mihaylova E, Martin S, Toal V. Holographic patterning of acrylamide-based photopolymer surface. Optics Express. 2005;**13**(13):4878-4889

[22] Lucchetta DE, Spegni P, Di Donato A, Castagna R. Hybrid surface-relief/volume one dimensional holographic gratings. Optical Materials. 2015;**42**:366-369

[23] Sabel T, de Vicente Lucas G, Lensen MC. Simultaneous formation of holographic surface relief gratings and volume phase gratings in photosensitive polymer. Materials Research Letters. 2019;**7**(10):405-411

[24] Roy D, Cambre JN, Sumerlin BS. Future perspectives and recent advances in stimuli-responsive materials. Progress in Polymer Science. 2010;**35**(1):278-301

[25] Colburn WS, Haines KA. Volume hologram formation in photopolymer materials. Applied Optics. 1971;**10**(7):1636-1641

[26] Tomatsu I, Peng K, Kros A. Photoresponsive hydrogels for biomedical applications. Advanced Drug Delivery Reviews. 2011;**63**(14):1257-1266

[27] Nguyen KT, West JL. Photopolymerizable hydrogels for tissue engineering applications. Biomaterials. 2002;**23**(22):4307-4314

[28] Lensen MC, Schulte V, Salber J, Dietz M, Menges F, Möller M. Cellular responses to novel, micropatterned biomaterials. Pure and Applied Chemistry. 2008;**80**(11):2479-2487

[29] Hahn BMS, Miller JS, West JL. Three-dimensional biochemical and biomechanical patterning of hydrogels for guiding cell behavior. Advanced Materials. 2006;**18**:2679-2684

[30] Berramdane K et al. Processing of holographic hydrogels in liquid media: A study by high-performance liquid chromatography and diffraction efficiency. Polymers. 2022;**14**(10):2089, 16 p

[31] Qi Y, Li H, Fouassier JP, Lalevée J, Sheridan JT. Comparison of a new photosensitizer with erythrosine B in an AA/PVA-based photopolymer material. Applied Optics. 2014;**53**(6):1052-1062

[32] Mikulchyk T, Martin S, Naydenova I. Investigation of the sensitivity to humidity of an acrylamide-based photopolymer containing N-phenylglycine as a photoinitiator. Optical Materials. 2014;**37**:810-815

[33] Qi Y, Gleeson MR, Guo J, Gallego S, Sheridan JT. Quantitative comparison of five different photosensitizers for use in a quantitative comparison of five different photosensitizers for use in a photopolymer. Physics Research International. 2012;**17**:11

[34] Tomal W, Ortyl J. Water-soluble photoinitiators in biomedical applications. Polymers. 2020;**12**(5):1-30

[35] Long S et al. High-performance photochromic hydrogels for rewritable information record. Macromolecular Rapid Communications. 2021;**42**(7):2000701

[36] Luo P-F, Xiang S-L, Li C, Zhu M-Q. Photomechanical polymer hydrogels based on molecular photoswitches. Journal of Polymer Science. 2021;**59**(20):2246-2264

[37] Rahman R. Photo - and Thermo - Responsive Poly (Ethylene Glycol) - Based Biomaterials: Synthesis, Characterization, Patterning and Application in Biological Studies. Berlin, Germany: Technische Universität; 2018

[38] García-amorós J, Velasco D. Recent advances towards azobenzene-based light- driven real-time information-transmitting materials. Beilstein Journal of Organic Chemistry. 2012;**8**:1003-1017

[39] Kuo C-T, Huang S-Y. The holographic investigation of diffusion of azo-dye in liquid crystals host. Applied Physics B: Lasers and Optics. 2007;**86**(4):699-702

[40] Zhang Z et al. Synthesis of poly(ethylene glycol)-based hydrogels via amine-michael type addition with tunable stiffness and postgelation chemical functionality. Chemistry of Materials. 2014;**26**(12):3624-3630

[41] Schulte VA, Diez M, Hu Y, Möller M, Lensen MC.Combined influence of substrate stiffness and surface topography on the antiadhesive properties of Acr-sP(EO-stat-PO) hydrogels. Biomacromolecules Dec 2010;**11**:3375-3383

[42] Lensen MC, Schulte VA, Diez M. In: Pignatello R, editor. Cell Adhesion and Spreading on an Intrinsically Anti-adhesive PEG Biomaterial, Biomaterials. Rijeka: IntechOpen; 2011

[43] Kelleher S, Jongerius A, Löbus A, Strehmel C, Zhang Z, Lensen MC. AFM Characterisation of elastically micropatterned surfaces fabricated by fill-molding in capillaries (FIMIC) and investigation of the topographical influence on cell adhesion to the patterns. Advanced Engineering Materials. 2012;**14**(3):B56-B65

[44] Kaastrup K, Sikes HD. Using photo-initiated polymerization reactions to detect molecular recognition. Chemical Society Reviews. 2016;**45**:532-545

[45] Yang R, Soper SA, Wang W. A new UV lithography photoresist based on composite of EPON resins 165 and 154 for fabrication of high-aspect-ratio microstructures. Sensors and Actuators A: Physical. 2007;**135**(2):625-636

[46] Sung J, Lee DG, Lee S, Park J, Jung HW. Crosslinking dynamics and gelation characteristics of photo- and thermally polymerized poly(ethylene glycol) hydrogels. Maternité. 2020;**13**(15):3277

[47] Sabel T, Zschocher M. Transition of refractive index contrast in course of grating growth. Scientific Reports. 2013;**3**(2552):1-7

[48] Sabel T. Spatially resolved analysis of Bragg selectivity. Applied Sciences. 2015;**5**(4):1064-1075

[49] Dimitrov O, Stambolova I, Vassilev S, Lazarova K, Babeva T, Mladenova R. Surface and morphological features of ZrO2 sol-gel coatings obtained by polymer modified solution. Materials Proceedings. 2020;**2**(6):1-8

[50] Sabel T. Volume hologram formation in SU-8 photoresist. Polymers. 2017;**9**(6):198

[51] Sabel-Grau T, Zhang Z, Rahman R, Lensen MC. Formation of diffraction gratings in optically patternable hydrogel films. Nano Select. 2022;**1**:5. DOI: 10.1002/nano.202200154

[52] Liu YJ, Dai HT, Leong ESP, Teng JH, Sun XW. Azo-dye-doped absorbing photonic crystals with purely imaginary refractive index contrast and all-optically switchable diffraction properties. Optical Materials Express. 2012;**2**(1):1667-1669

[53] Merino E, Ribagorda M. Control over molecular motion using the cis – trans photoisomerization of the azo group. Beilstein Journal of Organic Chemistry. 2012;**8**:1071-1090

Chapter 11

Digital Holographic Microscopy in Partially Coherent Illumination and Applications

Frank Dubois, Catherine Yourassowsky, Karim Zouaoui Boudjeltia and Jérôme Dohet-Eraly

Abstract

In this chapter, we describe several configurations of digital holographic microscopes operating with partially coherent illumination, including the implementation of color holography. The different ways of implementing partially coherent illumination are described and discussed, as well as the respective improvements they provide depending on the properties of the observed objects. Several significant applications in the biomedical and environmental fields are presented. In particular, we carry out researches on correlative quantitative phase-contrast-fluorescence imaging on blood with a special focus on the behavior of platelets in relation to relevant pathologies. In the environmental domain, we demonstrate the use of high-throughput digital holographic microscopy to monitor and analyze plankton and microalgal cultures.

Keywords: digital holographic microscopy, partially coherent illumination, coherent noise reduction, quantitative phase-contrast imaging, digital refocusing, environmental applications, biomedical applications

1. Introduction

Classical optical microscopy is highly limited by the small depth of focus due to the high magnification coupled with the high numerical apertures. With such a constraint, the analysis of a three-dimensional (3D) sample is complex and even impossible in the case of dynamic phenomena, which makes the extension of the depth of focus an important challenge. With the impressive recent developments of digital cameras, digital holography has provided an elegant and natural tool to counteract this limitation by enabling numerical refocusing at different depths within an experimental volume [1, 2]. In this context, digital holographic microscopy (DHM), which drastically increases the depth of focus (about $100\times$), became a very powerful tool to investigate numerous applications, in particular in the biomedical and environmental fields. Moreover, thanks to the wave nature of holography, DHM may provide the quantitative phase-contrast imaging (QPCI) that considerably extends the potential of this technology, especially in biological and environmental applications [3–6]. Several ways exist to implement DHM. The conceptually simplest implementation consists of using an interferometric configuration: it allows to record the interference pattern between the object beam, resulting from the transmission through—or the refection by—the object, and a separated reference beam [2, 7, 8]. Hereinafter, those configurations are called two-channel interferometers. Several other ways to proceed may be considered. Firstly, common-path interferometers [5, 9] allow to simplify the configuration or to improve mechanical robustness. However, other limitations may arise from the common-path configurations as, in some cases, the apparition of halo effects in the phase images or the difficulties to implement an illumination of partial coherence. Other configurations were also developed that aim to reach the same goals as, for instance, the implementation of the transport of the intensity equation in a usual microscope [10], the in-line holography [11] and the ptychography [12]. Those configurations are interesting but request more complex computations to get the results, which can make reconstruction ability or the quantitative phase recovery more difficult. However, with two-paths interferometers, the full complex amplitude, required for the refocusing process and the QPCI, is obtained with relatively simple and straightforward computation. Regarding the mechanical robustness of the two-channel configuration, our experience has shown that it is not actually an issue. The design of the compact optical system with usual optomechanical components indeed allows to build systems that are able to operate even in the harsh environment [13–15].

The most usual way to implement holography in a two-channel configuration is to use a laser source; it is then easy to have mutually coherent object and reference beams, giving interference patterns on the camera. However, coherent beams are very sensitive to any disturbance in their propagation, which leads to noisy results. This noise that alters the images, which is related to the speckle effect, is multiplicative and hence difficult to remove by post-processing. Additionally, the speckle disturbance is created everywhere in the optical setup and experimental cells from any small defect. For this reason, we developed several DHM configurations taking advantage of partially coherent illuminations. The use of such illuminations drastically reduces the noise in both the amplitude modulus and phase images.

It has to be noted that a lot of researches were also performed to develop digital holography to get holographic information with completely incoherent light, and in particular from fluorescent objects. Pioneer articles in this field are given by the references [16–18].

In Section 2, we present the DHM configurations, developed with partially coherent illumination, and the processing to obtain the amplitude modulus and the phase from the recorded hologram. Section 2 also describes how the fluorescence can be implemented to achieve multimodal DHM (see details in Section 4.1) and the extension of the technology to color DHM with partially coherent illumination. The benefit of using partially coherent illumination to drastically reduce the speckle noise is demonstrated in Section 3. In Section 4, we describe several valuable applications we implemented.

2. DHM configurations with partially coherent sources

When implemented in a DHM configuration, it is requested that the partial spatial coherence of the beam illuminating the sample is uniform over the complete field of view. Therefore, the mutual coherence function Γ has to be expressed, in a plane, as a function (γ) of the difference between the considered positions $\mathbf{x}_1 = (x_1, y_1)$ and $\mathbf{x}_2 = (x_2, y_2)$:

$$\Gamma(\mathbf{x}_1, \mathbf{x}_2) = \gamma(\mathbf{x}_1 - \mathbf{x}_2) \tag{1}$$

Such a mutual coherence function can be obtained from a spatially incoherent source or from a spatially highly coherent source according to the schemes in **Figure 1** [19].

On the left configuration in **Figure 1**, a fully spatially incoherent beam is illuminating an aperture $p(\mathbf{x})$. The aperture is placed at the front focal plane of a collimation lens. Assuming we have a quasi-monochromatic illumination, it results out of the lens a partial spatial coherence function that can be expressed by:

$$\gamma(\mathbf{x}_1 - \mathbf{x}_2) \propto (Fp)((\mathbf{x}_1 - \mathbf{x}_2)/\lambda f) \tag{2}$$

where F denotes the two-dimensional Fourier transformation, f is the focal length of the lens, and λ is the wavelength. The left configuration of **Figure 1** also allows the implementation of sources of reduced temporal coherence like light-emitting diodes (LEDs). However, the aperture removes a large part of the source intensity and the use of LEDs is more complex if the interferometric off-axis configuration is used.

It is also possible to obtain an equivalent partial spatial coherence illumination by the right configuration in **Figure 1**, in which a laser source is focused close to a rapidly moving ground glass placed at the front focal plane of a collimating lens. The ground glass generates a speckle field, whose width is adjusted by changing the position of the focused laser beam with respect to this ground glass and by the choice of the focal length of the collimating lens. It can be observed that, with time averaging, this source can be seen as a partially spatially coherent source, whose typical coherence width is equal to the average speckle size [20].

In the following, we describe two particular implementations of DHM with partially coherent illumination that we developed. The first one uses monochromatic illumination and can be coupled with a fluorescence module to achieve a multimodal instrument. It is shown in **Figure 2a**. A coherent beam emerging from a mono-mode laser diode is focalized by the lens ML1 close to the rough surface of the rotating ground glass (GG). The light transmitted by the GG is scattered and collimated by the lens L1. A first beam splitter BS1 divides the collimated beam into an object beam and a reference beam. The object beam passes through the sample and then through the microscope lens ML3. In a symmetric way, the reference beam passes through the microscope lens ML2, identical to ML3. Reference and object beams are then recombined by a second beam splitter (BS2) and transmitted by lens L2. Lenses ML3–L2 image one plane of the sample onto the camera sensor. The magnification G_s is then given by the ratio between the focal length of L2 and the focal length of ML3. Depending on the applications, the magnification is adjusted by changing the

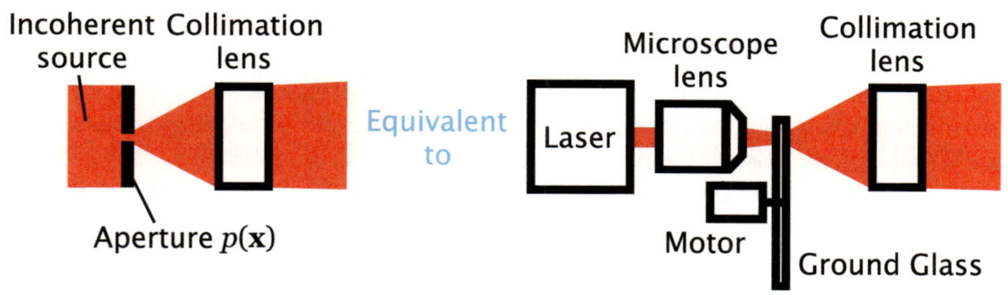

Figure 1.
Two ways to obtain a spatially partial mutual coherence function.

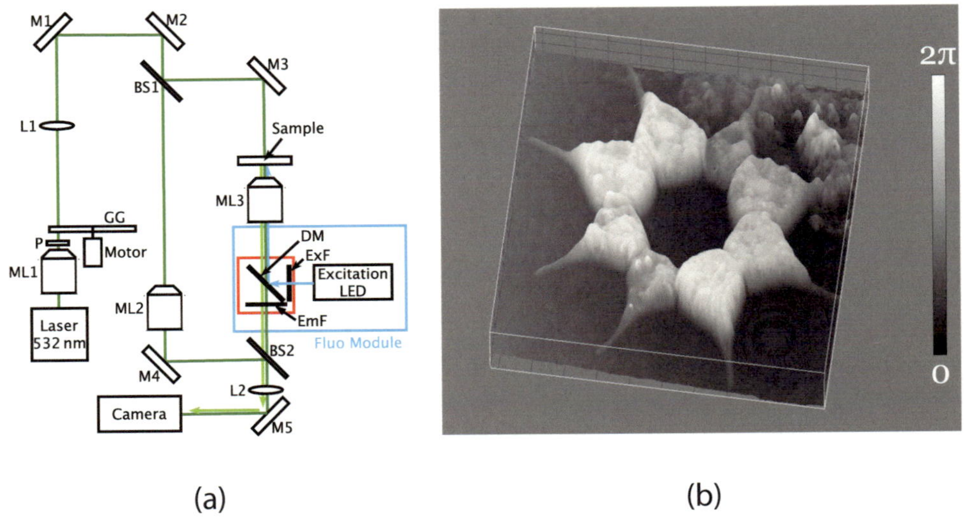

Figure 2.
Monochromatic DHM coupled with fluorescence to achieve a multimodal instrument. (a) diagram of the instrument. ML1-ML3: microscope lenses, P: polarizer, GG: ground glass, L1-L2: lenses, BS1-BS2: beam splitters, M1-M5: mirrors, EmF: emission filter, ExF: excitation filter, and DM: dichroic mirror. (b) 3D plot of the quantitative phase-contrast image of an alga Pediastrum *sp. Field of view: 80 × 80 μm, magnification × 100, NA: 1.3–oil immersion lens.*

couple of lenses ML2 and ML3 (typically we used microscope lenses ×10, ×20, ×40, ×63, and ×100 (immersion oil)).

In most of the applications, the interferometer is adjusted to have a slanted reference beam with respect to the object beam in such a way that a thin fringe pattern is recorded on the camera sensor. The computation of the full complex amplitude from every recorded hologram, involving the amplitude modulus and the phase, is obtained by using the Fourier method [21]. In the resulting phases, it happens that unwanted phase modulations appear, which can be originated from the experimental cell or from small misalignments of the interferometer. To compensate them, powerful correction techniques can be applied and are given in more detail in Section 4.2 [22]. An example of a quantitative phase-contrast image obtained, of an alga *Pediastrum* sp. at high magnification (×100), is shown in **Figure 2b**.

The fluorescence is implemented in the epifluorescence configuration, in such a way that the same sample is sequentially imaged by DHM and fluorescence modes on the same camera sensor. Applications of the combination of the DHM and fluorescence modes are given in Section 4.1.

In some applications, the color information is relevant. However, the configuration of **Figure 2a** is complex to adapt to several optical sources of different wavelengths. Indeed, in this case, the slant between the object and reference beams is identical for all the wavelengths, giving rise to fringe patterns with different inter-fringe spaces. This would make more difficult the processing with the Fourier method. In order to counter this limitation, we developed the color DHM instrument shown in **Figure 3** [23].

This color DHM is also based on a Mach–Zehnder configuration in which we added two afocal systems, L2–L3 and L4–L5, in the object and reference channels, respectively, between the microscope lenses and the lens in front of the camera sensor. In the object channel, a neutral density filter is placed in the back focal plane of L2. In the reference channel, a Ronchi grating is placed in the back focal plane of L4 and is

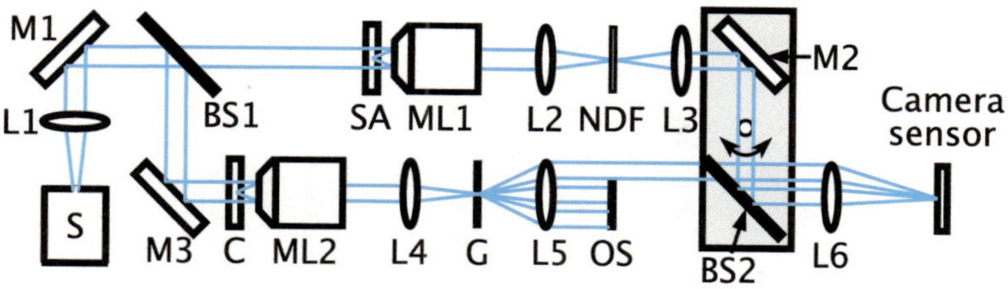

Figure 3.
Color off-axis DHM configuration. S: optical source, L1-L6: lenses, ML1-ML2: microscope lenses, BS1-BS2: beam splitters, M1-M3: mirrors, SA: sample, C: optional compensation plate, NDF: neutral density filter, G: grating, and OS: optical stop.

imaged on the camera sensor by L5–L6. In the back focal plane of L5, an optical stop is placed to keep only one of the first diffraction orders. In this way to proceed, the slant between the object and reference beams on the camera sensor is changing with the wavelength, in such a way that the same spatial frequency of the fringes is kept on the sensor. Therefore, this configuration is operating for any wavelength in the visible range without having to adjust the fringe spacing to implement the Fourier method. In order to operate in the red (R), green (G), and blue (B) wavelengths, the partially spatially coherent illumination is created by the combination of three laser beams into the system using a ground glass as described hereinabove. It has to be emphasized that the grating also enables the use of reduced temporal coherence sources, such as LEDs [23]. Indeed, if a reduced temporal coherence source is used in the configuration of **Figure 2a**, the slant between the object and reference beams makes the fringe contrast not uniform over the field of view. We can even expect to have no fringe at all in some areas of the field of view. With the configuration of **Figure 3**, however, due to the properties of the diffraction by a grating, this issue is solved. This configuration indeed provides a uniform fringe contrast with reduced temporal coherence illumination, accordingly, the optical paths between the object and reference beams are finely equalized. This equalization is adjusted, thanks to the rotating mount on which the mirror M2 and the beam splitter BS2 are placed.

An example where the color DHM can provide relevant information is the routine analysis of clinical cervical samples for the prevention and the checking of the development of cancerous cells. Analyses are usually performed on classical ThinPrep slides with the Papanicolaou stain. In this way, we performed the analysis of samples with the color DHM and recorded multispectral holograms (**Figure 4a**). From the recorded hologram, we extracted the intensity and the phase images in three wavelengths separately: 639 nm (R), 532 nm (G), and 488 nm (B). The combination of the RGB intensity images of stained slides gives color intensity images comparable to those obtained with classical microscopy used by histopathologists (**Figure 4b**). In addition to this RGB intensity image, DHM gives, simultaneously on the same sample, quantitative phase-contrast imaging. In **Figure 4c**, it can be seen that typical dysplastic cells (big nucleus and small cytoplasm) have the highest optical thickness.

One of the targets with the DHM is the refocusing of objects that have been recorded out of focus. For that purpose, the implementation in Fourier optics, of the convolutional form of the free space propagation is used [24]. The complex amplitude distribution t_d reconstructed at a distance d along the optical axis, by computing the Kirchhoff–Fresnel propagation integral in the paraxial approximation, is expressed by:

$$t_d(x', y') = \exp(jkd) F^{-1}_{x', y'} \exp\left(-j\pi\lambda d\left(\nu_x^2 + \nu_y^2\right)\right) F^{+1}_{\nu_x, \nu_y}\, t_0(x, y) \qquad (3)$$

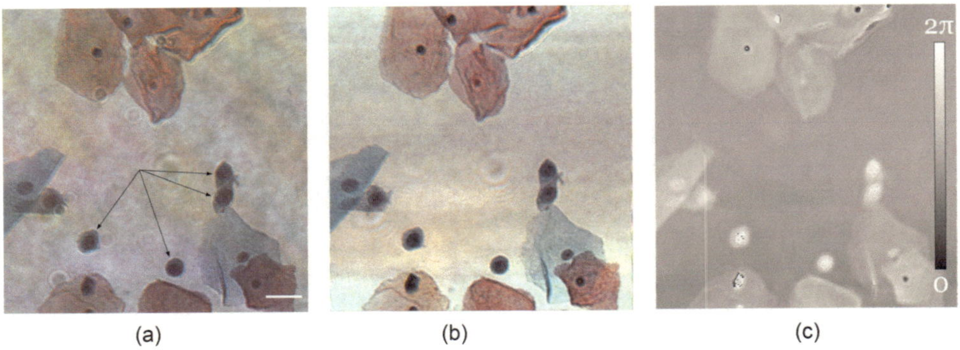

Figure 4.
Images obtained with the color DHM of a clinical cervical sample. (a) color hologram image, (b) color intensity image, and (c) phase image in the red wavelength (R). The arrows indicate typical dysplastic cells. Scale bar = 20 μm.

where t_0 is the complex amplitude of the optical field in the recorded plane, $k = 2\pi/\lambda$ is the wavenumber, (x, y) and (x', y') are the spatial variables in the focus plane and in the reconstructed plane, respectively, (ν_x, ν_y) are the spatial frequencies, $j = \sqrt{-1}$ and $F^{\pm 1}$ denote the direct and inverse two-dimensional continuous Fourier transformations. In its discrete form with a sampling distance Δ, Eq. (3) becomes:

$$t_d(s'\Delta, t'\Delta) = \exp(jkd) F^{-1}_{s',t'} \exp\left(\frac{-j\pi\lambda d}{N^2\Delta^2}\left(U^2 + V^2\right)\right) F^{+1}_{U,V}\, t_0(s\Delta, t\Delta) \tag{4}$$

where N is the number of pixels along the x and y directions; s, t, s', t', U, and V are integer numbers varying from 0 to $N-1$, and $F^{\pm 1}$ are here the direct and inverse discrete Fourier transformations.

The phase φ_{t_d} corresponding to the complex amplitude t_d is obtained by computing

$$\varphi_{t_d}(s'\Delta, t'\Delta) = \mathrm{Arctan}_{2\pi}(\mathrm{Im}[t_d(s'\Delta, t'\Delta)],\ \mathrm{Re}\,[t_d(s'\Delta, t'\Delta)]) \tag{5}$$

where $\mathrm{Arctan}_{2\pi}$ is the two-parameters arctangent function, and $\mathrm{Im}[z]$ and $\mathrm{Re}[z]$ denote the imaginary and real parts of z, respectively.

This holographic reconstruction does not indicate for which distance d the best focus of an object is reached. In order to dispose of criteria giving the best plane of each object present in the field of view, we use several methods based on the conservation of the amplitude integration [25–27] or the invariance of the modulus of the Fourier transform during the propagation [28].

3. Noise reduction by the use of partially coherent illumination

To study the illumination by a partially spatially coherent illumination, the left image of **Figure 1** may be considered. In this configuration, the illumination is performed by the set of mutually incoherent point sources located within the aperture. As the locations of those points are different, each of them gives rise, after the collimation lens, to plane waves with different propagation angles. Therefore, each object is illuminated by a set of mutually incoherent plane waves having different propagation directions. For an unfocused object, it results that the propagation up to the input focus plane of lens ML3 (in **Figure 2**) gives rise to a set of mutually incoherent diffraction patterns that are specifically shifted, due to the different propagation directions. Those shifts increase with the defocus distance d. It results that the digital holographic reconstruction creates a set of refocused objects at different locations in the reconstruction plane. This phenomenon results in a convolutional process [20], for an object defocused by a distance d, the complex amplitude $\hat{t}_d(\mathbf{x})$, extracted from the hologram in case of partial spatial coherence, is given by:

$$\hat{t}_d(\mathbf{x}) = A \iint \mathrm{d}\mathbf{x}' p\left(\frac{2f(\mathbf{x}-\mathbf{x}')}{dD}\right) t_d(\mathbf{x}') \tag{6}$$

where $\mathbf{x} = (x,y)$, A is a constant, p is the normalized pupil function equal to 1 within a circle of radius 1 and equal to zero outside, f is the focal length of the collimation lens, D is the diameter of the aperture in **Figure 1**, and $t_d(\mathbf{x}')$ is the complex amplitude that would have been obtained in case of fully coherent illumination.

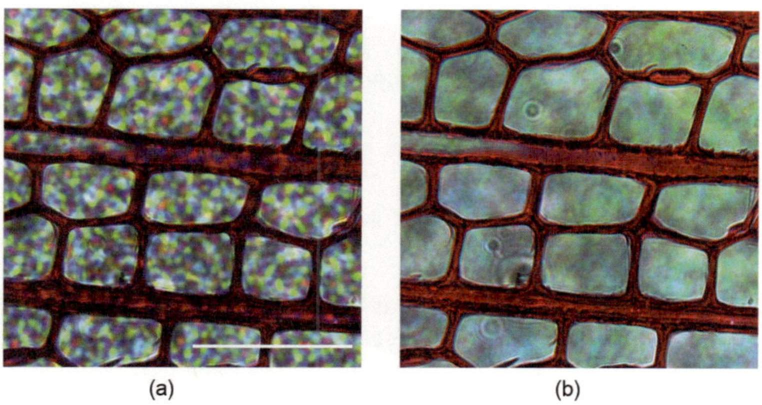

Figure 5.
Color images obtained with the color DHM with a noise slide located at 1 mm. Laser sources, lens magnification ×40, scale bar = 100 μm, equivalent aperture size of (a) 1 mm, (b) 5 mm.

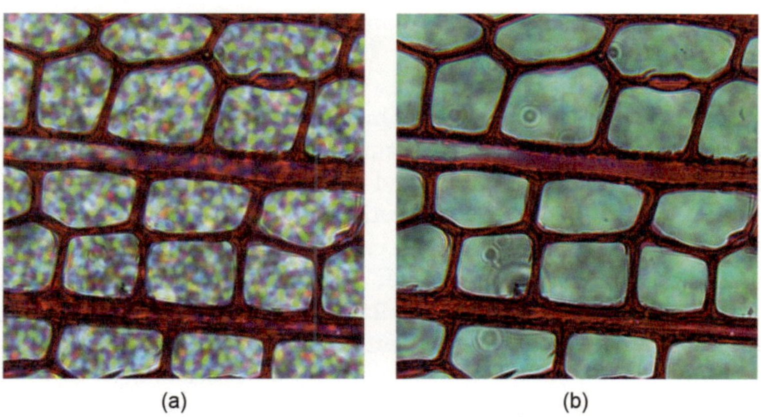

Figure 6.
Color image obtained with the color DHM with a noise slide located at 1 mm. LED sources, lens magnification ×40, (a) aperture size 1 mm, (b) aperture size 5 mm.

It results in an increasing loss of resolution when the defocus distance increases. The way to manage it consists to adjust the partial coherence by adapting D (left diagram in **Figure 1**) or the focusing distance of the laser beam with respect to the ground glass (right diagram in **Figure 1**), in such a way that the resolution requirement is filled for the longest defocus distance possible. The advantage of this phenomenon is the removal, by the convolution, of the defects of the experimental cell and of the optical system. Therefore, it is possible to implement a drastic reduction of the noise that happens with a fully coherent system. This noise reduction has been modeled and measured [29]. Examples of noise reduction obtained in color DHM are shown in **Figures 5** and **6**. The wood slice was placed in the color DHM configuration, and a contaminated slice with particles was placed at 1 mm of the wood slice. We recorded the pictures with the apertures of 1 mm and 5 mm with the LED illumination, or equivalently with the laser illumination. The focal length of L1 in **Figure 3** is f = 200 mm. Experiments were performed with laser and LED sources.

The comparison of the results in **Figures 5** and **6** shows that the reduction of the spatial coherence has a major impact on the reduction of the noise created by out-of focus disturbances. It is also to be noted that the reduction of the temporal coherence by replacing the lasers with LEDs (coherence length of 30 nm) gives, in this example, a weak reduction of the noise. This is due to the fact that the considered DHM configuration is in transmission and the noise-generating slide is relatively thin.

4. Applications

4.1 Multimodal DHM

The fluorescence mode in microscopy is of central importance in biology and biomedical applications, as specific fluorescent labeling allows us to tag specific structures and molecules in cells, tissues, and to perform cell analysis.

We were pioneers in the development of a multimodal DHM to combine the quantitative phase-contrast images provided by the recorded holograms and the fluorescence images of samples, labeled with fluorophores or that present an autofluorescence [19, 30, 31].

The quantitative phase-contrast imaging provides morphology and structural information on the sample. Fluorescence imaging provides information on molecules, thanks to the use of specific fluorophores. Therefore, the combination of both quantitative phase-contrast imaging and fluorescence imaging is of great interest, in particular in biomedical applications and cell biology. For that reason, this combination was afterward also performed by several authors [9, 32, 33].

One of the configurations we developed is presented in **Figure 2a**. In this setup, the fluorescence mode is obtained in reflection as in classical microscopy (epifluorescence mode), with the use of a classical fluorescence filter set selected for the fluorophore(s) to be studied. This filter set is inserted just below the microscope objective ML3 in the sample path (**Figure 2a**). A specific light source (LED) for microscope fluorescence mode is implemented for the excitation of the fluorophore(s) or the autofluorescence. The holograms and the fluorescence images are recorded with the same camera in a sequential mode. As the hologram is recorded in one shot, it allows to record the fluorescence signals with a very short delay.

The multimodal DHM working with partially coherent illumination allows to directly obtain the optical phase of the sample, the bright field image with reduced

DOI: http://dx.doi.org/10.5772/intechopen.105925

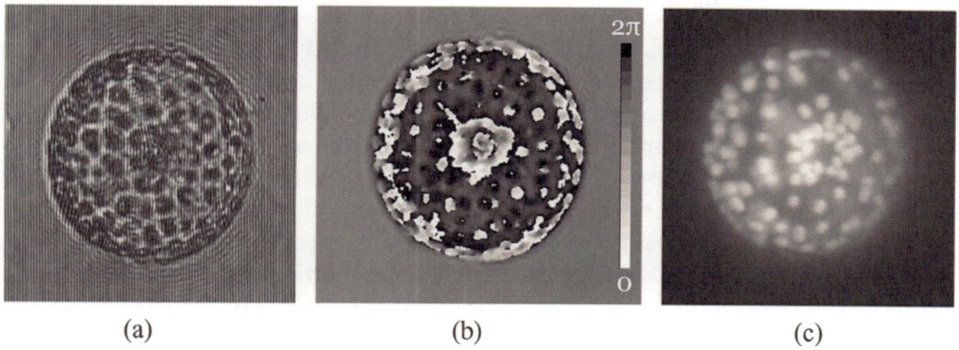

Figure 7.
Images of a phytoplankton organism obtained with the multimodal DHM. (a) the hologram, (b) the phase image, and (c) the chlorophyll auto-fluorescence. These species form colonies and the fluorescence images show the individual cell chloroplast.

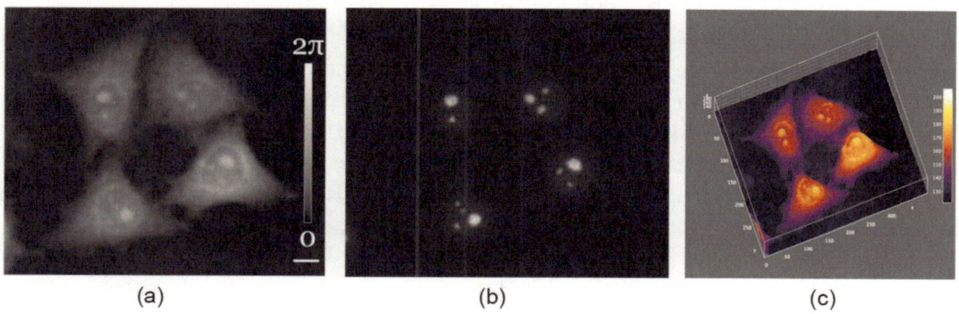

Figure 8.
Validation of the visualization of nucleoli inside the nucleus of HeLa cells, in the phase image. These HeLa cells have one nucleolar protein labeled with the green fluorescent protein (GFP). (a) the phase image, (b) the GFP fluorescent image, (c) the 3D-plot of the phase image (a), unit as gray levels in (a). Scale bar = 10 μm.

noise, and the fluorescence image. The combination of those images allows us to analyze the specimen in detail and brings precious information. An example obtained on a phytoplankton organism is presented in **Figure 7**.

Thanks to the overlay of the quantitative phase-contrast images and the fluorescent images, this DHM is a powerful tool to perform correlative quantitative phase-contrast-fluorescence microscopy. An example is presented in **Figure 8**, which shows the validation of the visualization of nucleoli inside the nucleus of HeLa cells in the phase image. The nucleoli are clearly visible on the phase image with their highest optical height. It is, therefore, possible to study the nucleoli thanks to the phase information, without fluorescence and in a noninvasive way.

Moreover, the recording of the quantitative phase information with low noise also enables the emulation of other classical microscopy modes. In particular, we emulated the differential interference contrast (DIC) mode, a classical tool to observe living cells [34].

4.2 In-flow analysis using DHM

The DHM configurations described in Section 2 are based on Mach–Zehnder off-axis interferometers. As explained above, this allows the one-shot acquisition of the hologram, that is, the extraction of the complex amplitude of the optical field, $t_0(\mathbf{x})$, from one single recorded image. The complex amplitude measured in one plane can be then numerically propagated in depth, that is, computed in other planes along the optical axis, giving $t_d(\mathbf{x})$, as described by Eq. (3) in Section 2; this process is known as the digital holographic reconstruction. This feature is of high interest for the fine analysis of dynamic phenomena, especially if the sample is thicker than the classical depth of field. The described technology is, therefore, well suited for the high-throughput, in-flow observation of objects in thick samples.

We developed in-flow DHM in partially coherent illumination, with specific algorithms for processing the holograms, resulting in high-quality refocused images of the objects of interest, regarding both optical phase and intensity [4, 35]. We used it in many applications, mainly in the environmental field, as illustrated in **Figure 9**, and in the biomedical field, as shown in **Figure 10**. Monochromatic ([4], **Figure 10**) or color RGB DHM ([35], **Figure 9**) were used in the different studies. The illumination was partially coherent, either both spatially and temporally (**Figure 9**) or only spatially while keeping a high temporal coherence (**Figure 10**).

The DHM configurations are those described in Section 2 above and in [4, 35]. The microscope objectives are Leica, magnification 40×, and numerical aperture 0.6. A flow chamber is placed in the object arm of the interferometer (position SA in **Figure 3**). The objects of interest are biological cells: either algal cells in **Figure 9**, or red blood cells in **Figure 10**. The cells are present in suspension in the sample, which is pulled into the flow chamber in such a way that the cells appear in the field of view of the microscope. A sequence of holograms is recorded in order to analyze a large number of cells.

In interferometric DHM, the knowledge of the complex amplitude, $t_0(\mathbf{x})$, extracted from the measured hologram, allows to undertake a powerful compensation of the aberrations in the field of view [35, 36]. One way exists to divide it by a blank complex amplitude, $b(\mathbf{x})$, which is the image containing all—and only—the aberrations due to the defects in the optical paths [35]:

$$t_{0c}(\mathbf{x}) = \frac{t_0(\mathbf{x})}{b(\mathbf{x})} \tag{7}$$

where $t_{0c}(\mathbf{x})$ is the corrected complex amplitude.

However, in the case of the in-flow imaging in DHM, the objects of interest are moving in the field of view, whereas the permanent defects and aberrations are static. As we showed in [4, 35], the blank image, $b(\mathbf{x})$, can then be computed, in a very

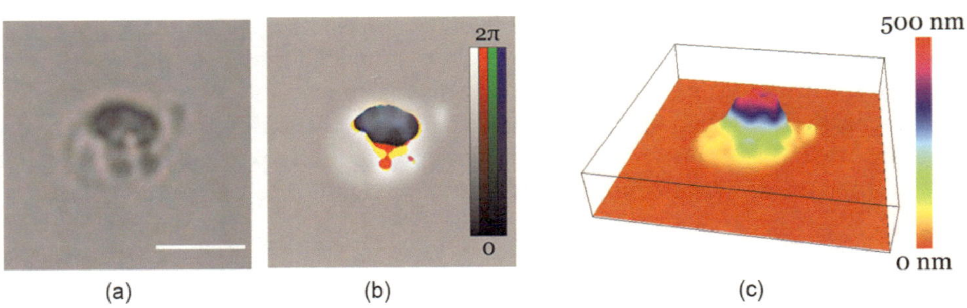

Figure 9.
Microplankton organism analyzed using color in-flow DHM. Automatically corrected and refocused images of: (a) the amplitude modulus, (b) the optical phase, and (c) the corresponding optical thickness. Scale bar = 10 μm.

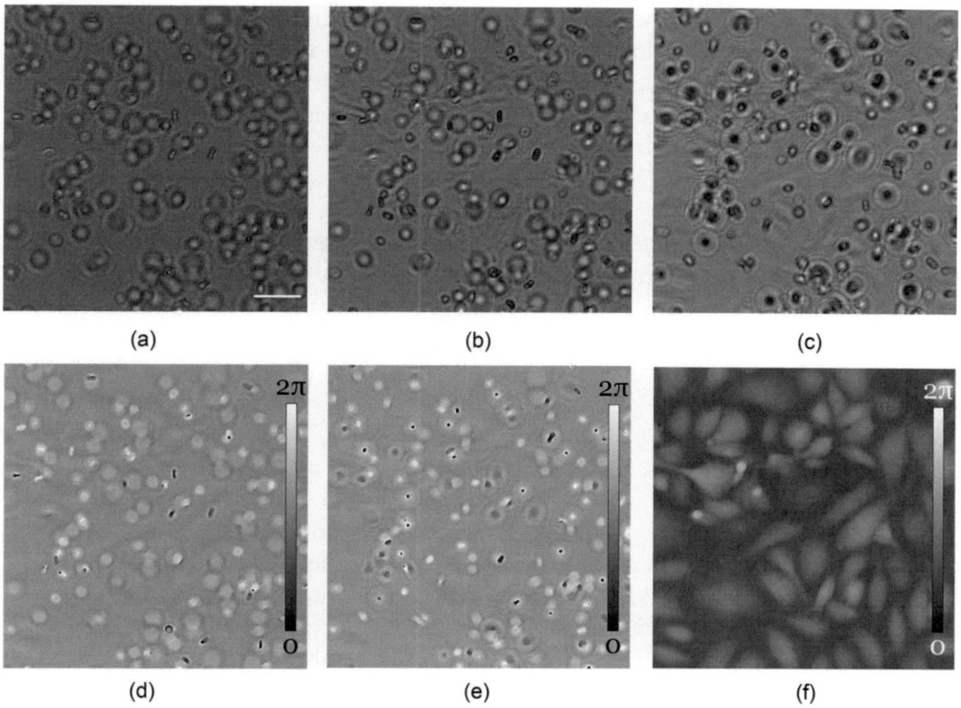

Figure 10.
Red blood cells observed behind an endothelial cell layer using in-flow DHM. Amplitude moduli of corrected images of red blood cells: (a) in the recorded plane, (b) and (c) reconstructed in two different parallel planes, (d) and (e) phase images corresponding to (b) and (c), respectively, and (f) computed phase image of the endothelial cell layer. Different refocused red blood cells can be seen in (b)–(e). Scale bar = 20 μm.

good approximation, as the averaged complex amplitude over all the images of the recorded sequence:

$$b(\mathbf{x}) = \frac{1}{N}\sum_{i=1}^{N} t_{0,i}(\mathbf{x}) \tag{8}$$

where $t_{0,i}(\mathbf{x})$ is the complex amplitude extracted from the i-th hologram of the recorded sequence ($i = 1,\ldots,N$) , and N is the number of recorded holograms in the sequence.

The blank, $b(\mathbf{x})$, is thus directly computed with the sequence of interest. It is not required to record a reference image separately. Two advantages result from this feature: (1) the blank well corresponds to the experiment itself, while a reference image recorded subsequently would be less close to the experimental conditions, and (2) the blank also contains the defects present on the walls of the flow chamber during the experiment.

The computation of the blank is efficient from about 100 averaged images. If the sequence experiment is very long, the blank $b_i(\mathbf{x})$ for the i-th image $t_{0,i}(\mathbf{x})$ can be computed with a moving average around this image:

$$b_i(\mathbf{x}) = \frac{1}{2P+1}\sum_{p=i-P}^{i+P} t_{0,p}(\mathbf{x}) \tag{9}$$

where $2P + 1$ is the number of averaged images, and this equation does not work for the P first and the P last images of the sequence. This allows a powerful correction, by reducing the effects of the instabilities of the illumination.

Dividing every image of the sequence by the computed blank provides the automatic correction of the defects and aberrations, which are permanent in the field of view. For the color in-flow DHM, the described method also automatically corrects the color balance in the images as illustrated in **Figure 9**. The method is very powerful, giving efficiently corrected complex amplitudes, resulting in high-quality images of both the optical phase and intensity, for every image of the sequence.

Once the images are corrected, every object of interest is automatically refocused as explained in Section 2. This allows to observe each object present in the image, even if it was recorded out of its focal plane. The whole depth of the flow chamber is thus analyzed.

4.2.1 An environmental application: the automatic analysis of plankton

We used the color in-flow DHM with LED illumination for the analysis of the microorganisms of plankton in pond and river water samples [35]. The algorithm described above was applied for the automatic correction of the permanent defects and aberrations, including the color balance, and for the automatic refocusing. **Figure 9** illustrates the study and shows an example of observed microorganisms.

4.2.2 A biomedical application: the analysis of human red blood cells through a layer of endothelial cells

We also present here another interesting application of in-flow DHM with partially coherent illumination: the *in vitro* observation of red blood cells inside a flow chamber, whose wall was plated by endothelial cells. This endothelial cell layer allows to mimic the real blood vessels. Observing the behavior of red blood cells close to this layer is therefore essential from a biomedical point of view. We used here the laser source with the rotating ground glass. The correction method described above, in Eqs. (7)–(9), is particularly efficient for such a study. Indeed, with this method, the computed blank gives an image of the endothelial cell layer, which remains static in the field of view, whereas the images of the red blood cells, pulled in the flow chamber, are rid of the modulations due to this layer. Results are shown in **Figure 10**.

4.3 Other blood applications

DHM is well suited for the analysis of blood cells, especially human blood cells [6, 37, 38]. Firstly, the quantitative phase analysis yields a powerful measurement tool for the accurate characterization of blood cells. The phase is indeed related to the optical thickness of the cells. Secondly, the holographic reconstruction in depth allows the extension of the depth of investigation, in comparison with the usual depth of field in classical optical microscopy. This allows to study the motion of blood cells in thick samples, of the same order of magnitude as real human blood vessels. Thirdly, the knowledge of both the quantitative phase and intensity makes easier the automatic processing of a large number of images. Therefore, statistically significant samples can be efficiently studied.

In addition to the in-flow analysis described in 4.2.2, we present here two other applications we performed, illustrating the potential of DHM in blood cell analysis.

4.3.1 Distribution of size and shape of red blood cells within one individual

All the red blood cells within one individual are not identical. Their size and shape are distributed over some range. Studying this distribution is important in biomedical research since it impacts platelet diffusion in blood flows. We used DHM for characterizing this distribution.

An example is shown in **Figure 11**, for one healthy individual (male, 27 years old). Microscope lenses Olympus, magnification 63×, and numerical aperture 0.7 were used. The illumination was made by a green laser source (532 nm) with rotating ground glass. The blood sample was diluted 2000× in the plasma of the donor, in order to avoid any osmotic shock, which would have resulted in a change in the shape and size of the red blood cells. Diluted blood is pulled into a flow chamber, before two-hour sedimentation. A sequence of holograms is then recorded by scanning the flow chamber. Holograms are processed using the method described in Section 4.2 for the compensation of defects and aberrations, and then refocusing. Biconcave red blood cells are then detected and segmented in each image using a method developed in [6], based on the normalized correlation product, denoted $\tilde{\circledast}_\Gamma$, which is defined by:

$$\left(g\,\tilde{\circledast}_\Gamma h\right)(\mathbf{x}) = \frac{|(g \circledast h)(\mathbf{x})|}{\|h\|\,\|g\|_\Gamma(\mathbf{x})} \tag{10}$$

where $g(\mathbf{x})$ is the image containing the red blood cells to be detected, $h(\mathbf{x})$ is a reference image containing a typical red blood cell in its center, $|z|$ denotes the modulus of z, $\circledast$ is the usual correlation product, $\|h\|$ is the Euclidean norm of h, computed by

$$\|h\| = \sqrt{(h \circledast h)(\mathbf{0})} \tag{11}$$

with $\mathbf{0}$ being the zero vector, and $\|g\|_\Gamma(\mathbf{x})$ is the moving norm of g over a domain Γ, which surrounds the red blood cell in h but centered in $\mathbf{x}$, and is given by:

$$\|g\|_\Gamma(\mathbf{x}) = \sqrt{\left(|g|^2 \circledast \gamma\right)(\mathbf{x})} \tag{12}$$

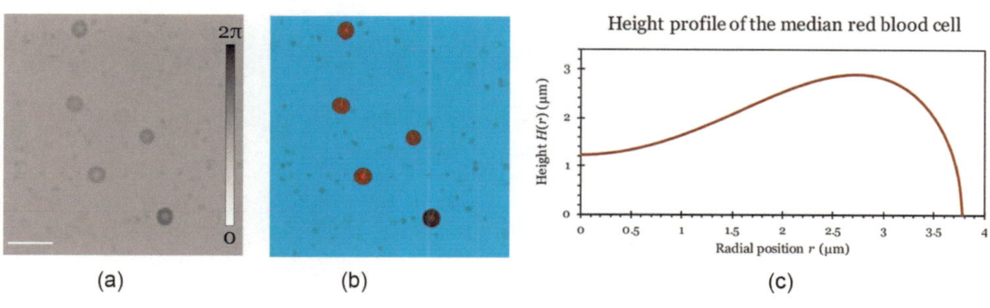

Figure 11.
Measurement of the statistical distribution of size and shape of the red blood cells within one individual. (a) optical phase of an image in the recorded sequence, (b) same image with the automatically detected, segmented, and measured red blood cells, and (c) red blood cell shape corresponding to the median of the red blood cell distribution in the individual (568 red blood cells analyzed). Scale bar = 20 μm.

with $\gamma(\mathbf{x}) = 1$ inside Γ and $\gamma(\mathbf{x}) = 0$ elsewhere. In practice, these expressions may be fast computed in the Fourier domain.

The normalized correlation product ranges from 0 to 1, exhibiting local maxima, close to 1, when a red blood cell is detected. Therefore, a fixed threshold applied to the image of the normalized correlation product allows to identify the position of each red blood cell. The segmentation is operated by applying a threshold to the phase image. Four parameters are then measured in each detected red blood cell: the mean radius, the minimal phase in the central part, the mean radius of the crest, and the mean phase value on the crest. From these four parameters, the parameters R, α, β, and δ in the following equation are assessed, commonly used for characterizing the height profile $H(r)$ of red blood cells in numerical simulations:

$$H(r) = 2R\sqrt{1 - \frac{r^2}{R^2}}\left[\alpha + \beta\frac{r^2}{R^2} + \delta\frac{r^4}{R^4}\right] \tag{13}$$

with r the radial position with respect to the center of the red blood cell. Results are given in **Figure 11**.

4.3.2 Statistical analysis of spread platelets from healthy and dialyzed donors

We also used DHM for an accurate analysis of the spreading of platelets, while comparing healthy volunteers and dialyzed patients [6]. Results are illustrated in **Figure 12**. The spreading occurred in glass slides surmounted by wells. For each of the 14 individuals of the study, seven healthy and seven dialyzed, six wells were used: two coated with laminin, two coated with collagen IV, and two without any coating, denoted as control. Platelet-rich plasma from each individual was poured into each well. Subsequent sedimentation allows the platelets to spread onto the slide, before washing and fixation. Using this experimental procedure, with the sedimentation step for the spreading of platelets, enhances the reproducibility. Each well was then analyzed using a color DHM working in partially spatially coherent illumination, made by three laser sources followed by a rotating ground glass. Forty holograms were recorded for each well. The processing of holograms, including the compensation of

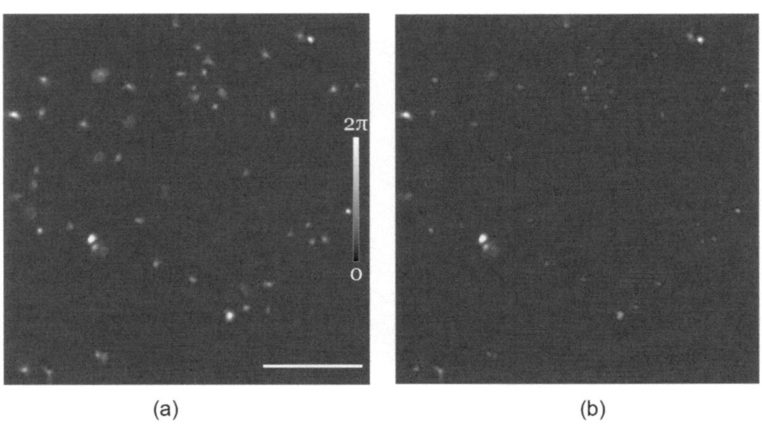

Figure 12.
Analysis of the platelet spreading using DHM, with the comparison between healthy and dialyzed donors. (a) example of a phase image obtained, and (b) the same image with the automatically detected and segmented single platelets, with the automatic rejection of platelets at the borders and platelets forming aggregates. Scale bar 50 µm.

aberrations, the refocusing, the detection, and the segmentation of platelets, was performed as described in 4.2 and 4.3.1. For every detected platelet, several parameters were measured, whose surface area S, optical volume V, and average optical height $\hat{h}$.

Whereas the surface area S covered by each platelet can also be measured in classical microscopy, the measurement of the average optical height $\hat{h}$, which provides information about the efficiency of the platelet spreading, is a huge advantage of DHM in this study. We hence performed a statistical analysis of the three-dimensional morphology of single spread platelets in both healthy volunteers and dialyzed patients, and demonstrated, for the first time to our knowledge, a significant difference between both.

5. Conclusions

In this chapter, we have described the advantages of digital holographic microscopy (DHM) working with partially coherent illumination. This illumination drastically reduces the speckle noise that alters the images in both the amplitude modulus and the phase images. This opens the investigation of numerous applications with instruments that are robust and easy to use. The instrumentation is accompanied by software allowing the hologram processing in an automated way.

We have depicted several DHM configurations that we developed as a color DHM and a multimodal fluorescence DHM. We have presented different valuable applications that we implemented and that show the high-quality image obtained, which demonstrate the high capabilities of the technology in the biomedical and the environmental fields.

Acknowledgements

The research was supported by the Walloon Region in the frame of the Decisiv and the HoloCancer projects and L'Institut d'encouragement de la Recherche scientifique et de l'Innovation de la Région de Bruxelles-Capitale (IRSIB-INNOVIRIS) in the frame of the Holoflow Impulse project.

Dr. Jérôme Dohet-Eraly was Aspirant and then Chargé de Recherches du Fonds de la Recherche Scientifique – FNRS (F.R.S.-FNRS, Belgium).

Conflict of interest

The authors declare no conflict of interest.

Author details

Frank Dubois[1*], Catherine Yourassowsky[1], Karim Zouaoui Boudjeltia[2] and Jérôme Dohet-Eraly[1]

1 Université libre de Bruxelles – Chimie physique, Microgravity Research Centre, Bruxelles, Belgium

2 Université libre de Bruxelles – Laboratoire de Médecine expérimentale, Charleroi, Belgium

*Address all correspondence to: frank.dubois@ulb.be

References

[1] Schnars U, Jüptner WPO. Digital recording and numerical reconstruction of holograms. Measurement Science and Technology. 2002;**13**:R85-R101. DOI: 10.1088/0957-0233/13/9/201

[2] Dubois F, Joannes L, Legros JC. Improved three-dimensional imaging with a digital holography microscope with a source of partial spatial coherence. Applied Optics. 1999;**38**: 7085-7094. DOI: 10.1364/AO.38.007085

[3] Marquet P, Rappaz B, Magistretti PJ, Cuche E, Emery Y, Colomb T, et al. Digital holographic microscopy: a noninvasive contrast imaging technique allowing quantitative visualization of living cells with subwavelength axial accuracy. Optics Letters. 2005;**30**: 468-470. DOI: 10.1364/OL.30.000468

[4] Yourassowsky C, Dubois F. High throughput holographic imaging-in-flow for the analysis of a wide plankton size range. Optics Express. 2014;**22**: 6661-6673. DOI: 10.1364/OE.22.006661

[5] O'Connor T, Shen J-B, Liang BT, Javidi B. Digital holographic deep learning of red blood cells for field-portable, rapid COVID-19 screening. Optics Letters. 2021;**46**:2344-2348. DOI: 10.1364/OL.426152

[6] Dohet-Eraly J, Zouaoui Boudjeltia K, Rousseau A, Queeckers P, Lelubre C, Desmet JM, et al. Three-dimensional analysis of blood platelet spreading using digital holographic microscopy: a statistical study of the differential effect of coatings in healthy volunteers and dialyzed patients. Biomedical Optics Express. 2022;**13**: 502-513. DOI: 10.1364/BOE.448817

[7] Zhang T, Yamaguchi I. Three-dimensional microscopy with phase-shifting digital holography. Optics Letters. 1998;**23**:1221-1223. DOI: 10.1364/OL.23.001221

[8] Cuche E, Bevilacqua F, Depeursinge C. Digital holography for quantitative phase-contrast imaging. Optics Letters. 1999;**24**:291-293. DOI: 10.1364/OL.24.000291

[9] Park Y, Popescu G, Badizadegan K, Dasari RR, Feld MS. Diffraction phase and fluorescence microscopy. Optics Express. 2006;**14**:8263-8268. DOI: 10.1364/OE.14.008263

[10] Barty A, Nugent KA, Paganin D, Roberts A. Quantitative optical phase microscopy. Optics Letters. 1998;**23**: 817-819. DOI: 10.1364/OL.23.000817

[11] Luo W, Zhang Y, Feizi A, Göröcs Z, Ozcan A. Pixel super-resolution using wavelength scanning scanning. Light: Science and Applications. 2016;**5**: e16060. DOI: 10.1038/lsa.2016.60

[12] Tian L, Liu Z, Yeh L-H, Chen M, Zhong J, Waller L. Computational illumination for high-speed *in vitro* Fourier ptychographic microscopy. Optica. 2015;**2**:904-911. DOI: 10.1364/OPTICA.2.000904

[13] Podgorski T, Callens N, Minetti C, Coupier G, Dubois F, Misbah C. Dynamics of vesicle suspensions in shear flow between walls. Microgravity Science and Technology. 2011;**23**:263-270. DOI: 10.1007/s12217-010-9212-y

[14] Callens N, Minetti C, Coupier G, Mader M-A, Dubois F, Misbah C, et al. Hydrodynamic lift of vesicles under shear flow in microgravity. Europhysics Letters. 2008;**83**:24002. DOI: 10.1209/0295-5075/83/24002

[15] Zegers I, Carotenuto L, Evrard C, Garcia-Ruiz J, De Gieter P,

Gonzales-Ramires L, et al. Counterdiffusion protein crystallisation in microgravity and its observation with PromISS (protein microscope for the international space station). Microgravity Science and Technology. 2006;**18**: 165-169. DOI: 10.1007/BF02870402

[16] Schilling BW, Poon T-C, Indebetouw G, Storrie B, Shinoda K, Suzuki Y, et al. Three-dimensional holographic fluorescence microscopy. Optics Letters. 1997;**22**:1506-1508. DOI: 10.1364/OL.22.001506

[17] Rosen J, Brooker G. Digital spatially incoherent fresnel holography. Optics Letters. 2007;**32**:912-914. DOI: 10.1364/OL.32.000912

[18] Rosen J, Brooker G. Non-scanning motionless fluorescence three-dimensional holographic microscopy. Nature Photonics. 2008;**2**:190-195. DOI: 10.1038/nphoton.2007.300

[19] Dubois F, Yourassowsky C. Method and device for obtaining a sample with three-dimensional microscopy. Patents. EP 01870147 (2001), EP1399730 B1, US 7009700 B2 (2006)

[20] Dubois F, Requena ML, Minetti C, Monnom O, Istasse E. Partial spatial coherence effects in digital holographic microscopy with a laser source. Applied Optics. 2004;**43**:1131-1139. DOI: 10.1364/AO.43.001131

[21] Kreis T. Digital holographic interference-phase measurement using the Fourier-transform method. Journal of the Optical Society of America. A. 1986;**3**:847-855. DOI: 10.1364/JOSAA.3.000847

[22] Zouaoui Boudjeltia K, Ribeiro de Sousa D, Uzureau P, Yourassowsky C, Perez-Morga D, Courbebaisse G, et al. Quantitative analysis of platelets aggregates in 3D by digital holographic microscopy. Biomedical Optics Express. 2015;**6**:3556-3563. DOI: 10.1364/BOE.6.003556

[23] Dubois F, Yourassowsky C. Full off-axis red-green-blue digital holographic microscope with LED illumination. Optics Letters. 2012;**37**:2190-2192. DOI: 10.1364/OL.37.002190

[24] Nazarathy M, Shamir J. Fourier optics described by operator algebra. Journal of the Optical Society of America. 1980;**70**:150-159. DOI: 10.1364/JOSA.70.000150

[25] Dubois F, Schockaert C, Callens N, Yourassowsky C. Focus plane detection criteria in digital holography microscopy by amplitude analysis. Optics Express. 2006;**14**:5895-5908. DOI: 10.1364/OE.14.005895

[26] Dohet-Eraly J, Yourassowsky C, Dubois F. Refocusing based on amplitude analysis in color digital holographic microscopy. Optics Letters. 2014;**39**:1109-1112. DOI: 10.1364/OL.39.001109

[27] Dubois F, El Mallahi A, Dohet-Eraly J, Yourassowsky C. Refocus criterion for both phase and amplitude objects in digital holographic microscopy. Optics Letters. 2014;**39**:4286-4289. DOI: 10.1364/OL.39.004286

[28] Dohet-Eraly J, Yourassowsky C, Dubois F. Fast numerical autofocus of multispectral complex fields in digital holographic microscopy with a criterion based on the phase in the Fourier domain. Optics Letters. 2016;**41**: 4071-4074. DOI: 10.1364/OL.41.004071

[29] Dohet-Eraly J, Yourassowsky C, El Mallahi A, Dubois F. Quantitative assessment of noise reduction with partial spatial coherence illumination in

digital holographic microscopy. Optics Letters. 2016;**41**:111-114. DOI: 10.1364/OL.41.000111

[30] Dubois F, Yourassowsky C, and Monnom O. Microscopie en holographie digitale avec une source partiellement cohérente. In: Faupel M, Smigielski P, Grzymala R, editors. Imagerie et Photonique pour les Sciences du Vivant et de la Médecine. Fontis Media; 2004. p. 287-302

[31] Dubois F, Yourassowsky C, Callens N, Minetti C, Queeckers P, Podgorski T, et al. Digital holographic microscopy working with a partially spatial coherent source. In: Ferraro P, Wax A, Zalevsky Z, editors. Coherent Light Microscopy, Imaging and Quantitative Phase Analysis. Berlin: Springer; 2011. p. 31-59

[32] Quan X, Nitta K, Matoba O, Xia P, Awatsuji Y. Phase and fluorescence imaging by combination of digital holographic microscopy and fluorescence microscopy. Optical Review. 2015;**22**:349-353. DOI: 10.1007/s10043-015-0042-4

[33] Nygate YN, Singh G, Barnea I, Shaked NT. Simultaneous off-axis multiplexed holography and regular fluorescence microscopy of biological cells. Optics Letters. 2018;**43**:2587-2590. DOI: 10.1364/OL.43.002587

[34] Dubois F, Yourassowsky C, Monnom O, Legros JC, Debeir O, Van Ham P, et al. Digital holographic microscopy for the three-dimensional dynamic analysis of in vitro cancer cell migration. Journal of Biomedical Optics. 2006;**11**:054032. DOI: 10.1117/1.2357174

[35] Dohet-Eraly J, Yourassowsky C, Dubois F. Color imaging-in-flow by digital holographic microscopy with permanent defect and aberration corrections. Optics Letters. 2014;**39**:6070-6073. DOI: 10.1364/OL.39.006070

[36] Colomb T, Cuche E, Charrière F, Kühn J, Aspert N, Montfort F, et al. Automatic procedure for aberration compensation in digital holographic microscopy and applications to specimen shape compensation. Applied Optics. 2006;**45**:851-863. DOI: 10.1364/AO.45.000851

[37] O'Connor T, Santaniello S, Javidi B. COVID-19 detection from red blood cells using highly comparative time-series analysis (HCTSA) in digital holographic microscopy. Optics Express. 2022;**30**:1723-1736. DOI: 10.1364/OE.442321

[38] Bianco V, Merola F, Miccio L, Memmolo P, Gennari O, Paturzo M, et al. Imaging adherent cells in the microfluidic channel hidden by flowing RBCs as occluding objects by a holographic method. Lab on a Chip. 2014;**14**:2499-2504. DOI: 10.1039/c4lc00290c

Chapter 12

Extended Lattice Light-Sheet with Incoherent Holography

Mariana Potcoava, Christopher Mann, Jonathan Art and Simon Alford

Abstract

Recently, lattice light-sheet instruments and imaging technology have shown a great improvement in exploring the dynamics of life at subcellular resolution. However, to reconstruct 3-dimensional structures the objective lens or the sample stage are invariably moved. Invention of incoherent holography lattice light-sheet (IHLLS) using FINCH provides quantitative information on the state and size of subcellular changes of neurons over the visible spectrum. The technique allows both faster three-dimensional amplitude as well as phase imaging without moving either sample stage or the detection objective. Amplitude and phase measurements provide intrinsic instrumental simplicity, larger scanning area, and higher resolution when compared to the original LLS schemes. IHLLS is built as an additional detection arm of an original lattice light-sheet (LLS) system. In this chapter, we review the evolution of IHLLS to address its feasibility and limitations.

Keywords: fluorescence, incoherent holography, lattice light-sheet

1. Introduction

Neurons in situ are inherently 3-dimensional structures, with processes that extend through the nervous system in all dimensions. Volumetric imaging solutions that have been applied to this include confocal imaging, multiphoton imaging, and light sheet methods [1, 2]. With conventional approaches it is not possible to image in the millisecond temporal range at multiple depths. Light sheet approaches have rather poor z-axis resolution of 4 to 10 μm and suffer from substantial light sheet divergence within refractive tissue. Lattice light-sheet (*LLS*) imaging [3, 4] goes some way to resolving those issues. Sheet scanning allows individual frames to be captured at rates limited by the camera frame rate, and the lattice structure allows both high z-plane resolution (close to diffraction limited; approximately 400 nm) and good tissue penetration (up to 100 μm) in brain slices. However, despite the substantial contribution of the LLS groups to the biomedical imaging, and that z plane scanning of the lattice sheet can be rapid using galvanometers, the goal of imaging biological cells using LLS systems without moving the detection objective or the sample stage has not yet been achieved.

To overcome these hurdles, we built an incoherent holography lattice light-sheet (IHLLS) [5–7] tool as a second detection module of the original LLS system, with the

 IntechOpen

capability to image quantitative phase information of biological samples without moving the detection objective or the sample stage. The IHLLS system was first developed for the excitation wavelength 488 nm and emission wavelength 520 nm, but it can be further developed into a color imaging system, called incoherent color holographic lattice light-sheet (ICHLLS).

Phase imaging of live cells and tissue opens several possibilities that have not been available to more established methods in microscopy [8–23]. The simplest example is the ability of holographic imaging to reconstruct 3-dimensional information by both effectively autofocusing and extending the depth of field from which valuable information can be gathered. Taking the nervous system as an example, in dendritic fields, neurons receive information rapidly and often synchronously at different locations at various 3-dimensional planes. Neurons also respond electrically to inputs. This activity is rapid (milliseconds duration) and spreads throughout their complex 3D structure. Interestingly, changes in the entire cell membrane follow rapidly from localized electrical changes, and these can be detected in phase images in transmission or reflection imaging modalities [24]. This opens the possibility of measuring excitability of nerve cells across the complex 3-dimensional structure of the neuron. At longer time scales, organelle and protein movement within cells change the local refractive index [25]. Since phase changes in reflection geometry do not include the refractive index difference between the cytoplasm and the medium, it implies that the movements restricted to structural changes, axonal transport, or vesicle recycling may be detected through the cell membrane dynamics using the reflection phase microscopy approach.

Synaptic transmission, which requires precise coupling between action potentials, Ca^{2+} entry and neurotransmitter release [26], is also fundamental to the function of the brain. Thus, understanding synaptic function is key to understanding how the brain works while similarly, understanding synaptic dysfunction is crucial to understanding diseases of the brain. One key to understanding synaptic function is to image the distributed spatiotemporal axonal, dendritic, and synaptic activity in 3-dimensional space simultaneously and with high resolution. Our long-term aim will be to record phase changes evoked by synaptic activity within small (e.g. pre- and postsynaptic structures of the brain) in the millisecond time periods during which this activity occurs. The combination of LLS excitation targeting specific neurons with holographic phase imaging of selective fluorescent markers, called IHLLS, to detect fast events, or of electrical or structural changes of phase, creates many new approaches in imaging neuronal activity.

In this chapter, we will review the fundamental concepts, design, calibration, optimization, and terminology of IHLLS systems and discuss their imaging capabilities. We will begin with an overview of the IHLLS principle. We will then focus on each imaging system, IHLLS 1L and IHLLS 2L, with one or two excitation wavelengths, as well IHLLS in two colors, 488 and 560 nm, called ICHLLS. Applications suitable for phase modulation using incoherent light of neuronal cells will be discussed. At the end of the chapter, we will draw our conclusions and recommendations.

2. Incoherent holographic lattice light-sheet (IHLLS)

IHLLS [5–7] is a new imaging technique. It is a noninvasive, noncontact, and cross-sectional imaging method, that provides the imaging capability to build the 3D complex amplitude volume of neuronal cells with resolution comparable to, or better than the resolution of the conventional LLS in dithering mode [3, 4] and with an extended FOV.

IHLLS operates as an extra detection module added on an existing LLS and utilizes the excitation technology of the LLS system, **Figure 1a**, in which excitation light is confined to a plane defined by a lattice of intersecting Bessel beams that self-reinforce as they project through tissue. Imaging detection uses the Fresnel incoherent correlation holography (FINCH) [27–31] principle. FINCH uses the self-interference property of the emitted fluorescent light to create Fresnel holograms of a 3D object in combination with the phase-shifting concept in which three or four interference patterns are created by a single channel on-axis interferometer. The beam splitter of the interferometer is replaced by a spatial light modulator (SLM), so that each spherical beam propagating from each 3D object's points is split into two spherical beams

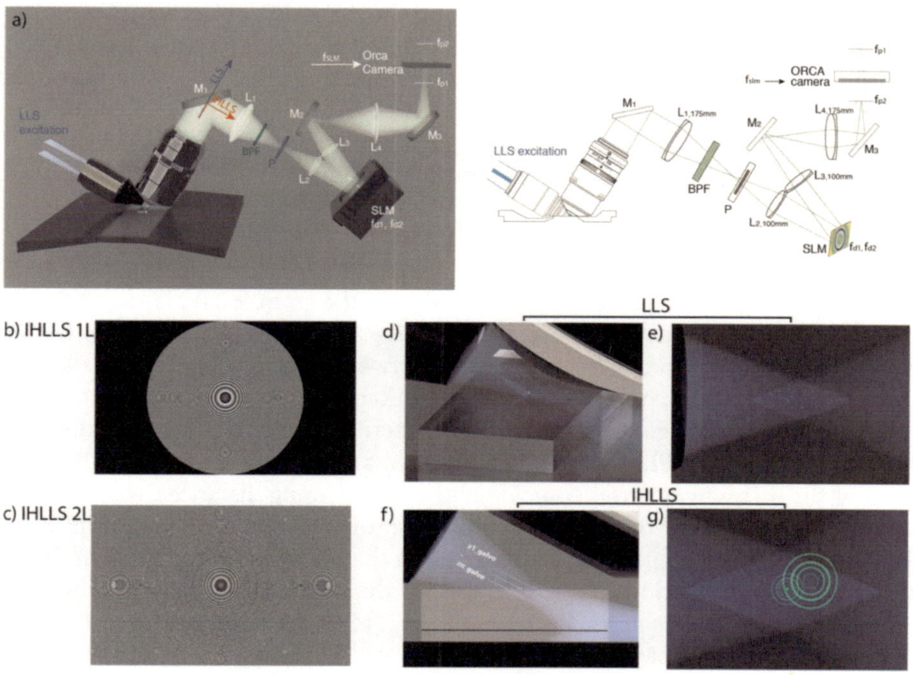

Figure 1.
The IHLLS system. (a) Schematics of the IHLLS systems with (b) one diffractive lens of focal length, f_{SLM} = 400 mm at the phase shift θ_1 = 0, and (c) two diffractive lenses with focal lengths f_{d1} = 220 mm and f_{d2} = 2356 mm, at the phase shift θ_1 = 0, superposed with a slight defocus to bring the objects in focus in the middle of the camera FOV; The system consists of a water immersed microscope objective MO (Nikon 25X, NA 1.1, WD 2 mm), lenses L_1 = L_4 with focal lengths 175 mm, L_2 = L_3 with focal lengths 100 mm; mirrors M_1, M_2, M_3; polarizer P; 520 nm center wavelength, 40 nm band pass filter BPF; spatial light modulator SLM. The light propagates through either pathway 1 (blue line in (a) for the original LLS or pathway 2 (red line in (a)) for IHLLS, depending on the orientation of sliding mirror. A collimated 30 Bessel beam is focused by an excitation objective lens (d, f) which generates a lattice light sheet. The vectors represent the x, y, z and s planes of the Bessel. z and x are moved by the z and x galvos. It excites only fluorophores in the focal plane (e) and in/off the focal plane (g) of the detection objective lens. While the z-galvo and z-piezo are moved along the z axis to acquire stacks in LLS (d-e), in IHLLS only the z-galvo is moved at various z positions (f-g), (Video_1 available at: https://bit.ly/3U4Iivy). For IHLLS, the size of the beam coming out the objective is diminished in half by the relay lens system, L_1 and L_2, to fit the size of the SLM. The SLM plane is optically conjugated with the objective back-focal-plane. The diffraction mask was positioned for all experiments on the anulus of 0.55 outer NA and 0.48 inner NA. The CMOS camera, tube lens, filter, and detection objective lens are used for fluorescence detection. The detection magnification M_{T-LLS}=62.5 and the illumination wavelength $\lambda_{illumination} = 488$ nm. The width of the light sheet in the center of the FOV is about 400 nm. x-axis is the direction of the x-galvo mirror motion, z-axis is the direction of the z-piezo mirror motion, and s-axis is the direction of excitation light propagation.

with different radii of curvature. The interference patterns from sample points are added incoherently, to further create Fresnel holograms. These holograms are numerically processed by in-house diffraction software.

So far, a short overview about the evolution of *IHLLS* devices was introduced. The following section will discuss various *IHLLS* systems and their technologies.

2.1 IHLLS systems and devices

We developed the IHLLS in two forms, with one diffractive lens and a constant phase with shared pixels, IHLLS 1L **Figure 1b**, **d**, **e**, or two diffractive lenses with randomly selected pixels, IHLLS 2 L (**Figure 1c**, **f**, **g**). The IHLLS 1L technique is used for calibration purposes and the IHLLS 2 L technique is used for the actual 3D sample imaging. The z-scanning principle in IHLLS 1L, same as in LLS, **Figure 1d**, **e**, is that both the z-galvanometric mirror (z-galvo) and the detection objective (z-piezo), synchronize in motion to scan the sample in 3D, but the axial resolution could be lower than the axial resolution in LLS due to the blurring effect of the constant phase lens added on the SLM that focusses to infinity. In IHLLS 2L, **Figure 1f**, **g**, the detection objective is kept fixed, but the z-galvanometric mirror moves to reach various depths in the sample and to increase the visibility of the Fresnel patterns.

2.1.1 IHLLS theoretical principle

The emitted light from a point object located in the front focal plane of the detection objective is collimated into a plane wave after passing the objective. This wave is transformed into two beams due to the two diffractive lenses with different focal lengths, f_{d1} and f_{d2}, uploaded on the SLM that focus at two focal points, f_{p1} and f_{p2}, **Figure 1a**. In the original setup of FINCH there are no optics between the microscope objective and the SLM or between the SLM and camera. Therefore, the focal lengths of the diffractive lenses coincide with their imaging distances, calculated from the SLM position toward the camera position. In this case the focal lengths of the diffractive lenses, f_{SLM}, f_{d1}, and f_{d2}, uploaded on the SLM, are different than their imaging planes measured from the SLM toward the camera. The beam is adjusted in size by using the two lens pairs, L_1, L_2 and L_3, L_4, to fit the SLM chip area. The SLM used here was a phase SLM (Meadowlark; 1920 × 1152 pixels, it was recalibrated to produce the desired focal lengths and phase shifts for a 520 nm wavelength and delivers a full 2π phase shift over its working range of 256 gray levels.

A hologram is obtained by the interference between the two positive spherical waves converging to the image points f_{p1}, and f_{p2} (relative to the SLM-plane), **Figure 1a**, located at -109 and 289 mm from the CMOS ORCA camera plane, and with interference efficiency $\tan(\varphi) \leq 0.04$.

Referring to eq. 16 from [29], $\tan(\varphi) \cong \frac{2R_0}{z_h} \leq \frac{\lambda}{2\delta_c}$, we get $0.03234 \leq 0.04$, where $R_0 = \max(R_{01}, R_{02}) = \max(11.393\, mm, 9.781\, mm) = 11.393\, mm$, R_{01},R_{02} are the radii of the two beams at the exit of lens TL_4, $z_h = 664\, mm$, is the distance from TL_4 to camera, $\lambda = 520\, nm$ the emission wavelength, and $\delta_c = 6.5\, \mu m$ is the camera pixel size.

The SLM transparency containing the two diffractive lenses has the expression: $\left[C_1 Q\left(-\frac{1}{f_{d1}}\right) + C_2 \exp(i\theta) Q\left(-\frac{1}{f_{d2}}\right)\right]$, where $Q(b) = \exp\left[i\pi b\lambda^{-1}(x^2 + y^2)\right]$ is a quadratic phase function with (x, y) being the space coordinates, θ is the phase shift of the SLM, and C_1,C_2 are constants. When the system works in IHLLS 2L mode, $C_1 = 0.5$, $C_2 = 0.5$. When $f_{d1} = \infty$ *(IHLLS 1L)*, the expression becomes:

$\left[C_1 + C_2 \exp(i\theta) Q\left(-\frac{1}{f_{SLM}}\right)\right]$, and $C_1 = 0.1$, $C_2 = 0.9$. Let us suppose the two waves, created by the diffractive lenses at the SLM, have the general expressions:

$$U_1(u, v) = A_1(u, v) \exp(i\phi_1(u, v)) \quad (1)$$

$$U_2(u, v) = A_2(u, v) \exp(i\phi_2(u, v) + \theta) \quad (2)$$

where (u, v) are the coordinates of the camera plane. With the assumption that the object is an infinitesimal object point, the intensity of the hologram at the sensor plane takes the following expression:

$$\begin{gathered} I_H(u, v) = |A_1(u, v)|^2 + |A_2(u, v)|^2 \\ +2 * A_1(u, v) A_1(u, v) \cos(\phi_1(u, v) - \phi_2(u, v) - \theta) \end{gathered} \quad (3)$$

Then, four phase-shifted holograms with phase shifts applied to one of the diffractive lenses of 0, $\pi/2$, π, $3\pi/2$, are recorded by the camera and superposed using the phase-shifting algorithm to build the complex hologram of the object point at the camera plane:

$$\begin{gathered} U(u, v) = A(u, v) \exp(i\phi(u, v)) = \\ \frac{1}{4}\left\{\left(I_{H(u, v, 0)} - I_{H(u, v, \pi)}\right) + i\left(I_{H(u, v, \pi/2)} - I_{H(u, v, 3\pi/2)}\right)\right\} \end{gathered} \quad (4)$$

where: $A(u, v)$ is the amplitude of the encrypted image, which is the product of the two amplitudes, $A(u, v) = A_1(u, v) * A_2(u, v)$, and the encrypted phase:

$$\varnothing = \phi_1(u, v) - \phi_2(u, v) = arctan \left[\frac{I_H(u, v, \pi/2) - I_H(u, v, 3\pi/2)}{I_H(u, v, 0) - I_H(u, v, \pi)}\right] \quad (5)$$

Therefore, the complex amplitude distribution of the object wave at the camera plane has the expression:

$$A_1(u, v) \exp(i\phi(u, v)) = U(u, v)/A_2(u, v) \quad (6)$$

The 3D image of the object is retrieved by using the angular spectrum method to reconstruct the complex amplitude of the object wave at any depth within the positive reconstruction distance z_r, found by the expression:

$$z_r = \pm \frac{\left(z_h - f_{p1}\right)\left(f_{p2} - z_h\right)}{f_{p2} - f_{p1}} \quad (7)$$

We used the following values for the dual lens FINCH: $f_{p1} = 555.185$ mm, $f_{p2} = 826.793$ mm. The two focal lenses were calculated using OpticStudio (Zemax, LLC) in a multiconfiguration system as described in our previous work [5]. The distance between the last lens in the system, L4, and the camera was calculated by the optical design software as being $z_h = 664$ mm and matches the value calculated with the expression $z_h = \frac{2f_{p1}f_{p2}}{f_{p2}+f_{p1}}$. Therefore, using Eq. (7), the reconstruction distance is $z_r \cong 63$ mm, (in terms of lens distances) above and below the middle position of the z-galvo scanning range. If we need to reach out to the most extreme parts of objects in the FOV of camera the reconstruction distance could increase to 8 mm. This distance corresponds to a *z-galvo* range of $z_{galvo} = \pm 40$ μm (in terms of object distances). This displacement is obtained for the LLS diffraction mask positioned on the anulus of outer NA of 0.55 and inner NA of 0.48, which generated a Bessel beam with a FWHM sheet length of 15 μm.

2.1.2 IHLLS optical design

We implemented the IHLLS approach in four main steps. *First*, we matched the effective pixel size in both configurations, conventional LLS and IHLLS 1L, **Figure 1f**, **g**, by using a USAF 1951 resolution target, **Figure 2a-d** and Opticstudio (Zemax, LLC) optical design, **Figure 2e-g**. The Opticstudio software simulated the optical components, beams propagation from the sample plane to the camera plane and calculated the correct distances between each sequential optical component. *Second*, we demonstrated the 3D imaging capabilities of polystyrene beads by using both conventional LLS and IHLLS 1L for the diffractive lens with f_{SLM} = 400 mm at phase $\theta = 0$ on the SLM, with x, z-galvo, and z-piezo motions, **Figure 3**, for the calibration purposes. *Third*, we have recorded and reconstructed IHLLS 2L bead holograms at various z-galvo scanning depths, from $-40\ \mu m$ to $+40\ \mu m$ in steps of $10\ \mu m$, without moving the z-piezo, to assess the instrument performance in comparison to those of the conventional LLS, **Figure 4a-h**. *Fourth,* volumetric imaging of nerve cells **Figures 5–7** at 488 nm excitation wavelength, and color imaging, **Figure 8**, at dual excitation wavelengths, 488 and 560 nm, are presented.

It was determined that for an emission wavelength of 520 nm, and with an overall transversal magnification set to 62.5, the distances should be $d_1 = 75\ mm$, $d_2 = 95.074\ mm$, $d_3 = 288.914\ mm$, $d_4 = 103.660\ mm$, $d_5 = 103.660\ mm$, $d_6 = 288.914\ mm$, and the distance between the lens TL_4 to the detector should be 664 mm (i.e., $d_7 = 664\ mm$), and the focal length of the single diffractive lens superimposed on the SLM was determined to be $f_{SLM} = 400\ mm$. For this step, the transversal magnification of 62.5 was checked by imaging the USAF 1951 resolution target, group 7, element 6, with both instruments using a white light source. An optimization of a multi-configuration optical system, **Figure 2b**, **c**, were performed to

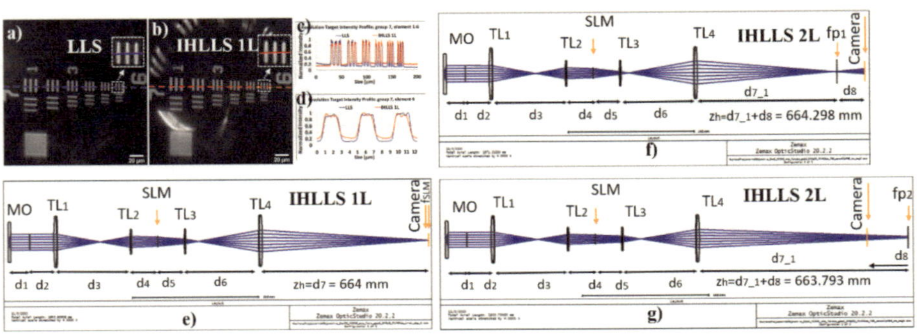

Figure 2.
Optical design of the IHLLS. (a) The LLS resolution target imaging (a, c (blue), d (blue)) with M_{T-LLS}=62.5; (b) The IHLLS resolution target imaging with one diffractive lens, IHLLS 1L, (b, c (red), d (red)) with $M_{T-IHLLS}$=62.5; (c), d, Cross-sections of group 7 and element 6 respectively; (e) Opticstudio simulation of IHLLS 1L, with the constraint $M_{T-IHLLS}$=62.5, to calculate the focal length (f_{SLM} = 400 mm) of a single diffractive lens uploaded on the SLM and the distances between each pair of optical components; (f), (g) These distances are kept constant for the simulation of IHLLS with two diffractive lenses, IHLLS 2L, to obtain the focal lengths of the two diffractive lenses on the SLM, f_{d_1} = 220 mm and f_{d_2} = 2356 mm.

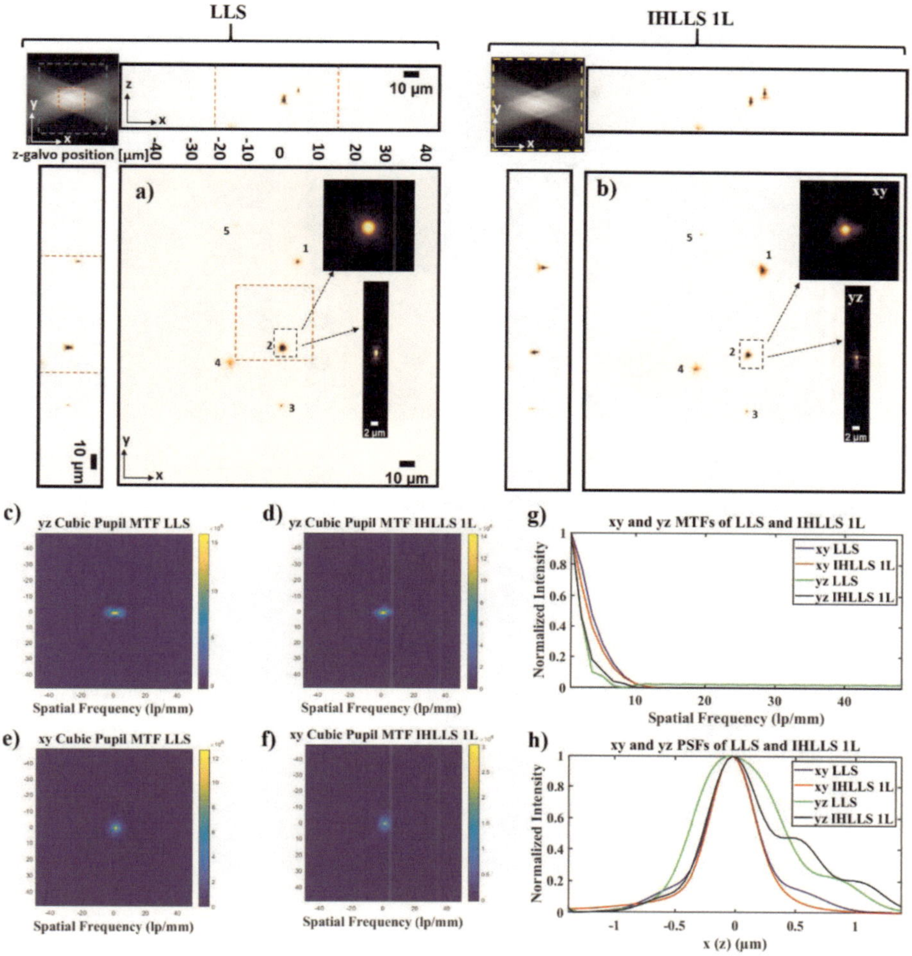

Figure 3.
Tomographic imaging of 0.5 µm, FOV 208 µm², in a conventional LLS (a) and incoherent LLS with only one diffractive lens (IHLLS 1L) of focal length 400 nm (b), without deconvolution. On the sides and above are shown the max projections through the volume (400 Z-galvo steps). The Bessel beams are displayed in the upper left corner of each xy-projection to show the orientation of the beams (FOV 208 µm²). The area enclosed inside the colored dashed rectangles are as follows: red- the scanning area for the original LLS (52 µm²), green – the extended area mentioned in [32] of 120 µm², and yellow – the actual scanning area for the LLS, IHLLS 1L, and IHLLS 2L. The bead #2 in the black dashed rectangle that is in the middle of the lattice sheet is considered for calculating the resolution for the two instruments. The transverse MTFs of the two imaging techniques are shown in (c, d) and the axial MTFs are shown in (e, f); (g) 1D xy and yz sections of the MTFs, (h) 1D xy and yz of the PSFs. The FWHM of the curves are blue- 0.530 µm, red 0.495 µm, green 0.8341 µm, and black 0.9004 µm.

calculate the focal lengths of the two diffractive lenses superimposed on the SLM to provide maximum overlap of the two beams at the plane of the detector but keeping fixed all the distances $d_1 \div d_7$ found in the previous step. After performing the optimization, the values of the two focal lengths were found to be $f_{d1} = 220\ mm$ and $f_{d2} = 2356\ mm$, which were used for the design of the two diffractive lenses. These two lenses focus at a distance of $d_{7\_1} = 555.185\ mm$ in the front of the camera, **Figure 2b**, and at $d_{7\_1} = 826.793\ mm$ behind the camera, **Figure 2c**, respectively. In implementation, the distance $d_{7\_1} + d_8$ may need to be tuned by ± 0.3 depending on tolerances and imperfections of optical parameters of other elements of the system

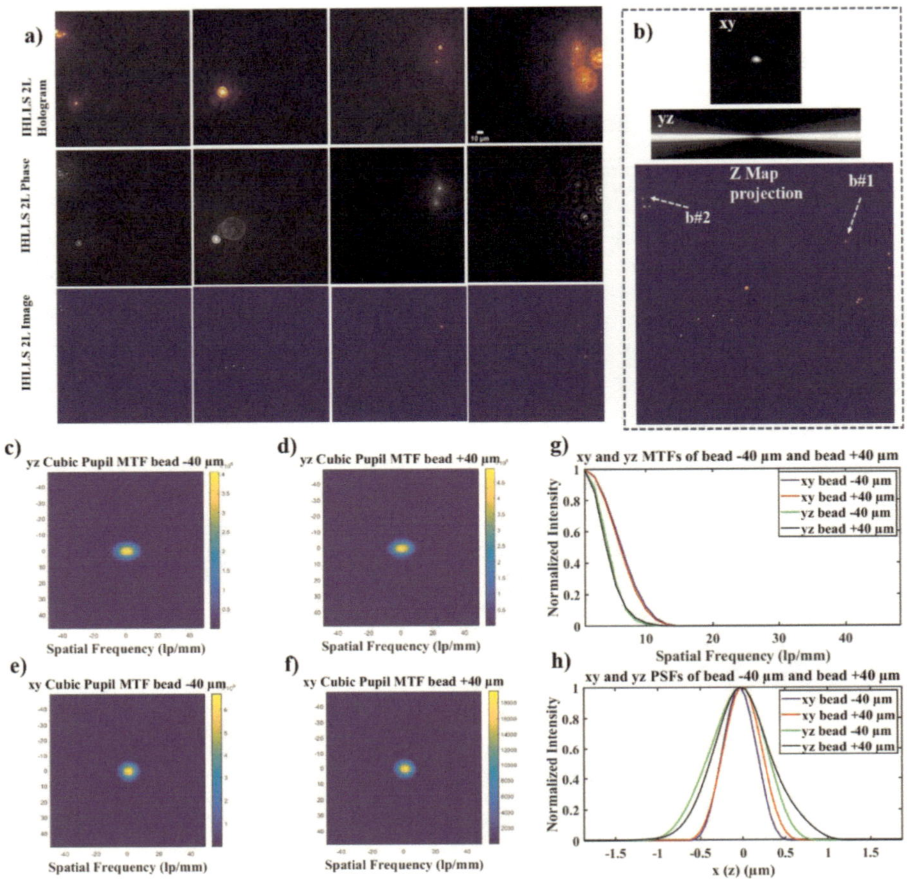

Figure 4.
IHLLS 2L beads volume reconstruction; (a) 500 nm beads holography ±40 μm, ± 30 μm, holograms (first row), phase maps (second row), and reconstructed images (third row); (b) The xy and yz cross-sections of a bead after the hologram reconstruction together with the z-max projection of all the best z-reconstructed planes. The IHLLS phase images contain the depth dependent phase information derived from the IHLLS holograms and the reconstructed IHLLS images show the complex holograms propagated to the best focal plane. The max projection of the reconstructed volume of the 500 nm beads sample contains the z-galvo levels ±40 μm, ± 30 μm, ± 20 μm, and 0 μm. The transverse MTFs of the two beads b#1 and b#2 are shown in (c, d) and the axial MTFs are shown in (e, f); (g) 1D xy and yz sections of the MTFs, (h) 1D xy and yz of the PSFs. The FWHM of the curves are blue-0.4534 μm, red 0.5118 μm, green 0.7663 μm, and black 0.7946 μm.

(e.g., tolerances of lenses, tolerances of phase resolution of the SLM, etc.). (i.e., $d_{7\_1} + d_8 = 664.298\ mm$, when calculating f_{d1}, and $d_{7\_1} + d_8 = 663.793\ mm$, when calculating f_{d2}.

2.1.3 IHLLS 1L calibration imaging

The first experiment was performed using the conventional LLS pathway where the z-galvo was stepped in $\delta z_{LLS} = 0.101\ \mu m$ increments through the focal plane of a 25x Nikon objective, which was simultaneously moved the same distance with a z-piezo controller for a displacement range of $\Delta z_{galvo} = 80\ \mu m$, **Figure 3a**, for scanning area of $208 \times 208\ \mu m^2$. The second set of images was obtained using the IHLLS 1L with focal length $f_{SLM} = 400\ mm$ displayed on the SLM, where both the z-galvo and

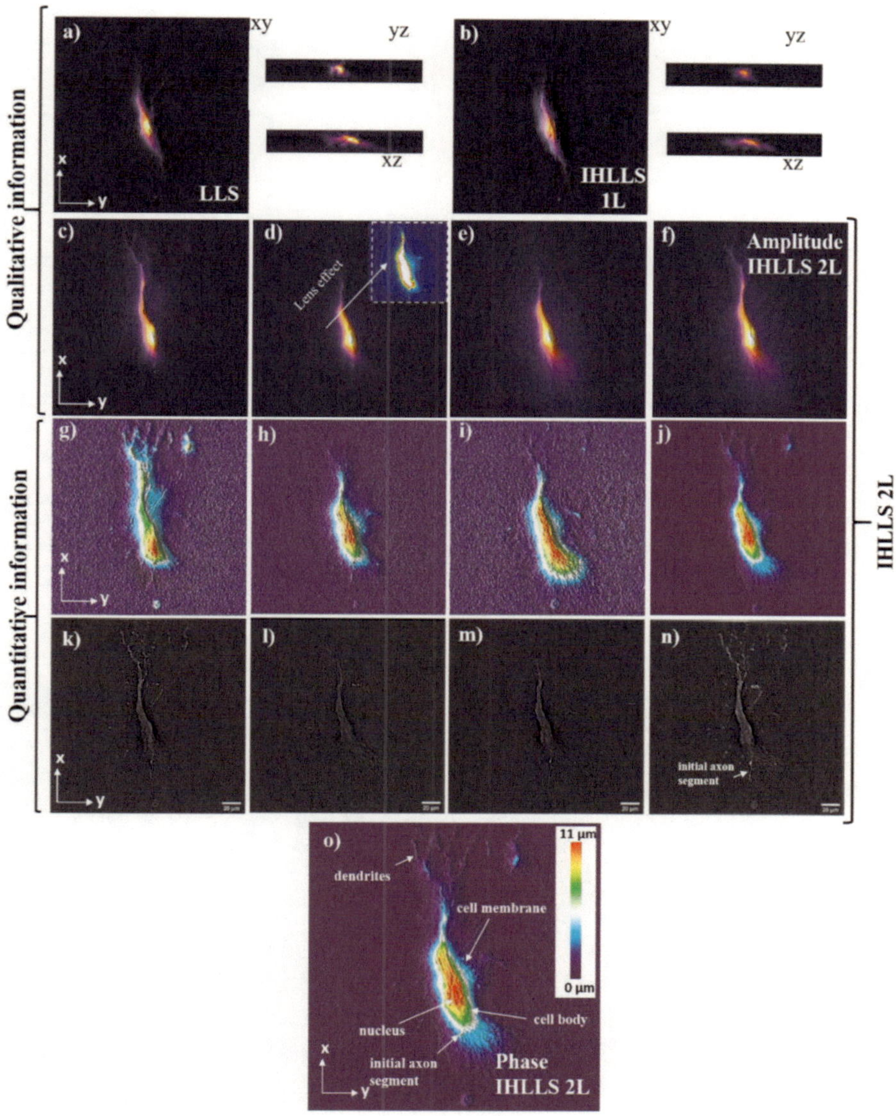

Figure 5.
Lattice light-sheet imaging of a lamprey spinal cord ventral horn neuron with dendrites; (a) Max projections through the volume (300 z-galvo steps) in a conventional LLSM without deconvolution; (b) Max projections through the volume (300 z-galvo steps) in an incoherent LLSM without deconvolution using IHLLS 1L; Amplitude reconstruction of a neuronal cell at three z-galvo positions: (c) + 30 µm, (d) 0 µm, (e) -30 µm, and (f) the superposition of all three; Phase reconstruction of a neuronal cell at z-galvo positions: (g) + 30 µm, (h) 0 µm, (i) -30 µm, and (j), (o) the superposition of all three; (k)-(n) Band-pass filter applied to the phase images from (g)-(j).

z-piezo were again stepped with the same $\delta z_{LLS} = 0.101\ \mu m$ increments through the focal plane of the objective for the same displacement $\Delta z_{galvo} = 80\ \mu m$, **Figure 3b**. The scanning area in a conventional LLS is at best $54 \times 54\ \mu m^2$ (red square in **Figure 3a** and in the upper left corner of **Figure 3a**), and it is too large for being illuminated by the Bessel beams. Therefore, to enlarge the scanned region these $54 \times 54\ \mu m^2$ areas can be moved in a mosaic-fashion by moving the sample. However, this requires a substantially longer acquisition time and image registration. It also prevents

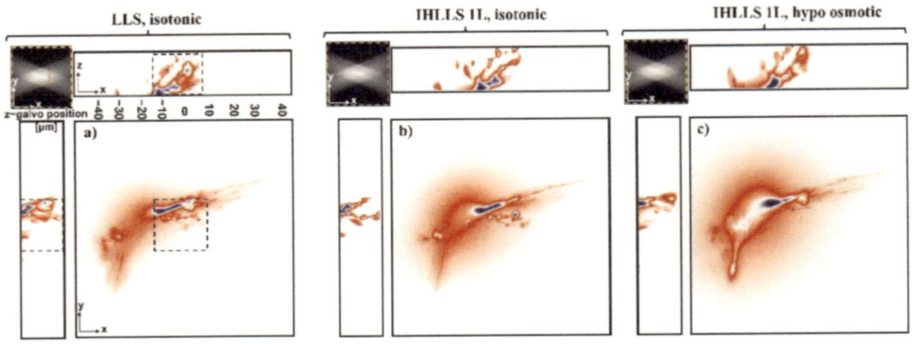

Figure 6.
Tomographic imaging of a lamprey spinal cord ventral horn neuron with dendrites, xy FOV *208 × 208 μm², 2048 × 2048 pixels,* yz, (xz) FOV *208 × 40 μm², 2048 × 400 pixels, in a conventional LLS (a) and incoherent LLS with only one diffractive lens (IHLLS 1L) of focal length 400 mm without deconvolution; recordings were initially in isotonic solution (b), and then in hypotonic solution to cause swelling (c). On the sides and above are shown the max projections through the volume (400 z-galvo steps). The Bessel beams are displayed in the upper-left corner of each xy-projection to show the orientation of the beams* (FOV *208 μm²). The area enclosed inside the colored dashed rectangles are as follows: black—the scanning area for the original LLS (52 μm²), and yellow—the actual scanning area for the LLS, IHLLS 1L, and IHLLS 2L.*

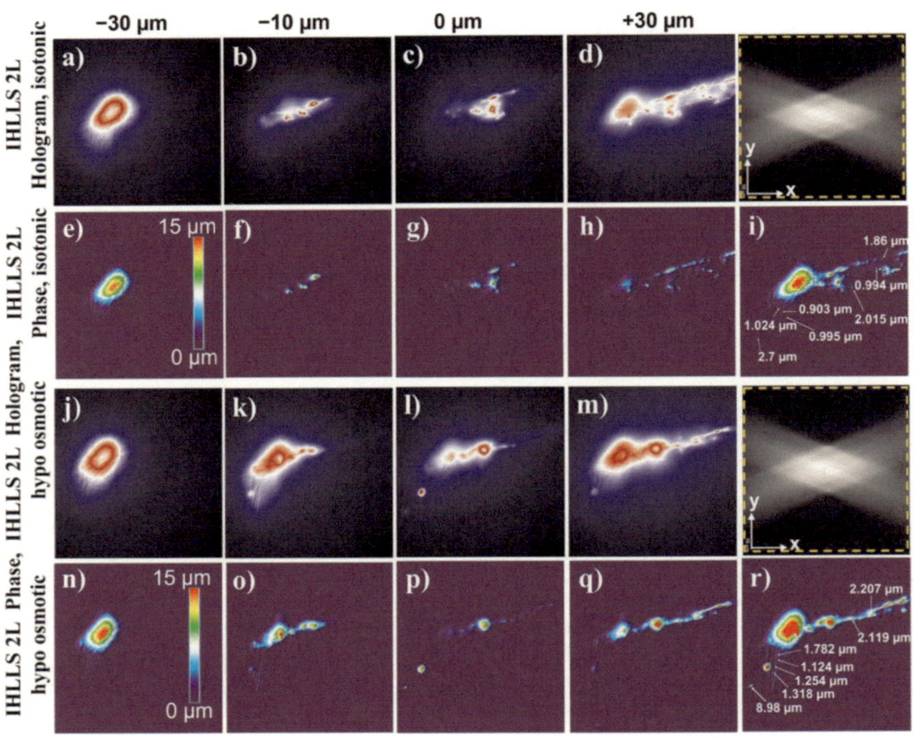

Figure 7.
IHLLS 2L imaging of a lamprey spinal cord ventral horn neuron with dendrites in a Ringer's solution (a–i) and hypotonic solution (j–r), FOV *208 × 208 μm², 2048 × 2048 pixels; (a–d, j–m) imaging holography at* −30 μm, − 10 μm, 0 μm, *and* +30 μm, *for the phase-shift θ = 0; (e–h, n–q) the corresponding reconstructed phase images; (i, r) the superposition of all reconstructed phase images at* ±40 μm, ± 30 μm, ± 20 μm, ± 10 μm,and 0 μm. *The Bessel beams are displayed in the right gray panel of each xy-projection to show the orientation of the beams* (FOV *208 × 208 μm²). The phase values of the reconstructed images were converted to the optical path length values in reflection mode.*

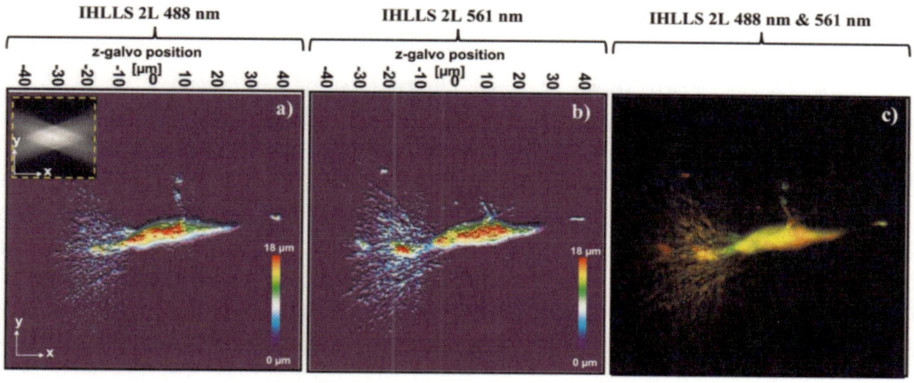

Figure 8.
IHLLS-2L max projection of the reconstructed phase at 9 z_{galvo} positions, z_{galvo} = ±40 μm, ±30 μm, ±20 μm, ±10 μm, and 0 μm, at (a) 488 nm, (b) 561 nm, and (c) combination of (a) and (b). The scanning area in both experiments was 208×208 μm². The scanning area in IHLLS-2L is at best 208×208 μm².

simultaneous use of other recording modalities such as electrophysiology. A recent more effective solution is to extend the scanning area to 170 × 170 μm^2 (green square in the upper left corner of **Figure 3a**). This was achieved by superposing a spherical phase profile on the illumination wavefront at the pupil plane which requires 17 tiling positions [33], but with a lower axial resolution than the original LLS. The IHLLS 1L, **Figure 3b,d**, performs better than LLS by scanning a bigger area, but the resolution is lower in the axial directions due to the blurring effect of one of the lenses that is focused to infinity. Resolving fine structure in an image depends on both the shape and the extent of the spatial frequency support of the optical transfer function (OTF). To evaluate the relative imaging performance of IHLLS 1L compared with conventional LLSM techniques, the OTFs of these methods were computed in the following way. We calculated the xy- and yz- point spread functions of the two systems and compared them to the bead # 2's transverse and axial images. The absolute values of the OTFs were calculated for the xy and yx cross-sections, **Figure 3c-f**, by taking the Fourier transform of the 2D PSFs distributions. The 1D OTFs are shown in **Figure 3g** and the corresponding PSFs in **Figure 3h**.

We calculated the FWHM in xy and yz directions for the two systems by fitting the PSFs from **Figure 3h** with a 1D Gaussian function. The calculations for the 500 nm beads show a xy FWHM of 530 nm (LLS), 495 nm (IHLLS 1L) and yz FWHM of 834 nm (LLS) and 900 nm (IHLLS 1L). The IHLLS performs better within the scanning areas of 54 × 54 μm^2 and it gets worse outside this area due to the blurring effect of the second lens with infinite focal length.

2.1.4 IHLLS 2L calibration imaging

A step-by-step initiative in the *IHLLS system* has been the implementation of dual diffractive lenses comprising randomly selected pixels (*IHLLS 2L*) for recording sample holograms, **Figure 1c**. Only the z-galvo was moved within the same $\Delta z_{galvo} = 40\ \mu m$ displacement range, above and below the reference focus position of the objective (which corresponds to the middle of the camera FOV), at $z_{galvo} = \pm 40\ \mu m \pm 30\ \mu m$, $\pm 20\ \mu m$,$\pm 10\ \mu m$, and $0\ \mu m$. This approach enables full complex-amplitude modulation of the emitted light for extended FOV and depth.

The results are summarized in **Figure 4(a)**, **(b)**, which depict the sample at four z-galvo planes, denoted $z_{galvo} = \pm 40\ \mu m$,and $\pm 30\ \mu m$. The z-galvo planes correspond to the same relative depths in the sample. The reconstruction distance calibration from the image plane to the sample plane is 20 mm in the IHLLS pathway for each $\pm 10\ \mu m$ displacement of the z-galvo mirror so, it means the conversion should be $80\ mm\ to \pm\ 40\ \mu m$.

The IHLLS 2L holograms ($\theta = 0$) for each z-galvo level are displayed in the upper row of **Figure 4a**, and the phase corresponding to each hologram displayed in the second row. The phase images contain the depth dependent phase information derived from the IHLLS holograms. Then, the complex hologram is propagated and reconstructed at the best focal plane using a custom diffraction Angular Spectrum method (ASM) routine programmed in MATLAB (MathWorks, Inc.). These IHLLS 2L reconstructed images are shown in the third row of **Figure 4a**. We chose the ASM and not the Fresnel reconstruction as the ASM can reconstruct the wave field at any distance from the hologram plane, without the minimum reconstruction distance requirement [9]. The max projection of all z-planes where the beads were found are displayed in **Figure 4b**. They show the complex holograms propagated to the best focal plane. We performed scanning at other three z-galvo levels, $\pm 10\ \mu m$, and $0\ \mu m$, but the results are not shown in **Figure 4**. Each column of **Figure 4a** results from images captured with the z-galvo positioned at a designated sample plane. Consequently, a sample plane at $z_{galvo} = +40\ \mu m$, contains the right layer of beads in this image, the sample plane at $z_{galvo} = -40\ \mu m$ contains the left layer and sample planes at $\pm 30\ \mu m \pm 20\ \mu m$, and $0\ \mu m$, are equidistant between the two layers. We also determined the distribution in transverse FWHM values for the 500 nm beads, **Figure 4**.

2.2 Applications of IHLLS

In this part of the chapter, we will provide information on potential use of *IHLLS systems in neuroscience*. LLS volumetric imaging, **Figure 5a**, can provide only morphological measurements in any sample volume, but not complex cellular parameters such as the cellular dry mass or refractive index (RI) [16]. These are features of cells that dynamically change in response to activity. As an example, the LLS is unable to account directly for potential contributions that might be provided to the surface area by subresolution features such as subdiffraction sized structure including microvilli [32], whose structure can be rapidly altered by Ca^{2+} entry [34–36] or the structure of cell membranes altered by voltage fluctuations [37]. Therefore, quantitative analysis of cells and membrane dynamics detected by using the IHLLS system will be of significance in understanding a range of cell functions.

2.2.1 Imaging neurons

For this step we chose a simple neuronal preparation in which a neuron could be readily labeled with fluorescent dye and maintained alive during imaging. Lamprey spinal cord ventral horn neurons have dendrites that are sufficiently large to cover the whole camera FOV of 208 μm^2. When performing tomographic imaging using the original LLS, **Figure 5a**, or IHLLS 1L (phase $\theta = 0$ in **Figure 5b**, the sample outside of the maximum area of $78 \times 78\ \mu m^2$ cannot be resolved, although the z-galvo mirror and the z-piezo stage objective were moved 300 steps in the range 60 μm. Deconvolution sharpening of the raw data is the standard solution for blur reduction and enhancement of fine image details in LLS image processing. When using the IHLLS 2L, **Figure 5c-f**, the max projection of only three reconstructed amplitude images from IHLLS holograms recorded at $z_{galvo} = \pm 30\ \mu m$, and $0\ \mu m$, could give similar results as the LLS, and within a larger resolved sample area. In addition to the amplitude images, the IHLLS 2L can provide access to the reconstructed phase information, **Figure 5g-j**, needed to understand the physiology and pathophysiology of various cells or other parts of samples under study. Applying further a band-pass filter to the above phase images other features could be identified, **Figure 5k-n**.

Within biological samples, axial resolution enhancement is possible if the phase difference of two interfering wavefronts is known. The combination of the two lenses with focal lengths, f_{d1} = 220 mm and f_{d2} = 2356 mm, gave a reconstruction distance with a maximum 80 *mm* above the reference plane (microscope objective position or the middle of the camera FOV) and 60 mm below the reference plane, but when converted to the z-galvo displacement the reconstructed distance was 30 μm and $-30\ \mu m$ respectively, which results in a 60 μm volume height. When the scanning range is increased to $\pm 40\ \mu m$, the reconstruction distance increases to about 250 *mm* above the reference plane and 80 mm below the reference plane, which correspond to z-galvo displacements of 40 μm and $-40\ \mu m$ respectively, and a 80 μm volume height.

3D perspective representation of the quantitative phase contrast image of the neuron is displayed in **Figure 5g-j**. Each pixel represents a quantitative measurement of the cellular optical path length (OPL) of the fluorescently labeled neurons and their subcellular compartments. The scale (at right) in **Figure 5o**) relates the OPL (in the color look-at table) to the morphology (in μm). The IHLLS 2L and LLS techniques have similar transverse performances, but the axial performance is poorer for the IHLLS 2L, when moving the z-galvo in 7 or 9 steps along the 80 μm ranges. Therefore, we need to increase the number of z-galvo increments to achieve a better axial performance.

2.2.2 LLS and IHLLS neuronal imaging at excitation wavelength 488 nm

Lamprey neurons were imaged in isolated spinal cords of newly transformed lampreys (*Petromyzon marinus*). Lamprey ventral horn neurons were fluorescently labeled with Alexa 488 hydrazide using previously published methods [6]. To ensure successful labeling of the neurons, the resultant epifluorescence was imaged during excitation with 470 nm LED illumination, and images were captured on an sCMOS camera (PCO AG, Kelheim, Germany). The chamber and spinal cord were then transferred onto the customized stage of the LLS microscope. The recording chamber was again continually superfused with cold, oxygenated Ringer (8–10°C) for the duration of the experiment. The labeled neurons were then imaged in this chamber using LLS and the osmotic potential of the superfusate changed by switching its input.

To examine the effects of applying IHLLS holography, we performed three experiments for this study for calibration purposes. The first was carried out using the conventional LLS pathway, where the *z-galvo* was stepped in $\delta z_{LLS} = 0.101$ μm increments through the focal plane of a 25x Nikon objective, which was simultaneously moved the same distance with a z-piezo controller for a displacement range of $\Delta z_{galvo} = 40$ μm, **Figure 6a**, for scanning area of 208×208 μm^2. The second set of images was obtained using the IHLLS 1L with focal length $f_{SLM} = 400$ mm, displayed on the SLM, where both the z-galvo and z-piezo were again stepped with the same

$\delta z_{LLS} = 0.101$ μm increments through the focal plane of the objective for the same displacement $\Delta z_{galvo} = 40$ μm. The intensity images using the IHLLS 1L mode were recorded only for the diffractive lens with phase shift $\theta = 0$.Initially, these images, obtained in both LLS and IHLLS 1L modes, were obtained in isotonic lamprey Ringer's solution (270mOsm, **Figure 6a**, **b**).

To expand the neurons as an initial test of the ability of the IHLLS system to record small changes in size, we repeated the experiment using the IHLLS 1L, where the Ringer's solution was modified with hypotonic solution (225mOsm), **Figure 6c**. The third step was achieved by combining the coherent properties of the Bessel beams with the incoherent properties of the fluorescent light emitted by each 3D point of the sample but made coherent in the self-interference process. This method was performed with both solutions, isotonic and hypotonic solutions, using the IHLLS pathway with two super-imposed diffractive lenses displayed on the SLM comprising randomly selected pixels (IHLLS 2L), where only the z-galvo was moved within the same $\Delta z_{galvo} = 40$ μm displacement range, above and below the reference focus position of the objective (which corresponds to the middle of the camera FOV), at $z_{galvo} = \pm 40$ μm $\pm$ 30 μm, $\pm$20 μm,$\pm$10 μm, and 0 μm. The two wavefronts interfere with each other at the camera plane to create Fresnel holograms. Four interference patterns were created using a phase shifting technique ($\theta = 0, \theta = \pi/2$, $\theta = \pi, \theta = 3\pi/2$) and further combined mathematically to obtain the complex amplitude of the object point at the camera plane. The results are summarized in **Figure 7** which depicts the samples in isotonic, **Figure 7a–i**, or hypotonic solutions, **Figure 7j–r**, at four z-galvo planes, denoted $z_{galvo} = \pm 30$ μm, − 10 μm, and 0 μm. The z-galvo planes correspond to the same relative depths in the sample. Four holographic images were recorded for each z-galvo position and combined mathematically to build the complex amplitude of each sample. The 3D field was reconstructed at various depth positions, and those images in focus at certain planes were chosen to build the tomographic structure of the neuron. We performed scanning at nine z-galvo positions and, therefore, nine phase images were selected to build the tomographic slice of all nine superposed images, **Figure 7i**, **r**. It is clear from this that the soma (**Figure 7e** compared to **Figure 7n**) and the dendrites (**Figure 7f–h** compared to **Figure 7o–q**) all showed an increase in volume. We have emphasized this by measuring the phase values of various points in the neuron's dendritic tree and calculating the optical path length that this represents, first in isotonic solution and then in hypo-osmotic solution. This treatment is expected to swell the neurons, including the dendritic tree, increasing the size of these structures. The analyses indicate that the technique can resolve sub-micrometer size changes, represented in depth by the color coding shown in **Figure 7**. For comparison, we measured the diameters of well separated structures in the xy plane, which are expected to be similar to the depth in these cylindrical structures. We have inserted these numbers in **Figure 7i**, **r**. The diameters measured increased from 1.41 $\pm$ 0.50 to 2.28 $\pm$ 1.04 μm. These are also clearly resolvable from the depth encoding.

2.2.3 LLS and IHLLS neuronal imaging at excitation wavelength 488 and 560 nm

Synchronous synaptic transmission, which requires precise coupling between action potentials, Ca^{2+} entry and neurotransmitter release [32], is fundamental to the function of the brain. Thus, understanding synaptic function is key to understanding how the brain works while similarly, understanding synaptic dysfunction is crucial to understanding diseases of the brain. One key to understanding synaptic function is to

DOI: http://dx.doi.org/10.5772/intechopen.107322

image the spatiotemporal axonal, dendritic, and synaptic activity in 3-dimensional space simultaneously and with high resolution. Our long-term aim will be to record phase changes evoked by synaptic activity within small (e.g. pre- and post-synaptic structures of the brain) in the millisecond time periods during which this activity occurs. Here, we describe a general approach to 2 color phase imaging in neurons which give high spatial and temporal resolution.

We have expanded our newly developed single wavelength IHLLS microscope, with one (IHLLS – 1L) or two diffractive lenses (IHLLS-2L) [1], to a two wavelengths optical design and imaging technique. These two excitation beams, 488 and 561 nm, will produce corresponding emission beams at 523 and 570 nm respectively.

The two-color technique still uses the self-interference properties of the emitted fluorescent light [24], in which three or four interference patterns are created using a phase shifting technique, to create Fresnel holograms of a 3D object. However, the spatial light modulation (SLM) optical design configuration has been modified to actively control the dual diffractive lenses phase-shifting at two colors sequentially. This is repeated at each z-galvo scanning level. The technique allows both faster three-dimensional amplitude and phase imaging without moving either the sample stage or the detection objective, for extended FOV (208×208 mm^2) and depth. The scanning depth is a function of two variables, the numerical aperture of the LLS diffraction mask annulus and the z-galvanometer mirror scanning range. Using an anulus of 0.55 outer NA and 0.48 inner NA, the scanning depth could reach up to 80 μm, using 9 z-galvo positions within the range Δz_{galvo} = 80 μm, at z_{galvo} = ±40, ±30, ±20, ±10, and 0 μm.

The optical setup of the IHLLS system at one wavelength is covered in [1]. The two-color system is driven by the same principle. The system calibration is done simultaneously with one diffractive lens, IHLLS-1L, of focal lengths, $f_{SLM\text{-}488}$ = 400 mm, and $f_{SLM\text{-}561}$ = 415 mm respectively. After that, we perform sample scanning using two diffractive lenses with randomnly selected pixels, for each wavelength, with focal lengths $f_{d1\text{–}488}$ = 220 mm, $f_{d2\text{–}488}$ = 2356 mm, and $f_{d1\text{–}561}$ = 228 mm and $f_{d2\text{–}561}$ = 2444 mm. The physical distances between each sequential optical component and the the focal distances for the two excitation wavelengths, 488 and 561 nm, were calculated using Opticstudio (Zemax, LLC). We designed a multi-configuration optical system with the condition that the height of the two beams generated by the two lenses was equal in size at the camera plane for a perfect overlap.

To examine the effects of applying ICHLLS holography, neurons fluorescently labeled live *in situ* in the central nervous system (lamprey spinal ventral horn neurons), **Figure 8**, were used as test samples. Fluorescent dyes, Alexa Fluor™ 488 and 568 nm hydrazide, were injected by microinjection into the neuron.

3. Conclusions

The use of IHLLS enhances lattice light sheet microscopy in several ways.

- The optical and mechanical designs of the incoherent arm expand the applicability of the lattice light-sheet system and could open entirely new imaging modalities in all light sheet imaging instruments. The use of IHLLS enhances lattice light sheet microscopy in several ways. In the original FINCH system, it was shown that using spatial incoherent light, such as fluorescent light, to form holograms, and producing images with better spatial resolution than conventional imaging, is possible, because the incoherent imaging systems break the Lagrange principle valid for classical imaging system. This occurs in such a way that it magnifies the gap between two spots more than it magnifies the spots themselves [38–40]. While we have indeed achieved higher resolution with IHLLS-2L than with IHLLS-1L or LLS, we must still do more work to optimize the system. We need to determine how many optimal galvanometric and hologram reconstruction steps are needed before we draw the same conclusion here. Another factor to consider is to investigate the Lagrange principle with a larger area SLM without the need to reduce and expand the beam size.

- In LLS focus of the emission objective must be maintained in the focal plane of the lattice sheet, requiring movement of a bulky objective, which takes time, but also at the high speeds of LLS imaging will apply fluid pressure movements across the sample. This will potentially distort images, but in addition will provide noise to transient measurements of cell surfaces caused by rapid activity, for example action potential firing in excitable cells. It is also difficult to resolve very small changes in surface structure at speed, even with volumetric imaging. Consequently, the ability to resolve rapid surface structure changes, without moving the objective lens and higher accuracies than traditional volumetric imaging approaches is valuable.

- We demonstrated the capacity of the IHLLS system to reconstruct 3D positions of beads as light point sources using a very reduced number of z-galvo mirror scanning planes when compared to the original LLS system using the dithered mode. We also showed the same or slightly improved resolution of the bead positions. The IHLLS system can therefore provide faster volumetric image acquisition and multi-plane imaging for probing the three-dimensional morphology and structure of biological samples. In this study we have demonstrated that IHLLS can be used to resolve artificially applied changes in neuron shape. We will seek in further studies to relate these changes to physiological activity.

- Although we eliminated the z-piezo objective motion needed to maintain the object focus, there are a few challenges in approaching this work. The conventional LLS system uses a very low dose of light and low exposure time without a polarizer in the detection path. The digital incoherent holography technique with two diffractive lenses uses two polarizers in the detection arm; therefore, the light dose and the exposure time must be increased from the middle of the z-galvo scanning range toward the two ends of the scanning range to maintain a similar signal efficiency as in the conventional LLS, but low enough to preserve the biological samples. For this reason, we opted out of using a double diffractive lens FINCH technique with randomly selected pixels which requires only one polarizer mounted before the SLM, to align the input beam to its active axis. The SLM was also positioned at 11° deflection angle to generate active beam steering at the highest efficiency. Another challenge in this configuration is the beam size of 17.6 mm in the back focal plane of the microscope objective (Nikon Apo LWD 25×1.1 W), which is too large for the SLM chip size (17.66×10.6 mm, Meadowlark Optics)), and we decided to place the SLM in a double 4f optic system configuration made by two lenses of 200 mm focal lengths and two lenses of 125 mm focal lengths. This combination has also helped to decrease the $z_{h_{min}}$ to about 664 mm.

Conflict of interest

There are no conflicts of interest.

Author details

Mariana Potcoava[1]*, Christopher Mann[2,3], Jonathan Art[1] and Simon Alford[1]

1 Department of Anatomy and Cell Biology, University of Illinois at Chicago, Chicago, USA

2 Department of Applied Physics and Materials Science, Northern Arizona University, Flagstaff, USA

3 Center for Materials Interfaces in Research and Development, Northern Arizona University, Flagstaff, USA

*Address all correspondence to: mpotcoav@uic.edu

References

[1] Keller PJ, Ahrens MB. Visualizing whole-brain activity and development at the single-cell level using light-sheet microscopy. Neuron. 2015;**85**(3):462-483

[2] Ji N, Freeman J, Smith SL. Technologies for imaging neural activity in large volumes. Nature Neuroscience. 2016;**19**(9):1154-1164

[3] Chen B-C, Legant WR, Wang K, Shao L, Milkie DE, Davidson MW, et al. Lattice light-sheet microscopy: Imaging molecules to embryos at high spatiotemporal resolution. Science. 2014; **346**(6208):1257998

[4] Liang Gao LS, Chen B-C, Betzig E. 3D live fluorescence imaging of cellular dynamics using Bessel beam plane illumination microscopy. Nature Protocols. 2014;**9**:1083-1101

[5] Potcoava M, Mann C, Art J, Alford S. Spatio-temporal performance in an incoherent holography lattice light-sheet microscope (IHLLS). Optics Express. 2021;**29**(15):23888-23901

[6] Potcoava M, Art J, Alford S, Mann C. Deformation measurements of neuronal excitability using incoherent holography lattice light-sheet microscopy (IHLLS). Photonics. 2021;**8**(9):383

[7] Rosen J, Alford S, Anand V, Art J, Bouchal P, Bouchal Z, et al. Roadmap on recent progress in FINCH technology. Journal of Imaging. 2021;**7**(10)

[8] Kim MK. Principles and techniques of digital holographic microscopy. SPIE Reviews. 2010;**1**(1):18005

[9] Kim MK, Yu LF, Mann CJ. Interference techniques in digital holography. Journal of Optics A: Pure and Applied Optics. 2006;**8**(7):S518-SS23

[10] Cuche E, Bevilacqua F, Depeursinge C. Digital holography for quantitative phase-contrast imaging. Optics Letters. 1999;**24**(5):291-293

[11] Mann C, Yu L, Lo CM, Kim M. High-resolution quantitative phase-contrast microscopy by digital holography. Optics Express. 2005;**13**(22):8693-8698

[12] Popescu G, Deflores LP, Vaughan JC, Badizadegan K, Iwai H, Dasari RR, et al. Fourier phase microscopy for investigation of biological structures and dynamics. Optics Letters. 2004;**29**(21): 2503-2505

[13] Marquet P, Rappaz B, Magistretti PJ, Cuche E, Emery Y, Colomb T, et al. Digital holographic microscopy: A noninvasive contrast imaging technique allowing quantitative visualization of living cells with subwavelength axial accuracy. Optics Letters. 2005;**30**(5): 468-470

[14] Alexander Khmaladze MK, Lo C-M. Phase imaging of cells by simultaneous dual-wavelength reflection digital holography. Optics Express. 2008;**16**: 10900-10911

[15] Barer R. Interference microscopy and mass determination. Nature. 1952; **169**(4296):366-367

[16] Popescu G, Park Y, Lue N, Best-Popescu C, Deflores L, Dasari RR, et al. Optical imaging of cell mass and growth dynamics. American Journal of Physiology - Cell Physiology. 2008; **295**(2):C538-CC44

[17] Müller P, Cojoc G, Guck J. DryMass: Handling and analyzing quantitative phase microscopy images of spherical, cell-sized objects. BMC Bioinformatics. 2020;**21**:226

[18] Muller P, Guck J. Response to Comment on "Cell nuclei have lower refractive index and mass density than cytoplasm" a comment on "How a phase image of a cell with nucleus refractive index smaller than that of the cytoplasm should look like?". Journal of Biophotonics. 2018;**11**(6):e201800033

[19] Schurmann M, Scholze J, Muller P, Guck J, Chan CJ. Cell nuclei have lower refractive index and mass density than cytoplasm. Journal of Biophotonics. 2016;**9**(10):1068-1076

[20] Barer R. Refractometry and interferometry of living cells. Journal of the Optical Society of America. 1957; **47**(6):545-556

[21] Belanger E, Levesque SA, Rioux-Pellerin E, Lavergne P, Marquet P. Measuring absolute cell volume using quantitative-phase digital holographic microscopy and a low-cost, open-source, and 3D-printed flow chamber. Frontiers in Physics (Lausane). 2019;7:172

[22] Rappaz B, Marquet P, Cuche E, Emery Y, Depeursinge C, Magistretti PJ. Measurement of the integral refractive index and dynamic cell morphometry of living cells with digital holographic microscopy. Optics Express. 2005; **13**(23):9361-9373

[23] Marquet P, Depeursinge C, Magistretti PJ. Exploring neural cell dynamics with digital holographic microscopy. Annual Review of Biomedical Engineering. 2013;**15**: 407-431

[24] Hodgkin AL, Huxley AF. A quantitative description of membrane current and its application to conduction and excitation in nerve. (Reprinted from Journal of Physiology, Vol 117, Pg 500-544, 1952. Bulletin of Mathematical Biology. 1990;**52**(1-2):25-71

[25] Hill RM, Barer J, Hill LL, Butler CM, Harvey DJ, Horning MG. An investigation of recurrent pine oil poisoning in an infant by the use of gas chromatographic-mass spectrometric methods. The Journal of Pediatrics. 1975; **87**(1):115-118

[26] Ramachandran S, Rodgriguez S, Potcoava M, Alford S. Single calcium channel nanodomains drive presynaptic calcium entry at lamprey reticulospinal presynaptic terminals. Journal of Neuroscience. 2022;**42**(12): 2385-2403

[27] Rosen J, Brooker G. Digital spatially incoherent Fresnel holography. Optics Letters. 2007;**32**(8):912-914

[28] Rosen J, Brooker G. Non-scanning motionless fluorescence three-dimensional holographic microscopy. Nature Photonics. 2008;**2**:190

[29] Katz B, Rosen J, Kelner R, Brooker G. Enhanced resolution and throughput of Fresnel incoherent correlation holography (FINCH) using dual diffractive lenses on a spatial light modulator (SLM). Optics Express. 2012; **20**(8):9109-9121

[30] Brooker G, Siegel N, Wang V, Rosen J. Optimal resolution in Fresnel incoherent correlation holographic fluorescence microscopy. Optics Express. 2011;**19**(6):5047-5062

[31] Bouchal P, Bouchal Z. Wide-field common-path incoherent correlation microscopy with a perfect overlapping of interfering beams. Journal of the European Optical Society: Rapid Publications. 2013;**8**:13011

[32] Aguet F, Upadhyayula S, Gaudin R, Chou YY, Cocucci E, He KM, et al. Membrane dynamics of dividing cells

imaged by lattice light-sheet microscopy. Molecular Biology of the Cell. 2016; **27**(22):3418-3435

[33] Ga L, Tang WC, Tsai YC, Chen BC. Lattice light sheet microscopy using tiling lattice light sheets. Optics Express. 2019;**27**(2):1497-1506

[34] Hardie RC, Minke B. The trp gene is essential for a light-activated Ca2+ channel in drosophila photoreceptors. Neuron. 1992;**8**(4):643-651

[35] Reuss H, Mojet MH, Chyb S, Hardie RC. In vivo analysis of the drosophila light-sensitive channels, TRP and TRPL. Neuron. 1997;**19**(6): 1249-1259

[36] Huber A, Sander P, Gobert A, Bahner M, Hermann R, Paulsen R. The transient receptor potential protein (Trp), a putative store-operated Ca2+ channel essential for phosphoinositide-mediated photoreception, forms a signaling complex with NorpA, InaC and InaD. The EMBO Journal. 1996;**15**(24): 7036-7045

[37] Ling T, Boyle KC, Zuckerman V, Flores T, Ramakrishnan C, Deisseroth K, et al. High-speed interferometric imaging reveals dynamics of neuronal deformation during the action potential. Proceedings of the National Academy of Sciences of the United States of America. 2020;**117**(19): 10278-10285

[38] Bouchal P, Kapitan J, Chmelik R, Bouchal Z. Point spread function and two-point resolution in Fresnel incoherent correlation holography. Optics Express. 2011;**19**(16): 15603-15620

[39] Rosen J, Siegel N, Brooker G. Theoretical and experimental demonstration of resolution beyond the Rayleigh limit by FINCH fluorescence microscopic imaging. Optics Express. 2011;**19**(27): 26249-26268

[40] Siegel N, Rosen J, Brooker G. Reconstruction of objects above and below the objective focal plane with dimensional fidelity by FINCH fluorescence microscopy. Optics Express. 2012;**20**(18):19822-19835

Chapter 13

Holography Cytometry: Imaging of Cells in Flow

Cindy X. Chen, Hillel B. Price and Adam Wax

Abstract

Holographic cytometry (HC) has been developed as an ultra-high throughput implementation of quantitative phase microscopy (QPM). While QPM has been well developed for studying cells based on endogenous contrast, few implementations have imaged cells in flow or provided high throughput measurements. Although QPI offers high resolution imaging, experiments are limited to examining a single cell at a time. The HC approach enables high throughput by imaging cells as they are flowed through microfluidic devices. Stroboscopic illumination is used in an off-axis interferometry configuration to produce holographic images of flowing cell samples without streaking artifact. The ability to profile large number of cells using individual images has been demonstrated in red blood cell and cancer cell samples. The large volume of data provides suitable training data for developing machine learning algorithms, producing excellent accuracy in classifying cell type. Analysis of the adherent cells to flow also produces diagnostically useful information in the form of biomechanical cell properties. Introduction of a new parameter, disorder strength, a measure of the variance of phase fluctuations across a cell, provides an additional window into the cell mechanical properties.

Keywords: high throughput cell screening, quantitative phase imaging, microfluidics, red blood cells, cancer cells

1. Introduction

Quantitative phase microscopy (QPM) has been developed as a means to visualize the structure, dynamics and function of biological cells without exogenous contrast agents. The approach achieves *nanometer depth sensitivity at millisecond time scales*, producing phase images of optically-transparent cells [1]. Several research groups have used QPM for cellular analysis, establishing feasibility; but there have been few successful implementations of a high throughput QPM system.

While there are many methods to obtain QPM images, the approach used in holographic cytometry (HC) is *off-axis interferometry,* where the reference field crosses the detector at an angle, producing high frequency spatial fringes. **Figure 1a** shows a typical interferogram containing such fringes. The 2D Fourier transform of the image in (a) is shown in **Figure 1b**. Due to the fringe, the imaging information is moved to higher spatial frequencies, shown as bright spots at the top left and bottom

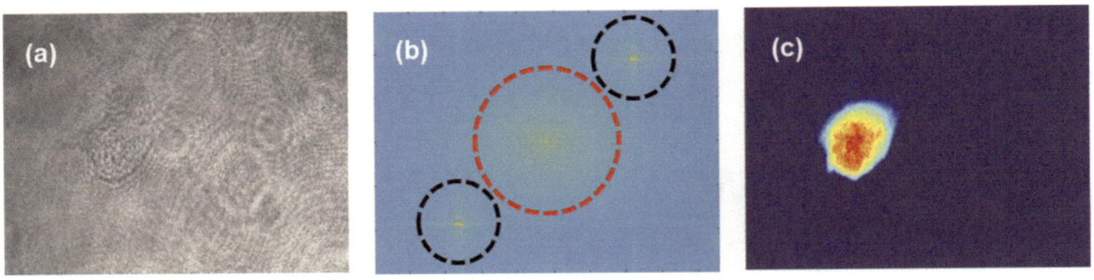

Figure 1.
Quantitative phase imaging of live breast epithelial (MCF10A) cell in growth media: (a) Single off-axis interferogram; (b) Spatial spectrum of this interferogram; and (c) Surface plots of the final unwrapped phase.

right. The complex field, containing phase delays due to the sample, is extracted by filtering the 2D Fourier transform (blue circle) to isolate one of the bright spots and then Fourier transforming back. **Figure 1c** shows the obtained image, producing phase contrast based on the phase argument of the complex field.

The phase due to a sample can produce an image based on optical path length (OPL). The change in optical path length is related to the measurement of optical phase ϕ as:

$$\Delta OPL(x, y) = \Delta n(x, y) \bullet h(x, y) = \frac{\Delta\phi(x, y)}{k} \tag{1}$$

Where n is the refractive index, h is the physical height of the cell and k is the wavenumber, equal to $\frac{2\pi}{\lambda}$ for wavelength λ. In earlier work with using machine learning to classify QPM images of malaria infected cells [2], a set of 23 morphological descriptors were defined based on OPL as well as geometrical features of a binary mask defining the cell shape. Additional parameters included metrics such as the gradient of the OPL and statistical measures of a histogram distribution of OPL values. This approach was highly accurate in discriminating stage of malaria infection, providing 99.7% accuracy in identifying the schizont stage.

Another approach for quantifying cell morphology based on QPM images is to sum up the OPL values across the cell to produce a metric termed Optical Volume (OV). To calculate OV, the following equation integrates across the OPL in each pixel [3]:

$$OV = \iint_{x,y} \Delta OPL(x, y) dxdy \tag{2}$$

In previous work establishing the utility of OV, we demonstrated that it is an *invariant* parameter for cells as they change orientation [4].

2. HC: high throughput QPM

While QPM has been widely used for cell analysis, typically only a handful of cells can be imaged at once due to the field-of-view (FOV) size limitation. Manual translation of sample stage to finding a good FOV and manual refocusing further restrict system throughput. Such limitations hinder QPM's ability to rapidly analyze large quantities of cells for disease diagnosis. For comparison, commercial imaging flow cytometry systems can capture 300–1000 cells/sec [5]. However, typical flow cytometry require a laborious immunophenotypic staining process and the lacking or overlapping of certain epithelial cell markers may compromise diagnostic judgment [6].

HC has been developed as a label-free, high throughput QPM modality that offers comparable acquisition rate (~420 cells/sec) to commercial imaging flow cytometers. **Figure 2** illustrates the classic Mach-Zehnder interferometry system setup used to capture off-axis holograms. Light from a laser diode is divided into sample and reference arms using a 1 × 2 fiber coupler. Light in the sample arm is collimated and directed to the sample. To enable high throughput of cell images, customized microfluidic channels are integrated into the QPM system. The illumination extends across multiple parallel channels in the microfluidic element. Light transmitted through the sample is imaged using a 20× microscope objective and a tube lens to create an image at the sensor plane. To enable imaging across a large number of channels a machine vision line scan array is used. This 2D sensor array (4096 × 96) captures the light from channels in one direction with interference fringes produced along the shorter dimension. The light from the reference arm is collimated and combined with the sample arm light using a beam splitter to produce an off axis interferogram.

Stroboscopic illumination is achieved using a 350 μs laser pulse, implemented using an acousto-optic modulator, to coincide with the duration of image acquisition, resulting in an overall framerate of 300fps. To avoid streaking effects, data acquisition rates are adjusted to ensure that the maximum distance traveled by cells during the pulse is less than diffraction-limited resolution.

Figure 3A and **B** shows the approach for creating the microfluidic element. A master mold is created using SU-8 photoresist comprising an array of parallel channels. Each channel is 40 μm wide and 30 mm long. Polydimethylsiloxane (PDMS) is cured on top of SU-8 mold for 2 hours at 80°C and subsequently plasma bonded onto glass coverslips. Inlets and outlets are punctured into PDMS slab, where metallic tubes are inserted for flow, shown in **Figure 3C**.

Interferograms of flowing cells are captured in video format and the complex field information describing red blood cell (RBC) height is reconstructed from the OPL data. Each complex field image is passed through digital refocusing algorithm [3, 4] which propagates the cell image to the plane by using an algorithm to determine the minimum variance in amplitude. A randomly selected empty frame with no cells

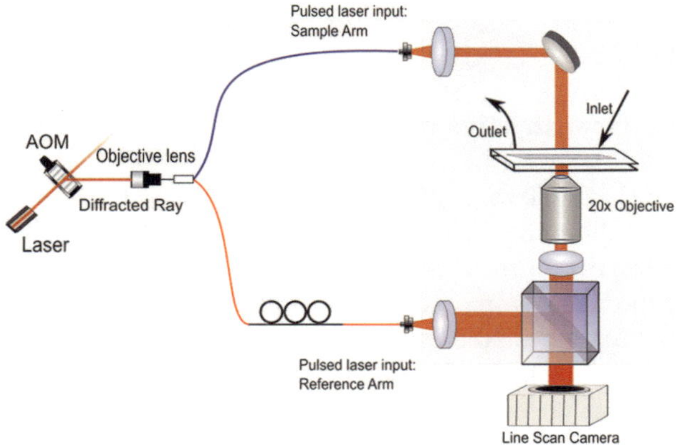

Figure 2.
Holographic Cytometry system diagram.

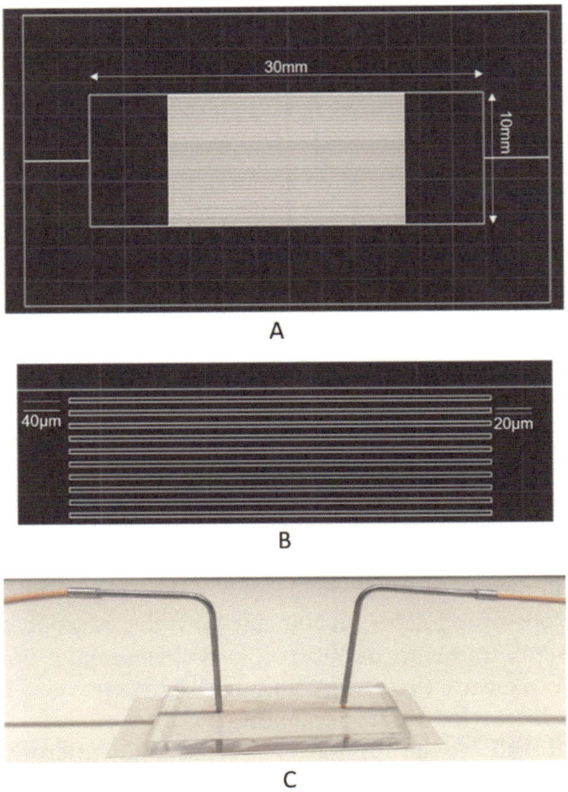

Figure 3.
(A) Overall mask design for the PDMS channels; (B) Magnified mask design showing the dimensions of the channels; and (C) PDMS channel.

apparent is used for background subtraction. Additional background noise is then removed by applying a 3rd-order polynomial fit to non-cell regions of the phase image. A segmentation algorithm is implemented so that single whole cells can be isolated for postprocessing classification, while partial cells and debris can be excluded.

3. Imaging RBCs

3.1 Flowing through constrictive microfluidic channels: storage lesion

RBCs are an ideal sample for QPM studies due to their almost uniform refractive index distribution and homogeneous structure. RBC membrane flexibility and deformability are important metrics that can be used to assess RBC's ability to transport gases within the body. Impaired deformability is often associated with diseases such as sickle cell disease or diabetes [7, 8]. In severe cases of the disease, poorly deformable RBCs may block blood flow to vital organs or promote splenic sequestration [7].

In one study [9], QPM was used to examine the biophysical deformability of RBCs from different storage periods, where mechanical stress was induced by flowing cells through custom-made constricted microfluidic channels. When squeezed through the

DOI: http://dx.doi.org/10.5772/intechopen.106581

channels, RBCs experience a water discharge that affects the overall cell height. **Figure 4** illustrates how the RBC deforms as it flows through the constricted channels and the optical volume (OV) values that are calculated from each time point. Note that the OV values appear to increase as the outflow of water results in a material with *higher* mass concentration.

Figure 5 shows that HC measurements detect a consistent increase in OV of RBCs in the same blood bag over longer storage times. Water efflux and increased hemoglobin concentration may be a possible explanation to the gradual OV changes seen over storage periods [9]. The increase in OV during squeeze is an outcome of increased cell height within confined space.

This study finds that water content exchange in RBCs under mechanical stress changes over time, showing a significant drop by day 29, as shown in **Figure 6**. This finding agrees with previous study which has shown that RBC stiffness significantly

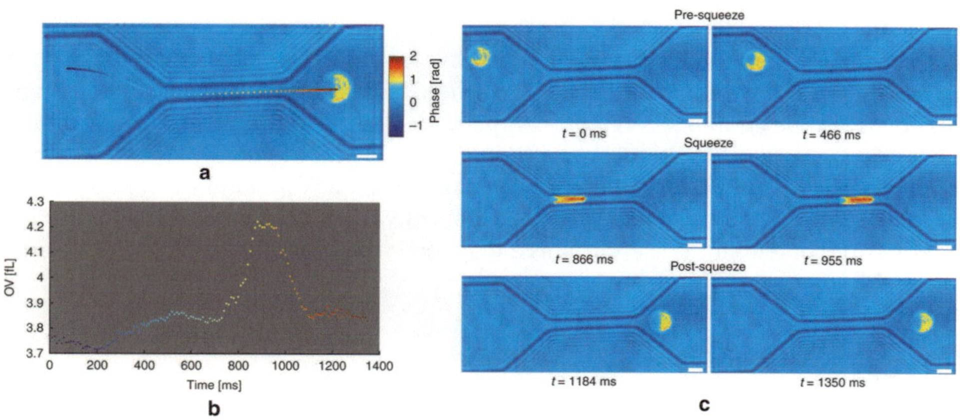

Figure 4.
(a) An RBC flowing through a channel while its position throughout the flow is color-coded to correspond to the temporal OV change shown in (b). (c) Three different segments (pre-squeeze, squeeze, and post-squeeze) of the transit through the constricting channel. Scale bar = 5 μm [9].

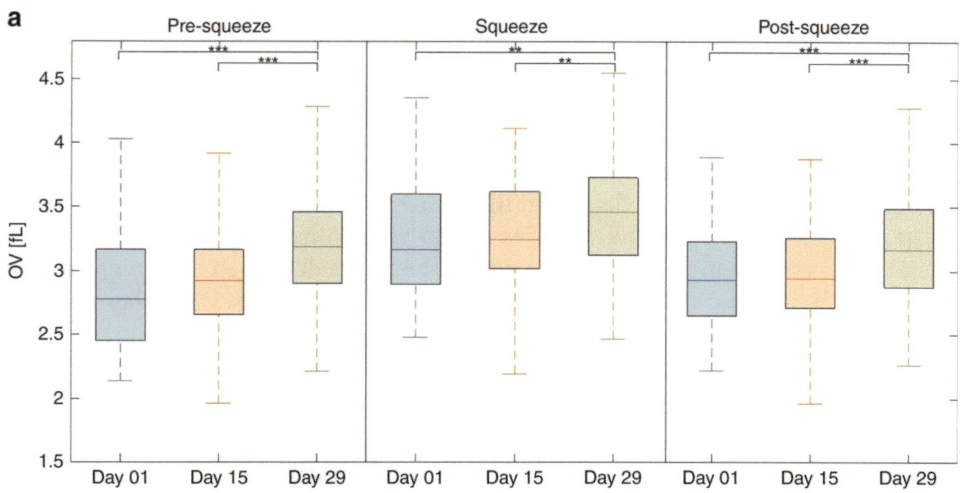

Figure 5.
*Box plots of RBC's OV flowing through the microchannels for sample 01, (**for P < 0.01; ***for P < 0.001) [9].*

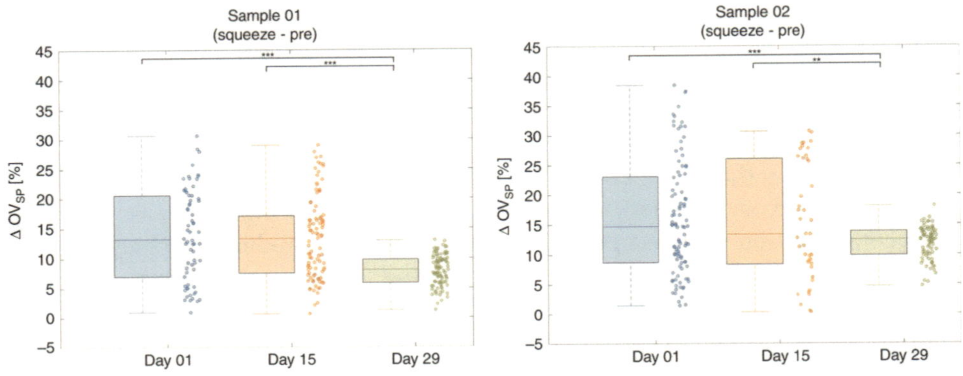

Figure 6.
Box plots and scatter plots of ΔOV_{SP}.

increases after week 4 [10]. The increased stiffness in RBC would explain the decrease in the volume change that is induced through mechanical confinement. Interestingly, this could also be an indication that, compared to the initial mixture of young and old RBCs that is expected to be encountered in circulating blood, there are now a greater proportion of old RBCs present within the blood bag. ΔOV_{SP} could thus, become a useful metric for evaluating aging RBCs. Coupled with constrictive microfluidics, QPM has the potential to be applied as a diagnostic tool for monitoring stored blood bag health.

3.2 High throughput RBC study

HC can provide cell morphology data without the need for time-consuming and costly staining processes and provides, otherwise inaccessible, 3D information on cell structure and dynamics. Its nanoscale sensitivity to membrane changes has been used to study the subtle biophysical changes in RBCs over different storage periods [11]. This previous study used 25 morphological descriptors, extracted from single cell phase images, to visualize cell structural variations at different time points. Notably, OV is a useful parameter that can differentiate between the different cell conditions.

Over 9 million single cell images were analyzed and a decreasing shift in OV is observed across blood bags from 3-week to 5-week storage, shown in **Figure 7**. One possible explanation to this phenomenon is that regular discocytes have gradually transformed to smaller-sized, deformed RBCs such as echinocytes, spheroechinocytes, and spherocytes, seen in previous studies [12–14]. The consistent decrease in OV over storage time can be potentially pinpointed to increase in deformed RBC population, which according to one study, increases from 4.9% on day 3 to 23.6% by day 42, in just a span of 5 weeks [12].

The extracted morphological parameters are passed through a logistic regression algorithm. The highest accuracy level achieved by the binary classification algorithm is 85.6%, shown in **Table 1**. These measures show that the analysis can discriminate the storage time even though each sample has a blend of young and new cells.

The HC system proves to be effective in characterizing millions of cells at an individual single image level. This study demonstrates the potential of HC system to be used as a screening tool that can provide rapid classification of cell types.

DOI: http://dx.doi.org/10.5772/intechopen.106581

Classification performance Avg ± std. (%)	Sample A	Sample B	Sample C	Sample D	Sample E
Week 01	90.8 ± 0.3	82.0 ± 0.4	83.2 ± 0.4	77.3 ± 0.3	71.6 ± 0.2
Week 05	81.5 ± 0.3	78.1 ± 0.3	75.3 ± 0.4	85.1 ± 0.3	80.4 ± 0.2
Accuracy	85.6 ± 0.1	79.5 ± 0.1	79.0 ± 0.1	82.6 ± 0.1	78.2 ± 0.1

Table 1.
Binary classification of cells from week 1 and 5 using logistic regression trained and tested with cells from the same sample [11].

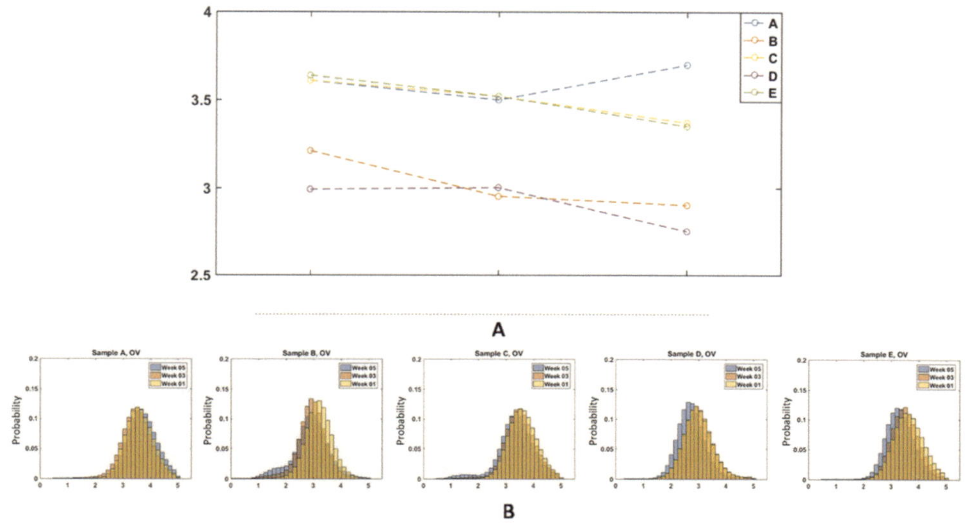

Figure 7.
(A) Line plots of optical volume for all samples over the storage time; and (B) Normalized histogram of optical volume of cells over the storage period for Sample A–E. Horizontal axis: femtoliters.

3.3 High throughput study of cancer cells

Cadmium is a group 1 carcinogen commonly found in food, water and surrounding environment [15]. Evidence show that chronic exposure to cadmium promotes cancer progression in epithelial cells [16]. In another HC study, it has been shown that HC is capable of differentiate between large number of carcinogen-exposed cells and normal healthy cells. Thousands of single cell images were collected from a Cd-treated cell line, normal epithelial cell line, and cancer cell line. This study aimed to evaluate QPI's diagnostic ability in identifying carcinogen-exposure at the single cell level.

The vast amount of cell data comprised individual phenotype profiles for each cell type, which when combined with machine learning and convolutional neural network (CNN), provides rapid and highly accurate classification on the carcinogenic status of cells. The system's high performance demonstrates its potential to be developed into an abnormal cell screening diagnostic tool.

Many morphological parameters exhibit similar trends – the Cd-treated cell line show intermediate parameter values that are in between healthy cells and cancer cells, this is especially the case for OV, shown in **Figure 8**. Since the Cd-treated cell line overlaps more with the cancer cell line than with the normal epithelial cell line, it is likely that there are more biophysical similarities between Cd-treated cells and cancer

25 parameters logistic regression			
	Sensitivity	Specificity	Mean accuracy
MCF10A/BT474	99.5 ± 0.49	99.0 ± 0.84	99.3 ± 0.51
MCF10A Cd/BT474	90.0 ± 2.87	91.38 ± 1.96	90.7 ± 1.52
MCF10A/MCF10A Cd	98.7 ± 1.11	98.3 ± 0.97	98.5 ± 0.50

Table 2.
Logistic regression results, trained with 25 morphological parameters [17].

Area, optical volume, major axis length, mean phase logistic regression			
	Sensitivity	Specificity	Mean accuracy
MCF10A/BT474	94.6 ± 1.92	89.0 ± 89.0	91.8 ± 1.34
MCF10A Cd/BT474	73.0 ± 2.45	70.4 ± 1.74	71.7 ± 1.41
MCF10A/MCF10A Cd	91.9 ± 3.06	92.4 ± 2.26	92.2 ± 1.84

Table 3.
Logistic regression results, trained with 4 morphological parameters [17].

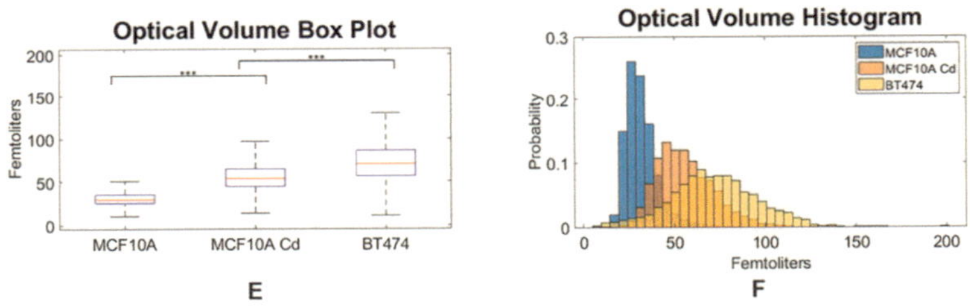

Figure 8.
(E) OV box plot; and (F) OV histogram [17].

cells. These similar traits may be monitored in future applications for identifying carcinogen exposure.

Large numbers of morphological parameters and single cell images are used as training set for logistic regression and CNN. Both algorithms exhibited extraordinarily high sensitivity (98.7%–99.5%) and high mean accuracy levels (98.5%–99.3%) for abnormal cell detection, shown in **Table 2**. Majority of misclassified cases occurred in the Cd/normal cell line binary classification. Thus, logistic regression results further validate that Cd-treated cells are more similar to cancer cells than they are to their healthy past selves.

It is also evident that not all morphological parameters contribute equally to perfecting the algorithm's performance. Among the 25 morphological parameters, 4 descriptors depicted the most phenotypic differences. **Table 3** shows the logistic regression performance with only 4 parameters as training set. With such a relatively high accuracy level (~92%) in identifying normal/abnormal cells, it is likely that HC can be utilized in real-time applications where only a few parameters will be enough to provide a preliminary rapid diagnosis.

Figure 9 demonstrates the classification performance of the CNN. The deep learning model achieved an accuracy level of up to 97.7%. It is also noted that majority of the misclassified cases rise from algorithm mistaking a Cd-treated cell for cancer cell or vice versa. The performance characteristic of the CNN again confirms the previous finding – morphological similarities between Cd-treated cells and cancer cells are dominant and apparent.

Overall, coupled with microfluidic channels and machine learning algorithms, HC provides a path to automated cytological sample diagnosis. Particularly, in low resource settings where labor-intensive and costly histopathological diagnostics are inaccessible, HC presents a low-cost label-free alternative.

4. Measuring cell mechanical properties

Knowing cellular mechanical properties can be useful when trying to determine health and disease states. Measuring mechanical properties often requires controlled probing of the cell and recording the response, e.g., atomic force microscopy (AFM). Physical probing can also cause damage [18] to the cellular environment, so less invasive methods like utilizing controlled flow of media over cells to induce a shear response can be attractive. The use of microfluidic channels and syringe pumps allow for fine control of shear imparted on a cell and stress testing. Off-axis holography quantitative phase imaging (QPI) has been shown to be able to record the cells' mechanical response over fine space and time scales [18–21]. QPI allows for recovery of cell elastic properties such as stiffness and shear modulus which can also be linked to cellular microstructure.

4.1 Phase deformation informs shear stiffness

When a cell is attached to the culture substrate, perturbing the cell and recording the response acts like a stress test. Fabricated flow chambers with well know physical boundaries combined with a syringe pump capable of constant volumetric flow allows for easy estimate of the wall shear stress, τ_s, imparted on the cell under laminar flow. The measured phase deformation, change in localized phase from before applying shear stress to when the cell reaches a steady state under shear, informs how cell height changes.

$$\Delta\phi(x, y, t) = \phi(x, y, t) - \phi(x, y, 0) \propto \Delta h(x, y, t)$$

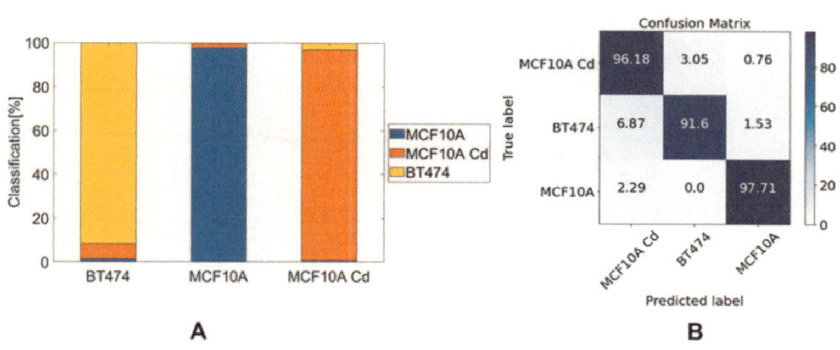

Figure 9.
Performance of CNN classification trained with three cell lines. (A) Stack box plot of classification results; and (B) Confusion matrix for CNN predictions (%) [17].

Combining measured cell height with cell position in x and y allows for the calculation of cell center-of-mass (COM). As shown in **Figure 10** tracking the 3D change in center of mass, $\vec{r}_{COM}(t)$, allows for the study of the cell's response to shear stress.

Using a Kelvin-Voigt model, the response of the cell at a steady state can be described in a differential equation like a 1D damped spring and provide information on the viscoelastic properties of the cell [19].

$$\left|\vec{r}_{COM}(t)\right| = \frac{\tau_s A_s}{k_s}\left(1 - e^{-\frac{tk_s}{\eta}}\right) \tag{3}$$

Where A_s is the estimated cell surface area, k_s is shear stiffness, and viscosity, η, acts as the damping element.

This has been shown to work experimentally in the 2016 publication by W. Eldridge et al. [19]. Here two groups of human skin cancer cells, A431; one in a normal environment as a control, the 2nd group treated with toxin cytochalasin D (CD); were observed in flow chambers. Before beginning flow, the cell's COM is constant, then once flow begins the phase displacement goes from negative to positive within the cell in the same direction as the flow, **Figure 10**. Under the constant volumetric flow rate the COM comes to a steady state. Cell displacement at steady-states was found to be

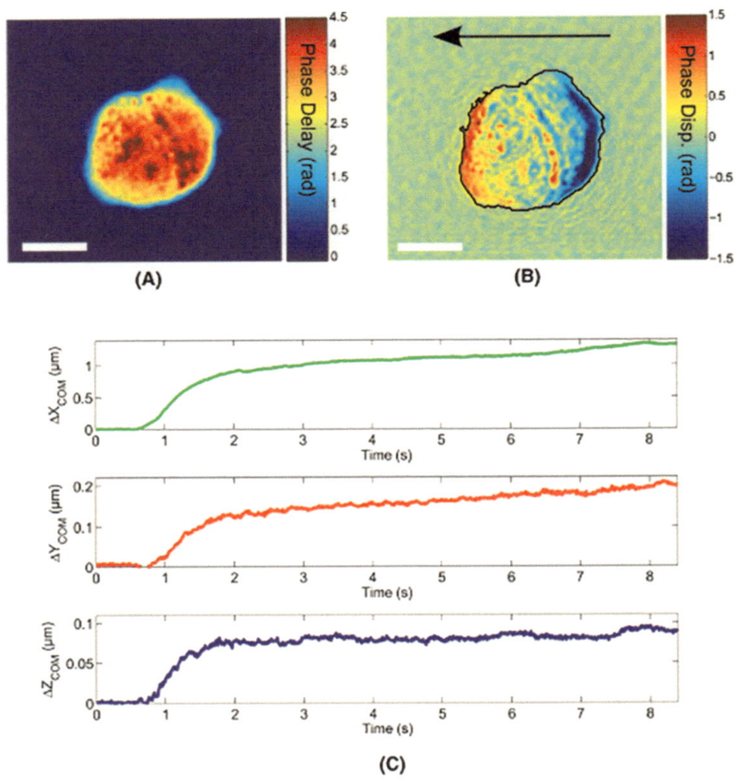

Figure 10.
(A) phase map of an A431 cell before starting the flow of media; (B) Phase deformation of the cell due to the shear stress cause by flow. The arrow shows the direction of flow, outline shows the x-y boundaries of observed phase deformation; and (C) Tracking cell COM in each direction do to shear, notice how the change in COM in the x direction is the largest since it best aligns with the direction of flow [19].

linearly related to the shear stress, matching predictions from the Kelvin-Voigt model. Agreeing with theory [22] and shown in **Figure 11**, cells dosed with CD were found to have a reduced shear stiffness when compared to control A431 cells because CD alters actin in the cell's cytoskeleton, and therefore microstructure, in turn altering its viscoelastic properties.

4.2 Disorder strength and microstructure

Disorder strength, L_D, a metric for quantifying a cell's degree of structural heterogeneity and nanoarchitecture, has been shown to be useful in differentiating cells in the early stages of carcinogenesis from normal cells even when the two types appear similar microscopically and histologically [23]. Originally measured using partial-wave spectroscopy, L_D has shown potential as a cancer diagnostic tool especially for pancreatic and colon cancer [23]. Work by W. Eldridge et al. [18] proves L_D can also be measured via QPI to provide information on cell stiffness and cytoskeletal organization with potential for high throughput measurements.

Measuring disorder strength via QPI takes advantage of the fine spatial sampling in quantitative phase images and local homogeneity in height over small length scales. Disorder strength calculated as the product of a cell's refractive index variance, $\langle \Delta n^2 \rangle$, and its spatial coherence length, l_c.

$$L_D = \langle \Delta n^2 \rangle l_c \tag{4}$$

Measurement of cell height, $h(x, y)$ at pixel (x, y) can be assumed to not vairy greatly over a small neighborhood e.g., 3x3 pixel region. Therefore, the local phase variance of a cell is proportional to its local refractive index variance, $\left\langle \Delta\phi(x,\ y)^2 \right\rangle \propto \left\langle \Delta n(x,\ y)^2 \right\rangle$. Normalizing the local phase variance by the local square mean phase, $\overline{\phi}^2(x, y)$, gives an equation that can be solved for refractive index variance.

$$\left\langle \Delta n(x,\ y)^2 \right\rangle \approx \frac{\left\langle \Delta\phi(x,\ y)^2 \right\rangle}{\overline{\phi}^2(x, y)} \overline{n}^2 \tag{5}$$

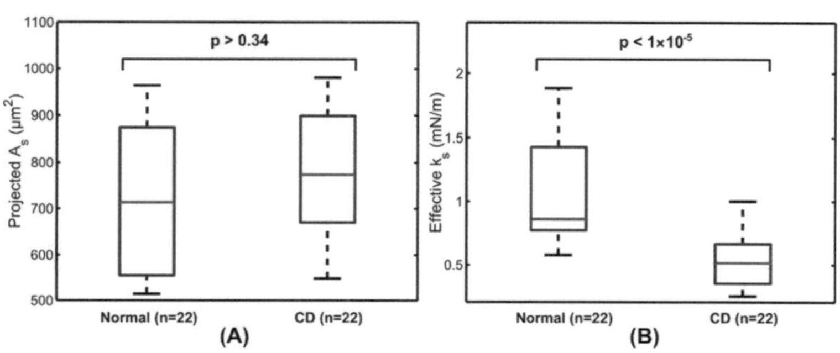

Figure 11.
(A) Plot showing the projected surface areas of each group of A431 cells is similar, highlighting the difficulty in differentiating groups just based on microscopically observable parameters like area; and (B) Whereas viscoelastic parameters like shear stiffness clearly differentiates between the normal and CD treated cells [19].

Spatial coherence length can be determined by fitting the falloff of the autocorrelation of the phase fluctuation to an exponential, allowing the phase disorder strength to be calculated via:

$$L_D = \left\langle \frac{\left\langle \Delta\phi(x,\ y)^2 \right\rangle}{\overline{\phi}^2(x,\ y)} \right\rangle \overline{n}^2 l_c \tag{6}$$

Experimentally L_D was measured for five different groups of cells: 1 group of HT-29 colon cancer cells, a 2nd group of HT-29 but with a CSK shRNA-mediated knockdown (referred to as CSK), 1 group of normal A431 human skin cancer cells, 2nd group of A431 cells that were treated with CD, and a single group of A549 human lung cancer cells. **Figure 12** shows phase maps, phase fluctuation maps, and mean L_D measurements for each group.

Similar to how shear stiffness was able to differentiate the two types of A431 cells in **Figures 11** and **12L** shows how L_D can also differentiate the CD-treated cells from the normal A431 cells. After performing the same shear stiffness analysis from Section 4.1 on these five groups of cells and plotting the mean L_D vs. the mean k_s, as shown in **Figure 13**, the negative correlation between the two metrics is clear. The fit constant, b, of 0.75 pN make sense when considering typical forces relating to cell microstructure [24, 25]. It is important to note that a single cell's mean L_D can calculated from a single QPI snapshot, whereas k_s measurement requires two images of the cell: one at rest and one after being perturbed by flow. The link between L_D and k_s showcases the potential of L_D for understanding more about a cell's viscoelastic properties without perturbation.

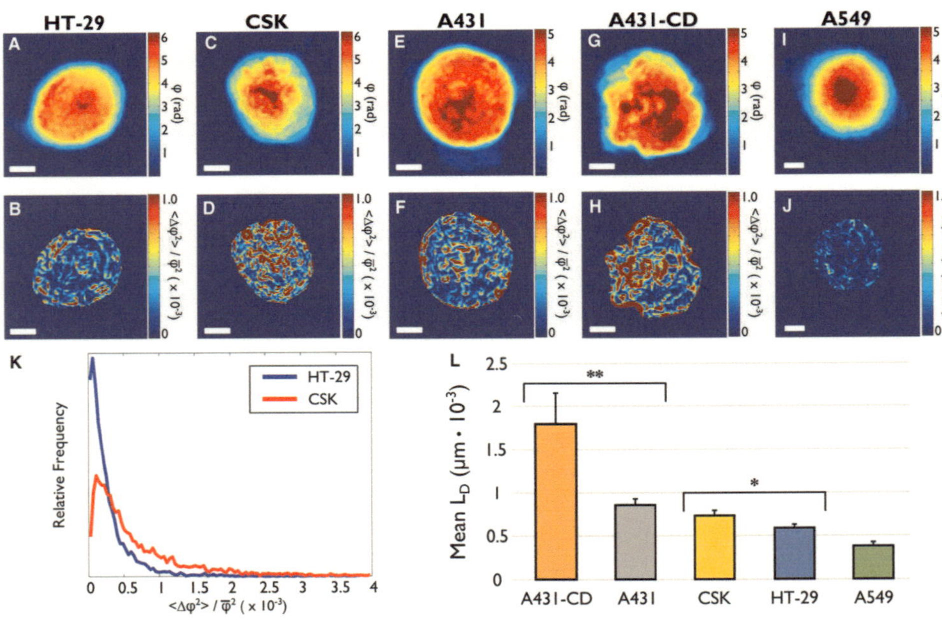

Figure 12.
(A–J) Example QPI images and corresponding phase fluctuation maps for each cell group with 5 μm scale bars; (K) Plots of the phase fluctuation probability density functions for the cells shown in B and D, highlighting the high mean phase fluctuation of the CSK cell when compared to HT-29;and (L) Mean measured L_D *for each group [18].*

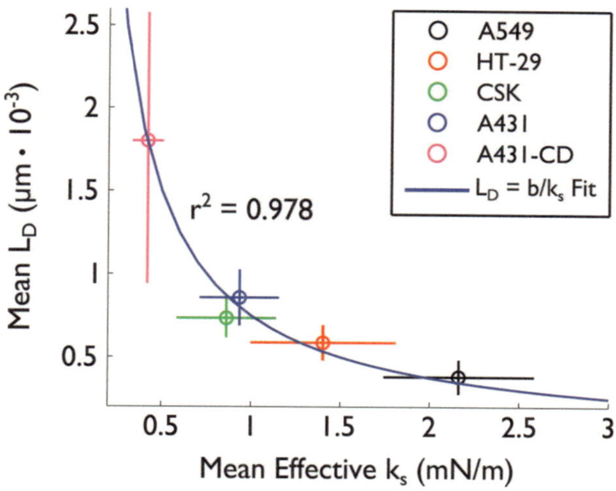

Figure 13.
Plot of mean disorder strength vs. mean shear stiffness for five groups of cells. Error bars show standard deviation for each parameter and show how all cell groups fall overlap with the negative correlation fit between L_D *and* k_s *[18].*

4.3 Shear modulus and elasticity

To truly understand cell elastic properties, shear stiffness or spring constant is not enough. Elastic moduli like the Young's modulus, E, or shear modulus, G, must be measured. These moduli are related according to:

$$E = 2G(1 + \nu) \tag{7}$$

where ν is the Poisson's ratio. Commonly the approximation $\nu \approx 0.5$ is used for cells [26], and therefore a linear relationship between the two moduli, $E = 3G$, can be assumed. Shear modulus is determined by the ratio of shear stress over shear strain. As shown in Section 4.1, the shear stress, τ_s, imparted on the cell from flowing media can be controlled with a microfluidic channel and a syringe pump. Shear strain can be treated as the transverse displacement, dX, divided by height. The shear stiffness analysis tracks the displacement of a cell's COM and QPI can measure localized cell height. To get a height estimate, h_e, of the whole cell the median of $h(x, y)$ over all values of x and y within the cell above a reasonable lower bound, in this case 5 μm, was used. **Figure 14** illustrates how h_e and G are calculated.

$$G = \frac{\tau_s}{dX} h(x, y) \approx \frac{\tau_s}{\left|\vec{r}_{COM}\right|} h_e \tag{8}$$

To experimentally determine G, two different types of human breast cancer cells were used: MCF7 and BT474. Each type was split into three groups based on concentration of treatment with CD: 0 μM (control), 0.2 μM, and 1.0 μM. To measure displacement, the COM of a cell was measured before flow at $t = 0\,s$ and again at $t = 8\,s$ after flow and reaching a steady state. To validate measured G, the same cell groups were also examined via AFM to measure E, results shown in **Figure 15**. In **Figure 15A** there is a negative trend in elasticity measured by AFM with increasing concentration of CD treatment. This agrees with the QPI measured shear modulus, as in **Figure 15B** not only within but across the two types of cells.

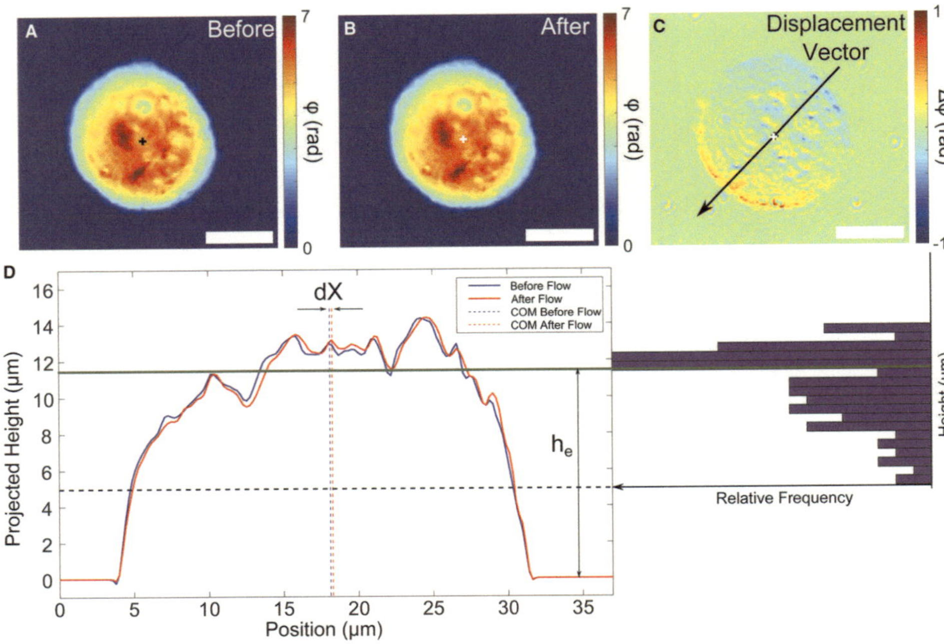

Figure 14.
(A, B) Phase map of BT474 before and after flow, respectively. Plus signs signify the COM; (C) Phase displacement map of the same cell shown in A and B; and (D) height profile of the cell in A and B showing the transverse displacement and height estimate (left), histogram showing how the height estimate was determined (right) [20].

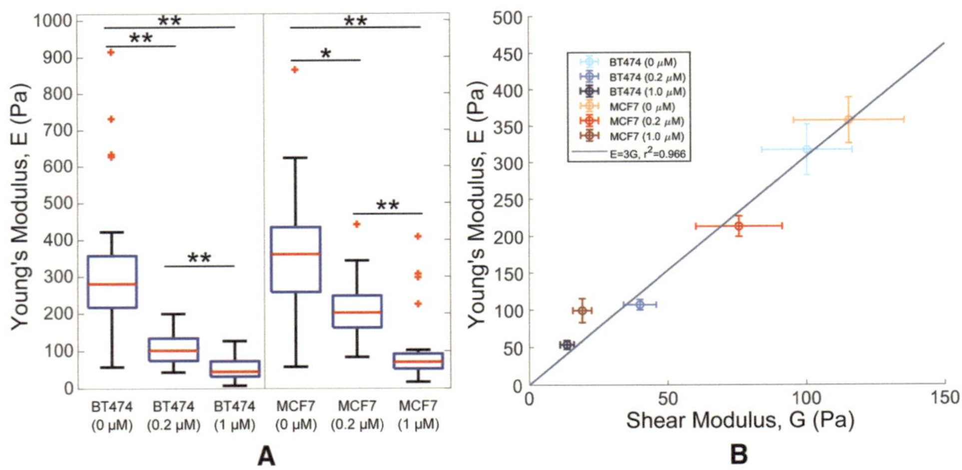

Figure 15.
(A) AFM measurements of Youngs modulus over the 6 cell groups; and (B) Plot of correlation between AFM measured E *and QPI measured* G*. Most cell groups follow the hypothesized* $E = 3G$ *relationship between the two moduli [20].*

The good agreement of AFM measured E and QPI measured G with the hypothesized $E = 3G$ relationship proves QPI's usefulness to measure elastic properties of a cell and shows it can be a reasonable alternative to AFM. One drawback is that to measure G, the cell must be imaged twice and be perturbed by flow. Disorder strength however can be measured with a single snapshot, without perturbation, and as shown

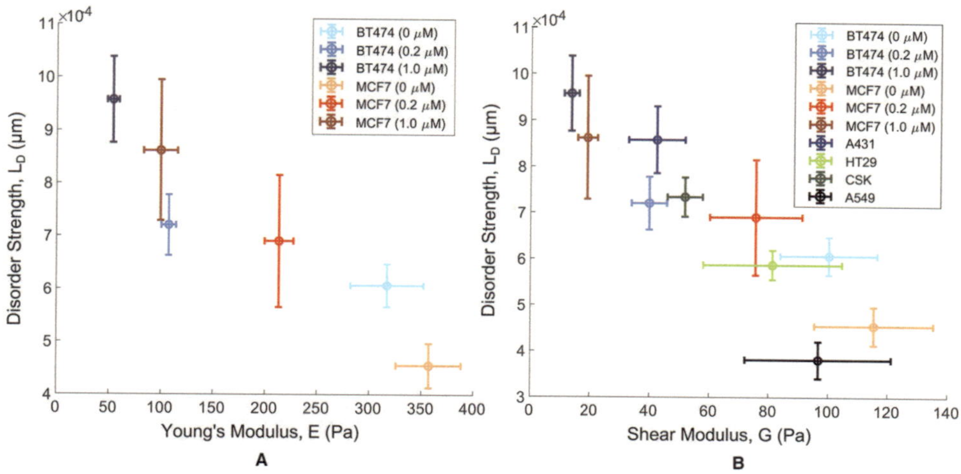

Figure 16.
(A) Plot of QPI measured disorder strength vs. AFM measured Young's modulus; and (B) Plot of QPI measured disorder strength vs. shear modulus. In both plots a clear negative correlation between disorder strength and elastic modulus is shown ref. [20].

in **Figure 13** is correlated with shear stiffness. Comparing measured L_D to E, **Figure 16A**, it can be seen that L_D is correlated with elastic moduli of a cell. This supports the findings in Section 4.2, that cell microstructure can be related to stiffness. There is still more research to be done in L_D, looking into other types of cells and their elasticity compared to cancer cells. When L_D is compared across multiple experiments and cell lines, **Figure 16B**, there is a negative correlation with shear modulus. Also, since L_D can be measured with single snapshot it has great potential to provide information on cell elasticity in high-throughput settings.

5. Discussions

With the use of microfluidics, QPM becomes a platform capable of high throughput measurements of cells at the individual level. Further, the platform allows observation of individual cells under unique mechanical conditions such as exposure to flow or mechanical constriction. To enable QPM to provide large volumes of imaging data, the HC setup has been implemented, which uses stroboscopic imaging to enable visualization of cell features without streaking due to motion. The high throughput scheme has been shown to provide the visualization of multi-million single cells within 8–16 parallel microfluidic channels. The analysis of imaging data is made feasible by composing a single vector for each cell, comprised of 25 morphological descriptors. Upon extracting these from each single cell image, a highly descriptive, phenotype profile is obtained for the given cytological specimen.

In the experiments discussed here, the HC system was demonstrated to provide insights on the deformability of RBCs under mechanical stress. The study shows that RBCs with different storage ages have distinct capabilities for exchanging water with their environment. This is observed as a refractive index increase which causes the increase in OV. The 29-day-old RBCs exhibit the smallest change in OV, and in turn less water exchange, due to channel constriction than other time points. This trend is consistent with the known properties of RBCs to become stiffer at longer storage times and thus are unable to exchange as much water content. The monitoring of water permeability could potentially be used as a predictive factor for stored blood bag quality control.

The high throughput capabilities of HC were demonstrated with experiments that examined millions of cells, generating significant volumes of data. Although direct analysis of the cell images can be instructive, the training time for direct application of machine learning algorithms can be substantial. Instead, the analysis is greatly streamlined using the 25 morphological descriptors. Studies of cultured mammalian cell lines show that distinct cellular characteristics are observed in the individual descriptors for varying phenotypes. In particular, the morphological descriptors easily distinguished the non-tumorigenic MCF-10A cells from the cancerous BT-474 cells. The Cd-treated cells displayed morphologies in between those of regular and cancer cells, suggesting that the heavy metal treatment induced potentially carcinogenic changes. Upon training with these data sets, machine learning models have been developed which exhibited high sensitivity and accuracy levels in identifying cell groups. These results suggest that HC can potentially be used for abnormal cell monitoring in disease diagnosis.

Measurement of cell mechanical properties in HC has been enabled by switching from a direct perturbation approach, in the form of a flow assay, to use of a descriptor that can be extracted from a single cell image. A cell's disorder strength L_D can be calculated from a single QPM image, using a simple determination of the phase variance, allowing for facile application of this measurement to many L_D has been shown to correlate with elastic moduli, which usually require multiple images and perturbation of an attached cell, limiting the throughput to just a handful of cells. Cellular microstructure and aL_D, showcasing its ability to differentiate cells of the same type but under different treatment. Combining the mechanical and L_D with the fast-paced single and multi-cell measurements of a high-throughput HC system potential L_D over time and length scales, as cells respond to mechanical or pharmacological stimuli. Also expanding disorder strength and cell mechanics measurements to a wider range of cell types.

6. Conclusions

In summary, we have presented a high throughput QPM approach termed holographic cytometry (HC). The approach uses microfluidic elements and stroboscopic illumination to obtain off-axis holographic images of cells in flow. Several experiments have been reviewed to showcase the high throughput capabilities of HC, including examination of RBCs and cancer cells, revealing excellent discrimination of phenotype based on the wealth of information obtained from morphological descriptors. Application to stored RBC's points to a potential utility in addressing the need to monitor the age of blood units. Experiments with cancer cells show that phenotype can potentially be discerned using this label free approach. Further development of the HC approach has enabled detailed studies of cell mechanical properties. The measurement of disorder strength provides a compelling window into cell structure that has been linked to essential mechanical properties such as stiffness and shear modulus.

Acknowledgements

This work has been supported through the Army Research Office, Grant W911NF-19-1-0306.

Conflict of interest

No conflicts of interest are declared for this work.

Notes/thanks/other declarations

The authors acknowledge Dr. Han Sang Park and Dr. Will J. Eldridge for their contribution to the foundational works presented in this chapter.

Author details

Cindy X. Chen*, Hillel B. Price and Adam Wax
Department of Biomedical Engineering, Duke University, Durham, NC, USA

*Address all correspondence to: xuewen.chen@duke.edu

References

[1] Shaked NT, Rinehart MT, Wax A. Dual-interference-channel quantitative-phase microscopy of live cell dynamics. Optics Letters. 2009;**34**(6):767-769

[2] Park HS, Rinehart MT, Walzer KA, Chi J-TA, Wax A. Automated detection of P. falciparum using machine learning algorithms with quantitative phase images of unstained cells. PLoS One. 2016;**11**(9):e0163045

[3] Rinehart MT, Park HS, Wax A. Influence of defocus on quantitative analysis of microscopic objects and individual cells with digital holography. Biomedical Optics Express. 2015;**6**(6): 2067-2075

[4] Park HS, Ceballos S, Eldridge WJ, Wax A. Invited article: Digital refocusing in quantitative phase imaging for flowing red blood cells. APL Photonics. 2018;**3**(11):110802

[5] Barteneva NS, Fasler-Kan E, Vorobjev IA. Imaging flow cytometry: coping with heterogeneity in biological systems. The Journal of Histochemistry and Cytochemistry. 2012;**60**(10):723-733

[6] Rushton AJ, Nteliopoulos G, Shaw JA, Coombes RC. A Review of Circulating Tumour Cell Enrichment Technologies. Cancers (Basel). 2021;**13**(5):6

[7] McMahon TJ. Red blood cell deformability, vasoactive mediators, and adhesion. Frontiers in Physiology. 2019;**10**:3

[8] Moon JS, Kim JH, Kim JH, Park IR, Lee JH, Kim HJ, et al. Impaired RBC deformability is associated with diabetic retinopathy in patients with type 2 diabetes. Diabetes & Metabolism. 2016; **42**(6):448-452

[9] Park HS, Eldridge WJ, Yang W-H, Crose M, Ceballos S, Roback JD, et al. Quantitative phase imaging of erythrocytes under microfluidic constriction in a high refractive index medium reveals water content changes. Microsystems & Nanoengineering. 2019; **5**(1):63

[10] Xu Z, Zheng Y, Wang X, Shehata N, Wang C, Sun Y. Stiffness increase of red blood cells during storage. Microsystems & Nanoengineering. 2018;**4**(1):17103

[11] Park H-S, Price H, Ceballos S, Chi J-T, Wax A. Single cell analysis of stored red blood cells using ultra-high throughput holographic cytometry. Cell. 2021;**10**(9):2455

[12] Roussel C, Dussiot M, Marin M, Morel A, Ndour PA, Duez J, et al. Spherocytic shift of red blood cells during storage provides a quantitative whole cell-based marker of the storage lesion. Transfusion. 2017;**57**(4):1007-1018

[13] Yoshida T, Prudent M, D'Alessandro A. Red blood cell storage lesion: Causes and potential clinical consequences. Blood Transfusion. 2019; **17**(1):27-52

[14] Cluitmans JC, Hardeman MR, Dinkla S, Brock R, Bosman GJ. Red blood cell deformability during storage: Towards functional proteomics and metabolomics in the Blood Bank. Blood Transfusion. 2012;**10 Suppl 2**(Suppl. 2):s12-s18

[15] World health organization international agency for research on cancer. IARC working group on the evaluation of carcinogenic risks to humans, cadmium and cadmium compounds. IARC Monographs on the Evaluation of Carcinogenic Risks to Humans; 1993;**82**:P121-P145

[16] McElroy JA, Shafer MM, Trentham-Dietz A, Hampton JM, Newcomb PA. Cadmium exposure and breast cancer risk. JNCI: Journal of the National Cancer Institute. 2006;**98**(12):869-873

[17] Chen CX, Park HS, Price H, Wax A. Automated classification of breast cancer cells using high-throughput holographic cytometry. Frontiers in Physics. 2021;**9**: 869

[18] Eldridge WJ, Steelman ZA, Loomis B, Wax A. Optical phase measurements of disorder strength link microstructure to cell stiffness. Biophysical Journal. 2017;**112**(4):692-702

[19] Eldridge WJ, Sheinfeld A, Rinehart MT, Wax A. Imaging deformation of adherent cells due to shear stress using quantitative phase imaging. Optics Letters. 2016;**41**(2): 352-355

[20] Eldridge WJ, Ceballos S, Shah T, Park HS, Steelman ZA, Zauscher S, et al. Shear modulus measurement by quantitative phase imaging and correlation with atomic force microscopy. Biophysical Journal. 2019; **117**(4):696-705

[21] Muñoz A, Eldridge WJ, Jakobsen NM, Sørensen H, Wax A, Costa M. Cellular shear stiffness reflects progression of arsenic-induced transformation during G1. Carcinogenesis. 2018;**39**(2):109-117

[22] Wakatsuki T, Schwab B, Thompson NC, Elson EL. Effects of cytochalasin D and latrunculin B on mechanical properties of cells. Journal of Cell Science. 2001;**114**(5):1025-1036

[23] Subramanian H, Pradhan P, Liu Y, Capoglu Ilker R, Li X, Rogers Jeremy D, et al. Optical methodology for detecting histologically unapparent nanoscale consequences of genetic alterations in biological cells. Proceedings of the National Academy of Sciences. 2008; **105**(51):20118-20123

[24] Kovar David R, Pollard TD. Insertional assembly of actin filament barbed ends in association with formins produces piconewton forces. Proceedings of the National Academy of Sciences. 2004;**101**(41):14725-14730

[25] Laakso JM, Lewis JH, Shuman H, Ostap EM. Myosin I can act as a molecular force sensor. Science. 2008; **321**(5885):133-136

[26] Nijenhuis N, Zhao X, Carisey A, Ballestrem C, Derby B. Combining AFM and acoustic probes to reveal changes in the elastic stiffness tensor of living cells. Biophysical Journal. 2014;**107**(7): 1502-1512

Chapter 14

Applications of Digital Holographic Interferometry in Heat Transfer Measurements from Heated Industrial Objects

Varun Kumar and Chandra Shakher

Abstract

Digital holographic interferometry (DHI) is used worldwide for many scientific and industrial applications. In DHI, two digital holograms; one in the reference/ambient state of the object and another in changed state of object are recorded by electronic imaging sensors (such as CCD/CMOS) as reference holograms and object holograms, respectively. Phase of object wavefronts in different states of the object is numerically reconstructed from digital holograms. The interference phase is reconstructed by subtracting the phase of reference hologram from the phase of object hologram, without performing any phase-shifting interferometry. Thus, no extra effort is needed in DHI for calculating the interference phase. Apart from direct reconstruction of interference phase from two digital holograms, the recent development, availability of recording devices at video rate, and high-performance computers make the measurements faster, reliable, robust, and even real-time. In this chapter, DHI is presented for the investigation of temperature distribution and heat transfer parameters such as natural convective heat transfer coefficient and local heat flux around the surface of industrial heated objects such as cylindrical wires and heat sinks.

Keywords: digital holography, interferometry, temperature, heat transfer parameters, convective heat transfer coefficient, heat flux

1. Introduction

The study of heat flow/heat transfer from industrial heated objects such as wires, cylinders, conductors, and heats sinks is involved in many scientific and industrial applications. The knowledge of surface temperature and temperature profile normal to the surface of heated objects is sought for the measurements of heat transfer parameters such as convective heat transfer coefficients and heat flux [1–3]. The temperature measurements by intrusive methods such as thermocouple, thermistor, hot wire anemometer, and resistance temperature detector (RTD) are not reliable because these contact type methods disturb the heat flow and also lack in providing full field temperature information [3]. In the past, many optical interferometric methods have been widely used for the measurements of temperature, temperature profile, and heat transfer studies. These methods being non-invasive and non-contact type in nature provide reliable measurement of temperature. Other advantages of optical methods are (1) these methods provide full field temperature measurements in contrast to contact-type methods which provide point-by-point measurements and (2) measurements are inertia free so rapidly time-varying phenomena can also be investigated [3, 4]. Careful use of optical methods leads to more accurate and precise temperature measurements.

In the past, most of the temperature and heat transfer studies such as natural convection, forced convections, heat flux in liquids, and gases from heated objects such as cylinders, vertical plates, horizontal plates, finned plates, compact channels, microprocessor chips, grooved channels, downward facing heater surfaces, rib-roughened vertical surfaces, and horizontal cylinders were performed using classical interferometry (Mach–Zehnder Interferometry [MZI] and Michelson Interferometry [MI]), and holographic interferometry (HI) [5–34]. Goldstein used MZI to study the aerodynamics and heat transfer [20] and Haridas et al. used both MI and MZI to measure the forced convection in compact channels [21]. In these studies it was concluded that MI provides a larger number of fringes (data points) in the temperature fields in comparison to MZI, thus MI is a better method of temperature field measurement in compact channels [21]. MI is more sensitive to temperature change than MZI but MI is not good for studying dynamic/rapidly varying phenomena as the light has to travel the test section/object field two times.

In the temperature and heat transfer studies using classical interferometry and HI mentioned in the above paragraph, the full field quantitative temperature field was not measured but the qualitative information on temperature was visualized. The temperature was calculated at each fringe position (i.e., isothermal line) using fringe count and fringe shift methods. The temperature profile was plotted by fitting the temperature data calculated at the location of each fringe. HI has several advantages in comparison to classical interferometry. HI does not require high-quality optical components in comparison to classical interferometry, because in HI, only the relative phase change of object wavefronts is evaluated. This leads to reduced costs of holographic interferometric equipment. HI provides the major advantage of greater experimental and technical simplicity. Thus, the time required for adjustment of equipment and taking measurement is also reduced considerably [26].

Furthermore, several optical interferometric methods such as classical interferometry [5, 6, 35, 36], holographic interferometry [22–24, 37], moiré deflectometry [38], speckle shearing interferometry [39], shearing interferometry [40, 41], Talbot interferometry (TI) [42, 43], Lau phase interferometry (LPI) [44], digital speckle pattern interferometry (DSPI) [45], holo-shear lens based lateral shear interferometry [46–48], and DHI [49–58] have been investigated to measure the temperature field inside the axi-symmetric gaseous flames and transparent hot objects. TI, Moiré deflectometry, lateral shear interferometry, and holo-shear lens-based interferometry are common path interferometric methods and hence these methods are relatively less sensitive to vibrations or external perturbations as compared to speckle shearing interferometry, HI, and DHI. In shearing interferometry and speckle shearing interferometry, prior information regarding temperature profile direction is required. Prior information regarding temperature profile direction is also needed, if TI and LPI are performed with linear grating. However, TI and LPI using circular grating do not require any prior information regarding temperature field direction [39]. Also, in TI and LPI with linear gratings, the alignment between the gratings is difficult to achieve.

LPI uses white light source, so the advantage of using LPI over TI is that it reduces coherent speckle noise in the interferogram, and thus provides high signal-to-noise ratio [39, 59]. TI, LPI, and holo-shear lens-based interferometry require high-quality optical components (such as gratings, lenses, and holo-shear lens) as compared to HI. The interference fringes in speckle shearing interferometry, and DSPI are not of good contrast because these fringes are formed due to speckles. Speckle filtering techniques are used to minimize the speckle noise to get usable information [60, 61].

In digital holography (DH), holograms are recorded in digitized form by using CCD or CMOS sensors and complex amplitude of object wave-front from recorded digital hologram is reconstructed by numerical methods using computers [62–64]. Thus precise quantitative reconstruction/evaluation of amplitude and phase of an object wave-front from single digital hologram is possible in DH using numerical reconstruction methods. In double exposure DHI, two digital holograms, one in the ambient/reference state of object and other in changed state of objects are recorded. The interference phase is directly calculated from two digital holograms by subtracting the numerically reconstructed phase of reference hologram from the phase of object hologram without performing any phase-shifting interferometry. Thus no extra efforts are required in DHI for the reconstruction of interference phase. Also in DH, the data acquisition rate is also high because by using CCD/CMOS sensor, signals are available at video frequencies. The data acquisition rate in DH is only limited by the frame rate of recording CCD/CMOS sensor. The reconstruction rate is also high ($\sim$1 Hz) and can be controlled by using fast computers and efficient reconstruction algorithms [65]. Hence, the processes of measurement become faster, reliable, and even almost real-time. The sensitivity of photographic plate is $\sim 10^5$ photons/μm^2, whereas the sensitivity of CCD can be $\sim$1 photon/μm^2 [66]. This leads to a large reduction in exposure time as well as simpler requirements on experimental set-up stability against external vibration and perturbations/disturbances. These parameters make the system simple, reliable, robust, more precise, accurate, and more suitable for industrial environment. Whereas, in conventional HI the time-consuming and tedious development process of recording materials makes the use of HI difficult in industrial environment.

In the past, DHI has been extensively used for many applications such as measurement of temperature [49–58, 67–69], heat transfer from surface of heated objects [52, 54], diffusion coefficient [70, 71], flow field analysis [72], observation of decomposition of crude oil under dispersant fluid [73], refractive index of liquids [74], heat conduction in glass [75], solution concentration variation in protein crystallization process [76], visualization of thermal gradients in bulb [77], and human skin temperature [78].

In this chapter, DHI in lenless Fourier transform (LLFT) configuration is presented for the investigation of temperature distribution and heat transfer parameters such as natural convective heat transfer coefficient and local heat flux around the surface of industrial heated objects such as cylindrical wires and heat sinks.

2. Theory/methodology

In DH, numerical methods are applied for hologram reconstruction in which diffraction of digital reference wave at the micro interference fringes of digital hologram is simulated using scalar diffraction theory. Several numerical reconstruction methods of hologram such as Fresnel reconstruction [62–64], convolution [79], and angular reconstruction method [80] have been developed. But, these reconstruction methods suffer from the following shortcomings (1) lateral resolution of the numerically reconstructed images is limited by the inherent coherent speckle noise. For getting high lateral resolution in reconstructed image, the speckle size should be minimized. (2) Due to low spatial resolution (~125 lines pair per mm for CCD with pixel size 4 μm) of CCD/CMOS sensor, the object needs to be placed at a minimum distance from the sensor to ensure that the angle between the object beam and reference beam does not exceed a certain value at all points of the surface of sensor so that sampling theorem is not violated which requires that fringe spacing of interference pattern in hologram is larger than double the pixel size [64]. (3) Most of holographic set-ups use plane reference waves for hologram reconstruction, and hence the angle between object and the reference beam varies over entire sensor chip and so does the spatial frequency. Thus, the full spatial bandwidth of sensor is not utilized in these set-ups. As the lateral resolution in the reconstruction relies on the complete evaluation of information captured by sensor, so full spatial bandwidth of the sensor should be utilized. Also, in Fresnel and convolution methods, several Fourier transforms and complex multiplications have to be performed. Therefore, to enhance the speed of digital image processing, computation of image reconstruction algorithm should be simpler and fast. All these problems as mentioned above are tackled by lensless Fourier transform digital holography (LLFTDH). Improved lateral resolution and use of the full spatial bandwidth of the CCD sensor can be achieved through LLFTDH. In LLFTDH, set-up, a spherical reference wave is used instead of plane reference wave and the point source of spherical wave is kept in the plane of object. This ensures that the angle between the interfering reference and object beams remains approximately same over the entire sensor chip. Thus, one can ensure that the sampling theorem is not violated over the full area of recording sensor and complete spatial-frequency spectrum of sensor is used at all point of sensor surface area. The entire spatial bandwidth of sensor can be utilized for complete evaluation of all the information one get from the sensor, if the ratio of object size and the distance between the object and the CCD target is optimized [3, 64]. Fresnel reconstruction method is used to derive the reconstruction algorithm of LLFTDH. The digitized form of complex amplitude of object wavefront is expressed as [63].

$$\begin{aligned} O(m\Delta X_I, n\Delta Y_I) = \frac{i}{\lambda d} \exp\left(-i\frac{2\pi}{\lambda}d\right) \exp\left[-i\pi\lambda\, d\left(\frac{m^2}{M^2\Delta X^2} + \frac{n^2}{N^2\Delta Y^2}\right)\right] \\ \times IFFT\left\{E_R(p,\ q)H(p,\ q) \exp\left[-i\frac{\pi}{\lambda d}\left(p^2\Delta X^2 + q^2\Delta Y^2\right)\right]\right\}. \end{aligned} \tag{1}$$

In Eq. (1), M × N is the total number of pixels on the CCD/CMOS sensor, $m, p = 0,1,2,3,\ldots \ldots \ldots M-1$ and $n, q = 0,1,2,3,\ldots \ldots \ldots N-1$, and ΔX, ΔY are pixel size on the sensor chip in X and Y directions, respectively. "λ" is the wavelength of laser light, "d" is the distance between object and recording plane (imaging sensor), $E_R(p, q)$ is the digitized reference wave, and $H(p, q)$ is the digital hologram. ΔX_I and ΔY_I are the pixel sizes in the reconstructed image. Pixel sizes in image plane are connected with the pixel sizes of the recording/hologram plane as [81, 82].

$$\Delta X_I = \frac{\lambda d}{M\Delta X}; \Delta Y_I = \frac{\lambda d}{N\Delta Y}. \tag{2}$$

In LLFT configuration of DH, the object and point source of spherical wave are kept in same plane. Thus in this configuration, the spherical phase factor associated with the Fresnel diffraction of the transmitted wave through hologram is eliminated by using of a spherical reference wave $E_R(p, q)$ with the same average curvature that was used during the recording [63].

$$E_R(p, q) = (\text{const.}) \exp\left[i\frac{\pi}{\lambda d}\left(p^2\Delta X^2 + q^2\Delta Y^2\right)\right]. \tag{3}$$

By substituting $E_R(p, q)$ in Eq. (1), a simpler algorithm of complex amplitude of object wavefront reconstruction in LLFTDH is obtained and expressed as [63].

$$O(m\Delta X_I, n\Delta Y_I) = \frac{i}{\lambda d} \exp\left(-i\frac{2\pi}{\lambda}d\right) \exp\left[-i\pi\lambda\, d\left(\frac{m^2}{M^2\Delta X^2} + \frac{n^2}{N^2\Delta Y^2}\right)\right] \times IFFT\{H(p,\ q)\}. \tag{4}$$

In Eq. (4), inverse fast Fourier transform (IFFT) of only single term, that is, recorded digital hologram is evaluated apart from some multiplicative constant. Also an autofocus image is obtained without tuning the distance "d". Thus, this method is simpler and faster as compared to other reconstruction methods such as Fresnel reconstruction and convolution method, in which combination of several Fourier transform and complex multiplications need to be evaluated. The simpler and faster reconstruction algorithm of LLFTDH enhances the possibility of real-time applications [3]. The spatial phase distribution from reconstructed object wavefront is evaluated as [3].

$$\varphi(m\Delta X_I, n\Delta Y_I) = \arctan\frac{\text{Im}[O(m\Delta X_I,\ n\Delta Y_I)]}{\text{Re}\,[O(m\Delta X_I,\ n\Delta Y_I)]}. \tag{5}$$

In the above equation, the operators "Re" and "Im" provide the real and imaginary part of complex functions, respectively. In double exposure DHI, two digital holograms, one in Ref. or ambient state of object, and other in changed state of object are recorded. Phases of two different states of the object wavefronts are calculated using Eq. (5). Let $\varphi_1(m, n)$ be the phase of object wavefront in ambient state and $\varphi_2(m, n)$ be the phase of changed state of object. The interference phase, which is the phase difference between ambient and changed state of object, is calculated by modulo 2π subtraction as [83].

$$\Delta\varphi(m\Delta X_I, n\Delta Y_I) = \begin{cases} \varphi_1(m\Delta X_I,\ n\Delta Y_I) - \varphi_2(m\Delta X_I,\ n\Delta Y_I) & \text{if } \varphi_1(m\Delta X_I,\ n\Delta Y_I) \ge \varphi_2(m\Delta X_I,\ n\Delta Y_I), \\ \varphi_1(m\Delta X_I,\ n\Delta Y_I) - \varphi_2(m\Delta X_I,\ n\Delta Y_I) + 2\pi & \text{if } \varphi_1(m\Delta X_I,\ n\Delta Y_I) < \varphi_2(m\Delta X_I,\ n\Delta Y_I). \end{cases} \tag{6}$$

From this equation, one can directly evaluate the interference phase from the digital holograms. This phase is wrapped between 0 and 2π. A continuous phase difference is obtained by applying appropriate phase unwrapping algorithm [84]. The physical parameters of interest like deformation, refractive index, temperature, etc. are calculated from the unwrapped phase difference.

3. Applications of LLFTDHI in heat transfer measurements

3.1 Measurement of heat transfer parameters along the surface of heated wire

Measurement of natural convective heat transfer from heated wires/cylinders is used in several scientific and industrial applications such as quality control of electrical wires, to find ideal insulation properties of current carrying conductors, vertical tubes of HVAC (Heating, Ventilation and Air Conditioning), in resistive heating of electronic components, space shuttle launch pads, waste nuclear rods stored in repositories, refrigerating coils, and hot radiators [3, 57, 85]. Surface temperature and temperature gradient normal to the surface of the heated object is needed for the calculation of heat transfer parameters such as local convective heat transfer coefficient "h_c" and heat flux "Q(y)". DHI is very suitable for the measurement of the heat transfer parameters. Phase difference of the order of λ/30 (0.209 radians) could be measured using DHI in our lab. Hence, the measurement accuracy of local "h_c" and local Q(y), which essentially depend on the phase difference measurement from DHI system is expected to be high [3].

In this section, an application of LLFTDHI is presented for the measurement of the natural local (h_c) and Q(y) along the surface of electrically heated tungsten wire of different diameters and different heating conditions. The schematic of the LLFTDHI experimental set-up used for the measurement of heat transfer parameters is shown in **Figure 1(a)** and photograph of spring loaded mount is shown in **Figure 1(b)**. Photograph of experimental set-up while conducting the experiment is shown in **Figure 1(c)**. In the experiment, a 30 mW He-Ne laser (Make - Spectra Physics, λ = 632.8 nm) is used as a light source. Beam splitter divides laser beam into object and reference beams. Object beam illuminates the air around the wire placed in vertical position as can be seen in **Figure 1(c)**. The interference pattern formed by the superposition of spherical reference beam and object beam was captured by CCD camera (Lumenera's Infinity3-1 M). First, a digital hologram of ambient air is recorded without heating the wire as a reference hologram. A known amount of current and voltage is applied across the wire using a variable dc power supply and second, digital hologram is recorded in steady state convective flow of air around the wire.

Phase difference maps of heated air and ambient air around the wire were calculated from the two digital holograms by using Eq. (6), and unwrapped phase difference map was calculated using the Goldstein phase unwrapping method [84]. The phase change $\Delta\varphi(X)$ along a line at distance X from axis of wire is given as [52].

$$\Delta\varphi(X) = \varphi_2(X) - \varphi_1(X) = \frac{2\pi}{\lambda}\int_0^L [n_r(r) - n_0(r)]dZ \quad (7)$$

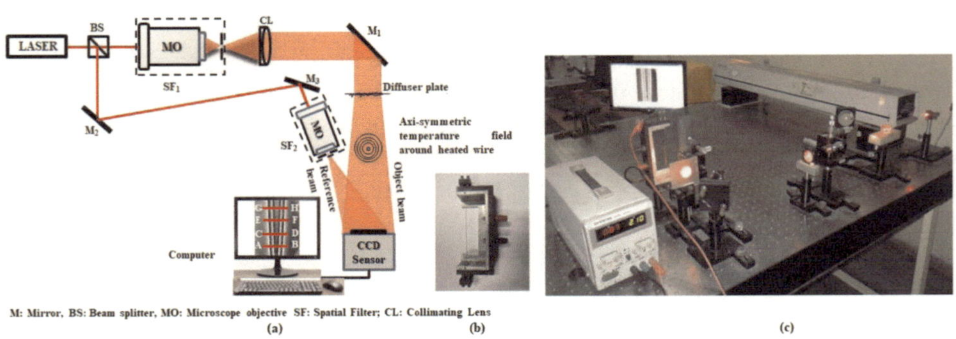

Figure 1.
(a) Schematic of experimental set-up for the measurement of temperature and heat transfer parameters, (b) photograph of spring-loaded mount used to clamp the wire in vertical position, and (c) photograph of experimental set-up [3, 58].

Here r is the radial distance, z is the direction of propagation of light, λ is the wavelength of light used in the experiment, L is the distance traveled by light in heated air, and $n_r(r)$ and $n_0(r)$ arerefractive indices of heated air and ambient air, respectively (see **Figure 2**).

Experiment was conducted in a closed room with stable air, refractive index of heated air around wire can be considered cylindrically symmetrical. For the axisymmetric objects, the evaluated phase difference map can be transformed into a radial distribution of refractive index difference distribution by using inverse Abel transforms as [52].

$$\Delta n(r) = \frac{-\lambda}{2\pi^2}\int_r^R \frac{d(\Delta\varphi)/dX}{(X^2 - r^2)^{½}} dX \tag{8}$$

where R is the radial distance up to which hot air is present, outside the R, refractive index is "n_o" and Y is along the axis of symmetry (see **Figure 2**). After calculating refractive index difference data, temperature distribution can be obtained by using Lorentz–Lorenz formula [39, 52, 86].

$$T = T_0\left[\left(\frac{n - n_0}{n_0}\right)\left(\frac{3PA + 2RT_0}{3PA}\right) + 1\right]^{-1}, \tag{9}$$

where T_o is the room temperature and n_0 is the refractive index of ambient air in the laboratory. P is the atmospheric pressure and A is the molar refractivity of air. R is the universal gas constant. The local natural "h_c" and local Q(y) can be calculated as [52, 54].

$$h_c = -\frac{K_s}{(T_s - T_0)}\left(\frac{\partial T}{\partial r}\right)_s; \qquad Q(y) = -K_s\left(\frac{\partial T}{\partial r}\right)_s \tag{10}$$

where T_s, and T_0 are surface temperature of the heated object (wire/heat sink), and ambient temperature, respectively. K_s is the thermal conductivity of air at the surface temperature of heated wire, and $\left(\frac{\partial T}{\partial r}\right)_s$ is the derivative of temperature normal to surface of wire.

Initially, an experiment was conducted on three different heating conditions of tungsten wire of diameter 0.4 mm, when applied power across the wire were (I) P_1 = 70 Watt, (II) P_2 = 46.8 Watt, and (III) P_3 = 37 Watt. Different reconstruction steps involved in the temperature profile calculation from the wrapped phase difference map in case when P = 46.8 W, are shown in **Figure 3**. From **Figure 3(d)**,

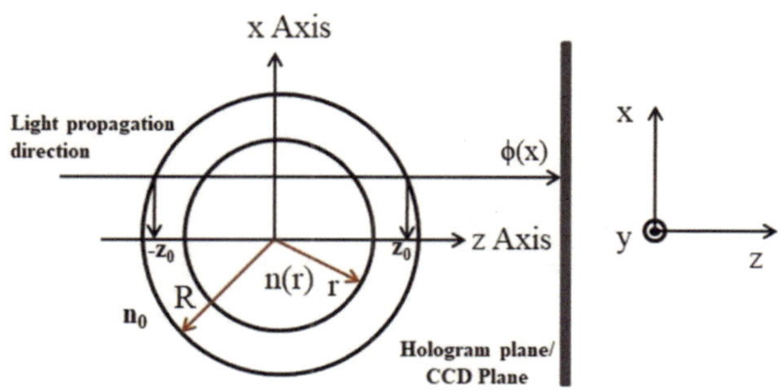

Figure 2.
Top view of the cross-section of an axisymmetric refractive index distribution [3].

experimentally calculated surface temperature of wire is found to be 477 K. The surface temperature of wire was also verified with the temperature measured by thermocouple. The thermocouple readings were also in the range 475–479 K (Avg. Temperature 477 K $\pm$ 2 K). Temperature profile along the lines AB, CD, EF and GH shown in **Figure 3(d)** are used to calculate the natural local "h_c".

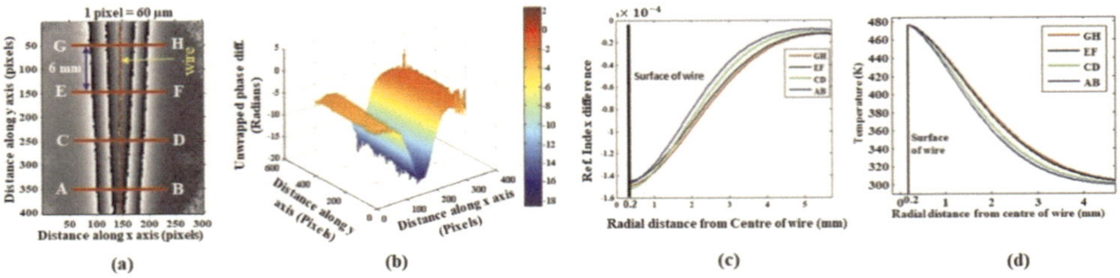

Figure 3.
*(a) Wrapped phase difference map of hot air and ambient air when applied power P = 46.8 W; (b) 3D unwrapped phase difference map corresponding to **Figure 3(a)**; (c) refractive index difference distribution along the lines AB, CD, EF, and GH; and (d) temperature profile of hot air along the AB, CD, EF, and GH.*

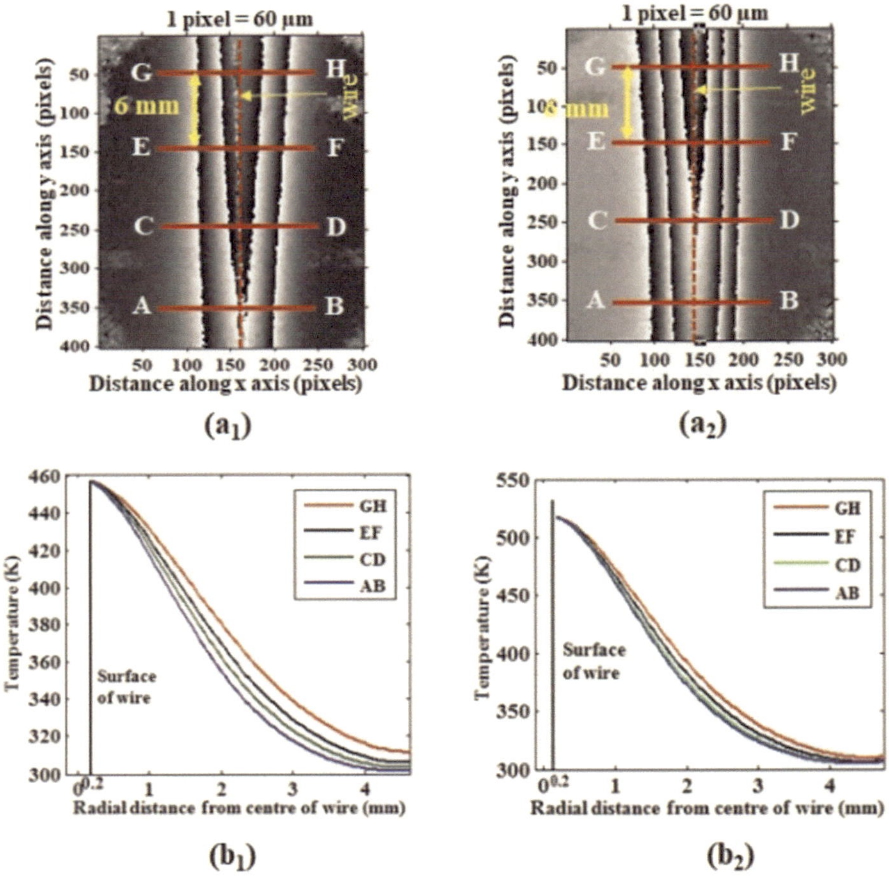

Figure 4.
Wrapped phase difference map of heated air and ambient air when applied power across the wire is (a_1) P = 37 W, (a_2) P = 70 W; (b_1–b_2) temperature profiles corresponding to (a_1) and (a_2) at four different heights of wire.

Experimental results corresponding to power P = 70 W and P = 37 W are shown in **Figure 4**. Experimentally calculated values of "h_c" along the surface of wire at four different heights for three different heating conditions are compared in **Table 1**. The results indicate that "h_c" decreases as height increases from bottom to top of wire. The reason for the increasing value of "h_c" is the broadening of isothermal line around the wire. Experimentally measured values are within range as given in [14]. Results indicate that measured values of h_c increases as the power applied across of wire of equal length and diameter increases.

A comparison of temperature measured by DHI and thermocouple is shown in **Figure 5**, and a good agreement is observed between the two measurements. Deviations in the temperature measured by DHI and thermocouple are found to be within 0.82–5.2%. The value of "h_c" measured with thermocouple is 13.9 W/(m^2 K), while with DHI is 13.30 W/(m^2 K). The difference between the two values is 4.31%.

Further, an experiment was conducted on wires of different diameters (d_1 = 0.4 mm, d_2 = 0.6 mm, d_3 = 0.78 mm) of equal length with a constant power (70 W) applied across the wire. Results are shown in **Figure 6**. A comparison of experimentally calculated values of local "h_c" and Q(y) using DHI for wires of three

Height	Applied input power = 70 W, T_w = 517 K, wire diameter d_1 = 0.4 mm	Applied input power = 46.8 W, T_w = 477 K, wire diameter d_2 = 0.4 mm	Applied input power = 37 W, T_w = 457 K, wire diameter d_3 = 0.4 mm
	h_c (W/m^2 K)	h_c(W/m^2 K)	h_c (W/m^2 K)
Along AB	13.94	11.55	10.87
Along CD	13.30	10.50	9.56
Along EF	12.20	8.93	8.41
Along GH	10.84	8.60	6.92

Table 1.
Experimentally calculated values of local natural convective heat transfer coefficient for three heating conditions of wire.

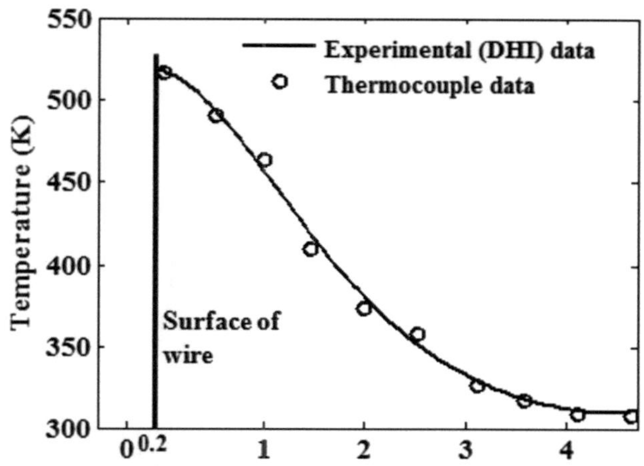

Figure 5.
Comparison between the temperature profile measured by DHI and thermocouple [3].

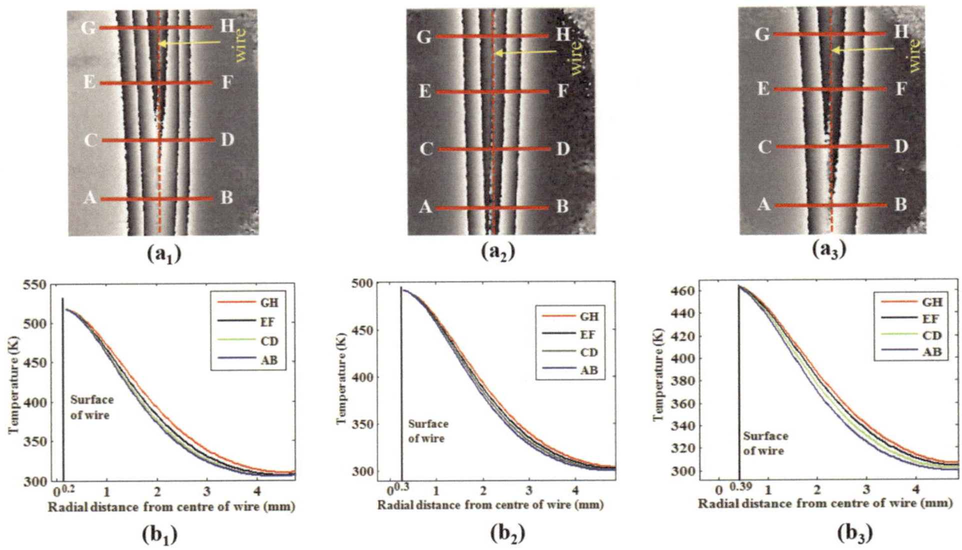

Figure 6.
(a_1–a_3) Wrapped phase difference map of heated air and ambient air around the wire for different diameters (d_1 = 0.4 mm, d_2 = 0.6 mm, d_3 = 0.78 mm) of wire with equal power applied across the wire of equal length; (b_1–b_3) temperature profiles corresponding to (a_1–a_3) at four different heights of wire.

different diameters are shown in **Table 2**. Experimentally measured values are within range as given in literature [14]. Results indicate that measured values of local natural "h_c" and Q(y) decreases as the diameter of wire increases. The reason for this decreasing trend of "h_c" and Q(y) is that when same electrical power is applied across the wire of different diameters of same length, surface temperature of wire decreases as diameter of wire increases. This happens because of increasing surface area of wire. The thermal conductivity "K_s" of air depends on temperature and it decreases as surface temperature of wire (T_s) decreases. The temperature gradient also decreases because of its dependence on the temperature difference (T_s-T_o). Both "h_c" as well as "$Q(y)$" depend on the value of temperature gradient normal to surface of wire, and K_s. Hence, local natural "h_c", and Q(y) decreases. **Table 2** also shows a decreasing trend in the values of local "h_c", and $Q(y)$ with the increase in height from the bottom of wire.

Height	Applied input power = 70 W, T_w = 517 K, wire diameter d_1 = 0.4 mm		Applied input power = 70 W, T_w = 491 K, wire diameter d_2 = 0.6 mm		Applied input power = 70 W, T_w = 465 K, wire diameter d_3 = 0.78 mm	
	h_c (W/m^2 K)	Q (W/m^2)	h_c (W/m^2 K)	Q (W/m^2)	h_c (W/m^2 K)	Q (W/m^2)
Along AB	13.94	3136.5	12.38	2463.62	10.28	1778.44
Along CD	13.3	2992.5	10.78	2145	8.80	1522
Along EF	12.2	2745	10.12	2013	8.37	1448
Along GH	10.84	2439	9.37	1864	7.72	1335.56

Table 2.
Calculated values of convective heat transfer coefficient and local heat flux using digital holographic interferometry (DHI) for wires of three different diameters.

3.2 Heat Flow performances of heat sinks

Electronics circuits, components/devices are highly demanded in every aspect of modern world ranging from toys, appliances to high performance computers. In recent, with increasing demand in processing, manufacturer of electronic circuits/ equipment are forced to increase the performance and functionality of electronics components/ICs chips in addition to reducing their sizes. In high performance electronic components /ICs/computers, rate of heat generation is comparatively high. This elevated heat generation from high performance electronic components/ devices produces high temperature level in the electronics components/devices, which may lead to their failure due to overheating. Operating temperature of high performance components/devices/computers plays a vital role in maintaining trouble free functioning. So, there is requirement to keep these electronic devices in reasonable temperature range suitable for their functioning by enhancing the transfer of generated heat. The heat sink as a source of heat transfer plays an important role in effective thermal management strategies, which makes the optimization of heat sink design a useful area of investigation. Heat sink are used in effective thermal management for increased reliability and proper functioning of such devices, while micro-channel heat sinks are used in miniaturized heat transport systems like compact heat exchangers, fuel cell powered mechanisms, components of advanced propulsion systems, micro-electromechanical systems (MEMS) applications, and cooling system for fusion reactor, rocket nozzles, avionics, hybrid vehicle power electronics and systems, etc. [3, 54, 56, 85]. The temperature distribution surrounding the heat sink must be measured for investigating the heat flow performance of heat sinks.

This section presents the investigations about heat flow performance from plate fin heat sink using LLFTDHI. Visual inspection of reconstructed phase difference maps of heated air and ambient air around the heat sink provide qualitative information about the variation of temperature and heat dissipation process. Quantitative information of temperature distribution is obtained from the relationship between the digitally reconstructed phase difference map of ambient air and heated air. The effect of fin spacing on the heat flow performance of heat sink is experimentally studied in case of natural heat convection. From experimental data, heat transfer parameters such as local heat flux and convective heat transfer coefficients are also calculated. Same LLFTDHI set-up is used in this application. **Figure 7** shows a photograph of DHI set-up used to study the heat flow performance of plate fin heat sink. In experiment, aluminum heat sink is attached with a load resistor in its bottom to heat the heat sink. A known value of voltage and current (5 V and 2.4 A) is applied across the load resistor to heat the heat sink.

During the heating process, several digital holograms are recorded at different time interval. Reconstructed wrapped phase difference maps of hot and ambient air surrounding the heat sink at different times are shown in **Figure 8(a)–8(h)**. Wrapped phase difference maps at time 75 and 90 min. are same. It indicates that that after 75 minute temperature surrounding the heat sink has reached in steady state. It is assumed that the refractive index of air along the direction of propagation of light ray is uniform, then the relationship between the unwrapped phase difference data $\Delta\varphi(x, y)$ and refractive index change $\Delta n(x, y)$ can be expressed as [54].

$$\Delta\varphi(x, y) = \frac{2\pi}{\lambda} L \Delta n(x, y) = \frac{2\pi}{\lambda} L[n(x, y) - n_o] \tag{11}$$

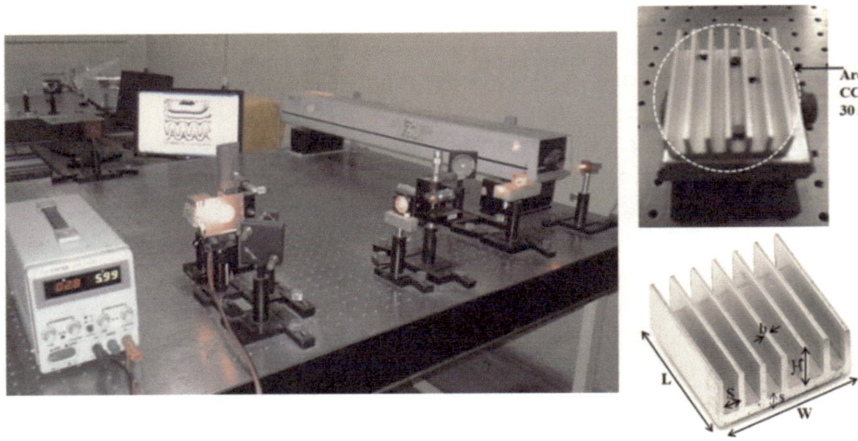

Figure 7.
Photograph of experimental setup, and heat sink with its dimensional representations.

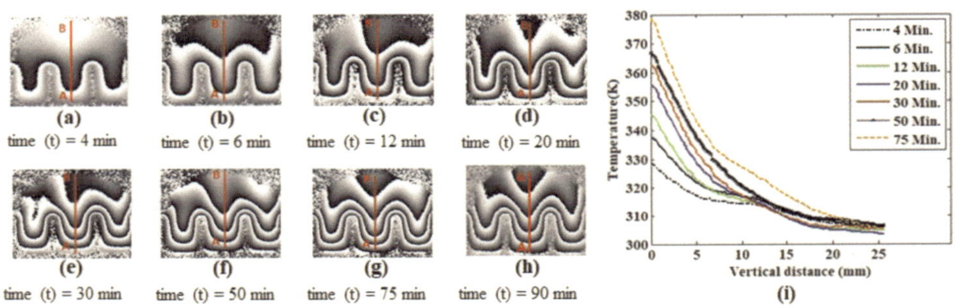

Figure 8.
(a)–(h) Wrapped phase difference maps of air surrounding the heat sink at different times during heating process, (i) temperature profile along line AB at different times in vertical direction corresponding to ***Figure 8(a)–(g).***

where "L" is the geometrical path length that the laser light travels in temperature field, "λ" is the wavelength of laser light, and n_0 is the refractive index of ambient air. Refractive index of hot air $n(x, y)$ can be related to temperature distribution $T(x,y)$ using Gladstone-Dale relation and ideal gas equation through a relation $T(x, y) = \frac{uPK}{R[n(x,\ y)-1]}$ [54]. Here "u" is the molar mass of air, "P" is atmospheric pressure, K is Gladstone–Dale constant, and R is universal gas constant. Substituting this relation in Eq. (11), temperature distribution can be obtained from phase difference map as [54]

$$T(x, y) = \left[\frac{\Delta\varphi(x,\ y)\lambda R}{2\pi LuPK} + \frac{1}{T_0}\right]^{-1} \tag{12}$$

In Eq. (12), T_0 is the ambient temperature. Thus, one can calculate the temperature distribution by knowing T_o and phase difference distribution $\Delta\varphi(x, y)$. In the experiment, ambient temperature T_0 was measured using K type (Chromel–Alumel) thermocouple with multi-logger. Ambient temperature was 305.5 K. Temperature distribution was reconstructed from the phase difference map using Eq. (12). **Figure 8 (i)** shows the reconstructed temperature profiles along the line AB at different times. Same trend was observed in the temperature profiles drawn along the central line AB at different times. Temperature gradient was observed to be maximum in steady state. It reveals that heat dissipation capability of heat sink is maximum in the steady state.

Further an experiment was conducted to study the effect of channel width/fin spacing (S) on heat flow performance of the plate fin heat sink. For this purpose, heat sinks with different fin spacing (Number of channels = 6, S_1 = 3.25 mm; Number of channels = 5, S_2 = 5 mm; Number of channels = 3, S_3 = 9 mm) were chosen and all other dimensional parameters of heat sink were kept same. An equal power of P = 16.8 W was applied across the load resistor for all the three heat sinks. In steady state of hot air, experimentally reconstructed phase difference maps of hot air and ambient air for the heat sinks of three different fin spacing are shown in **Figure 9(a_1)–9(a_3)**. The reconstructed 2D temperature distribution around the heat sink of three channel widths corresponding to **Figure 9(a_1)–9(a_3)** are shown in **Figure 9(b_1)–9(b_3)**. The color variations shown in **Figure 9(b_1)–9(b_3)** indicate that the temperature distribution/variation between the adjacent fins. Temperature is observed to be almost constant (i.e., almost negligible temperature gradient) in the case of heat sink of smaller fin spacing (S_1 = 3.25 mm), and temperature gradient is the highest in case of the heat sink of largest channel width (S_3 = 9 mm). It indicates that heat is accumulated between the adjacent fins of a heat sink with smaller fin spacing and it is more difficult to diffuse the heat than that of heat sinks with larger fin spacing in the case of natural convection.

Temperature profiles along the line AB for the heat sinks of different fin spacing are shown in **Figure 9(c)**. The temperature gradient along AB is observed to be highest for heat sink of wider fin spacing/channel width S_3 = 9 mm than that of the heat sinks of lower channel widths (i.e., S_1 = 3.25 mm, and S_2 = 5 mm). This is the reason why natural local convective "Q" and "h_c" increases with increase in fin spacing of heat sink.

Table 3 shows a comparison of experimentally calculated values of local convective "Q" and "h_c" for the heat sinks of three different channel widths. From the **Table 3**, it can be observed that heat sink with wider channel width S_3 = 9 mm has higher value of local convective "Q" and "h_c" than the heat sinks of lower channel widths (i.e., S_2 = 5 mm and S_1 = 3.25 mm). Therefore, for better heat dissipation, there

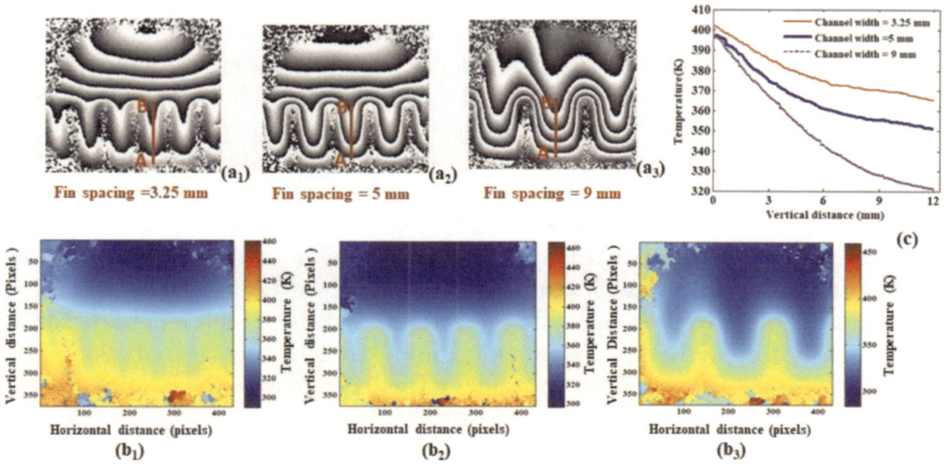

Figure 9.
(a_1–a_3) Reconstructed wrapped phase difference map of air field when S_1 = 3.25 mm, S_2 = 5 mm, and S_3 = 9 mm, respectively; (b_1–b_3) reconstructed temperature distribution corresponding to (a_1–a_3); (c) temperature profiles along the line AB corresponding to (a_1–a_3), respectively.

Fin spacing or channel width S (mm)	Dimensions of heat sink Length L = 50 mm, Width W = 32 mm, Height H = 15 mm, base thickness h = 3.5 mm, fin thickness b = 1.5 mm Room temperature = 32.5°C, applied power across load resistor = 16.8 W			
	Calculated base temperature T_b (K)	Thermocouple value of T_b (K)	Local convective heat flux (w/m^2) $Q(x) = -K_s\left(\frac{dT}{dy}\right)_s$	Experimentally calculated h_c W/(m^2 K)
S_1 = 3.25	402	404	182.9	1.89
S_2 = 5	397.2	400	243.6	2.64
S_3 = 9	396.3	399	305	3.34

Table 3.
Experimentally calculated values of natural local heat flux and local convective heat transfer coefficient for heat sinks with different channel widths [54].

should be an optimization between the no. of channels and fin spacing so that heat does not accumulate inside the fins.

4. Conclusions

In conclusion, this chapter demonstrates LLFTDHI technique for the investigation of temperature distribution, and measurement of heat transfer parameters around the heated industrial objects such as wires/conductors and heat sinks. The LLFT configuration of DHI offers following advantages (1) due to a simpler hologram reconstruction algorithm, it provides faster reconstruction than other counterpart methods such as Fresnel and Convolution approach, (2) it can be implement almost in real time, and (3) LLFTDH provides improved lateral resolution in the reconstructed image because entire spatial bandwidth can be utilized for complete evaluation of information from sensor by optimizing the ratio of object size and distance between object and sensor. The LLFTDHI system is investigated for the following scientific and industrial applications (1) Measurement of temperature profile, natural local convective heat transfer coefficient and heat flux along the surface of electrically heated wire. These investigations are helpful to study the heat transfer characteristics of wires/conductors, and finding ideal insulation properties of wires/conductors. (2) The heat dissipation performances of heat sinks. Heat dissipation capabilities of heat sinks are determined by qualitative and quantitative estimations of temperature distribution and heat transfer parameters. These investigations can be helpful in making the proper choice of the heat sink in the electronic circuit, and are also required in making electronic circuits of high reliability for critical industrial as well as for space application. Based on investigated applications, it can be concluded that LLFTDHI provides a feasible and effective method for temperature measurements and heat transfer studies.

Author details

Varun Kumar[1,2]* and Chandra Shakher[1]*

1 Laser Applications and Holographic Laboratory, Centre for Sensors, Instrumentation and Cyber Physical System Engineering (SeNSE), Indian Institute of Technology Delhi, New Delhi, India

2 Department of Physics and Photonics Science, National Institute of Technology Hamirpur, Himachal Pradesh, India

*Address all correspondence to: varunphy@gmail.com and cshakher@sense.iitd.ac.in

References

[1] Hauf W, Grigull U. Optical methods in heat transfer. In: Hartnett JP, Irvine TF. Jr. (eds.). Advances in Heat Transfer. Vol. 6. New York: Elsevier; 1970 Jan 1. pp. 133-366

[2] Naylor D. Recent developments in the measurement of convective heat transfer rates by laser interferometry. International Journal of Heat and Fluid Flow. Jun 1 2003;**24**(3): 345-355

[3] Kumar V. Digital holographic interferometry for the measurement of temperature profile, heat dissipation and contouring [Doctoral dissertation], IIT Delhi; 2016

[4] Norris JK. Interferometric measurement of free convective heat transfer coefficients in the transitional and turbulent regimes [Doctoral dissertation], Georgia Institute of Technology. 1973

[5] Wilkie D, Fisher SA. Measurement of temperature by Mach-Zehnder interferometry. Proceedings of the Institution of Mechanical Engineers. Jun 1963;**178**(1):461-470

[6] Goldstein RJ. Optical measurement of temperature. In: Eckert ERG, Goldstein RJ, editors. Measurement Techniques in Heat Transfer, AGARDograph, no. 130. Slough, London: Technivision Services [distributed by Technical Press]; 1970 Jan 1. pp. 177-228

[7] Kennard RB. Temperature distribution and heat flux in air by interferometry. Temperature: Its Measurement and Control in Science and Industry. 1941;**1**:685-706

[8] Eckert ER, Soehngen EE. Studies on heat transfer in laminar free convection with the Zehnder-Mach interferometer. Air Materiel Command Wright-Patterson AFB OH; 1948 Dec 27.

[9] Goldstein RJ, Eckert ER. The steady and transient free convection boundary layer on a uniformly heated vertical plate. International Journal of Heat and Mass Transfer. Aug 1 1960;**1**(2–3): 208-218

[10] Ostroumov GA. Unsteady heat convection near a horizontal cylinder. Soviet Physics-Technical Physics. Jan 1 1956;**1**(12):2627-2641

[11] Vest CM, Lawson LM. Onset of convection near a suddenly heated horizontal wire. International Journal of Heat and Mass Transfer. 1972;**15**: 1281-1283

[12] Duarte DN. Direct temperature gradient measurement using interferometry. Experimental Heat Transfer. Oct 1 1999;**12**(4):279-294

[13] Goharkhah M, Ashjaee M, Madanipour K. Investigation of the accuracy of different methods of interferogram analysis for calculation of local free convection heat transfer coefficient on axisymmetric objects. Experimental Thermal and Fluid Science. 2009 Nov 1;**33**(8):1188-1196

[14] Fatehi S, Madanipour K, Parvin P. Measurement of convective heat transfer coefficient around a vertical hot wire in different heights using Michelson interferometer based on Abel transform. Applied Physics Express. 2011;**1**:124-132

[15] Kwak CE, Song TH. Natural convection around horizontal downward-facing plate with rectangular

grooves: experiments and numerical simulations. International Journal of Heat and Mass Transfer. Mar 1 2000; **43**(5):825-838

[16] Wright JL, Jin H, Hollands KG, Naylor D. Flow visualization of natural convection in a tall, air-filled vertical cavity. International Journal of Heat and Mass Transfer. Mar 1 2006;**49**(5–6): 889-904

[17] Sajith V, Sobhan CB. Digital interferometric measurement of forced convection heat transfer in a miniature rectangular channel. Experimental Heat Transfer. Oct 2 2008; **21**(4):314-333

[18] Sajith V, Sobhan CB. Characterization of heat dissipation from a microprocessor chip using digital interferometry. IEEE Transactions on Components, Packaging and Manufacturing Technology. Jun 19 2012; **2**(8):1298-1306

[19] Ahadi A, Saghir MZ. An extensive heat transfer analysis using Mach Zehnder interferometry during thermodiffusion experiment on board the International Space Station. Applied Thermal Engineering. Jan 25 2014;**62**(2): 351-364

[20] Goldstein RJ. Interferometer for aerodynamic and heat transfer measurements. Review of Scientific Instruments. Oct 1965;**36**(10): 1408-1410

[21] Haridas D, Reddy GR, Sobhan CB. Digital interferometric measurement of forced convection fields in compact channels. International Journal of Optomechatronics. Jan 2 2015;**9**(1): 9-34

[22] Vest CM. Holographic Interferometry. New York: Wiley; 1979

[23] Farrell PV, Springer GS, Vest CM. Heterodyne holographic interferometry: concentration and temperature measurements in gas mixtures. Applied Optics. May 1 1982;**21**(9):1624-1627

[24] Reuss DL. Temperature measurements in a radially symmetric flame using holographic interferometry. Combustion and Flame. Jan 1 1983;**49**(1–3):207-219

[25] Sweeney DW, Vest CM. Measurement of three-dimensional temperature fields above heated surfaces by holographic interferometry. International Journal of Heat and Mass Transfer. Dec 1 1974;**17**(12):1443-1454

[26] Mayinger F, editor. Optical Measurements: Techniques and Applications. Berlin Heidelberg GmbH, Germany: Springer; 1994

[27] Herman C, Kang E. Comparative evaluation of three heat transfer enhancement strategies in a grooved channel. Heat and Mass Transfer. Sep 2001;**37**(6):563-575

[28] Ambrosini D, Paoletti D, Tanda G. Measurement of free convection heat transfer coefficient along a rib-roughened vertical surface. In 16th International Symposium on Transport Phenomena. Prague, Czech Republic: ISTP; 2005

[29] Ambrosini D, Tanda G. Comparative measurements of natural convection heat transfer in channels by holographic interferometry and schlieren. European Journal of Physics. Dec 22 2005;**27**(1):159

[30] Herraez JV, Belda R. A study of free convection in air around horizontal cylinders of different diameters based on holographic interferometry. Temperature field equations and heat

transfer coefficients. International Journal of Thermal Sciences. Mar 1 2002; **41**(3):261-267

[31] Tauscher R, Mayinger F. Visualization of flow temperature fields by holographic interferometry – optimization of compact heat exchangers. Honolulu, USA: Proceedings of PSFVIP-2; 1999 May 16

[32] Herman C, Kang E. Experimental visualization of temperature fields and study of heat transfer enhancement in oscillatory flow in a grooved channel. Heat and Mass Transfer. Jan 2001;**37**(1): 87-99

[33] Manickam S, Dhir V. Holographic interferometric study of heat transfer associated with a single vapor bubble sliding along a downward-facing heater surface. In: Heat Transfer Summer Conference 1 Jan 2003 (Vol. 36940). Las Vegas, Nevada, USA: HTSC; pp. 317–327

[34] Spagnolo GS, Ambrosini D. Sandwich holography for determining the convective heat transfer coefficient. Journal of Optics A: Pure and Applied Optics. 2000;**2**(1):39

[35] Karaminejad S, Askari MH, Ashjaee M. Temperature field investigation of hydrogen/air and syngas/air axisymmetric laminar flames using Mach–Zehnder interferometry. Applied Optics. Jun 20 2018;**57**(18): 5057-5067

[36] Wang Q, Meng H, et al. Optical fiber temperature sensor based on a Mach-Zehnder interferometer with single-mode-thin-core-single-mode fiber structure. Review of Scientific Instruments. Jan 1 2020;**91**(1):015006

[37] Zhu J, Huang S, Lv W, Zhou H. Study on the measurement of temperature field using laser holographic interferometry. Frontiers in Energy. Mar 2011;**5**(1):120-124

[38] Keren E, Bar-Ziv E, Glatt I, Kafri O. Measurements of temperature distribution of flames by moiré deflectometry. Applied Optics. Dec 15 1981;**20**(24):4263-4266

[39] Shakher C, Nirala AK. A review on refractive index and temperature profile measurements using laser-based interferometric techniques. Optics and Lasers in Engineering. Jun 1 1999;**31**(6): 455-491

[40] Stella A, Guj G, Giammartini S. Measurement of axisymmetric temperature fields using reference beam and shearing interferometry for application to flames. Experiments in Fluids. Jul 2000;**29**(1):1-2

[41] Singh P, Faridi MS, Shakher C. Measurement of temperature of an axisymmetric flame using shearing interferometry and Fourier fringe analysis technique. Optical Engineering. Feb 2004;**43**(2):387-392

[42] Shakher C, Daniel AP. Talbot interferometer with circular gratings for the measurement of temperature in axisymmetric gaseous flames. Applied Optics. Sep 1 1994;**33**(25): 6068-6072

[43] Khramtsov PP, Penyazkov OG, Shatan IN. Temperature measurements in an axisymmetric methane–air flame using Talbot images. Experiments in Fluids. Feb 2015;**56**(2):1-9

[44] Thakur M, Vyas AL, Shakher C. Measurement of temperature and temperature profile of an axisymmetric gaseous flames using Lau phase interferometer with linear gratings. Optics and Lasers in Engineering. Oct 1 2001;**36**(4):373-380

[45] Pandey PK, Kumar M, Kumar V, Shakher C. Measurement of temperature and temperature profile of wick stabilized micro diffusion flame under the effect of magnetic field using digital speckle pattern interferometry. Optical Engineering. Jan 2017;**56**(1):014106

[46] Kumar V, Shakher C. Measurement of temperature and temperature profile of candle flame using holo-shear lens and Fourier fringe analysis technique. Optical Engineering. Aug 2015;**54**(8): 084105

[47] Kumar V, Rastogi V, Agarwal S, Shakher C. Holo-shear lens based interferometer for measurement of temperature distribution and fluctuation of temperature in micro flame. In: Holography, Diffractive Optics, and Applications IX 18 November 2019; (Vol. 11188). Hangzhou, China: SPIE; pp. 244–250

[48] Kumar V, Rastogi V, Agarwal S, Shakher C. Investigation of temperature profile and temperature stability of micro diffusion flame under the influence of magnetic field by use of a holo-shear lens-based interferometer. Optical Engineering. Jun 2020;**59**(6): 064107

[49] Sharma S, Sheoran G, Shakher C. Digital holographic interferometry for measurement of temperature in axisymmetric flames. Applied Optics. 2012;**51**(16):3228-3235

[50] Kumar V, Kumar M, Sharma S, Shakher C. Measurement of temperature profile around heated wire using digital holography. In: Osten W, editor. Fringe. Berlin, Heidelberg: Springer; 2013. pp. 497-502

[51] Doleček R, Psota P, Lédl V, Vít T, Václavík J, Kopecký V. General temperature field measurement by digital holography. Applied Optics. Jan 1 2013;**52**(1):A319-A325

[52] Kumar V, Kumar M, Shakher C. Measurement of natural convective heat transfer coefficient along the surface of a heated wire using digital holographic interferometry. Applied Optics. Sep 20 2014;**53**(27):G74-G83

[53] Kumar V, Kumar M, Shakher C. Temperature measurement of axi-symmetric butane diffusion flame under the influence of upward decreasing gradient magnetic field using digital holographic interferometry. In SPECKLE 2015: VI International Conference on Speckle Metrology 24 August 2015; (Vol. 9660). Guanajuato, Mexico: SPIE; pp. 99–106

[54] Kumar V, Shakher C. Study of heat dissipation process from heat sink using lensless Fourier transform digital holographic interferometry. Applied Optics. Feb 20 2015;**54**(6):1257-1266

[55] Agarwal S, Kumar V, Shakher C. Temperature measurement of wick stabilized micro diffusion flame under the influence of magnetic field using digital holographic interferometry. Optics and Lasers in Engineering. Mar 1 2018;**102**:161-169

[56] Kumar V, Shakher C. Visual and quantitative investigation on heat flow performance from heat sinks using digital holographic interferometer. In: Holography, Diffractive Optics, and Applications IX 18 November 2019; (Vol. 11188). Hangzhou, China: SPIE; pp. 230–237

[57] Kumar V, Shakher C. Digital holography for local heat flux measurement along the surface of heated wire. In CLEO: QELS_Fundamental Science 5 May 2019; San Jose, California,

United States: Optica Publishing Group; pp. JW2A-1

[58] Rastogi V, Agarwal S, Kumar V, Shakher C. Holographic optical element based digital holographic interferometer for the study of macro flames, micro flames and their temperature instability. Optics and Lasers in Engineering. Nov 1 2019;**122**:29-36

[59] Quan C, Thakur M, Tay CJ. Lau phase interferometry with a vibrating object. Optik. Jan 15 2006;**117**(1):9-14

[60] Kaufmann GH, Galizzi GE. Speckle noise reduction in television holography fringes using wavelet thresholding. Optical Engineering. Jan 1996;**35**(1): 9-14

[61] Sharma A, Sheoran G, Jaffery ZA. Improvement of signal-to-noise ratio in digital holography using wavelet transform. Optics and Lasers in Engineering. Jan 1 2008;**46**(1):42-47

[62] Schnars U, Jüptner WP. Digital recording and numerical reconstruction of holograms. Measurement Science & Technology. Aug 7 2002;**13**(9):R85

[63] Schnars U, Jüeptner W. Digital Holography: Digital Hologram Recording, Numerical Reconstruction, and Related Techniques. Berlin Heidelberg GmbH, Germany: Springer; 2005

[64] Wagner C, Seebacher S, Osten W, Jüptner W. Digital recording and numerical reconstruction of lensless Fourier holograms in optical metrology. Applied Optics. Aug 1 1999;**38**(22): 4812-4820

[65] Jeong K, Turek JJ, Nolte DD. Speckle fluctuation spectroscopy of intracellular motion in living tissue using coherence-domain digital holography. Journal of Biomedical Optics. May 2010;**15**(3): 030514

[66] Yu X, Hong J, Liu C, Kim MK. Review of digital holographic microscopy for three-dimensional profiling and tracking. Optical Engineering. Apr 2014;**53**(11):112306

[67] Wetzel M, Herman C. Accurate measurement of high-speed, unsteady temperature fields by holographic interferometry in the presence of periodic pressure variations. Measurement Science & Technology. Jun 1 1998;**9**(6):939

[68] Hossain MM, Shakher C. Temperature measurement in laminar free convective flow using digital holography. Applied Optics. Apr 1 2009; **48**(10):1869-1877

[69] Imbe M. Radiometric temperature measurement by incoherent digital holography. Applied Optics. Feb 10 2019;**58**(5):A82-A89

[70] Sheoran G, Anand A, Shakher C. Lensless Fourier transform digital holographic interferometer for diffusivity measurement of miscible transparent liquids. Review of Scientific Instruments. May 6 2009; **80**(5):053106

[71] Agarwal S, Kumar M, Kumar V, Shakher C. Analysis of alcohol-water diffusion process using digital holographic interferometry. In: SPECKLE 2015: VI International Conference on Speckle Metrology 24 August 2015; (Vol. 9660). Guanajuato, Mexico: SPIE; pp. 135–141

[72] Desse JM, Picart P, Tankam P. Digital three-color holographic interferometry for flow analysis. Optics Express. Apr 14 2008;**16**(8):5471-5480

[73] Gopalan B, Katz J. Turbulent shearing of crude oil mixed with dispersants generates long microthreads and microdroplets. Physical Review Letters. Feb 1 2010;**104**(5):054501

[74] Hossain MM, Mehta DS, Shakher C. Refractive index determination: an application of lensless Fourier digital holography. Optical Engineering. Oct 2006;**45**(10):106203

[75] Wang Q, Zhao J, Jiao X, Di J, Jiang H. Visual and quantitative measurement of the temperature distribution of heat conduction process in glass based on digital holographic interferometry. Journal of Applied Physics. May 1 2012; **111**(9):093111

[76] Zhang Y, Zhao J, Di J, Jiang H, Wang Q, Wang J, et al. Real-time monitoring of the solution concentration variation during the crystallization process of protein-lysozyme by using digital holographic interferometry. Optics Express. Jul 30 2012;**20**(16): 18415-18421

[77] Desse JM, Picart P. Stochastic digital holography for visualizing inside strongly refracting transparent objects. Applied Optics. Jan 1 2015;**54**(1):A1-A8

[78] Rastogi V, Kumar V, Dubey SK, Khan GS, Shakher C. Noncontact temperature measurement of human hand skin using volume phase holographic optical element based digital holographic interferometer. Optics and Lasers in Engineering. Apr 1 2022;**151**: 106886

[79] Kris TM. Frequency analysis of digital holography with reconstruction by convolution. Optical Engineering. 2002;**41**(8):1829-1839

[80] Matsushima K, Schimmel H, Wyrowski F. Fast calculation method for optical diffraction on tilted planes by use of the angular spectrum of plane waves. Journal of the Optical Society of America A. Sep 1 2003;**20**(9):1755-1762

[81] Kumar V, Khan GS, Shakher C. Phase contrast imaging of red blood cells using digital holographic interferometric microscope. In Third International Conference on Applications of Optics and Photonics 22 August 2017; (Vol. 10453). Faro, Portugal: SPIE; pp. 678–684

[82] Kumar V, Shakher C. Testing of micro-optics using digital holographic interferometric microscopy. In 2016 4th International Conference on Photonics, Optics and Laser Technology (PHOTOPTICS) 27 February 2016; Rome, Italy: IEEE; pp. 1–6

[83] Kumar V, Shakher C. Characterization of micro-lenslet array using digital holographic interferometric microscope. In: Ribeiro PA, Raposo M, editors. Optics, Photonics and Laser Technology. Springer Nature, Switzerland AG: Springer, Champions; 2018. pp. 25-45

[84] Goldstein RM, Zebker HA, Werner CL. Satellite radar interferometry: Two-dimensional phase unwrapping. Radio Science. Jul 1988; **23**(4):713-720

[85] Mudawar I. Two-phase microchannel heat sinks: theory, applications, and limitations. Journal of Electronic Packaging. 2011;**133**(4): 041002

[86] Farrell PV, Hofeldt DL. Temperature measurement in gases using speckle photography. Applied Optics. Apr 1 1984;**23**(7):1055-1059

Chapter 15

Two-Rail Photonic Qubit Utilizing the Quantum Holographic Imaging Idea

Kamil Wereszczyński and Krzysztof Cyran

Abstract

We present the novel approach to physical implementation of qubits with the technology of photonic chips. Proposed multi-rail qubit model, called QBell, utilizes hyper-entanglement to work in Decoherence Free Subspace on physical layer. This makes this solution robust and can result in increasing fidelity of quantum circuit used in this model. We elaborate the two-rail case. We define the QBell and discuss its internal structure. We construct also one- and two-qubit gates to make the model comprehensive and ready to implement. Proposed model utilizes the early-stage ideas for optical quantum computation, but by using the polarization and position entanglement as the resource of computation allows to avoid the general problem of them, like heralded photon technique. The technology of photonic chips allows to brake other limitations that are pointed in the text. The presented model was inspired by quantum holographic imaging and uses the holographic technique for implementing the z-rotation operation. The final product will be the photonic quantum processor using multi-rail qubits. It will find the application in many domains (e.g., medical) on earth and in the space.

Keywords: quantum computing, quantum holographic imaging, qubit, Bell states, photonic chips, quantum gates, robust quantum systems, photonic quantum processors

1. Introduction

Photonic qubits (quantum bits), which means qubits utilizing the quantum optical phenomena, was widely elaborated at the turn of XX and XXI centuries. The name *photonic* is a modern name, since in that time authors tell about optical qubits. However, this name belongs currently to ion-trapped qubits using standard energetic structure of atoms, e.g., Bruzewicz et al. [1]. The first one-qubit operation was described by Simon and Mukunda in [2] in the context of transformation between polarization and position logic of qubits. The early studies of this area assumed that the photonic qubits should utilize the linear optics phenomena, since the general name of such solutions is LOQC (Linear Optics Quantum Computing). The first *protocol* of quantum computing using optical circuit was elaborated by Knill, Laflamme, and

Milburn and named after author's initials *KLM* protocol, published in several articles [3–5]. Finally, Myers and Laflamme published a review and tutorial of *KLM theory* [6]. The fundamental problem with this scheme is that it uses as the basic building block so-called *nonlinear sign* NS gate. It is *probabilistic* gate, which means that it returns the correct result not always but with some probability depending on the specific implementation, e.g., Knill, Laflamme, and Milburn implementation success probability is only equal to $\frac{3-\sqrt{2}}{7} \sim 0.2265$! To inform which result is correct, one has to measure additional output. When on this output, there appears a photon, the result is correct. Currently such photons are called *heralded* [7]. Knill in [8] proved theoretically that the upper bounds of NS gate success probability are equal to 1/2 and for conditionally NS gate with two modes equal to 3/4. The problem with such "heralded" gates is that in case of high frequency of changes between correct and not correct results, the detection of the moment of correct result is very hard. Hence, the fidelity of such gates was quite low—at the level of 84%—Gasparoni et al. [9].

There was some improving KLM scheme protocols. The main improvement was include to linear systems nonlinear elements for generating the numbers of entangled photons. Yoran-Reznik [10] proposed a protocol based on "entanglement chains." This is important effort for us, since it was the first time, when hyper-entanglement was used for computation. However, he still uses the probabilistic gates and heralded photons. There is also a class of protocols based on the *cluster states*, also known as *multipartite entanglement* [11]. It was, e.g., Nielsen [12] or Browne-Rudolph [13] protocol.

All those ideas were yet treated as "demonstratory" or even "simulation" of quantum computers. We think that we have no right to name it "simulation" since in those circuits there are real quantum effect and such a protocols are widely used in quantum security and quantum communication solutions. The problem that there was only demonstration of quantum computers lays in two facts. Firstly, they are not programmable—if you have new algorithm you have to rebuild the setup on the optical table, which is not practical in production application. Secondly, the number of devices needed to grow exponentially with the growth of qubit number. This is even harder limitation in practical solutions. Therefore, at about 2005, the number of works on linear optics application for quantum computing decreased rapidly, also by arising realistic superconducting and ion-trap solution in the area of quantum processors.

However, the idea of using quantum optics for computation is very tempting because it does not need extreme cooling like in mentioned superconducting solution (cooling below 1 K) or ion-trap (cooling to 10–35 K, depending on the components). The worst case for quantum dots-based optical current is temperature on the level of 213 K, which is much affordable. The most solution does not require cooling at all and potentially can work in any temperature. Therefore, when the tool for implementation appears, several solutions were developed. This tool is the technology of programmable photonic chips announced by Bogaerts et al. in [14]. There have been developed some quantum processors based on similar technology like Xanadu (Madsen et al. [15]) or QuiX (Taballione et al. [16], deGoede et al. [17]).

When we moving in the area of quantum holography, we have to pay attention on the nomenclature. That is because this notion is used for holographic data storage system [18] or even in psycholgy and social science [19]. In this chapter we will use the *quantum holography* notion as the tool for generating holographic images. There are two similar works of Defienne et al. [20] and Toepher et al. [21]. We will define works that inspired us in the elaboration of proposition of two-rail qubit, which will be described in more detail in Section 2. At this point it is enough to say that it uses the hyperentangled photons, like in Yoran-Roznik protocol of linear optics quantum computation.

Our **contribution** is the definition of the novel type of qubits that are consisting on more than one rail, to create more robust circuit with high fidelity. In this work we present the definition of two-rail qubit called *two-rail Bell's qubit, QBell*. We discuss its internal structure, define and show how to construct one qubit gates for QBell needed for generation of universal gate. We also show how to physically implement controlled gate. This all constructions avail physical implementation of such qubits, especially on the photonic chips.

Our **motivation** in long-term perspective is to create multi-rail robust qubits with high fidelity, which allows to create robust photonic quantum circuit, where it will be possible to use one gate multiple times. The possibility of creating programmable quantum photonic chips is already available. Together with multiple usage of gates they will break, in long term, the mentioned limitation of photonic quantum computing. This work is one first step on this road.

2. Quantum holographic imaging

Two-rail optical qubit proposed by us is inspired by quantum holographic imaging, described by Defienne et al. in [20]. Like in classical case, it relies on extracting the differences in phase of light reflected from the object being imaged. Unlike the classical method in presented technique, the hyperentanglement is utilized in the channels of polarization and space.

Deffiene uses the output of SPDC (Spontaneous Parametric Down Conversion) type I process, which produces the states $|VV\rangle + |HH\rangle$[1]. This formula describes the state, which is entangled in polarization channel, but there is no information about space entanglement. In quantum computing we pursue to narrow laser beam as possible, and we consider existing the photon on the "first" or "second" beam, therefore we can easily rewrite the state in four-slot bra or ket describing the state in Fock space. However, in the holographic imaging application, there are wider beams used, which illuminates the area of a given object and the group of pixels in the EMCCD camera playing the role of detector in the whole system. Therefore the position of the photon is given by its momentum $\mathbf{k} = (\mathbf{k_x}, \mathbf{k_y})$. The position entanglement means that if two photons have the same k_y component, one of them has k_x momentum component and the second one $-k_x$. we will denote this situation that one photon has $\mathbf{k}$ momentum while the second one $-\mathbf{k}$. Now, the hyperentangled state can be rewritten as: $|V\rangle_{\mathbf{k}}|V\rangle_{-\mathbf{k}} + |H\rangle_{\mathbf{k}}|H\rangle_{-\mathbf{k}}$. Since we mentioned about consideration of an area illuminated by a beam of hyperentangled photons, the final state equation is given by:

$$|A\rangle = \sum_{\mathbf{k}}[|V\rangle_{\mathbf{k}}|V\rangle_{-\mathbf{k}}+|H\rangle_{\mathbf{k}}|H\rangle_{-\mathbf{k}}] \qquad (1)$$

The procedure of obtaining the quantum holographic image is shown schematically on the **Figure 1**. The procedure starts with producing the hyperentangled state

[1] In fact this is not normalized state. The normalized should be divided by square root of 2. We will follow this not normalized notation in this section. However, in the sections describing the qubit notion, we will use the normalized states to be formal correct.

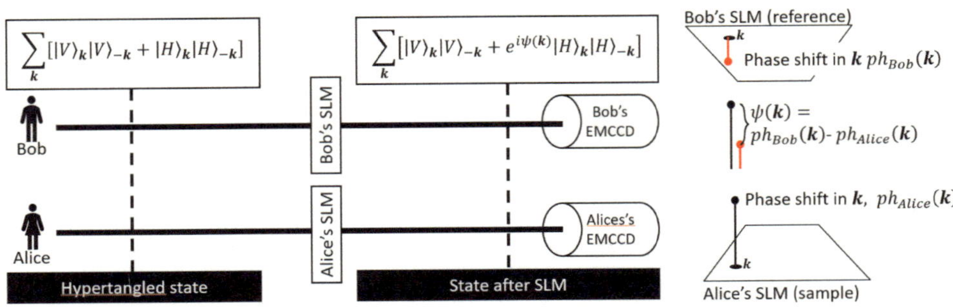

Figure 1.
The general scheme of quantum holography imaging. On the picture, there are shown the states before and after going through SLM (Spatial Light Modulator). On the right side there is schematically shown the meaning of $\psi(\mathbf{k})$ *function.*

$|A\rangle$, e.g., using the SPDC. Alice transmits its beam through the pretreated sample. On the scheme it is signed "Alice's SLM" (Spatial Light Modulator), since in Defienne experiment this device was used for creating the images in phase domain. However, it can be any sample that we would to examine, e.g., medical preparation. Bob transmits its beam through another SLM. It plays the role of phase correction device. Before the examination, instead of sample there should be placed a device that has the exactly the same phase in all places. Because of imperfection of the instruments on the Bob's and Alice's cameras, the pictures can differ. Using Bob's SLM we can mitigate these differences changing the phase shifts in all positions. We can consider this procedure as the setup calibration process. The pattern arised on the Bob's SLM we call *reference pattern*. Having the sample and reference pattern, we can determine the function $\psi(\mathbf{k})$, which is the difference between the phase shift of sample and reference SLM, which is shown on the right part of the scheme. The state after SLM is given by an equation:

$$|A'\rangle = \sum_{\mathbf{k}} \left[|V\rangle_{\mathbf{k}} |V\rangle_{-\mathbf{k}} + e^{i\psi(\mathbf{k})} |H\rangle_{\mathbf{k}} |H\rangle_{-\mathbf{k}} \right] \tag{2}$$

At the end the beams are detected after passing polariser rotated by 45 degrees. The intensity correlations are measured between $\mathbf{k}$-th pixel on Bob's camera and $-\mathbf{k}$ on Alice's one. The procedure is repeated several times for different homogeneous phase shift on Bob's SLM. In the Deffiene experiment, the iterations was made for phase shifts: $0,\pi/2,\pi$ and $3\pi/2$. Basing on those intensity correlation images, the original object can be reconstructed.

There are many possible application of such a technique, including medical (e.g, histopathology) preparations. Authors point the robustness of the system, which is very crucial for quantum computation as well.

In next sections we will show how to adopt this idea, especially the phase and polarization entanglement, in a quite similar way to the method described above.

3. Foundations of quantum computation

In the area of non-imaging optics, prominent place is taken by implementation of quantum computation. Quantum bit, called *qubit*, is the smallest piece of quantum information, just like bit is the realization of classical one. While bit can be in one of two states (0 or 1) during whole its life, qubit can be in the state that is superposition of 0 and 1. Two states, equivalent to 0 and 1 are denoted in bracket notation as $|0\rangle,|1\rangle$, as well as the qubit itself. Then the superposition for qubit $|q\rangle$ is described by equation: $|q\rangle = \alpha|0\rangle + \beta|1\rangle$, where $\alpha,\beta \in \mathbb{C}$ and $|\beta|^2 + |\beta|^2 = 1$, called *unit condition*. This representation of qubit is, generally speaking, a vector in Hilbert space[2]. At first glance, one can think it has four degrees of freedom (DoF), since four real numbers are needed to describe such system. However, if we include the unitary condition we can convince ourselves that: $\alpha = cos\left(\frac{\vartheta}{2}\right), \quad \beta = e^{i\phi} sin\left(\frac{\vartheta}{2}\right)$, thus it has two DoF: ϕ,ϑ. This formula is the general equation of qubit. It defines the representation of qubit on a *Bloch Sphere* in polar coordinates, considering unit length radius. On the north pole of the Bloch sphere, there is a representation of $|0\rangle$ and on the south pole of $|1\rangle$. Every other position is the superposition.

The bit state can be changed by a gate, which we identify as the logic (or Boolean) function. Similarly, the state of the qubit can be changed by a quantum gate, which is *operator* in Hilbert space. Because of unit condition, the gate has to be unitary, since unitary operators save the norm. There are plenty of different types of gate, e.g., rotation around x,y and z axis of Bloch sphere, X, Y, Z Pauli gates, phase shift gate, etc. However, for physical implementation ability to change the state by changing the coordinates on the Bloch Sphere is enough. There could be other set of operations depended on physical implementation, certainly. For example, IBM-Q uses rotation around the Z and X axis of Bloch sphere. Using such a rotation the general form of quantum gate can be constructed:

$$U(\theta, \varphi, \lambda) = \begin{bmatrix} cos\left(\frac{\theta}{2}\right) & -e^{i\lambda} sin\left(\frac{\theta}{2}\right) \\ e^{i\varphi} sin\left(\frac{\theta}{2}\right) & e^{i(\varphi+\lambda)} cos\left(\frac{\theta}{2}\right) \end{bmatrix} \tag{3}$$

The multiqubit states can be constructed by tensor product of single qubits. In that case the gates are the tensor products of one qubit gates as well. The tensor products create the interference, which is the second quantum phenomenon used for computation. Nevertheless, there are states that cannot be expressed by tensor product of single qubit states. Such a states are called *entangled state*. Along the tensor product the system $|s\rangle$ made of S qubits has 2^S coordinates. They can be understood as something like 2^S parallel quantum cores for computation. The entanglement creates dependencies between such "cores" and interference allows to affect the chosen subset of them concurrently.

Measurement of a quantum system effects with a collapse of qubit to the so called *eigen states*, which are $|0\rangle$ and $|1\rangle$. It is not a determined process but each of eigenstate appearance is defined by probabilities equal to $\langle q|0\rangle = |\alpha|^2$ and $\langle q|1\rangle = |\alpha|^2$. The probability distribution function of eigen-state appearance is one of the possible results of quantum computation called *quantum sampling*. In multiqubit case there appears 2^S possible outputs, which would make such a procedure inapplicable in practical solutions. However the entanglement allows to limit the number of output states to the reasonable size.

Quantum computation can be divided into four parts: (1) initialization, which means to set up the qubit to state $|0\rangle$, (2) evolution of the qubit state with an evolution

[2] In fact it is a ray in Hilbert space, since the global phase is not distinguishable physically

operator made of gates by composition and tensor product, (3) measurement, and (4) interpretation of the result.

To create one qubit:

1. We have to know what is the physical meaning of being in eigen state $|0\rangle$ or $|1\rangle$ for qubit.

2. We have to create such a setup that will cover all possible rotations on a Bloch Sphere. In other words, the setup has to implement the general quantum gate (Eq. (3)). It is obvious it implies that qubit can be superposed.

3. We have to be able to measure the qubit and to decide what state the qubit had collapsed to after measurement.

4. In case of multiqubit state we have to implement:

   a. the interference.

   b. the entanglement of qubits.

In this chapter we show how to build qubit and multiqubit systems using the linear and nonlinear optics. We will also discuss the possible practical implementation.

4. The concept of two-rail optical qubit

Two rail qubit employs the idea described in the Section 2. The process starts like in the holographic case—we create two streams of hyperentangled photons using nonlinear unit. In the next section we will discuss which materials can be used for efficient implementation of this process. The state after this process is (see [22]):

$$\frac{|HV\rangle + e^{i\delta}|VH\rangle}{\sqrt{2}} \tag{4}$$

In the equation above $e^{i\delta}$ is a local phase, which will be eliminated during qubit calibration process. Moreover this equation tell us, that there two possible states that system will collapse to, with equal (0.5) probability. Hence the state is in the superposition of those states. In both cases two photons generated by nonlinear element are on the separate rails and have different orthogonal polarizations (H and V). To establish qubit we will call the first one *control rail* TR and the second one *computation rail* CR. The photons on those two rails are entangled, since if on the TR there is horizontally polarized photon then on the CR is the one vertically polarized, and vice-versa.

Now, we will define the two rail qubit, using 4 points defined in the Section 3.

1. In our proposition we will use the polarization entanglement and temporal correlations of photon as a resource for creation the qubit. As was mentioned, we have only got two possible configuration of photons, initially: $|HV\rangle$ and $|VH\rangle$, which we consider as $|0\rangle$. Moreover this configuration is $|\Psi^+\rangle$ Bell state.

However, it is easy to change polarization between Bell states, e.g., with half plates, the changes of phase are much more complicated. This operation produces the Bell states except of local phase. Therefore we will call them Bell-like states. Finally, we will consider $\Psi^{\pm}$-like as $|0\rangle$ and $\Phi^{\pm}$-like as $|1\rangle$.

2. Polarization rotator and phase modulator will be used for implementation of one qubit gates.

3. The measurement relies on splitting each rail of the qubit by polarization beam splitter and each of four rails will be measured by photo detectors. Then the time coincidence of photon detected will be determined. We will take into considerations only those detections that cover one of the Bell state. The rest of them are considered as noise.

4. The multiqubit systems will consists of $2n$ rails. Therefore the tensor product of them appears naturally on the output set of detectors.

    a. In the virtue of above the interference appears in the system spontaneously.

    b. The entanglement between qubits, will be established basing on the CNOT (Controlled NOT) gates designed for one-rail qubits in the polarization channel.

4.1 The representation of the states

For the formal description of two-rail qubit we use the Fock space (or second quantization) formalism owing to the nature of the light. Firstly we establish the symbols for creation and annihilation operators. There four pairs of such operators:

$$\begin{array}{cccc} h^{\dagger}_{tr},h_{tr} & v^{\dagger}_{tr},v_{tr} & h^{\dagger}_{cr},h_{cr} & v^{\dagger}_{cr},v_{cr} \\ \text{horizontal,TR} & \text{vertical,TR} & \text{horizontal,CR} & \text{vertical,CR,} \end{array} \tag{5}$$

where † means the creation operator and no upper index means annihilation one. The acting of operators from TR group on the state, creates the photon with H or V polarization on the control rail. Similarly, operators from CR group creates the photon on the computation rail. The description of state in Fock space is defined as follows:

$$|\Psi\rangle = |H_{TR}V_{TR},H_{CR}V_{CR}\rangle, \tag{6}$$

where X_Y means the number of photons with polarization X (horizontal or vertical) on the rail Y (TR or CR). For example, $h^{\dagger}_{tr}v^{\dagger}_{cr}|0\rangle = |10,01\rangle$, which represents the state $|HV\rangle$ in Hilbert space.

4.2 The qubit

We can define the operator defining the output of SPDC process (Eq. 4), using the second quantization:

$$\hat{S} = \frac{h^{\dagger}_{tr}v^{\dagger}_{cr} + v^{\dagger}_{tr}h^{\dagger}_{cr}}{\sqrt{2}} \tag{7}$$

Indeed, if this operator acts on the vacuum we obtain:

$$\hat{S}|0\rangle = \frac{1}{\sqrt{2}}\left(h_{tr}^{\dagger}v_{cr}^{\dagger}|0\rangle + v_{tr}^{\dagger}h_{cr}^{\dagger}|0\rangle\right) = \frac{1}{\sqrt{2}}\left(h_{tr}^{\dagger}|00,01\rangle + v_{tr}^{\dagger}|00,10\rangle\right)$$
$$= \frac{|10{,}01\rangle + |01{,}10\rangle}{\sqrt{2}} = \frac{|HV\rangle + |VH\rangle}{\sqrt{2}}.$$

From the above equation we learn that the SPDC's process output is the superposition of two eigen-states only: $|10{,}01\rangle$ and $|01{,}10\rangle$. It means that after measurement, the time-correlated two photons are detected on the different rails and has mutually orthogonal polarization. Moreover one of them has the same polarization as the input to the SPDC process. Now, we can formally define our proposition for an implementation of photonic qubit as follows.

Definition 4.1 *The two-rail Bell's qubit, QBell* is the system of two hyperentangled photons, which is composed of, so-called *Bell*-like states:

$$|\Psi_l^+\rangle = \frac{|10{,}01\rangle + e^{i\lambda_1}|01{,}10\rangle}{\sqrt{2}} \quad |\Phi_l^+\rangle = \frac{|10{,}10\rangle + e^{i\lambda_2}|01{,}01\rangle}{\sqrt{2}}$$
$$|\Psi_l^-\rangle = \frac{|10{,}01\rangle - e^{i\lambda_3}|01{,}10\rangle}{\sqrt{2}} \quad |\Phi_l^-\rangle = \frac{|10{,}10\rangle - e^{i\lambda_4}|01{,}01\rangle}{\sqrt{2}},$$

where $\lambda_j \in \mathbb{R}^3$ is local phase and can be any number from the range $[-2\pi, 2\pi]$. The Bell-like state $|\Psi_l^{\pm}\rangle$ is considered as the qubit eigen state $|0\rangle$, and $|\Phi_l^{\pm}\rangle$ as qubit eigen state $|1\rangle$. For convenience, we will denote the QBell states with Q in index behind the bra-ket: $|0\rangle_Q, |1\rangle_Q$.

The first part of Bell-like states' component we call the *leading*, e.g. $|10{,}01\rangle$ for $|\Psi_l^{\pm}\rangle$. The second part we call *complementary* state. If we denote the leading state as $|A\rangle$, then the complementary state we denote $|\overline{A}\rangle$.

Note that leading part always begins with horizontally polarized photon, while the complementary—with vertically polarized one.

Definition 4.2 Two QBells components $|q\rangle_Q$ and $|p\rangle_Q$ are considered as *similar with* phase exclusion $|q\rangle_Q \sim |p\rangle_Q$ (or just similar) if and only if:

$$\forall \delta \in \mathbb{R}\ \exists n,m \in \mathbb{R}\ s.t.: |q\rangle_Q = \frac{|A\rangle + e^{in\delta}|\overline{A}\rangle}{\sqrt{2}} \wedge |p\rangle_Q = \frac{|A\rangle + e^{im\delta}|\overline{A}\rangle}{\sqrt{2}}$$

The relation defined above, is the equivalence relation, certainly, which can be proved using elementary methods, hence we omit them. Therefore it divides the set of QBells' components into equivalence classes. Finally each class can be considered as the QBell's component itself. Therefore we can use it for deeper look on the coefficient *inside* the QBell (see **Figure 2A**). Namely the coefficient of the leading part is always real and equal to $1/\sqrt{2}$, and the coefficient of the complementary part can be placed on the complex plain on a circle with a center in the $(0, 0)$ and passing through the coefficient of the leading point.

Corollary 4.1 Acting on the Bell-like eigen-state with an operator changing the phase of leading state changes the global phase of this eigen-state.

---

[3] Phase λ is always real number, since in the coefficient we have complex exponent $e^{i\lambda}$.

Proof: Indeed, let us denote Bell-like state as $|X_l\rangle = \frac{|A\rangle + e^{in\lambda}|\overline{A}\rangle}{\sqrt{2}}$ and consider that a state $|Y\rangle = \hat{O}_\alpha|X_l\rangle = \frac{e^{i\alpha}|A\rangle + e^{i\varepsilon}|\overline{A}\rangle}{\sqrt{2}}$, where $\alpha, \varepsilon \neq 0$. In that case we can write:

$$|Y\rangle = e^{i\alpha}\frac{|A\rangle + e^{i\cdot 1\cdot(\varepsilon-\alpha)}|\overline{A}\rangle}{\sqrt{2}} = e^{i\alpha}|X_l\rangle, n = 1, \delta = \varepsilon - \alpha$$

Looking on the **Figure 2B**, we see that the global phase rotates both leading and complementary coefficients around the circle of complementary coefficients (dotted line on picture B) on the complex space. As the effect the new state does not fulfill the QBell condition, that the leading coefficient has to be equal to $1/\sqrt{2}$. In algebraic sense it means that we factor out the global phase. In geometric representation we can imagine that it is a rotation (picture B on mentioned figure) of complex space by the phase in order to place the leading coefficient on $1/\sqrt{2}$ location up again. It means that the global phase of the given eigen-state becomes *local phase* of QBell.

Corollary 4.2 Acting on the Bell-like state with an operator changing the complementary phase does not change the QBell state.

Proof: Indeed, let us denote Bell-like state as $|X_l\rangle = \frac{|A\rangle + e^{in\lambda}|\overline{A}\rangle}{\sqrt{2}}$ and consider that a state $|Y\rangle = \hat{O}_\alpha|X_l\rangle = \frac{|A\rangle + e^{i\alpha}e^{i\varepsilon}|\overline{A}\rangle}{\sqrt{2}}$, where $\alpha, \varepsilon \neq 0$. In that case:

$$|Y\rangle = \frac{|A\rangle + e^{i\cdot 1\cdot(\varepsilon+\alpha)}|\overline{A}\rangle}{\sqrt{2}} = |X_l\rangle, n = 1, \delta = \varepsilon + \alpha$$

In this case, shown on the **Figure 2D** we change the local phase of one of the QBell's component, but not the QBell state itself. The original and changed component's state are similar in the light of def. 4.2. Therefore, they represents the same component, so the QBell state stays untouched.

On the left hand, there is shown two types of Bell-like states. The potential existence of the photon on the rail is marked with a black dot with label describing which polarization state it concerns. The correlated states from control and computation rails

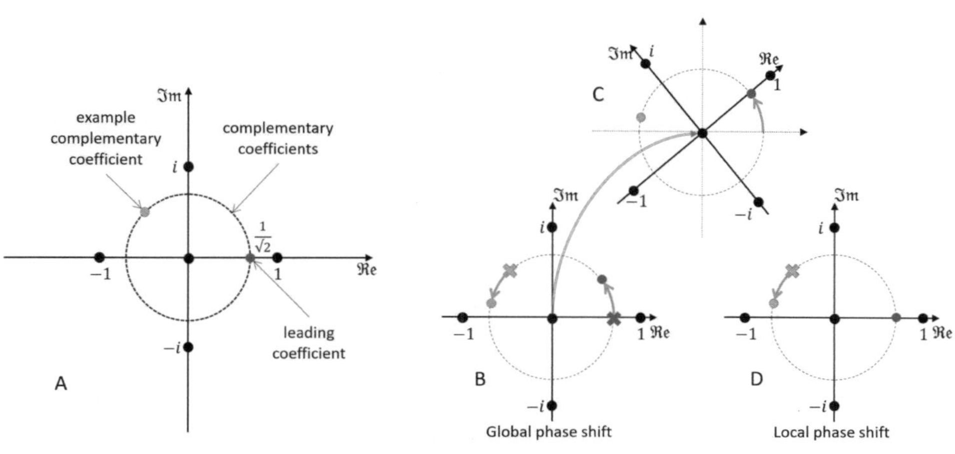

Figure 2.
The location on complex space of the leading and complementary coefficients and the local and global phase on the state.

are marked with a line. The same color of the lines means the same Bell-like state. Solid line is the leading state while the dashed one complementary state.

On the right hand, is the process of measurement. First the horizontal and vertical polarized photons are distributed on separate lines by polarized beam splitters. Nevertheless the entanglement is hold, which is shown by correlation lines. On the most right side the long rectangles symbolize the detectors. The circles inside them are the results of measurement- filled is the detection of photon while empty is the lack of detection. Solid fill means the detection of leading state, while dashed one - the complementary state. Each column through the detectors means the time correlation of measurement. The first two of them describes the situation that we interpret the result as Φ-like state, hence $|0\rangle_Q$, the last two - as Ψ-like state, hence $|1\rangle_Q$.

Initialization. This qubit obtained in SPDC process is in the $|0\rangle_Q$ state already. Since the initialization state occurs automatically, so we do not need to make any operations to generate it, in opposite to other solutions.

4.3 The measurement

The measurement is performed by avalanche photo detectors preceded by division both rails using polarization beam splitter. In such a way we can check how many photons had received each polarization on each rail, during the wave function collapse caused by measurement. On theoretical level we consider two photons and their polarization-space configuration, like it is shown on the **Figure 3**. There are two photo-detectors for TR rail, marked with TRH for horizontal polarized photon and TRV for vertical polarized photon. The detectors for CR rail are marked similarly: CRH i CRV. Let us consider that our qubit is in state.

$$|q\rangle_Q = \alpha|0\rangle_Q + \beta|1\rangle_Q \tag{8}$$

It means that on physical, optic layer the system is in the state:

$$\frac{\alpha|1010\rangle + \alpha e^{i\delta_1}|0101\rangle}{\sqrt{2}} + \frac{\beta|1001\rangle + \beta e^{i\delta_2}|0110\rangle}{\sqrt{2}}$$

Thus, the probability that detectors TRH and CRH click in the same time is equal to $\frac{|\alpha|^2}{2}$ as well as TRV and CRV, which can be seen on mentioned **Figure 3** where the clicks of detectors at the same time are arranged in columns. Hence we can say that the probability of event that the photons with the same polarization on separate rails was measured is equal to $|\alpha^2|$. It means that we measured Φ-like state. Hence our QBell is in $|0\rangle_Q$ state with probability $|\alpha^2|$, which is consistent with our assumption from Eq. (8). Such configuration of detectors' clicks we will call *alternating*. Using the same line of reasoning, we conclude that probability of the event that photons on the TR and CR rails has different polarizations, is equal to $|\beta|^2$. Hence the probability of obtaining QBell in state $|1\rangle_Q$ is equal to $|\beta|^2$ as well, which is also consistent with the same assumption. This configuration of detector clicks we will call *straight*.

In the real application, in current *state of the art*, it is not possible to generate *separate in space and time* hyperentangled photon pairs efficiently in the process of SPDC. It's because of very low efficiency of the process even in the sources considered as strong. Therefore we use extremely narrow laser pulses, nano- or even femtosecond long, for powering the SPDC sources, nevertheless photon count in such an impulse is

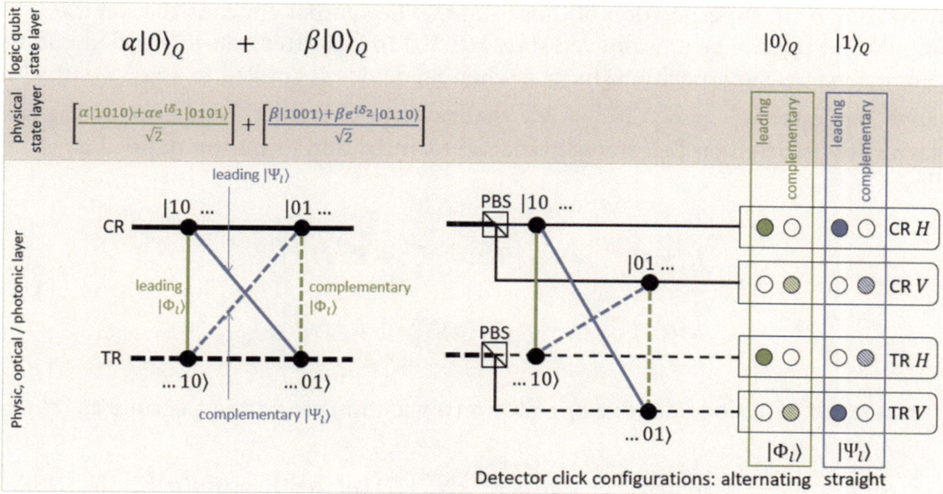

Figure 3.
Scheme of the two-rail Bell's qubit, QBell and its measurement. There are shown three layers of abstraction. The most bottom one contains the physical optical current. The middle one—the state of the current in the Fock space. The upper one is the QBell logic, which is an interpretation of the current state in the perspective of QBell definition or measurement results.

on the order of 10^{12}. Therefore, a detection of time correlated photo diodes clicks is one of the key point of the quantum computing systems.

4.4 One qubit quantum gates

For practical implementation of the complete set of one qubit quantum gates the universal $U(\theta, \varphi, \lambda)$ has to be physically available. One of available forms of such a gate is as follows.

$$U(\theta, \varphi, \lambda) = R_z\left(\varphi + \frac{\pi}{2}\right)R_x(\theta)R_z\left(\lambda - \frac{\pi}{2}\right) \tag{9}$$

From equation above we see that we need to implement rotations around z and x axis of Bloch sphere.

There are two building blocks of the gates: *polarization rotator* and *phase modulator*.

4.5 Rotation around x axis of Bloch sphere

In the terms of second quantization, *polarization rotator* $\hat{R}(\alpha)$, where α is the rotation angle, has got the input creation operators $R_i^{h\dagger}$,$R_i^{v\dagger}$ and output creation operators $R_o^{h\dagger}$,$R_o^{v\dagger}$. The relation between them is defined as follows.

$$\begin{aligned} R_i^{h\dagger}(\alpha) &= cos\,(\alpha)R_o^{h\dagger} - isin(\alpha)R_o^{v\dagger} \\ R_i^{v\dagger}(\alpha) &= cos\,(\alpha)R_o^{v\dagger} - isin(\alpha)R_o^{h\dagger} \end{aligned} \tag{10}$$

We apply the polarization rotator to the control rail. One photon on it can be in states: $|10, \cdot\cdot\rangle$ in case of horizontal polarization and $|01, \cdot\cdot\rangle$ in case of vertical polarization or in superposition of those states. The symbol $\cdot\cdot$ means that on the computation rail can be any allowed state (01,10) In fact after non-linear element, it is in homogeneous superposition. In case when no device is applied to a rail we use “no change” operator $N_i^{h\dagger} = N_o^{h\dagger}, N_i^{v\dagger} = N_o^{v\dagger}$ that saves polarization. Hence the acting of polarization rotation on Bell states in second quantization terms are defined as follows.

$$\begin{aligned} \hat{R}(\alpha)|\Psi_l^+\rangle &= \frac{1}{\sqrt{2}}\left[R_i^{h\dagger}(\alpha)N_i^{v\dagger} + R_i^{v\dagger}N_i^{h\dagger}\right] \\ \hat{R}(\alpha)|\Phi_l^+\rangle &= \frac{1}{\sqrt{2}}\left[R_i^{h\dagger}(\alpha)N_i^{h\dagger} + R_i^{v\dagger}N_i^{v\dagger}\right], \end{aligned} \tag{11}$$

As a result of action of above operator on a vacuum, we obtain the state as follows.

$$\begin{aligned} \hat{R}(\alpha)|\Psi_l^+\rangle &= \frac{1}{\sqrt{2}}\left[\left(cos(\alpha)R_o^{h\dagger} - isin(\alpha)R_o^{v\dagger}\right)N_o^{v\dagger} + \left(cos(\alpha)R_o^{v\dagger} - isin(\alpha)R_o^{h\dagger}\right)N_o^{h\dagger}\right]|0\rangle \\ &= \frac{1}{\sqrt{2}}\left[cos(\alpha)|10,01\rangle - isin(\alpha)|01,01\rangle + cos(\alpha)|01,10\rangle - isin(\alpha)|10,10\rangle\right] \\ &= \frac{1}{\sqrt{2}}\left[cos(\alpha)(|10,\ 01\rangle + |01,\ 10\rangle) - isin(\alpha)(|10,\ 10\rangle + |01,\ 01\rangle)\right] \\ &= cos(\alpha)|\Psi_l^+\rangle - isin(\alpha)|\Phi_l^+\rangle \end{aligned}$$

Similar reasoning for $\hat{R}(\alpha)$ acting on remaining Bell states, will lead us to the result

$$\begin{aligned} |0\rangle_Q = |\Phi_l^\pm\rangle &\xrightarrow{\hat{R}} cos(\alpha)|\Phi_l^\pm\rangle - isin(\alpha)|\Psi_l^\pm\rangle = cos(\alpha)|0\rangle_Q - isin(\alpha)|1\rangle_Q \\ |1\rangle_Q = |\Phi_l^\pm\rangle &\xrightarrow{\hat{R}} -isin(\alpha)|\Psi_l^\pm\rangle + cos(\alpha)|\Phi_l^\pm\rangle = -isin(\alpha)|0\rangle_Q + cos(\alpha)|1\rangle_Q \end{aligned}$$

The above formula can be rewritten in the context of Pauli matrices σ_x,σ_y and σ_z, using famous Euler formula:

$$\hat{R}(\alpha) = e^{-i\sigma_x\alpha} \tag{12}$$

In virtue of above, we say that polarization rotation makes the rotation around x by angle 2α on Bloch sphere where north pole is $|\Psi^+\rangle$ and south pole is $|\Phi^+\rangle$. In the light of the definition of QBell this Bloch sphere also represents this qubit. Hence polarization rotation is rotation around x axis of QBell's Bloch sphere by an angle 2α:

$$R_x(\alpha) = \hat{R}\left(\frac{\alpha}{2}\right). \tag{13}$$

4.6 Rotation around z axis of Bloch sphere

Phase modulator $\hat{M}(\gamma)$ placed on the TR applies the *phase shift* operation, as follows.

$$\hat{M}_i^\dagger = e^{-i\gamma}\hat{M}_o^\dagger \Leftrightarrow \hat{M}_{i,TR}^\dagger = e^{i\gamma}\hat{M}_{o,CR}^\dagger \tag{14}$$

The above equivalence is correct because γ is a *local phase* considered as difference of phase between two rails, similar like in holography method and Mach-Zehnder interferometer. Global phase $\left(e^{i\phi}(\alpha|0\rangle + \beta|1\rangle)\right)$ is ignored since it cannot be measured[4]: $\left|e^{i\phi}\alpha\right|^2 = e^{i\phi}e^{-i\phi}|\alpha|^2 = |\alpha|^2$. Thus, phase modulator represents the *phase shift* gate P_γ. It is obvious that phase modulator does not change the polarization, hence it is not possible to change the type of Bell state:

$$
\begin{aligned}
|\Psi^\pm\rangle &\xrightarrow{P_\gamma} e^{-i\gamma}|\Psi^\pm\rangle \\
|\Phi^\pm\rangle &\xrightarrow{P_\gamma} e^{-i\gamma}|\Phi^\pm\rangle
\end{aligned} \tag{15}
$$

Therefore and from 4.1, the phase modulator itself change the global phase of QBell, not a local one, which is needed for z rotation operator.

In order to obtain the rotation around z axis on the Bloch Sphere we use the more general circuit called by us *General Phase Shift circuit*, GPSh and market with $\mathcal{G}_{\zeta,\gamma}$, shown on **Figure 4**. We start the evolution with the polarization separation phase, which we divide the control rail into two sub-rails (red and yellow) with polarizing beam splitter (PBS). It works in such a way that horizontally polarized photons go straight (red trajectory on the mentioned figure) an vertically are reflected (yellow trajectory). After

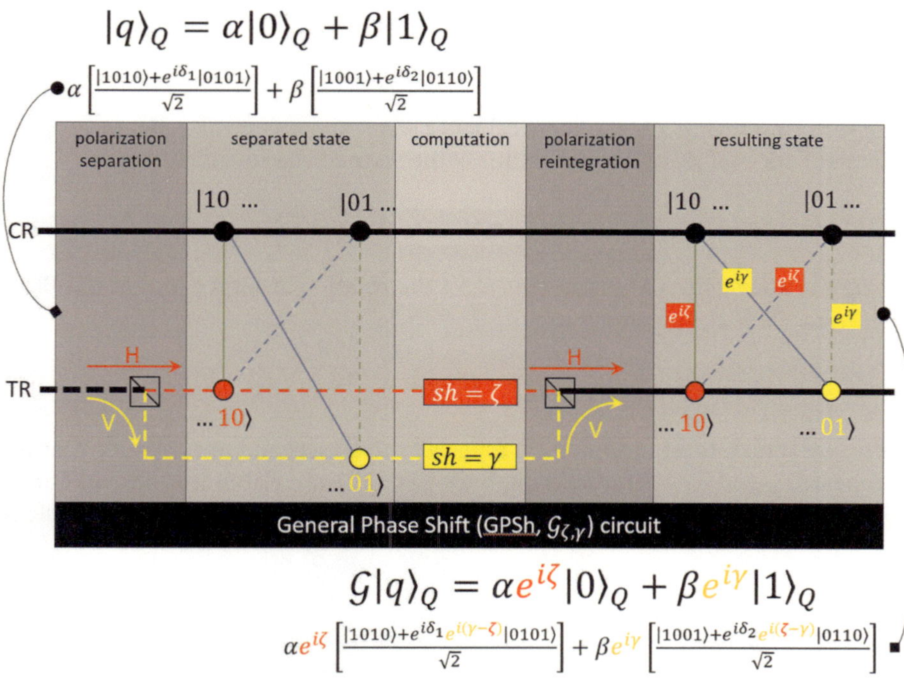

Figure 4.
General Phase Shift Circuit (GPSh) diagram. The polarization separation and integration is made with two Polarization Beam splitters. The trajectory of horizontally polarized photon is market with red color, while vertically polarized with yellow one. In resulting state we see that phase shifts done for vertical (ζ)/ horizontal (γ) polarized photons propagates on both $|\Psi\rangle$,$|\Phi\rangle$ Bell-like states. However ζ affects the leading part of $|\Phi\rangle$ representing $|0\rangle_Q$ on logic layer, while γ its complementary part. In the case of $|\Psi\rangle$ phase shift propagation occurs in opposite way.

[4] Therefore the state of qubit is not represented by *vector* $|q\rangle$ but by *ray* $e^{i\phi}|q\rangle, \phi \in \mathbb{R}$ in Hilbert space.

this operation, the system state, called *separated state* is roughly the same as before it. Nonetheless the horizontally and vertically TR polarization are now on the different rails. On the TR rail, there is indeed one photon in superposition of horizontal and vertical polarization, in general case. Thus after separation there is a superposition of the photons position on red and yellow sub-rail. Moreover the red sub-rail represents the horizontal eigen-state, while the yellow one - vertical eigen state. Indeed, we change the polarization basis into two sub-rail spatial basis. It allows us to apply any operations including phase shifts to each polarization eigen-state separately. Which we actually do in the computation phase with,red" adjustable phase shifter (with phase ζ) applied to horizontally polarized sub-rail and the "yellow" one with shift γ. Then, the polarization reintegration phase changes the basis again to polarization one. Because of photon (on TR and CR) entanglement, phase shift affect the photon on CR *crosswise* against the entangled relation. It can be seen on the scheme since, in the resulting state, one red label is on the green relation and other one on the blue line. Yellow labels behave similarly. Furthermore, if we apply with GPSh to the input state $|q\rangle_Q = \alpha|0\rangle_Q + \beta|1\rangle_Q$, which is rewritten in physical layer on the top of the scheme, we obtain:

$$\alpha\left[\frac{e^{i\zeta}|10,10\rangle + e^{i\delta_1}e^{i\gamma}|01,01\rangle}{\sqrt{2}}\right] + \beta\left[\frac{e^{i\gamma}|10,01\rangle + e^{i\delta_2}e^{i\zeta}|01,10\rangle}{\sqrt{2}}\right]$$
$$= \alpha e^{\zeta}\left[\frac{|10,10\rangle + e^{i\delta_1}e^{i(\gamma-\zeta)}|01,01\rangle}{\sqrt{2}}\right] + \beta e^{i\gamma}\left[\frac{|10,01\rangle + e^{i\delta_1}e^{i(\zeta-\gamma)}|01,10\rangle}{\sqrt{2}}\right]$$

The resulting formula is the physical layer state, shown on the scheme as well. In the virtue of cor. 4.2 we can say that this is the state of the form:

$$|q'\rangle_Q = \mathcal{G}_{\zeta,\gamma}|q\rangle_Q = \alpha e^{i\zeta}|0\rangle_Q + \alpha e^{i\gamma}|1\rangle_Q \tag{16}$$

Using the above circuit we can construct the rotation around z axis of the Bloch sphere. In fact it is simply equal to:

$$R_z(\alpha) = \mathcal{G}_{-\frac{\alpha}{2},\frac{\alpha}{2}} = e^{-i\sigma_z\alpha} \tag{17}$$

Now, we return to equations: 3, 9, especially to the second one. We see that there only appears R_z and R_x operators, which we have already elaborated above. The physical structure of the U Gate is presented on the **Figure 5**. The above consideration

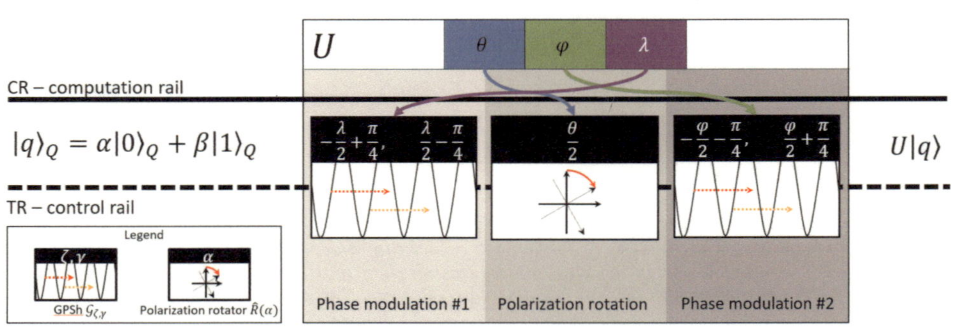

Figure 5.
The scheme of the universal gate for QBell using the Eq. (9). The circuit consists of three phases. Two phase modulations sandwiches the polarization rotation phase. The execution of phases is in opposite direction to the equation, since operators are applied from right to left.

should be treated as the constructive proof, that proposed QBell qubit can be implemented physically. This is because w have shown how to construct the universal gate. It mean that all one-qubit gates can be constructed using it. However it does not mean that a particular gate cannot be implemented in other way, e.g. the phase shift gate $P_\alpha = \mathcal{G}_{0,0.5\alpha}$. In practical solution, there each gate implementation should be carefully optimized from the perspective of the number of devices needed, however this issue goes beyond the scope of this chapter.

4.7 Two QBell systems

Not entangled states of QBell are obtained in very simple way. If we build two QBell qubits one next to other and we perform measurement (which means we detect time-correlations of clicks on eight detectors), we obtain four possibilities:

1st QBell detectors	2nd QBel detectors
Alternating	Alternating
Alternating	Straight
Straight	Alternating
Straight	Straight

If we add another one, the number of options will grow to eight. Thus, we see that it fulfills the pattern of tensor product. Therefore we can say, that we create the interference (which comes down to tensor product) of two QBell just by putting them one next to other and detect the time correlation of detectors' click.

For the purpose of generating $CNOT_Q$ gate, we assume that we dispose *Toffoli* gate for one rail qubits. It is realistic assumption since such circuits are implemented using linear optics systems already, like Huang et al. in [23]. Toffoli circuit work in three qubits systems. It changes polarization of the photon of the third qubit if and only if both photons of the first and second qubit has, for the sake of argument, horizontal polarization[5]. The first two qubits are called *controlling* while the last one *controlled*. Of course in case of photons polarization superposition, this operation is applied to polarization eigen-states building it. On the **Figure 6** there is a scheme of $CNOT_Q$ gate acting on the logic level on the two QBells' system. Let us assume that the first QBell is controlling qubit, while the second one is controlled. We start with the tensor product of two QBells, which is shown on the left part of this scheme by two independent patterns of QBell states. Then there is the $CNOT_Q$ gate shown on the second part of the scheme. First we act with Toffoli gate where controlling rails are CR_1 and TR_1 and controlled TR_2. It results in changing the polarization of the photon on the TR_2 in case photons on CR_1 and TR_1 are vertically polarized. It means that the first QBell is in the $|1\rangle_Q$ state. In that case, the second QBell state will be changed from $|0\rangle_Q$ to $|1\rangle_Q$ and opposite. It occurs because the polarization of TR_2 is changed, while the polarization of CR_2 is not. If the first QBell has different polarizations, nothing happen. In that point one can think, that one Toffoli circuit fully implements the $CNOT_Q$ gate, but it is not true. Namely if first QBell is in complementary state, nothing will happen as well. At this point we have two options. Within the first one, shown on the scheme,

5 There is possible the complementary solution, where these photons has vertical polarization, certainly.

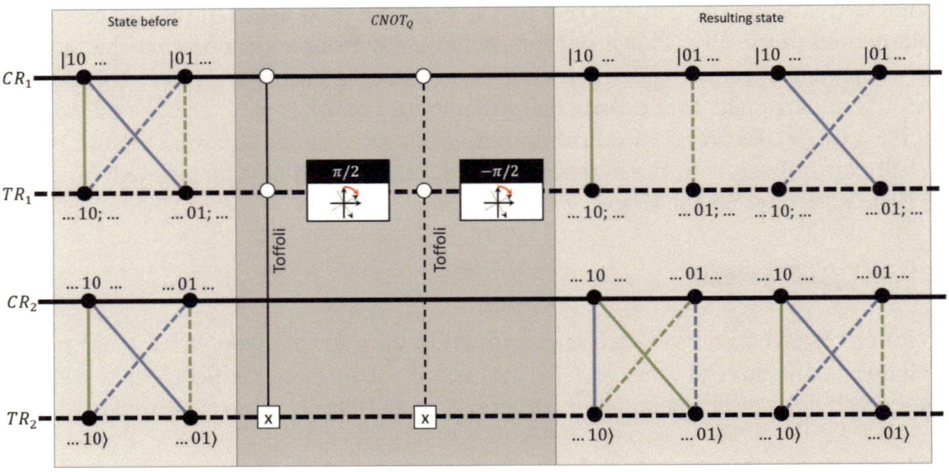

Figure 6.
$CNOT_Q$ gate implementation on the logic level of two QBells system, using Toffoli one-rail qubit gates.

we change the polarization of the first QBell with the polarization rotator with angle $\pi/2$, which acting was described in previous sections. Thus, if both photons was polarized vertically, now they are polarized horizontally. So we can use the same Toffoli gate once again. It changes the polarization of the second QBell in the same way as for its leading part. Finally we turn back the polarization to left the state of the first QBell untouched. The second option is to use the vertical-sensitive Toffoli gate, which changes the controlled state if the controlling qubits has the vertical polarization.

Now, we have the state, shown schematically in the last part of mentioned figure as the, resulting state". We see there that for both "green states" (leading and complementary), the colors on the second QBell are changed, while in case of "blue states" the second QBell looks the same. It means that with operation described above, in the system arises a correlation, that is an entanglement between two QBells.

As we can see above, we use hyper-entanglement as the internal structure of proposed qubit. Thus, we utilize the notion of multipartite entanglement (see Szalay [11]). The system of n qubits is the simple system of the first type consisting of biparities (**Figure 7**). The CNOT and in fact all controlled gates, generates the clusterization of second type (Szalay, ibidem), since Toffoli gate connect CR, TR from the first qubit and TR from the second one. So there is no connection for CR from the second qubit. Hence this is not full entanglement, thus the set $\{CR1, TR1, CR2, TR2\}$ is not a clique, so it is

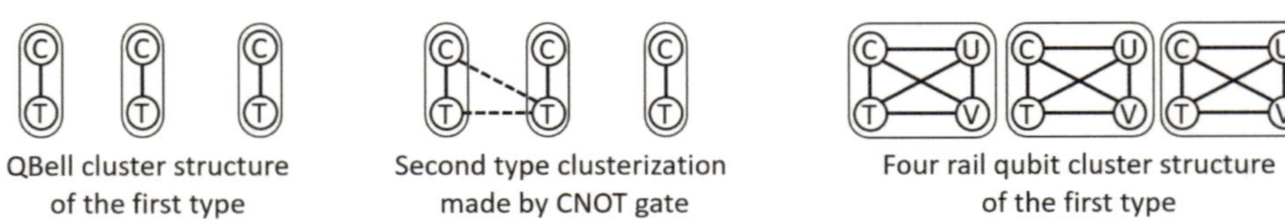

Figure 7.
The multipartite look at the QBell and possible four-rail solution. On the left side simple three QBels structure with no controlled gate applied. In the middle the second type of clusterization arising by CNOT gate application on the first two qubits. At the right side, there is a simple structure of four-rail qubits utilizing GHZ states for $|0\rangle_Q$ and alternating polarization arrangement of $|1\rangle_Q$.

not coarser (see Szalay, ibidem) then sets $\{CR1, TR1\}$,$\{CR2, TR2\}$. Generally we can say, that n-rail systems creates the n-partite structure of the first type and the controlled gate will cover it with the structure of the second type.

5. Discussion and conclusions

In this chapter we have defined the qubit named QBell inspired by quantum holography imaging. We also defined all operation needed for quantum computation, articulated in Section 3. Namely, we have defined how to implement the rotation around x and z axis on the Bloch Sphere for QBell. We pointed how to build the universal, one qubit gate using those two rotations. Furthermore, we defined how to create the interference of two QBells and how to generate entanglement on the level of QBell. In the light of above, we are permitted to announce that we have constructed the two-rail Bell's qubit, defined in def. 4.1 and we have constructively shown that it is physically implementable.

If we look at the internal structure of QBell once again, we see that all the time of QBell living, we shell ensure that there are two, time-correlated photons on two separate rails, one photon on one rail. But this is only subspace of Hilbert space representing any arrangement of two photons on two rails. Because of hyperentanglement, the system has the feature, that any external distortion of phase affects all eigenstates of QBell in the same way. It means that this is so called *Decoherence Free Subspace* DFS, described by Lidar et al. in [24]. This fact is also confirmed by Deffiene et al. (ibidem). Therefore QBell is robust in the perspective of dephasing type of decoherence.

Due to taking into consideration only time-correlated photons, there are also excluded the influence of noise induced by other reactions with environment, like scattering, damping, time de-correlation, devices imperfection, etc. In that case we lowering the efficiency but we exclude many incorrect results. Which also increase the robustness of our solution.

The next advantage we can see in a case of more than two rails. Let us consider $|1\rangle_Q$ as the GHZ-like (Greenberger-Horne-Zeilinger) state, which is an extension of Bell Φ-like state in case of two-rail QBell. $|1\rangle_Q$ as the superposition of the states consisting of alternating horizontally and vertically polarized photons, starting from horizontal polarization (leading part) and vertical polarization (complementary part), which is an extension of Bell Ψ-like state. In that case any external influence touching the polarization has to concern a half of rails. The probability of such an event decreases with the growth of the rails. We can say that the computation subspace has always four dimension, while the dimension of the space where it is embedded grows. Indeed, it is an exponential growth. Therefor the mentioned probability also decreases exponentially. This advantage creates the robustness of proposed solution in the polarization channel as well.

This work opens the wide area of research on multi-rail qubits working in the DFS areas. The possibility of its implementation on photonic chips can lead to mature solutions in the area of photonic quantum processors braking the limitations of quantum optics implemented without this technology.

There are many applications of such technology, terrestrial and space. Quantum processors, with all needed apparatus can be fitted in a few components of nano satellites. It makes the idea of using quantum processors on earth orbit affordable, which is not possible in a case of superconducting and ion-trap implementation due to cooling modules size and weight.

Except of common hopes connected with quantum computation technology, like quantum advantage (or supremacy), photonic solution has some additional advantages. Namely they can be connected with other optical devices without in-between silicon devices. It can speed-up the computations in the area of quantum communication and security. It avail also straight connections with quantum sensors like quantum holographic cameras that was described in the beginning of this chapter.

Recapitulating, in this chapter we have presented the theoretical study on QBell, however it has a potential for great impact for the practical implementation. This, practical aspects of its physical implementation in photonic chips technology will be the subject of a work of our team in the future.

Acknowledgements

This publication was supported by the Department of Graphics, Computer Vision and Digital Systems, under research project for young scientists (Rau6, 2022) and statue research project (Rau6, 2022), Silesian University of Technology (Gliwice, Poland).

Author details

Kamil Wereszczyński† and Krzysztof Cyran*†
Department of Computer Graphics, Vision and Digital Systems, Silesian University of Technology, Gliwice, Poland

*Address all correspondence to: krzysztof.cyran@polsl.pl

† These authors contributed equally.

References

[1] Bruzewicz CD, Chiaverini J, McConnell R, Sage JM. Trapped-ion quantum computing: Progress and challenges. Applied Physics Reviews. 2019;**6**(2):021314

[2] Simon R, Mukunda N. Minimal three-component SU(2) gadget for polarization optics. Physics Letters A. 1990;**143**(4-5):165-169

[3] Knill E, Laflamme R, Milburn G. Thresholds for Linear Optics Quantum Computation. arXiv. 2000

[4] Knill E, Laflamme R, Milburn GJ. A scheme for efficient quantum computation with linear optics. Nature. 2001;**409**(6816):46-52

[5] Knill E. Quantum computing with realistically noisy devices. Nature. 2005; **434**(7029):39-44

[6] Myers CR, Laflamme R. Linear Optics Quantum Computation: An Overview. arXiv. 2005

[7] Thekkadath GS, Mycroft ME, Bell BA, Wade CG, Eckstein A, Phillips DS, et al. Quantum-enhanced interferometry with large heralded photon-number states. npj Quantum Information. 2020; **6**(1):89

[8] Knill E. Bounds on the probability of success of postselected nonlinear sign shifts implemented with linear optics. Physical Review A. 2003;**68**(6):064303

[9] Gasparoni S, Pan JW, Walther P, Rudolph T, Zeilinger A. Realization of a photonic controlled-NOT gate sufficient for quantum computation. Physical Review Letters. 2004;**93**(2):020504

[10] Yoran N, Reznik B. Deterministic linear optics quantum computation with single photon qubits. Physical Review Letters. 2003;**91**(3):037903

[11] Szalay S. Multipartite entanglement measures. Physical Review A. 2015;**92**: 042329

[12] Nielsen MA. Optical quantum computation using cluster states. Physical Review Letters. 2004;**93**(4):040503

[13] Browne DE, Rudolph T. Resource-efficient linear optical quantum computation. Physical Review Letters. 2005;**95**(1):010501

[14] Bogaerts W, Pérez D, Capmany J, Miller DAB, Poon J, Englund D, et al. Programmable photonic circuits. Nature. 2020;**586**(7828):207-216

[15] Madsen LS, Laudenbach F, Askarani MF, Rortais F, Vincent T, Bulmer JFF, et al. Quantum computational advantage with a programmable photonic processor. Nature. 2022;**606**(7912):75-81

[16] Taballione C, Anguita MC, de Goede M, Venderbosch P, Kassenberg B, Snijders H, et al. 20-Mode Universal Quantum Photonic Processor. arXiv. 2022

[17] de Goede M, Snijders H, Venderbosch P, Kassenberg B, Kannan N, Smith DH, et al. High Fidelity 12-Mode Quantum Photonic Processor Operating at InGaAs Quantum Dot Wavelength. arXiv. 2022

[18] Xiao L, Jianying H, Mingjie Z, Tiangui D, Hui L, Yuhong R. Optical holographic data storage—The time for new development. Opto-Electronic Engineering. 2019;**46**(3):180642

[19] Raković D. On extended quantum-holographic framework for

consciousness and FREE WILL. International Journal of Prenatal and Life Sciences. 2019;**3**(3):4-5

[20] Defienne H, Ndagano B, Lyons A, Faccio D. Polarization entanglement-enabled quantum holography. Nature Physics. 2021;**17**(5):591-597

[21] Töpfer S, Gilaberte Basset M, Fuenzalida J, Steinlechner F, Torres JP, Gräfe M. Quantum holography with undetected light. Science Advances. 2022;**8**(2):eab14301

[22] Magnitskiy S, Frolovtsev D, Firsov V, Gostev P, Protsenko I, Saygin M. A SPDC-based source of entangled photons and its characterization. Journal of Russian Laser Research. 2015;**36**(6):618-629

[23] Huang HL, Bao WS, Li T, Li FG, Fu XQ, Zhang S, et al. Deterministic linear optical quantum Toffoli gate. Physics Letters A. 2017;**381**(33): 2673-2676

[24] Lidar DA, Birgitta Whaley K. Decoherence-free subspaces and subsystems. In: Beig R, Englert BG, Frisch U, Hänggi P, Hepp K, Hillebrandt W, et al., editors. Irreversible Quantum Dynamics. Berlin, Heidelberg: Springer; 2003

Chapter 16

Mixed Reality Applications in Business Contexts

Anett Mehler-Bicher, Lothar Steiger and Dirk Weitzel

Abstract

Mixed reality is becoming increasingly relevant in business. In the corporate environment, such as logistics or maintenance, the use of data glasses allows extensive possibilities for process optimization and quality assurance. In the area of construction, virtual models either as augmentation of reality or mapped in virtual reality offer new approaches to experience ability. The goal of this paper is to show the manifold possibilities of mixed reality in the enterprise environment. For this purpose, selected application scenarios with corresponding realization stages will be shown and analyzed regarding their added value.

Keywords: mixed reality, augmented reality, virtual reality, data glasses, augmentation of reality, application scenarios, realization stages

1. Introduction

Mixed reality as an umbrella term for augmented reality (AR) and virtual reality (VR) encompasses technologies that have gained significant importance in recent years and are increasingly being used commercially. Even though theoretical principles were developed as early as the beginning of the 1990s, the increased computing power only makes widespread use possible today.

The media often focus on corresponding opportunities; the potential that arises from use in the corporate environment is often still missing from the discussion.

Mixed reality is not a passing hype. The relevance is made clear by studies of various market research institutes and will continue to increase in the discussion around the metaverse.

It is often difficult to translate new technologies into application scenarios. Based on various application scenarios, answers to possible uses and corresponding added values emerge. Decision-makers need to get ideas for applications and business models that they can develop further to achieve added value for their company.

The aim of this article is to present mixed reality, to differentiate augmented reality and virtual reality from each other, to describe technical prerequisites, to show the current state of development, to present suitable application scenarios and realization stages, and to derive opportunities and risks in use to generate competitive advantages.

After the introduction, the basics of mixed reality are presented, followed by realization stages and application scenarios, which are illustrated by applications in the corporate environment. Afterwards, criteria are presented as to when augmented or virtual reality is recommended in which implementation stage. Finally, a conclusion and outlook follow.

2. Basics of mixed reality

Mixed reality (MR) as an umbrella term subsumes augmented reality (AR) and virtual reality (VR) and corresponding gradations. While virtual reality is understood to mean the representation and simultaneous perception of reality and its physical properties in a real-time computer-generated, interactive, virtual environment, and the real environment is consequently switched off, augmented reality aims to enrich the existing real world with additional computer-generated objects. In contrast to virtual reality, no entirely new worlds are created, but the existing reality is supplemented with a virtual reality [1].

There is no uniform definition of MR in the literature [2]; mostly the "reality-virtuality continuum" is referred to. This postulates a steady transition between real and virtual environments (**Figure 1**) [3].

The left region of the continuum defines environments composed only of real objects and includes all aspects observed when a real scene is viewed by a person or through any medium such as a window, camera, etc. The right-hand domain, on the other hand, defines environments that consist only of virtual objects such as corresponding computer game simulations [2].

Within this framework, mixed reality is defined as an environment in which real and virtual objects are combined in any way in a representation, i.e., lying between the two extreme points of the continuum environments [3]. In Augmented Reality, the real component predominates, whereas in augmented virtuality, the virtual component predominates. A distinction between augmented reality and augmented

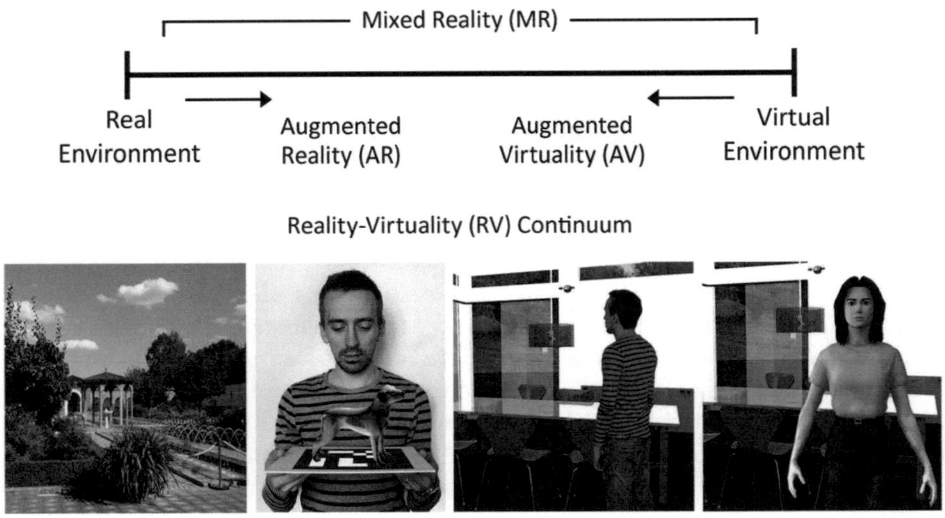

Figure 1.
Reality-virtuality-continuum.

virtuality is rarely made in practice today. Likewise, the terms augmented reality and mixed reality - more rarely also enhanced reality - are often used synonymously [1].

Mixed reality applications require corresponding devices or displays for their use. The displays used differ depending on the type of mixed reality - augmented reality or virtual reality. Virtual reality can usually only be used with data glasses. In addition to classic devices such as smartphones and tablets, data glasses are also becoming increasingly relevant for augmented reality. Also, pseudo-holographic displays are of interest for innovative augmented reality solutions, which enable new application scenarios but are often less flexible in use than the classic devices [4].

3. Realization stages and application scenarios in mixed reality

3.1 Realization stages

Ref. [5] distinguish according to different realization stages to classify different application purposes. Level 1–3 are presented because of their practical relevance [1].

- Realization stage 1: Visualization.

  With the help of MR, computer-generated objects can be projected into the real or virtual environment. In this stage, the user is a passive observer of an MR scene.

- Realization stage 2: Guidance & Control.

  In this stage, prefabricated scenarios of stage 1 are put into a logical sequence; the sequence is controlled by the change of the real viewing object.

- Realization stage 3: Interaction

  Stage 3 differs from stage 2 in the way that the MR scenario does not run according to a predefined pattern, but the sequence of substeps is controlled by the user himself. This can be done voice- or gesture-controlled. Control via input devices such as mouse, keyboard, or touchscreen, but also via facial expression or gesture control are conceivable and depend on the respective situation.

3.2 Application scenarios

In case of VR, there is no necessity to differ between applications scenarios. According to A. Mehler et. al., [1] the following application scenarios can be distinguished in relation to AR/MR:

- Living Mirror

  In living mirror, a camera detects the viewer's face and places three-dimensional objects on the face or head in the correct position. The projection is usually done via a large screen or a beamer so that a mirror effect is created.

- Living Print

  This scenario is based on the recognition of a print medium and corresponding augmentation. Here, a distinction is made between different print media, be it collectible or greeting cards (living card), brochures or leaflets (living brochure), or packaging materials (living object). Further possibilities exist in the augmentation of books (living book) or games (living game print-based).

- Living game mobile

  Mobile devices are the basis of living game mobile; e.g., augmented games are brought to bear on smartphones.

- Living architecture

  A typical application in the field of architecture arises when a viewer wants to "experience" an impression of a room or an entire building by determining its representation himself through movements such as turning his head or walking through a real room and further actions such as speech or gestures.

- Living poster

  A living poster is understood to be an advertising message in public space that is enhanced with manipulative information elements using augmented reality.

- Living presentation

  Trade show booths and presentations must become increasingly spectacular and interesting to be perceived at all in times of information overload. AR technology can be used to achieve this goal. In addition, it is possible to display and even interact with real objects that cannot be "presented" live due to their sheer size or complexity.

- Living meeting

  Due to increasing globalization, more and more meetings take place as tele- or videoconferences. Augmented reality can be used to enrich tele- and videoconferences so that they almost seem like real meetings.

- Living environment

  All AR applications that use mobile systems to enhance real environments or facilities with additional information of any kind, such as text, 2D objects, 3D objects, video, and audio sequences, are called Living Environment.

The goal is basically a timely information acquisition (time-to-content) by the user simply by the fact that the camera captures one or more objects and thereby corresponding additional information is provided. In the case of the living environment, a combination of several sensors is possible and often desired.

The list of application scenarios is not necessarily complete, as further application possibilities may arise because of technical developments. The above scenarios are open regarding their intended use [1].

3.3 Combination of realization stages and application scenarios

If we combine AR/MR application scenarios with realization stages, we see that the living environment in particular is suitable for the third realization stage (**Table 1**).

Application scenario	Realization stage		
	Visualization	Guidance & Control	Interaction
Living Mirror	suitable	suitable	
Living Print	suitable	suitable	
Living Poster	suitable	suitable	
Living Architecture	suitable	suitable	suitable
Living Presentation	suitable	suitable	
Living Meeting	suitable	suitable	suitable
Living Environment	suitable	suitable	suitable

Table 1.
Combination of application scenarios and realization stages in AR/MR.

Static elements for generating AR/MR animations such as living print are more suitable for the first two realization levels. living game mobile is excluded because there are no economic opportunities for use here [1].

4. Application areas of mixed reality in business contexts

4.1 Overview

Mixed reality with a focus on augmented reality can be used in almost all areas of everyday life; many applications have a rather playful character. The use of mixed reality leads to significant advantages in many areas such as production, construction, and logistics. Typical examples are that designers can collaborate with real and virtual colleagues on the same three-dimensional model [1].

Many applications in production, construction, or logistics are also possible in the corporate environment. Possible applications of MR in the industry are mainly:

- Optimization and acceleration of industrial processes
- Immediate display and validation of planning status
- Collision control
- Evaluation of the design of new components on the existing product
- Virtual guidance directly on the real product
- Increased understanding of complex products or processes
- Simplified target/actual comparison
- Improvement of team-internal communication

Communication with customers can also be improved. Conventional advertisements are given an interactive component through the integration of mixed reality, which offers an additional benefit. Product catalogs can be enriched accordingly and offer added value to the customer. More complex solutions are virtual studios, such as those used by television studios. These are virtual sets in which real actors are combined with virtual reality in real-time.

The more the technology develops, the more complex and interesting application scenarios can be realized. In particular, these include electronic devices that exist only virtually but respond to real touch, enabling artificial sensory enhancements such as x-ray vision and open-world computer games. mixed reality supports complex and difficult tasks; these include especially applications in construction or maintenance of machines.

- Maintenance area
  - By displaying important additional information on demand, difficult tasks can be performed more easily, safely, and with higher quality.
  - E.g., based on an exploded view, a maintenance technician can display the machine to be repaired in 3D, move, and rotate it as desired, so that he gets a better impression and can perform the repair more quickly.
  - Another option is to show the mechanic labels for individual parts of the device as well as corresponding work instructions.
- Design
  - MR offers the possibility of simulating constructed objects in space in the design department.
  - Digital design data can be efficiently matched with existing real geometries. This also enables the broad use of digital validation methods when combining digital data with real prototypes or designs.

Further industrial applications with complex tasks have already been partially realized:

- Navigation
  - Navigation supplemented by mixed reality is conceivable in many different application areas:
  - In the maintenance of complex industrial plants
  - For operations in the field of disaster management
  - In the automotive sector, MR is used in cockpit displays that provide the driver with graphical navigation and traffic information in live images
  - In aircraft, MR is used in conjunction with head-mounted or head-up displays to increase safety and efficiency in poor visibility and weather conditions. Use in military aircraft was one of the first industrial applications of MR.

- As a complement to navigation Google began offering its users the live MR navigation in Google Maps. Pedestrians can use *live view* instead of the normal 2D map view to see directional arrows as well as street names and additional information [6].

- Visualization

  Although mixed reality is very often used for representation, there are still some application areas whose main goal is to better visualize certain aspects.

  - Display and interactive analysis of maps and terrain features, e.g., for the extraction of mineral resources or in the field of geoinformatics [7].
  - Representation of destroyed historical buildings or future architectural projects [8].
  - Enrichment of museums and exhibitions by displaying virtual objects [9].

- Simulation

  - E.g., to enrich existing flight and driving simulators with mixed reality elements.

- Collaboration

  - E. g., to support virtual teams in their collaboration on simulated 3D models or conferences with real and virtual participants [10].

One example of the use of MR is education/training in the form of edutainment. Learners can view learning material on the computer via webcam and receive information elements on the computer that provide further explanations and information on the topic under consideration. Sound sequences that go beyond this are also possible as part of the animation. This approach gives learners the feeling of tangible proximity as well as individual and interactive help. This helps to reduce or eliminate learning barriers. The aim of integrating 3D animations based on MR is above all to simulate a haptic perception for the learners and to strengthen their interest.

The application examples already realized clearly show that the aspect of communication in the form of visualization is in the foreground. Complex issues are simplified and made easier to understand or experience by visualizing them.

Increasing computing power enables real-time application scenarios in a variety of ways - especially in the areas of user support and entertainment. Operating system interfaces, especially for the consumer sector, can be enriched with MR elements. Program windows and icons can be displayed as virtual elements in real space and operated by facial expressions or gestures. Thinking further, conventional screens or device control panels can be replaced by new types of devices and corresponding multimedia applications that use mixed reality. These include, for example, pseudo-holographic virtual screens, virtual holodecks, or virtual surround cinema, as already known from various science fiction series. Using MR enhancements such as X-ray vision, it is also possible to display occluded targets [11].

In recent years, MR solutions based on natural feature recognition and face tracking are gaining more and more importance:

- Natural feature recognition allows tracking without the use of artificial markers and enables the recognition of completely unknown environments and thus new application possibilities. The further development of this technology will enable mobile devices to recognize natural environments and enrich them with virtual objects.

- Face tracking, for example, opens new forms of communication with viewers. The viewer is recorded - essentially reduced to the face - and age- and gender-specific characteristics are recognized, evaluated, and interpreted, resulting in a corresponding individual response. Possible application scenarios arise in advertising. While posters currently reflect static information, it is possible to develop interactive posters based on mixed reality that react smartly to their counterparts. Based on age- and gender-specific characteristics, the poster reacts individually and offers suitable advertising messages.

Mixed reality has become a strategic product of hardware manufacturers such as Apple and Microsoft. Meta (Facebook) is increasingly investing in MR (VR as well as AR) technology. The so-called Metaverse is intended to enable users to communicate in virtual and augmented spaces. However, this requires glasses that support both VR and AR scenarios and are affordable for end consumers.

The possibility of gesture and voice control will allow further fascinating opportunities in MR, as it supports easy and intuitive handling of MR applications. Meta is already experimenting with a bracelet that measures brain waves to control MR applications.

4.2 Illustration of application examples in business contexts

Based on the three realization stages, concrete examples of MR applications in the corporate context are described below. A distinction is made between AR and VR in order to illustrate the broad range of possibilities. In the case of augmented reality, these examples are usually living environment solutions.

- Realization stage 1: Visualization.

    - Augmented reality

      Displaying technical information during machine maintenance or product visualizations, such as in ref. [12], are typical applications since an integration in real situations is possible.

    - Virtual reality

      Visualized virtual tours through computer-generated buildings, technical facilities, or operation- or process-visualizations, e.g., in ref. [13], are suitable, especially buildings and facilities that are not yet realized or cannot be used for training purposes.

- Realization stage 2: Guidance & control
  - Augmented reality

    Instructions for the maintenance of a machine or picking, as in ref. [14], represent common application areas for this realization level in the enterprise context.
  - Virtual reality

    Virtual tours through buildings or technical facilities under guidance, if necessary, with appropriate explanation or instruction, as well as training of complex activities, as shown in ref. [15], are typical application possibilities.
- Realization stage 3: Interaction
  - Augmented reality

    Interactive instructions for the maintenance of a machine (including the recognition of next steps or errors in the process) (see, e.g., [16]) or collaborative work are typical applications.
  - Virtual reality

    Virtual tours of buildings or facilities that can be changed through interaction, interactive product design, or collaborative work can be realized especially with tools like Mozilla Hubs, Google Spatial, or other tools for generating VR spaces. These tools provide the first approaches to implementing metaverse aspects.

4.3 Criteria for deciding between augmented and virtual reality

AR and VR have many similarities in terms of the levels of realization, yet they differ in the application possibilities. Therefore, the first thing to check is whether you want to implement an AR or VR application. Decision criteria are primarily simulation of situations, their availability as well as their hazard potential, necessity of embedding, effort, and costs. A corresponding classification is given in **Table 2**.

If one considers the repair of a plant, then AR lends itself to this. The necessity of embedding, in reality, is given; the functionality, i.e., an everyday situation, is to be achieved. Likewise, the availability of the plant is given and there is no potential danger. The situation is different, e.g., for a flight simulator. In this virtual world,

Criterium	Augmented reality	Virtual reality
Simulation of situations	Everyday situations	Exceptional situations
Availability of situations	High	Exceptions, rarely occur in reality
Hazard potential of situations	None	Low to high
Necessity of embedding in reality	Yes	None
Effort	Low to medium	High
Costs	Low to medium	High

Table 2.
Decision criteria.

mainly exceptional situations are to be practiced, which rarely occur, but then have a high hazard potential.

Once a decision has been made between AR and VR, it must be clarified which realization stage is required: Is visualization (stage 1) sufficient, is guidance & control (stage 2) needed, or is an application for interaction (stage 3) necessary? The higher the realization stage, the greater the development, training effort, and the associated costs.

Once a decision has been made in favor of AR, a suitable application scenario must be selected depending on the realization stage.

Applications in the AR and VR area stand and fall with the generated added value. Of course, the implementation of an AR or VR application requires corresponding development effort, which should not be underestimated. However, it is much more important that as much creativity as necessary and as much pragmatism as possible flow into the conception of the application so that the application later delivers the necessary added value.

5. Conclusions

MR applications are on the verge of a commercial breakthrough and are becoming increasingly interesting for companies. Since 2015, various providers have been presenting AR as well as VR glasses; in some cases, second-generation glasses are already available and are becoming increasingly suitable for everyday use.

Despite the similarities that AR and VR have in terms of realization stages, there must first be clarity within the company as to which technology is to be pursued. Once this decision has been made, the next steps are to select the realization stage and, if necessary, the application scenario.

Both technologies covered with the umbrella term MR offer a multitude of opportunities, but at the same time have risks and limitations. Two main problems today are the lack of appropriate hardware as well as the lacking integration of corresponding applications into the corporate IT landscape. With the increased appearance of corresponding MR applications, solutions to this will be developed successively.

However, it is important today that companies do not miss the trend and may no longer be marketable or competitive. This will pose major challenges for small and medium-sized companies. The current discussion related to the Metaverse shows the relevance of the topic very impressively.

Author details

Anett Mehler-Bicher*, Lothar Steiger and Dirk Weitzel
Mainz University of Applied Sciences, Mainz, Germany

*Address all correspondence to: anett.bicher@hs-mainz.de

References

[1] Mehler-Bicher A, Steiger L. Augmented Reality - in Theorie und Praxis. München: DeGryter; 2022

[2] Milgram P, Takemura H, Utsumi A, Kishino F. Augmented reality: A class of displays on the reality-virtuality continuum. Telemanipulator and Telepresence Technologies. SPIE. 1994;**2351**:282-292

[3] Milgram P, Kishino F. A taxonomy of mixed reality visual displays. In: IEICE Transactions on Information and Systems, Special Issue on Network Reality. 1994

[4] Christou A, Gao Y, Navaraj W, Nassar H, Dahiya R., 3D Touch Surface for Interactive Pseudo-Holographic Displays. 2020

[5] Porter M, Heppelmann J. Why every organization needs an augmented reality strategy. Harvard Business. 2017;**2017**:1-20

[6] Google. Use Live View on Google Maps. 2021. [Online]. Available from: https://support.google.com/maps/answer/9332056?hl=en&co=GENIE.Platform%3DAndroid. [Accessed: 26 June 2022]

[7] VR Owl | VR & AR Agency. Geopark the Hondsrug – Nature Route Enhancement. 2019. [Online]. Available from: https://www.vrowl.io/portfolio/unesco-global-geopark-the-hondsrug. [Accessed: 12 July 2022]

[8] Instructables. How to Make a Time Lapse With Your Android Phone. 2019. [Online]. Available from: https://www.instructables.com/How-to-Make-a-Time-Lapse-With-Your-Android-Phone/. [Accessed: 14 July 2022]

[9] Metropolitan Museum of Art. Roam the Galleries of the Metropolitan Museum via Augmented Reality. 2021. [Online]. Available from: https://hyperallergic.com/614503/roam-the-galleries-of-the-metropolitan-museum-via-augmented-reality/. [Accessed: 24 June 2022]

[10] Microsoft. Mirosoft mesh. 2021. [Online]. Available from: https://news.microsoft.com/innovation-stories/microsoft-mesh/. [Accessed: 24 June 2022]

[11] Tesla. Tesla Full Self-Driving's cool augmented reality view previewed by noted hacker. 2020. [Online]. Available from: https://www.teslarati.com/tesla-fsd-augmented-reality-video/. [Accessed: 25 June 2022]

[12] Preality. Preality: Official Trailer 2018. 2018. [Online]. Available from: https://www.youtube.com/watch?v=BIleg-nykvI. [Accessed: 29 June 2022]

[13] Kompetek Interaktiv. Showreel 2022. 2022. [Online]. Available from: https://www.youtube.com/watch?v=5uRfkooUnRk. [Accessed: 13 July 2022]

[14] Simplifier. Connect Smart Glasses and SAP with the Simplifier. 2013. [Online]. Available from: https://www.youtube.com/watch?v=ZWsBHISOqjA. [Accessed: 28 June 2022]

[15] BMC Medical Education. Pre-graduation medical training including virtual reality during COVID-19 pandemic: A report on students' perception. 2020. [Online]. Available from: https://bmcmededuc.biomedcentral.com/articles/10.1186/

s12909-020-02245-8. [Accessed: 12 July 2022]

[16] Vuforia. Augmented Reality | Transforming How Industrial Workforces Communicate, Learn and Work. 2020. [Online]. Available from: https://www.youtube.com/watch?v=Fwikx1TOidE. [Accessed: 28 June 2022]

Chapter 17

Beam Implementation with a Translucent Twisted-Nematic Liquid Crystal Display

Kavan Ahmadi

Abstract

This chapter describes an efficient approach to generating light beams with arbitrary intensity profile and phase distribution. Accordingly, a fast method is described to characterize liquid crystal displays based on the Mach-Zehnder interferometer and fringe analysis in the Fourier domain. Then, the double-pixel hologram Arrizón's approach is reviewed. This approach is able to generate an on-axis computer-generated hologram into a low-resolution twisted-nematic liquid crystal for encoding arbitrary complex modulations. Furthermore, a fast algorithm to map holographic cells based on the k-nearest neighbor (k-NN) classifier is introduced in order to generate computer-generated holograms faster than the conventional calculation. Finally, two beam profiles are produced with the described approach and assessed at the entrance pupil and the depth of focus of a high-NA microscope objective.

Keywords: Twisted-nematic liquid crystal display, computer-generated hologram, beam shaping, highly focused beam, machine-learning algorithms

1. Introduction

Liquid crystal display devices are a kind of spatial light modulators (SLMs). SLMs are able to relate electronic data to spatially modulated coherent light. In particular, twisted nematic liquid crystal spatial light modulators (TNLC-SLMs) are kind of relatively low-cost electro-optics devices widely used in many branches of optical information processing, such as digital holography [1, 2], spatially-variant polarized beams [3], coherent diffraction imaging [4], generating vector beams [5–7], pattern recognition and optical correlators [8], Fresnel lenses [9, 10], and optical cryptosystem [11–13].

On the other side, highly focused beams and their properties have been investigated in many fields, such as nonlinear optics, super-resolution microscopy, tomography, and optical tweezers [14–21]. Tightly focused beams attract much attention because of the non-neglectable component of the electric field in the direction of propagation.

Since SLMs change the properties of light, such as amplitude, phase, and polarization, it is necessary to find the proper operating conditions to control the SLM's response. The twisted angle and the birefringence of a TNLC are two main parameters that control the modulation. There are several proposed methods and configurations to find these parameters to introduce the Jones matrix of TNLC [22–26]. On the contrary, Martín-Badosa et al. proposed a method to characterize liquid crystal displays (LCDs) based on the fringe analysis obtained by a Mach-Zehnder interferometer configuration, without the necessity of finding the physical properties of LCD [27]. Nevertheless, their approach is based on counting the displacement of the fringes and is time-consuming. Wang et al. proposed a faster method to characterize SLM based on extracting the phase values of the Fourier spectra of the interference pattern directly [28]. They indicated that analyzing the phase values in the frequency domain is sufficient to obtain the imposed modulated phase. Their mathematical approach is based on Fresnel diffraction considering the transfer function. However, this approach can be modified by adding a 4-f imaging system to the Mach-Zehnder interferometer.

The double-pixel hologram Arrizón's approach is a cell-oriented computer-generated hologram (CGH) encoding. In the cell-oriented method, each encoding point is split into a couple of holographic cells [29]. He proposed a modification of the previous works regarding a double-phase CGH method with an on-axis reconstruction field [30–32]. His approach improved the signal-to-noise ratio (SNR) in the reconstruction plane using two pixels of SLM to encode one holographic cell [33]. Since then, he generalized his approach to producing a more symmetric and high-SNR-signal domain using four pixels of SLM, the so-called modified Double-Pixel Hologram (DPH) [34]. Although his approach was suggested for encoding phase-only SLMs, he adjusted the method to encoding complex modulation with a transmission TNLC as a low-resolution spatial light modulator [35–37].

This chapter aims to present a modification of the method proposed by Martín-Badosa et al. and Wang et al. to characterize LCDs and review the DPH Arrizón's approach to generating an on-axis CGH proper for encoding TNLCs. Since the codification algorithm is time-consuming, a fast algorithm using the k-NN classifier is presented. The application of this codification is extended to provide vector beams using a high-NA microscope objective.

The following text is organized as follows: Section 2 briefly describes the physical and optical properties of a TNLC display. Section 3 describes the characterizing process based on the modified Mach-Zehnder interferometer and fringe analysis in the Fourier domain. Section 4 reviews the DPH Arrizón's approach. Section 5 introduces a fast algorithm to generate DPHs applying the k-NN machine learning algorithm. Section 6 demonstrates the experimental setup to produce and analyze highly focused beams with arbitrary beam profiles.

2. Twisted nematic liquid crystal

A twisted nematic liquid crystal display is constructed by sandwiching a nematic liquid crystal between two transparent glass plates. Different voltages impose an external electric field through the medium using electrodes connected to each glass plate. Nematic molecules inside the medium have a helical structure parallel to their elongated direction with an optical axis. In this regard, TNLCs are a sort of birefringent medium with ordinary and extraordinary refractive indices (n_o,n_e). The birefringence of the medium is altered by applying different voltages resulting in tilting nematic molecules in the direction of the applied electric field. Hence, a TNLC can electrically be controlled to be used as optical wave retarders, modulators, and switches. In the absence of the applied electric field, nematic molecules are appropriately oriented with respect to each other in the plane, parallel to the surface of glasses. In the particular case that the input beam is linearly polarized parallel to the direction of the liquid crystal director (practically is unknown to users), the beam keeps its state of polarization traveling through TNLC but rotating as much as its twisted angle (α), which is usually equal to 90°. In this condition, the TNLC acts as a polarization rotator. If the direction of polarization of the input beam is oriented concerning the LC director, the beam experiences phase retardation as $2\pi(ne - no)d/\lambda$, where d is the thickness of LC display. In the presence of the applied electric field, the nematic molecules tilt to be aligned with the direction of the applied field. The amount of this tilt angle (θ) depends on the applied voltages causing different birefringence. So, LC becomes a variable retarder with retardation $\Gamma = 2\pi[n(\theta) - n_0]d/\lambda$. The retardation varies monotonically from 0 (when the molecules are not tilted, $\theta = 0$) to $\Gamma_{max} = 2\pi[n_e - n_0]d/\lambda$ (when molecules are tilted 90°, $\theta = 90$). In summary, based on the physical configuration of the experimental setup, an input beam traveling through nematic liquid crystal cells might face only-twisted, twisted and tilted, or only-tilted nematic molecules. The twisted angle, the birefringence, and the director of LC should be obtained to introduce the Jones matrix of TNLC display (for more information, see [38]).

3. Characterizing TNLC display

Regarding specific situations, including the state of polarization, the wavelength of input beams and applied voltages, three different modulations are mainly interesting: amplitude-only, phase-only, and complex amplitude-phase. Applying different voltages to SLM results in different degrees of birefringence that can be practically obtained by displaying different gray values on the LCD, ranging from 0 to 255. Accordingly, the phase and intensity of the input beam alter passing through the SLM differently. Hence, the effect of a TNLC display in the optical system can be described by

$$E_t = A \exp(-j\varphi), \tag{1}$$

where A stands for the transmitted amplitude and φ represents the imposed phase to the transmitted beam. In this regard, characterizing LCDs aims to find the Eq. (1) for each gray value ranging from 0 to 255.

Martín-Badosa et al. applied a Mach-Zehnder interferometer for characterizing LCDs. With the same optical setup and a different mathematical description, Wang et al. proposed a faster and more convenient way to obtain phase modulation. Since the digital holography approach used in this work is based on DPH Arrizón's approach, the Mach-Zehnder interferometer should be accordingly modified in order to eliminate the off-axis diffraction orders caused by the codification algorithm (it is explained in Section 4). Therefore, the optical setup used in this work is modified by adding a 4-f spatial filtering system. Accordingly, the mathematical analysis based on this experimental setup is presented.

As shown in **Figure 1**, the optical setup consists of a Mach-Zehnder interferometer plus a 4-f spatial filtering system. The coherent beam provided by a pig-tailed laser (Thorlabs LP520-SF15A) with $\lambda = 514$ nm is collimated and linearly polarized by means of the collimator lens and the linear polarizer (LP1), respectively. Then, the

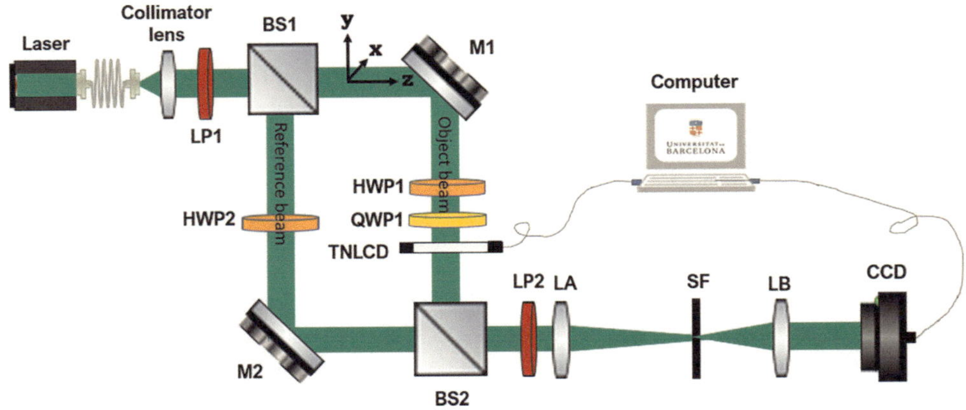

Figure 1.
The experimental setup. LP, BS, HWP, QWP, M, SF, and CCD stand for the linear polarizer, beam splitter, the half-wave plate, the quarte-wave plate, mirror, spatial filter, and charged-coupled device, respectively.

beam is divided into two arms of the interferometer by the first beam splitter (BS1). The right arm of the interferometer, which the object beam passes through, includes a half-wave plate (HWP1), a quarter-wave plate (QWP1), and the transmissive TNLCD (Holoeye HEO 0017 with a resolution of 1024 × 768 pixels and a pixel pitch of 32 μm).

The left arm of the interferometer, which the reference beam passes through, includes a half-wave plate (HWP2). Subsequently, the interference occurs when the two beams reach the second beam splitter (BS2). Then the interferometric pattern passing through the second linear polarizer (LP2) reaches the CCD camera's sensor plane through the 4-f imaging system.

Regarding the experimental setup, LCD is placed at the back focal plane of lens A (LA), and the CCD camera is placed at the front focal plane of lens B (LB). Besides, the diffraction orders (except zero-order) caused by the pixelated structure of LCD and the digital holography approach are removed using the spatial filter (SF) placed at the common focal plane of LA and LB. Mirror M2 is properly tilted in such a way that the fringes are aligned in the y-direction, as shown in **Figure 2a**. HWP2 is used to adjust the contrast intensity of the interference pattern. The orientation of axes of linear polarizers, LP1 and LP2, and fast axes of HWP1 and QWP1 with respect to each other define the modulated characteristic of the TNLC display. Hence, the desired modulation can be obtained by some practical attempts.

The fringe pattern recorded by the CCD can be expressed mathematically in the following form:

$$g(x,y) = a(x,y) + b(x,y)\cos\left[2\pi f_0 x + \phi(x,y) + \varphi_m\right], \qquad (2)$$

where $\phi(x,y)$ and φ_m are the phase of the object beam and the modulated phase imposed by SLM, respectively. Furthermore, $a(x,y)$ represents possible nonuniform background, $b(x,y)$ represents the local contrast of the pattern, and f_0 is the spatial-carrier frequency. As explained in [39], the fringe pattern can be rewritten in the following form:

$$g(x,y) = a(x,y) + c(x,y)\exp\left(j2\pi f_0 x\right) + c^*(x,y)\exp\left(-j2\pi f_0 x\right), \qquad (3)$$

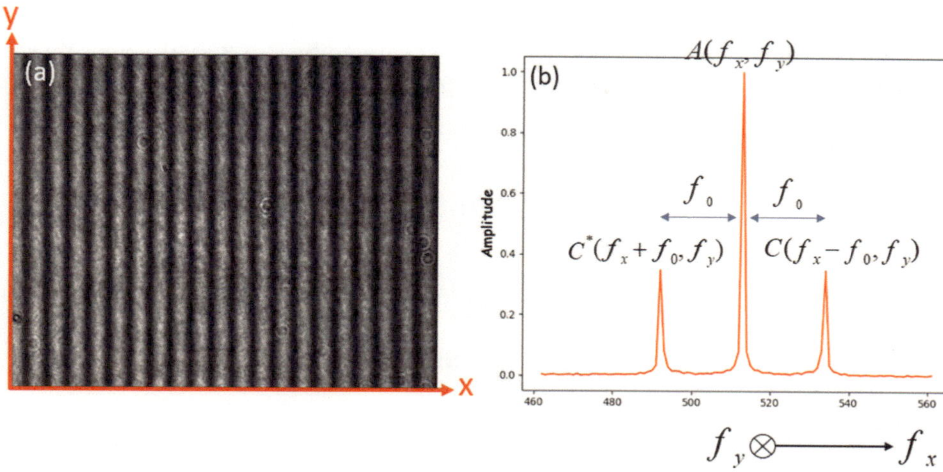

Figure 2.
(a) The interferometric pattern recorded by CCD. (b) The corresponding Fourier spectra.

with

$$c(x,y) = (1/2)b(x,y)\exp[j\phi(x,y)]\exp(j\varphi_m). \tag{4}$$

The Fourier transform of Eq. (3) gives

$$G(f_x, f_y) = A(f_x, f_y) + C(f_x - f_0, f_y) + C^*(f_x + f_0, f_y), \tag{5}$$

where the capital letters denote the Fourier spectra, f_x and f_y are the spatial frequencies in the x- and y-direction, respectively. In fact, $A(f_x, f_y)$ is the zero-order of the interference and $C(f_x - f_0, f_y)$ and $C^*(f_x + f_0, f_y)$ are ± 1 interference orders, respectively.

Since the spatial variations of $a(x,y)$, $b(x,y)$, and $\phi(x,y)$ are slow compared with spatial frequency f_0, the Fourier spectra are separated by carrier frequency f_0, as shown in **Figure 2b**.

Once the configuration of the experiment, for instance, wavelength and the optical path difference between the reference and object beam, remains constant, the position of the peaks in the frequency domain remains unchanged. Thereby, the modulated phase can be obtained by analyzing either C or C^*. Complex value C can be rewritten as follows:

$$C(f_x, f_y) = \mathrm{FT}\{(1/2)b(x,y)\exp[j(\phi(x,y) + 2\pi f_0 x)]\exp(j\varphi_m)\} = |C|\exp(j\Phi)\exp(j\varphi_m), \tag{6}$$

where FT denotes the Fourier transform, and $|C|$ and Φ are the amplitude and phase of C in the frequency domain, respectively. Since the experimental parameters remain constant during the whole process of the measurement, the phase of $C(f_x, f_y)$

varies only with φ_m, that is, the modulated phase imposed by SLM. As a result, if we split the object beam into two parts which are separated by a gray reference value (zero) and variable gray value ranging from 0 to 255, subtracting the phase of $C(f_x, f_y)$ corresponding with each part gives the phase shift as follows:

$$C(f_x, f_y)_{rg} = |C| \exp(j\Phi) \exp(j\varphi_{rg}), \tag{7}$$

$$C(f_x, f_y)_{vg} = |C| \exp(j\Phi) \exp(j\varphi_{vg}), \tag{8}$$

$$\Delta\varphi = \arg\left[C(f_x, f_y)_{vg}\right] - \arg\left[C(f_x, f_y)_{rg}\right] + \varphi_0. \tag{9}$$

The subscripts rg, vg, and $\Delta\varphi$ indicate the reference, the variable gray values, and the phase shift, respectively. φ_0 is selected in which $\Delta\varphi = 0$ when the image loading on LCD has zero value. In the following, the experimental processes are expressed in two main stages: phase modulation and amplitude modulation.

3.1 Phase modulation

The gray-level images should be synthesized in such a way that the images are divided into two parts corresponding with the reference part (zero value) and the variable part (ranging from 0 to 255), as explained previously. These images loading on SLM cause different phase shifts between the variable and reference part resulting in a displacement of fringes. The amount of this phase shift depends directly on the orientation axes of retarders and linear polarizers and can be obtained mathematically by means of Eq. (9). The first row of **Figure 3** demonstrates three examples of the gray-level images loading on the LCD, whereas the second row indicates the corresponding interference patterns recorded by the CCD camera. The fringes were experimentally obtained by setting the axes of LP1 and LP2 at 0 and 90 degrees,

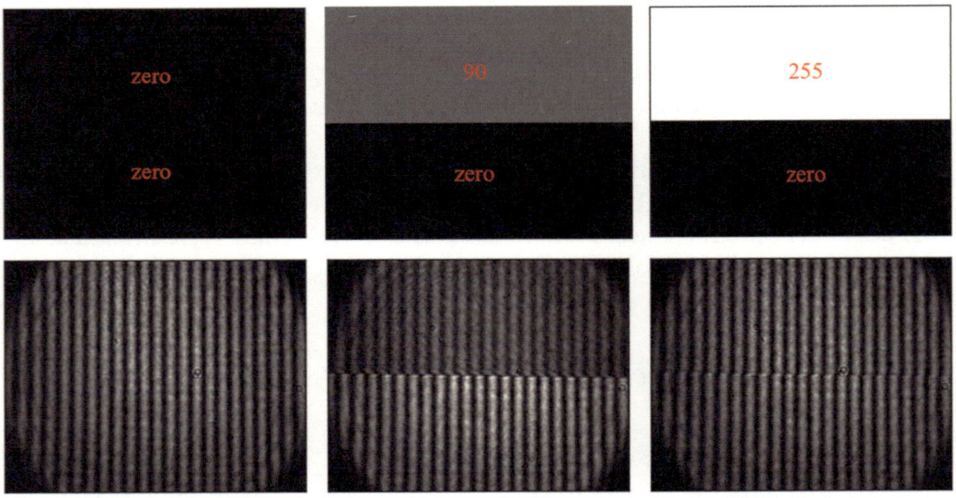

Figure 3.
The first row demonstrates three examples of gray-level images loaded on the LCD. The second row indicates the corresponding interference patterns recorded by the CCD camera.

respectively, with respect to the x-axis. In addition, the fast axes of HWP2, HWP1, and QWP1 were rotated −20, −63, and 45 degrees, respectively, with respect to the x-axis.

Then, two equal regions of the reference and variable part of each interference pattern should be selected for calculating the 2D fast Fourier transform, as indicated by the red rectangles in **Figure 4b**. As a result, the modulated phase can be obtained by applying Eq. (9) to the known frequency $C\left(f_x,f_y\right)$ for each gray value. Note that the frequencies are shown in **Figure 4c** that the first order of interference occurring remains constant for the variable and reference parts and all 256 interference patterns.

In this practical case, the maximum achievable phase imposed by SLM is around 1.75π. However, this value depends on the TNLC physical properties, the optical wavelength, and the polarization state of the incoming light. The obtained phase curve versus gray values is shown in **Figure 5a**.

3.2 Amplitude modulation

With the same configuration without changing the direction axes of linear polarizers or retarders, the amplitude curve can be obtained as follows: First, blocking the left arm of the Mach-Zehnder interferometer. Then, providing 256 gray-level images in which the entire images have the same gray value. Loading gray-level images on the LCD and subsequently recording the intensity pattern by the CCD camera. The modulated amplitude versus each gray value can be obtained by calculating the square root of the mean value of the recorded intensities and normalizing them to the maximum obtained amplitude. Also, this process can be done by using an intensity detector. The experimentally obtained amplitude curve versus gray values is shown in **Figure 5b**. Finally, Eq. (1) can be obtained for each gray value by combining the amplitude and phase responses of the SLM in consequence of the present

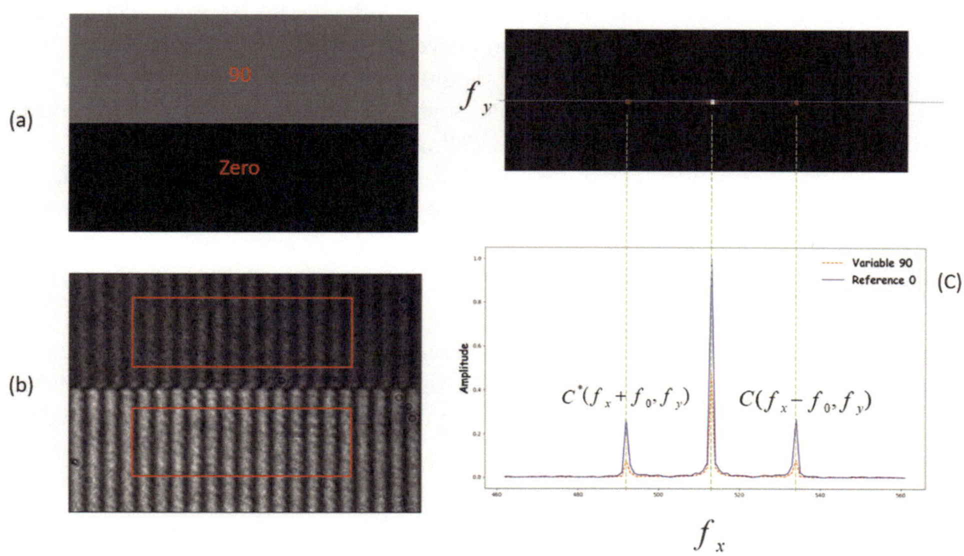

Figure 4.
(a) An example of synthesized gray-level image loading on LCD. (b) The recorded fringe pattern. The red selected rectangles have been used for calculating the 2D Fourier transform. (c) The Fourier domain.

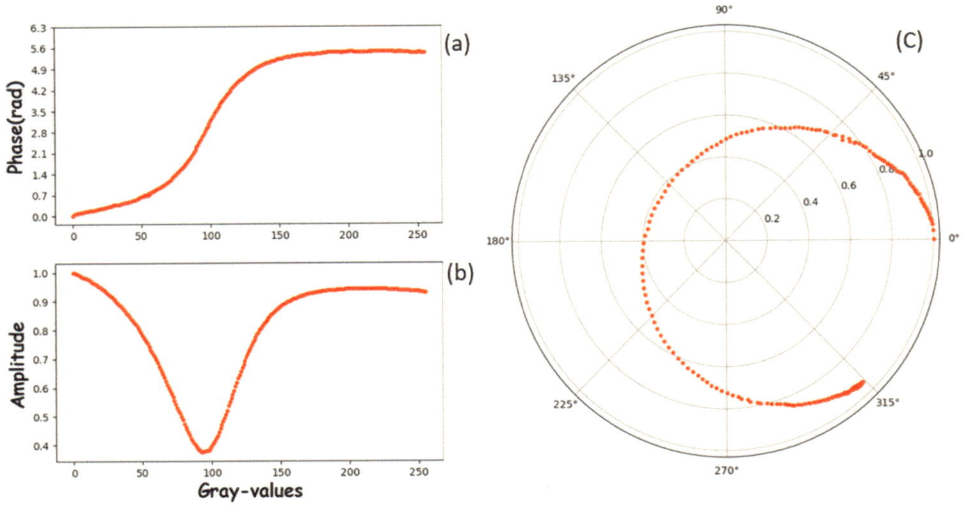

Figure 5.
(a) The phase modulation. (b) The amplitude modulation. (c) The polar plot of the phase-mostly modulation.

configuration. This process is called the characterization or calibration of SLM. The polar plot shown in **Figure 5c** indicates the obtained phase-mostly modulation curve.

4. Computer-generated hologram: Double-pixel Arrizón's approach

Very often, TNLCs provide a coupled amplitude-phase modulation, as indicated in **Figure 5**. This section aims to review the DPH Arrizón's approach to expanding the accessible modulations beyond the restricted SLM response. Then, we review the modified DPH approach by applying 4 pixels to encode one holographic cell adapted to the experimentally obtained modulation curve. As a result, the approach is able to generate an on-axis computer-generated hologram with the optimum reconstruction efficiency, maximum signal bandwidth, and high SNR suitable for encoding arbitrary complex modulation into a low-resolution TNLC display.

Considering the obtained phase-mostly modulation curve shown in **Figure 5**, assume each complex modulation point belongs to the modulation curve as follows:

$$M_g = |M_g| \exp\left(j\psi_g\right), \tag{10}$$

where subscript g denotes gray values, which is an integer value between 0 and 255. In addition, the pixelated structure of the display is considered as a matrix with $N \times M$ arrays (pixels). The modulation M_{nm} in the (n,m) the pixel can be defined as follows:

$$M_{nm} = |M_{nm}| \exp\left(j\psi_{nm}\right). \tag{11}$$

To encode a desired complex modulation value

$$q_{mn} = |q_{nm}| \exp\left(j\tau_{nm}\right), \tag{12}$$

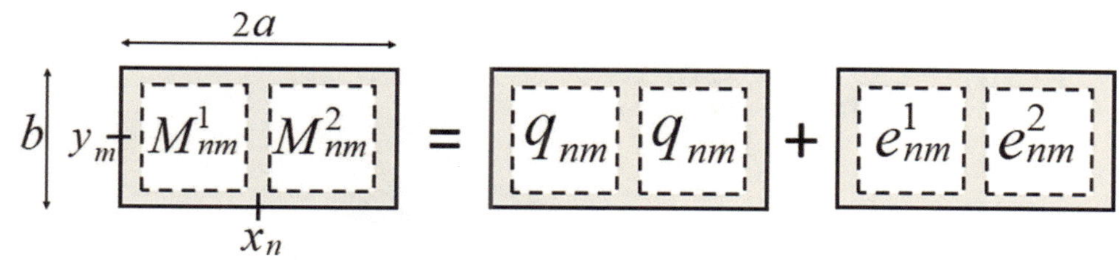

Figure 6.
The DPH configuration.

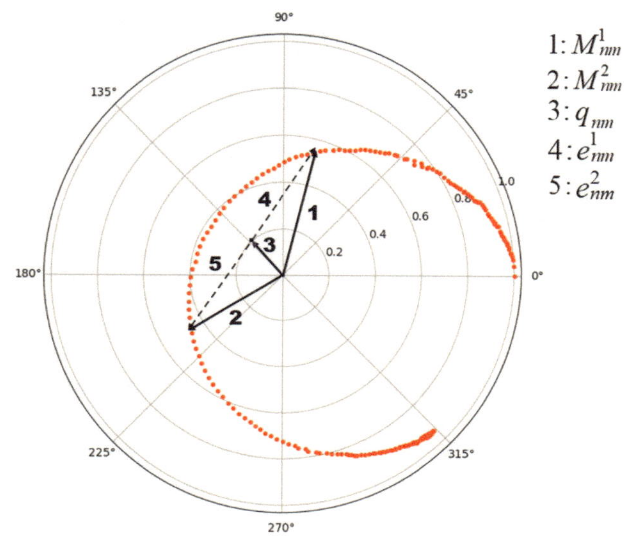

Figure 7.
The codification algorithm.

Arrizón employed the holographic double pixel shown in **Figure 6**, whose pixels have the complex modulation M^1_{nm} and M^2_{nm} that belong to the modulation curve. As shown in **Figure 6**, the holographic cell is equal to a double pixel with the encoded modulation q_{nm} plus an error double pixel, with the modulation values e^1_{nm} and e^2_{nm}. The conditions required to produce an on-axis signal reconstruction with a null contribution of the error term at the zero frequency are

$$q_{nm} = \left(M^1_{nm} + M^2_{nm}\right)/2, \tag{13}$$

$$e^2_{nm} = -e^1_{nm}. \tag{14}$$

To explain this, considering the modulation points on the modulation curve (M_g) as a vector with the origin of the polar plot shown in **Figure 7**, encoded modulation points (q) are obtained by the average of the superposition of vectors M^1 and M^2. As a result, we can access the modulation points (q) beyond the restricted SLM responses (M_g). Consequently, the modulation errors are $e^1_{nm} = M^1_{nm} - q_{nm}$ and $e^2_{nm} = M^2_{nm} - q_{nm}$, which lead to Eq. (14).

Regarding the experimental modulation curve, all possible complex values that can be obtained with this codification algorithm are shown with the green points in **Figure 8**. However, only those complex values that fall inside the blue circle with a

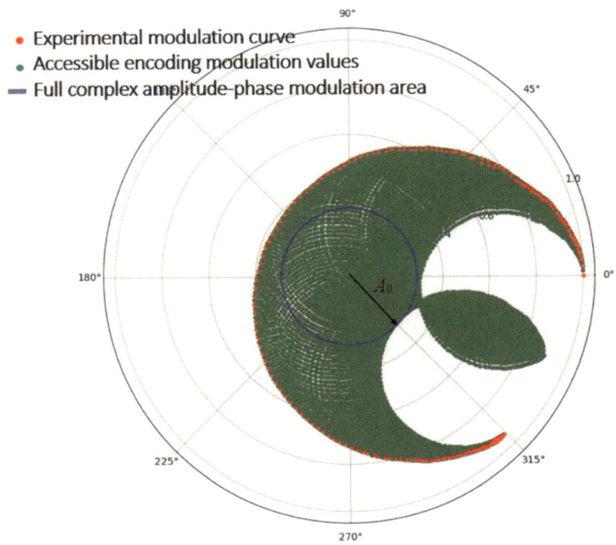

Figure 8.
The red and green points show the experimental modulation curve and all possible complex values using the DPH approach. The green points inside the blue circle with the radius A_o = 0.29 are those accessible complex values to encode a complex function.

radius $A_0 = 0.29$ can encode a complex function with an amplitude ranging from 0 to A_0 and a phase ranging from 0 to 2π.

To go through Arrizón's approach in more detail, assume the transmittance of the CGH that can be displayed on the LCD is

$$h(x,y) = \sum_{n,m} M_{nm} w(x - np, y - mp), \tag{15}$$

where p is the pixel pitch and $w(x,y) = \text{rect}(x/a)\text{rect}(y/b)$. Considering the CGH is intended to encode the spatially quantized complex function

$$q(x,y) = \sum_{nm} q_{nm} w(x - np, y - mp), \tag{16}$$

where q_{nm} is defined in Eq. (12) in which $|q_{nm}| \leq 1$. Assuming the spectrum of $q(x,y)$ denoted $Q(u,v)$ is centered at the zero frequency $(u,v) = (0,0)$. Hence, the CGH transmittance must be related to the encoded complex modulation $q(x,y)$ by the following expression:

$$h(x,y) = A_0 q(x,y) + e(x,y), \tag{17}$$

then, the Fourier transform of Eq. (17) gives

$$H(u,v) = A_0 Q(u,v) + E(u,v). \tag{18}$$

The error spectrum, $E(u,v)$, should be negligible within the largest possible band centered at the zero frequency to obtain a high SNR. So, Arrizón proposed an error function as follows:

$$e(x,y) = l(x,y)g(x,y), \tag{19}$$

$$l(x,y) = \sum_{n,m} l_{nm} w(x - np, y - np), \tag{20}$$

$$g(x,y) = \sum_{n,m} g_{nm} w(x - np, y - np). \tag{21}$$

He demonstrated that the optimal choice $l(x,y)$ is the binary grating with discrete modulation $l_{nm} = (-1)^{n+m}$. In this regard, the Fourier transform of the error function contributed by the noise field is given mainly by four off-axis replicas of the function $G(u,v)$ centered at the spatial frequency coordinates $(1/2p, 1/2p)$, $(-1/2p, 1/2p)$, $(1/2p, -1/2p)$, and $(-1/2p, -1/2p)$. Therefore, the reconstructed field with a zero noise contribution places on the optical axis while symmetric off-axis error contributions occur far enough from the optical axis at the Fourier plane, which should be removed using a 4-f spatial filtering system.

According to Eqs. (17) to (21), $g(x,y)$ is specified by its discrete modulation l_{nm}, which is related to the CGH modulation by the formula

$$M_{nm} = A_0 q_{nm} + (-1)^{n+m} g_{nm}. \tag{22}$$

Since both function $q(x,y)$ and $g(x,y)$ have on-axis spectrum bands, their variation should be negligible when the increment $(\Delta n = 1)$ in x is of the order of the pixel pitch. To satisfy this condition, Arrizón proposed to establish the discrete function g_{nm} such that both complex vectors,

$$M_{nm}^1 = A_0 q_{nm} + g_{nm}, \tag{23}$$

$$M_{nm}^2 = A_0 q_{nm} - g_{nm}, \tag{24}$$

belong to the SLM modulation curve. Note that Eqs.(23) and (24) are the general forms of Eqs. (13) and (14). The constant value A_0 is a maximum possible amplitude (for instance, the radius of the blue circle shown in **Figure 8**) to fulfill complex amplitude-phase modulation, in which the average of each pair of modulation points (M_{nm}^1, M_{nm}^2) on the modulation curve should be an interior point inside the circle. As a result, the pair of modulation points (M_{nm}^1, M_{nm}^2) always exist. Besides, the maximum CGH efficiency is related to the maximum possible valueA_0.

Since the set of the modulation points is finite and discrete, we should find the nearest accessible complex value denoted q_{nm}^a to the desired complex value $A_0 q_{nm}$. Thus, practically we select the pair modulation points (M_{nm}^1, M_{nm}^2) in such a way that its middle point has the minimum Euclidean distance from the desired complex value corresponding with each holography cell as follows:

$$\varepsilon_{nm} = \min \left| A_0 q_{nm} - q_{nm}^a \right|. \tag{25}$$

The remaining issue is to select the position of M_{nm}^1 and M_{nm}^2 on the modulation curve. In this regard, there are two possibilities as follows:

- Selection 1: M_{nm}^1 performs a clockwise rotation (smaller than 180°) of the radial line containing q_{nm}^a, where M_{nm}^1 has a smaller phase than M_{nm}^2 as demonstrated in **Figure 9a**.

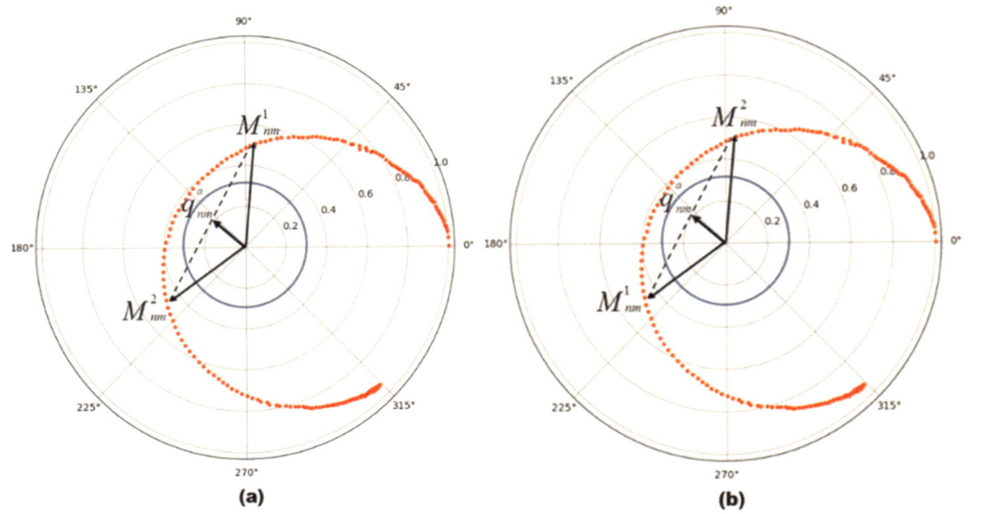

Figure 9.
(a) Selection 1 (b) selection 2.

- Selection 2: M_{nm}^1 performs a counter-clockwise rotation of this radial line, where M_{nm}^2 has a smaller phase than M_{nm}^1, as indicated in **Figure 9b**.

On one side, he showed that the appropriate position-selection of the pair modulation points to encode a discrete modulation $q_{nm} = |q_{nm}| \exp(j\tau_{nm})$, where both the modulus $|q_{nm}|$ and the phase τ_{nm} are soft or quasi-continuous functions, is a way explained in selection 1, which leads to

$$M_{nm} = \begin{cases} M_{nm}^1 & (n+m)\text{even} \\ M_{nm}^2 & (n+m)\text{odd} \end{cases}. \tag{26}$$

This configuration is shown in **Figure 10**.

On the other side, he demonstrated that the appropriate position-selection for the pair modulation points to encode complex functions of the type $q_{nm} = r_{nm} \exp\left(j\tau_{nm}^0\right)$, where both r_{nm}, which is a real factor, and τ_{nm}^0 are quasi-continuous functions, is as follows:

$$M_{nm} = \begin{cases} \begin{cases} M_{nm}^1 & (n+m)\text{even} \\ M_{nm}^2 & (n+m)\text{odd} \end{cases} & r_{nm} \geq 0 \\ \begin{cases} M_{nm}^2 & (n+m)\text{even} \\ M_{nm}^1 & (n+m)\text{odd} \end{cases} & r_{nm} < 0 \end{cases} \tag{27}$$

This configuration is shown in **Figure 11**. The display plane is divided into two areas: Area U corresponds $r \geq 0$ and area D corresponds $r < 0$.

This encoding algorithm is also appropriate for encoding continuous functions of the type $F(r, \theta) = R(r) \exp(jt\theta)$ where t is the topological charge, (r, θ) are polar coordinates, and $R(r)$ is a real function of a radial coordinate. In addition, if we intend to encode the complex function $F_{nm} = |r_{nm}| \exp(j\theta_{nm})$ the property of this function

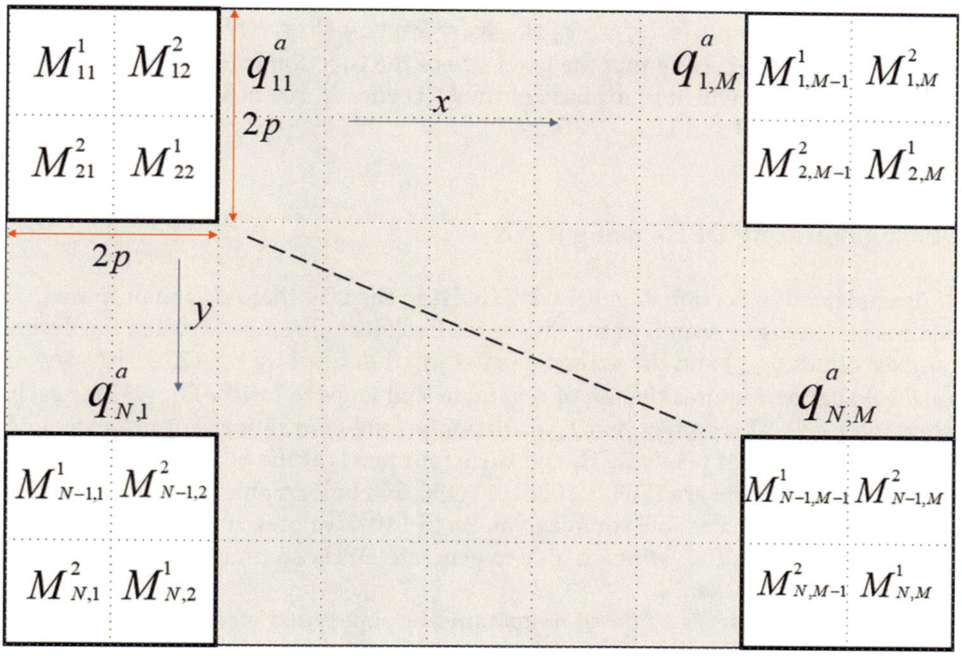

Figure 10.
The representation of the modified DPH at the LCD plane applying selection 1.

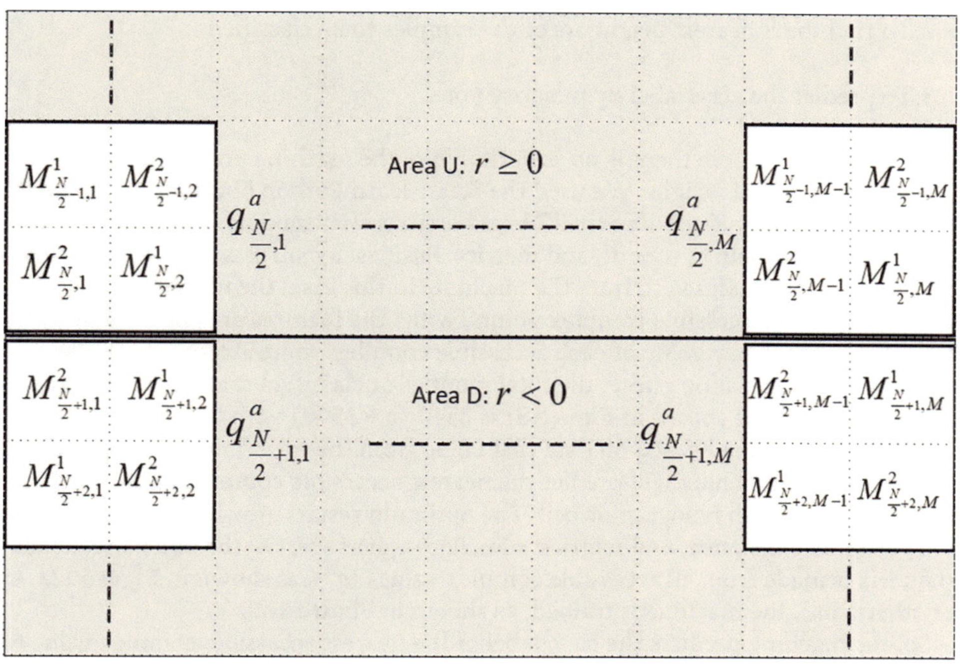

Figure 11.
The representation of the modified DPH at the LCD plane applying selections 1 and 2. The order of the distribution of values M^1_{nm} and M^2_{nm} in the transition area from U to D is changed.

should be defined as $r_{nm} = \left(x_{nm}^2 + y_{nm}^2\right)^{1/2}, x_n = 2npy_n = 2mp$, $-N/2 \leq n < N/2$, and $-M/2 \leq m < M/2$. Note that the pixel size of the function is twice the pixel size of the applied LCD, in which four pixels of the LCD encode one holographic cell, as shown in **Figures 10** and **11**.

5. Fast generating DPHs using k-NN

As explained in Section 4, generating CGHs in the way that Arrizón proposed requires an extensive search of the minimum Euclidean distance between the desired complex values (q_{nm}) and the accessible ones (q_{nm}^a) defined by Eq. (25). The conventional calculation requires the use of several nested loops to satisfy Eq. (25) for each holographic cell. This conventional calculation is time-consuming. For instance, to generate a 768 × 1024 pixels CGH, and since four pixels at the SLM plane provide one holographic cell, there are (768 × 1024)/4 = 196,608 holographic cells that should be mapped among all accessible complex values. In [40], we presented a method using the k-NN classifier, which is able to generate DPHs 80 times faster than the conventional calculation.

The k-NN classifier is a type of nonparametric supervised machine learning algorithm [41]. Nonparametric models are characterized by memorizing the training datasets instead of learning them. This technique aims to train a dataset to label them into different known classes based on defined features. This algorithm can be described in three main steps:

1. To select the optimum number of k based on a distance metric.

2. To find the k nearest neighbors of the samples to be classified.

3. To predict the class label by majority vote.

In those cases where there is no majority vote, the machine predicts the class label based on the defined weight. We used the Scikit-learn Python library to implement the k-NN algorithm. Regarding the KNeighborsClassifier module, three parameters should be determined: k, weight, and metrics. Besides, a matrix with a-samples and b-features should be defined to train the machine. In this case, the number of samples is the number of all accessible complex points, while the features are chosen based on the real and imaginary value of each accessible complex point. According to the experimental modulation curve, the total number of classified samples (green points inside the blue circle shown in **Figure 8**) is 3540 (a = 3540) with two features (b = 2). The machine is trained based on data that come from the experimental modulation curve, while the machine will predict the nearest accessible complex values to the desired ones for each holographic cell. The optimum results are obtained by choosing k = 1, weights = distance, and metrics = Euclidean distance. On the one hand, a look-up matrix is made from all accessible complex values (q^a), as shown in **Figure 12a**. On the other hand, the machine is trained, as shown in **Figure 12b**.

So, the machine predicts the class label of the nearest accessible complex value to the desired one. According to the predicted label and look-up matrix, the pair of gray-level (M_g^1,M_g^2) will be distributed to the corresponding holographic cell, as shown in **Figures 10** and **11**.

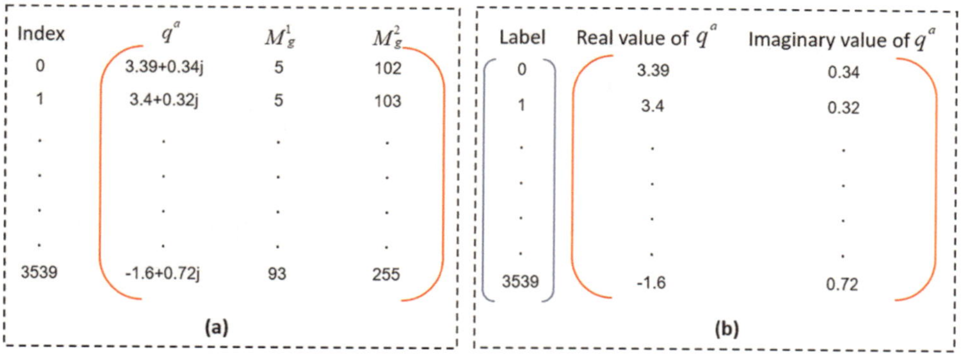

Figure 12.
(a) The provided look-up matrix with 3540 rows and 3 columns. (b) the configuration of labeled training dataset with 3540 samples and two features.

Here, two beam profiles are considered to generate their CGHs. The first one is a (1,1)-Hermite-Gaussian (HG_{11}) with the wave equation given by

$$HG_{11} = \frac{4xy}{w_0{}^2} \exp\left(-\left(\frac{r}{w_0}\right)^2\right) circ\left(\frac{r}{R}\right), \tag{28}$$

where $r = \sqrt{x^2 + y^2}$, w_0 is the beam waist ($w_0 = R/2$), and R is the radius of the circular beam support. The first row of **Figure 13** indicates the amplitude and phase of HG_{11}, respectively, whereas the second row demonstrates the nearest accessible values predicted by the k-NN classifier according to the experimental modulation curve.

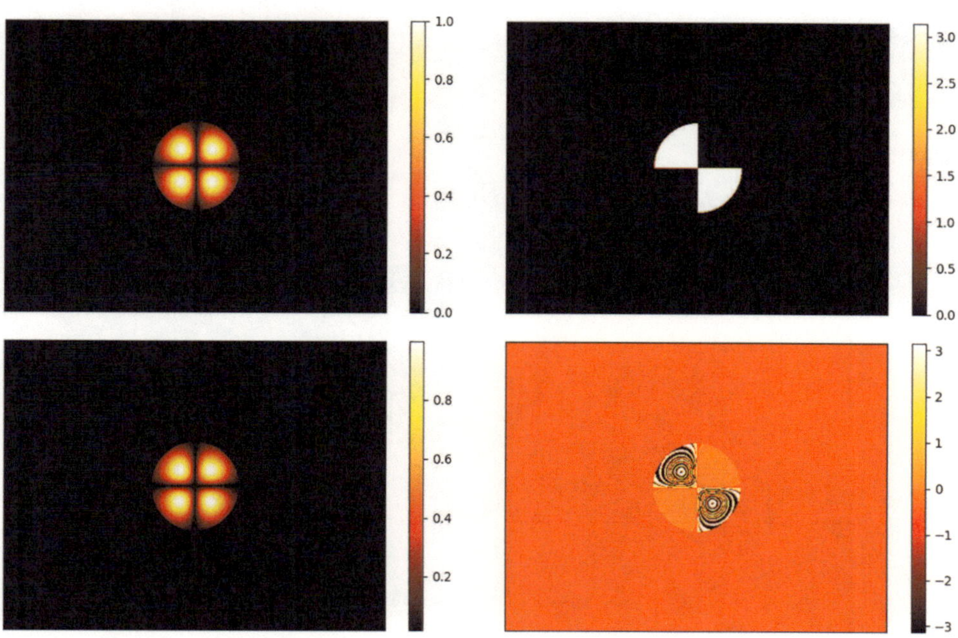

Figure 13.
The first row shows the distribution of the amplitude and phase of HG11, respectively. The second row shows their nearest values predicted by the k-NN classifier according to the experimental modulation curve.

The second beam profile is a (0,1)-Laguerre-Gaussian (LG_{01}), in which the complex wave equation is given by

$$LG_{01} = \left[\left(\frac{2x}{w_0}\right) + j\left(\frac{2y}{w_0}\right)\right] \exp\left(-\left(\frac{r}{w_0}\right)^2\right) circ\left(\frac{r}{R}\right) \tag{29}$$

The amplitude and phase distribution of LG_{01} and their nearest accessible values predicted by the machine are shown in the first and second rows of **Figure 14**. The corresponding CGHs for HG_{11} and LG_{01} are shown in **Figure 15**.

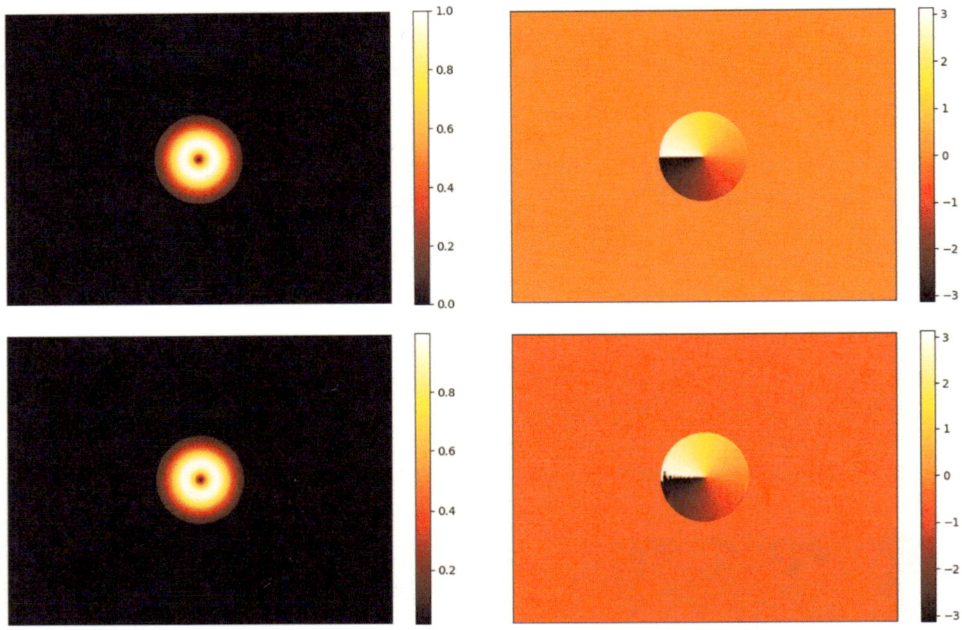

Figure 14.
The first row shows the numerical distribution of the amplitude and phase of LG_{01}*, respectively. The second row shows their nearest values predicted by the k-NN classifier.*

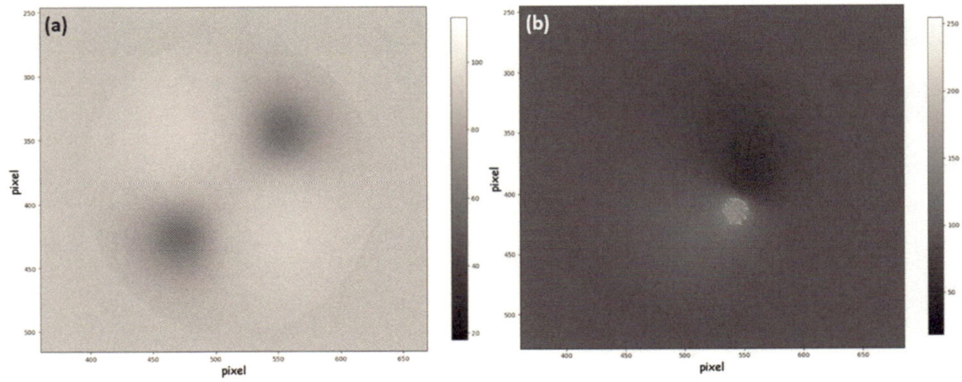

Figure 15.
The CGHs correspond with (a) HG_{11} *and (b)* LG_{01}*. The practical part of CGHs is selected for illustration purposes.*

Pixels	250 × 250	512 × 512	768 × 1024	1152 × 1920
HG_{11}	0.16	0.64	1.92	5.47
LG_{01}	0.38	1.54	5.53	16.39

Table 1.
Processing time (in seconds) for generating CGHs using the k-NN classifier.

Table 1 indicates the required time to generate CGHs corresponding with HG_{11} and LG_{01} with four different resolutions. Numerical calculations have been carried out using Python 3.7.5 and the Scikit-learn library, and a Laptop with CPU i7-4510U (2 GHz) and 6 GB RAM. Besides, the processing time is obtained by the timeit module, and the results are averages of 10 runs. Note that the k-NN classifier can be accelerated using the RAPID cuML library performing on GPUs. As reported in [42], performing k-NN using the RAPID cuML library on GPUs is 600 times faster than performing k-NN applying the Scikit-learn library on CPUs. As a result, this proposed approach can generate real-time double-pixel computer-generated holograms.

6. Experimental results

The experimental setup (sketched in **Figure 1**) is modified in order to generate and analyze the beam at the entrance pupil and focal plane of a highly focusing system, as shown in **Figure 16**.

The left arm of the Mach-Zehnder interferometer is blocked by an obstacle. A vortex retarder (ThorLab, WPV10L-532) or a QWP (R) is added after LP2 to provide a radially or circularly polarized beam, respectively. In order to provide a radial polarization, the fast axis of the vortex retarder is placed parallel to the y-axis. In order to provide a circular polarization, the fast axis of QWP is rotated 45° with respect to the x-axis.

The beam is separately imaged at the entrance pupil of microscope objective MO1 (Nikon Plan Fluorite N40X-PF with NA = 0.75) and at the sensor plane of CCD1 by

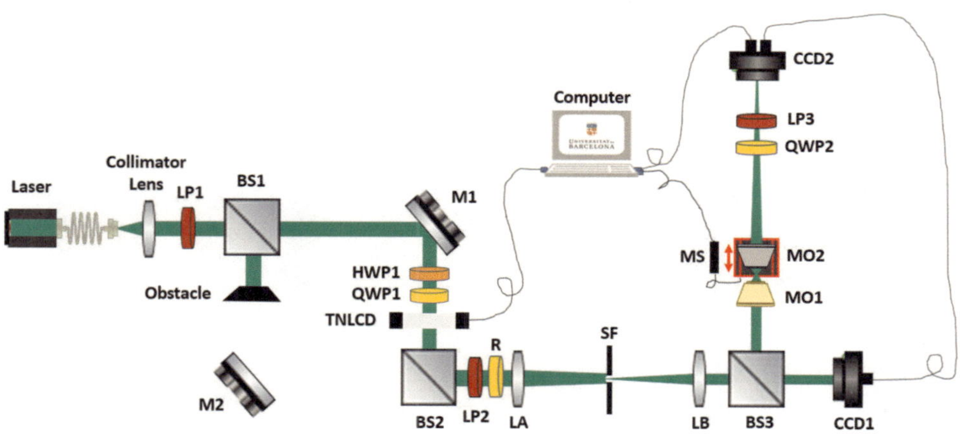

Figure 16.
The sketch of the experimental setup to capture the beam profile at the SLM and focal plane. R stands for a retarder that can be a QWP or vortex for providing circularly or radially polarized beam, respectively. MO stands for microscope objective. MS stands for the movable stage.

means of the 4f-system and BS3. Microscope objective MO2 (Nikon with NA = 0.8) is mounted on a movable stage driven by a motorized device (Newport LTA-HL) with uni-directional repeatability of ±100nm. MO2 is used to scan different planes close to the focal plane of MO1 and image them to the sensor plane of the CCD2 camera. Note that MO2 has a larger NA than MO1 to collect the entire beam. Furthermore, the actual magnification of the imaging system provided by MO2 is obtained by imaging a USAF target placed in the front of MO2, resulting in a 100x and spatial sampling of 37.5 nm. LP3 and QWP2 are used to record a set of six polarimetric images.

We recently used this experimental setup to estimate the longitudinal component of a highly focused beam using a phase retrieval algorithm and Gauss's theorem. The method is able to retrieve transverse and longitudinal components of a highly focused electromagnetic field (for more details, see [43]).

Since the intensity pattern of a beam at the focal plane strongly depends on its polarization state at the entrance pupil of MO1 [44], two different polarization states have been considered to compromise the experimental results with the numerical ones. The numerical calculations have been implemented by applying the focused field calculation method introduced in [16]. **Figure 17** shows the intensity patterns recorded by CCD1 regarding HG_{11} and LG_{01} beams.

Figure 18 indicates the intensity patterns of the circularly polarized HG_{11} beam at the focal plane of MO1. The first row indicates the Stokes images, which were obtained numerically, whereas the second row shows the intensity measurement of

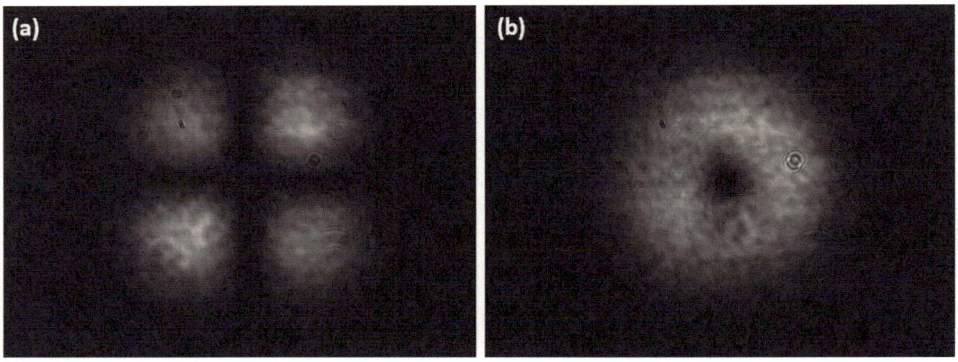

Figure 17.
The intensity patterns recorded by CCD1 correspond to (a) HG_{11} *beam and (b)* LG_{01} *beam.*

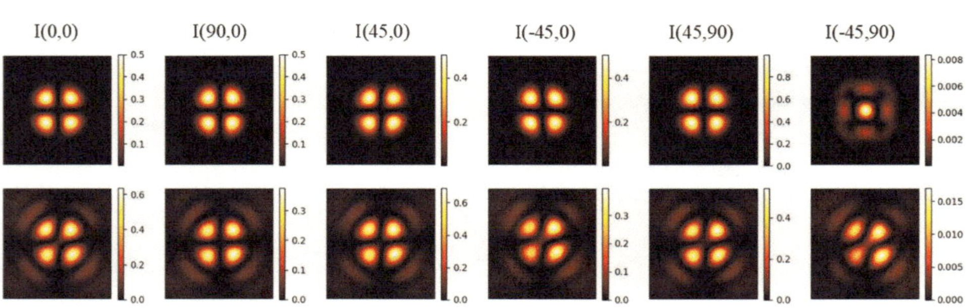

Figure 18.
The stokes images correspond to the circularly polarized HG_{11} *beam at the focal plane. The first row shows the numerical results, while the second row demonstrates the recorded intensity by CCD2. The size of each image is 3 μm.*

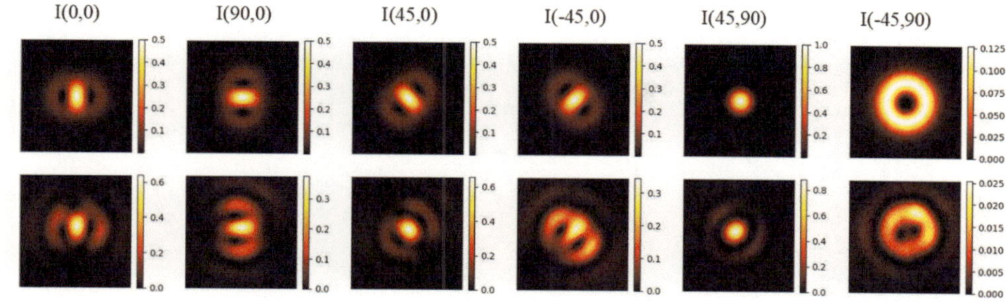

Figure 19.
The stokes images correspond to the radially polarized LG_{01} *beam at the focal plane. The first row shows the numerical results, while the second row demonstrates the recorded intensity by CCD2. The size of each image is* $3\ \mu m$.

the Stokes images, which were recorded by CCD2. The Stokes images are denoted by $I(\theta, \delta)$, where θ and δ are the rotation angles of the axis of LP3 and the phase delay introduced by means of QWP2 with respect to the x-axis, respectively. Moreover, the polarimetric images are normalized by the maximum intensity of the transverse components of the electromagnetic field.

In a similar way, **Figure 19** indicates the Stokes images correspond to the radially polarized LG_{01} beam. As results show, the obtained Stokes images are in excellent agreement with the numerical ones. However, the state of polarization is altered slightly due to the imperfection of applied retarders.

7. Conclusions

This chapter provided all the necessary steps to generate complex beams with arbitrary intensity and phase distribution using a translucent TNLC display in one frame. Characterizing a TNLC-SLM accompanied by the DPH Arrizon's approach has been widely reviewed, and the k-NN classifier has been applied to generate CGHs faster than conventional calculation. Two wave functions have been experimentally assessed at the SLM and the focal plane of a high-NA microscope objective. Finally, the experimental setup has been described in order to generate focused vector beams and measure the corresponding Stokes images.

Acknowledgements

The author acknowledges support from the PredocsUB program.

Author details

Kavan Ahmadi
University of Barcelona, Barcelona, Spain

*Address all correspondence to: k1ahmadi@ub.edu

References

[1] Maria-Luisa C, Albertina C, Arrizón V. Phase shifting digital holography implemented with a twisted-nematic liquid-crystal display. Applied Optics. 2009;**48**:6907-6912

[2] Maluenda D, Juvells I, Martínez-Herrero R, Carnicer A. A digital holography technique for generating beams with arbitrary polarization and shape. Proceedings of SPIE, Optical System Design. 2012:**8550**:84403Q. DOI: 10.1117/12.979926

[3] Davis JA, McNamara DE, Cottrell DM, Sonehara T. Two-dimensional polarization encoding with a phase-only liquid-crystal spatial light modulator. Applied Optics. 2000;**39**:1549-1554

[4] Shen C, Tan J, Wei C, Liu Z. Coherent diffraction imaging by moving a lens. Optics Express. 2016;**24**:16520-16529

[5] Rong Z-Y, Han Y-J, Wang S-Z, Guo C-S. Generation of arbitrary vector beams with cascaded liquid crystal spatial light modulators. Optics Express. 2014;**22**:1636-1644

[6] Martínez-Herrero R, Maluenda D, Juvells I, Carnicer A. Experimental implementation of tightly focused beams with unpolarized transversal component at any plane. Optics Express. 2014;**22**: 32419-32428

[7] Martínez-Herrero R, Maluenda D, Juvells I, Carnicer A. Synthesis of highly focused fields with circular polarization at any transverse plane. Optics Express. 2014;**22**:6859-6867

[8] Moreno I, Campos J, Gorecki C, Yzuel MJ. Effects of amplitude and phase mismatching errors in the generation of a Kinoform for pattern recognition. Japanese Journal of Applied Physics. 1995;**34**:6423-6432

[9] Lin S, Li C, Kuo C, Yeh H. Fresnel lenses in 90° twisted-Nematic liquid crystals with optical and electrical controllability. IEEE Photonics Technology Letters. 2016; **28**(13):1462-1464. DOI: 10.1109/ LPT.2016.2555699

[10] Lin C, Huang H, Wang J. Polarization-independent liquid-crystal Fresnel lenses based on surface-mode switching of 90° twisted-Nematic liquid crystals. IEEE Photonics Technology Letters. 2010;**22**(3):137-139. DOI: 10.1109/LPT.2009.2036738

[11] Maluenda D, Carnicer A, Martínez-Herrero R, Juvells I, Javidi B. Optical encryption using photon-counting polarimetric imaging. Optics Express. 2015;**23**:655-666

[12] Wu CW, Thompson G, Wright SL. Multiple images viewable on twisted-nematic mode liquid-crystal displays. IEEE Signal Processing Letters. 2003; **10**(8):225-227. DOI: 10.1109/ LSP.2003.814394

[13] Cheng C-J, Chen M-L, Tu H-Y. Polarization encoding for multi-channel optical encryption using twisted Nematic liquid crystal displays. In: Pacific Rim Conference on Lasers and Electro-Optics. Vol. 2005. Tokyo: CLEO- Technical Digest; 11-15 Jul 2005. pp. 981-982. DOI: 10.1109/CLEOPR. 2005.1569632

[14] Dorn R, Quabis S, Leuchs G. Sharper focus for a radially polarized light beam. Physics Review Letters. 2013;**91**:233901

[15] Davidson N, Bokor N. High-numerical-aperture focusing of radially polarized doughnut beams with a parabolic mirror and a flat diffractive lens. Optics Letters. 2004;**29**:1318-1320

[16] Leutenegger M, Rao R, Leitgeb RA, Lasser T. Fast focus field calculations. Optics Express. 2006;**14**:11277-11291

[17] Kozawa Y, Sato S. Sharper focal spot formed by higher-order radially polarized laser beams. Journal of the Optical Society of America. A. 2007;**24**: 1793-1798

[18] Wang H, Shi L, Lukyanchuk B, Sheppard C, Chong CT. Creation of a needle of longitudinally polarized light in vacuum using binary optics. Nature Photonics. 2008;**2**:501-505

[19] Lerman G, Levy U. Effect of radial polarization and apodization on spot size under tight focusing conditions. Optics Express. 2008;**16**:4567-4581

[20] Hao X, Kuang C, Wang T, Liu X. Phase encoding for sharper focus of the azimuthally polarized beam. Optics Letters. 2010;**35**:3928-3930

[21] Khonina SN, Volotovsky SG. Controlling the contribution of the electric field components to the focus of a high-aperture lens using binary phase structures. Journal of the Optical Society of America. A. 2010;**27**: 2188-2197

[22] Moreno I, Bennis N, Davis JA, Ferreira C. Twist angle determination in liquid crystal displays by location of local adiabatic points. Optics Communications. 1998;**158**:231-238

[23] Davis JA, Moreno I, Tsai P. Polarization eigenstates for twisted-nematic liquid-crystal displays. Applied Optics. 1998;**37**:937-945

[24] Ponce R, Serrano-Heredia A, Arrizón VM. Simplified optimum phase-only configuration for a TNLCD. Proceedings of SPIE. 2004; **5556**:206-213

[25] Pezzaniti JL, Chipman RA. Phase-only modulation of a twisted nematic liquid-crystal TV by use of the eigenpolarization states. Optics Letters. 1993;**18**:1567-1569

[26] Chiang Y, Chou T, Chen S, Chao C. Effects of low viscosity liquid on the electro-optical properties of inverse twisted Nematic liquid crystal display. IEEE Transactions on Electron Devices. 2017;**64**:1630-1634

[27] Matín-Badosa E, Carnicer A, Juvells I, Vallmitjana S. Complex modulation characterization of liquid crystal devices by interferometric data correlation. Measurement Science and Technology. 1997;**8**:764-772

[28] Wang H, Dong Z, Fan F, Feng Y, Lou Y, Jiang X. Characterization of spatial light modulator based on the phase in Fourier domain of the hologram and its applications in coherent imaging. Applied Sciences. 2018;**8**:1146. DOI: 10.3390/app8071146

[29] Wu YH, Chavel P. Cell-oriented on-axis computer-generated holograms for use in the Fresnel diffraction mode. Applied Optics. 1984;**23**:228-238

[30] Hsueh CK, Sawchuk AA. Computer-generated double-phase holograms. Applied Optics. 1987;**17**:3874-3883

[31] Florence JM, Juday RD. Full-complex spatial filtering with a phase mostly DMD. In: Wave Propagation and Scattering in Varied Media II. San Diego, CA, United States: SPIE 1558; 1991. DOI: 10.1117/12.49655

[32] Mendlovic D, Shabtay G, Levi U, Zalevsky Z, Emanuel M. Encoding technique for design of zero-order (on-axis) Fraunhofer computer-generated holograms. Applied Optics. 1997;**36**:8427-8434

[33] Arrizón V. Improved double-phase computer-generated holograms implemented with phase-modulation devices. Optics Letters. 2002;**27**:595-597

[34] Arrizón V, Sánchez-de-la-Llave D. Double-phase holograms implemented with phase-only spatial light modulators: Performance evaluation and improvement. Applied Optics. 2002;**41**: 3436-3447

[35] Arrizón V. Complex modulation with a twisted-nematic liquid-crystal spatial light modulator: Double-pixel approach. Optics Letters. 2003; **28**(15):1359-1361. DOI: 10.1364/ ol.28.001359

[36] Arrizón V. Optimum on-axis computer-generated hologram encoded into low-resolution phase-modulation devices. Optics Letters. 2003;**28**: 2521-2523

[37] Arrizón V, González LA, Ponce R, Serrano-Heredia A. Computer-generated holograms with optimum bandwidths obtained with twisted-nematic liquid-crystal displays. Applied Optics. 2005; **44**:1625-1634

[38] Saleh BEA, Teich MC. Electro-optics. In: Saleh BEA, Teich MC, editors. Fundamentals of Photonics. 2nd ed. Hoboken, New Jersey: Wiley & Sons; 2007. pp. 696-736

[39] Takeda M, Ina H, Kobayashi S. Fourier-transform method of fringe-pattern analysis for computer-based topography and interferometry. Journal of the Optical Society of America. 1982; **72**:156-160

[40] Ahmadi K, Maluenda D, Carnicer A. Fast mapping of double-pixel holograms using k-nearest neighbors. In: OSA Imaging and Applied Optics Congress. Washington, DC United States: OSA Technical Digest: DW5E-7. 19-23 Jul 2021. DOI: 10.1364/DH.2021.DW5E.7

[41] Raschka S. Python Machine Learning. Birmingham: Packt Publishing; 2015

[42] Deotte C. Accelerating k-Nearest Neighbors 600x using RAPIDS cuML [Internet]. 2005. Available from: https:// medium.com/rapids-ai/accelerating-k-nearest-neighbors-600x-using-rapids-cuml-82725d56401e

[43] Maluenda D, Aviñoá M, Ahmadi K, Martínez-Herrero R, Carnicer A. Experimental estimation of the longitudinal component of a highly focused electromagnetic field. Scientific Reports. 2021;**11**:17992. DOI: 10.1038/ s41598-021-97164-z

[44] Novotny L, Hecht B. Propagation and focusing of optical fields. In: Novotny L, Hecht B, editors. Principales of Nano-Optics. New York: Cambridge University Press; 2006. pp. 45-87

Chapter 18

Correlations in Scattered Phase Singular Beams

Vanitha Patnala, Gangi Reddy Salla and Ravindra Pratap Singh

Abstract

We discuss about the correlations present in the scattered phase singular beams and utilize them for obtaining the corresponding mode information. We experimentally generate the coherence vortices using the cross-correlation functions between the speckle patterns and validate them with the exact analytical expressions. We also explore their propagation characteristics by considering their geometry along with their divergence. We utilize the autocorrelation measurements of speckle patterns for obtaining the mode information. Further, we study the correlations present in scattered perfect optical vortices which lead to a new class of coherence functions, Bessel-Gauss coherence functions, and utilized for generating the non-diffracting random fields, i.e. propagation invariant fields. We utilized these correlation functions, which are order-dependent although the speckle patterns are order-independent, for encrypting the information which has higher advantage than normal random optical fields.

Keywords: phase singular beams, scattering, speckles, correlation function, optical encryption

1. Introduction

Phase singular beams or optical vortices are well known due to their applications in guiding the particles, for coding larger information per photon, for transferring spatial structure to the materials, and for enhanced accuracy in metrological measurements [1–10]. These beams have ring-shaped intensity distribution along with helical wavefronts and have phase singularity at the center [11–16]. These beams carry an orbital angular momentum (OAM) of $m\hbar$ per photon due to its azimuthal phase, where m is the order or topological charge defined as number of helices completed in one wavelength. The propagation of these light beams through various media such as turbid media [17, 18], turbulence atmosphere [19, 20], and under water [21, 22] have attracted lot of interest in recent days for utilizing them for communication applications [23–30]. The vortices can be generated using computer generated holography [31, 32] along with the help of spatial light modulator [33, 34], spiral phase plate [35, 36], and using an astigmatic mode converter [37]. Some advanced techniques have been introduced for generating vortex beams through laser cavity and using materials [38, 39]. After including the polarization to the spatial mode of light beam, we get the vector vortices which have been studied extensively for sensing and communication applications [40, 41]. For sensing the magnetic field, these beams will pass through the materials that have magnetic field-dependent properties [42].

The field distribution of an optical vortex beam in polar coordinates can be expressed mathematically as [43]:

$$E(r, z) = E_0 r^m \exp(im\phi) \exp\left(-\frac{r^2}{\omega(z)^2}\right) \therefore r^2 = x^2 + y^2 \quad (1)$$

where E_0 is the field amplitude, $\omega(z)$ is the beam width at propagation distance z, and m is topological charge. The wavefront, phase profile, and the intensity distribution of vortices have been shown in **Figure 1**.

The vortices can be observed in all the random optical fields, known as speckles, which have random temporal and spatial coherence properties [44, 45]. These patterns can be obtained upon the propagation of coherent random waves through an inhomogeneous media such as ground glass plate (GGP) [46, 47]. This speckle is due to the superposition of many scattered waves originating from the inhomogeneities of the medium [48]. The size of the speckles can be carried by changing the width and wavelength of the beam, and the distribution of speckles can be changed by varying

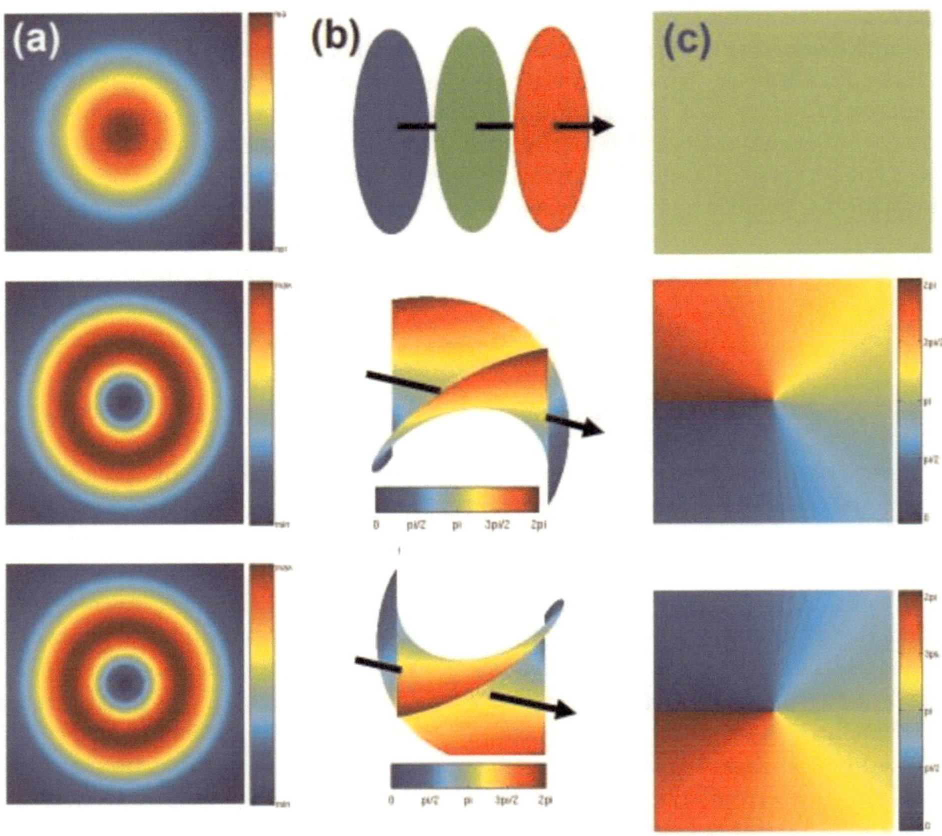

Figure 1.
The intensity distribution (left), wavefronts (middle), and the phase profiles (right) of optical vortex (OV) beams with m = 0 *(top),* m = +1 *(middle), and* −1 *(bottom).*

Figure 2.
The speckle patterns generated by the scattering of an optical vortex of order +1 (left), order +2 (middle), and order +3 (right) through the ground glass plate.

the field distribution incidenting on the rough surface [49–52]. The speckle patterns obtained by the scattering of optical vortices of orders $m = 1$–3 have been shown in **Figure 2**.

The phase singularities have also been observed in correlation functions and named as coherence vortices [47, 53–55]. The singularities have been verified both theoretically and experimentally using the interferometric techniques. The intensity correlation between the speckle patterns has attained a lot of interest due to their applications in speckle imaging and encryption applications [56–60]. These correlations have been used for finding the roughness of the surface and the effect of turbulence on the spatial modes [61, 62]. The roughness of the surface can be characterized by assuming the delta-correlated random phase screen and well described using a Gaussian correlation function.

In this chapter, we consider the correlations present in the scattered phase singular beams, normal optical vortices, and perfect optical vortices (POVs) for obtaining the information about the spatial mode. We discuss about the coherence vortices which can be obtained through the cross-correlation present in the speckle patterns corresponding to two optical vortices of different orders. We present the intensity distribution and propagation characteristics of coherence vortices by considering the cross-correlations and utilize the autocorrelation measurements for obtaining the mode information. Then we study the correlations present in scattered perfect optical vortices which lead to a new class of coherence functions, Bessel-Gauss functions, and utilized for generating the non-diffracting random fields. We utilized these correlation functions, which are order-dependent although the speckle patterns are order-independent, for encrypting the information which has higher advantage than normal random optical fields.

2. Cross-correlations present in scattered optical vortices: realization of coherence vortices

The phase singularities have been studied extensively in coherent light beams, and in recent days, partially coherent phase singularities have gained a considerable interest due to their robustness against atmospheric propagation [63–65]. The vortices present in partially coherent fields are known as coherence vortices as they can be realized in correlation functions [66–70]. These coherence vortices have been utilized for many applications such as free-space optical communication, remote sensing, and optical imaging [71–75]. The correlation between the two optical random fields plays an important role in obtaining the various types of coherence functions and their usage in applications, such as optical communication and for producing the physical unclonable functions (PUFs) for cryptography [60, 76–78]. The coherence vortices can be observed in the intensity correlation between two speckle patterns obtained by scattering the coherent vortices of different orders [55, 79]. The coherence vortices can be formulated with mutual coherence function between two speckle patterns corresponding to the vortices of orders m_1 and m_2 and is given by [55, 80]:

$$\tilde{\Gamma}_{m_1,m_2} = A \int r_1^{|m_1|+|m_2|} e^{[i(m_1-m_2)\phi_1]} e^{[-2r_1^2/\omega_0^2]} e^{[-2i\vec{k}\cdot\vec{r}_1]} d\vec{r}_1 \tag{2}$$

One can clearly observe the phase singularity with order $m = m_1 - m_2$ where m is the order of the coherence vortex.

For realizing the singularities in coherence functions, we need to scatter the coherent vortex beams through a rough surface such as GGP. The coherent optical vortices can be generated using a computer-generated hologram displayed on a spatial light modulator. After selecting the required vortex beam by an aperture, we scatter these beams through the GGP, and the corresponding speckle patterns are recorded using a CCD camera. We now find the cross-correlation function between two speckle patterns corresponding to optical vortices of different orders using MATLAB software. **Figure 3** shows the speckle patterns along with the determined coherence functions for different values of m_1 and m_2. It is clear from the figure that the autocorrelation between the speckle patterns provides the coherence function of order

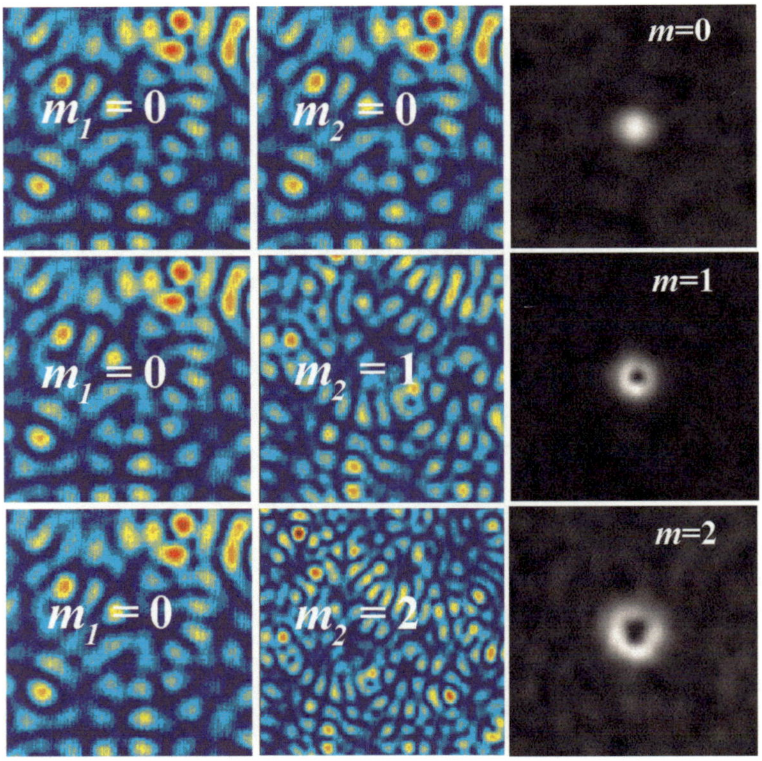

Figure 3.
The recorded speckle patterns and the corresponding cross-correlation functions, coherence vortices (here $m = m_2 - m_1$).

0. The cross-correlation between the speckle patterns corresponding to two different orders provides the higher-order coherence functions.

The coherence vortices have been characterized through their geometry by considering similarly as that of coherent vortex beams. **Figure 4** shows the intensity distribution of an optical vortex and its line profile along the center for order m = 1. We characterized the optical vortices by considering them as thin annular rings and using the parameters inner and outer radii r_1, r_2 as shown in figure. These are the nearest (inner) and farthest (outer) radial distances from center at which the intensity falls to $1/e^2$ (13.6%) of the maximum intensity observed at $r= r_0$ [46].

Figure 5 shows the variation of inner and outer radii of the coherence vortex ($m = m_2 - m_1$) of order 2 obtained by considering the cross-correlation between two speckle patterns of different values of m_1 and m_2 (with constant m) at the propagation distance of z = 20 cm. We considered the combinations of (m_1, m_2) = (0,2), (1,3), (2,4), (3,5), (4,6), (5,7), (6,8) where the difference (m_1-m_2) is constant.

From the figure, we observe that the inner and outer radii for all the combinations mentioned earlier are constant and independent of the input vortex beams considered for scattering. From this, we confirm that the intensity distribution of coherence

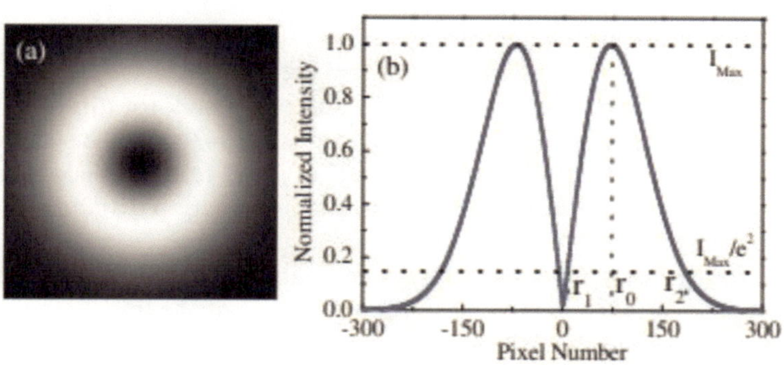

Figure 4.
(a) Intensity distribution and (b) line profile through its center for an optical vortex of order 1.

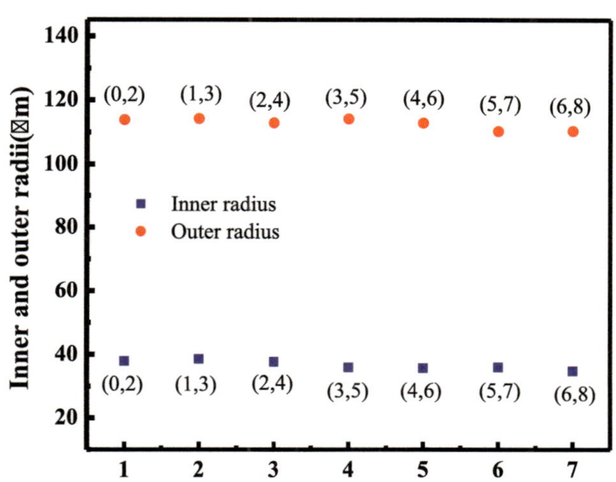

Figure 5.
Variation of inner and outer radii of the coherence vortex of order 2 with different combinations of input vortex beams.

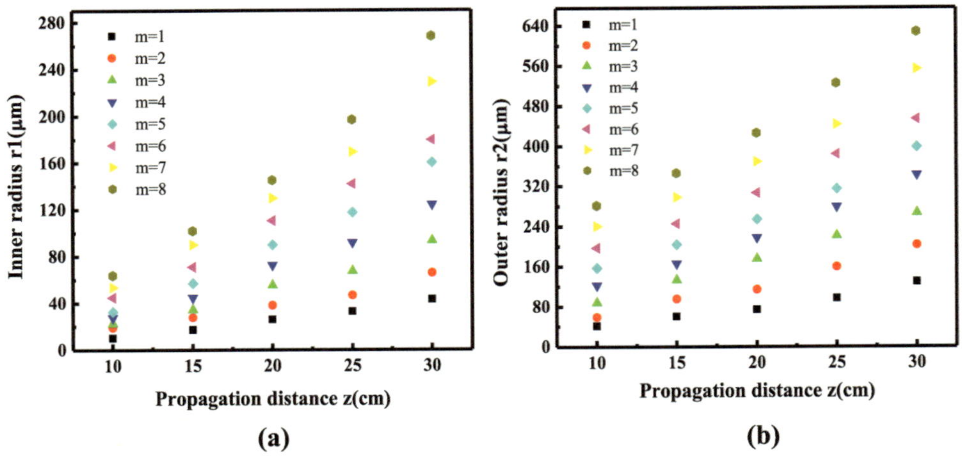

Figure 6.
Variation of inner (a) and outer (b) radii for coherence vortices of order m = 1–8 *with the propagation distance* z.

vortices depends only on the order difference but not on the individual orders of the optical vortices considered for scattering.

Now, we study the propagation characteristics of these coherence vortices. **Figure 6a** and **6b** show the variation of inner and outer radii for different orders m = 1–8 with respect to the propagation distance from z = 10–30 cm. The speckles have been recorded from z = 10–30 cm at an interval of 5 cm. It is observed that the inner and outer radii from figure vary linearly with the propagation distance for all orders and increase with order as shown in **Figure 6**.

We consider the rate of change of inner and outer radii with propagation distance as divergence and can be obtained by finding the slope of the line drawn between inner or outer radius and the propagation distance [79, 81, 82]. The slope has been determined using the linear fit to the experimental data. The variation of divergence with the order by considering inner and outer radii has been shown in **Figure 7**. It is clear from the figure that the divergence increases linearly with order (m). One can utilize the inner and outer radii at the source plane and their divergence for

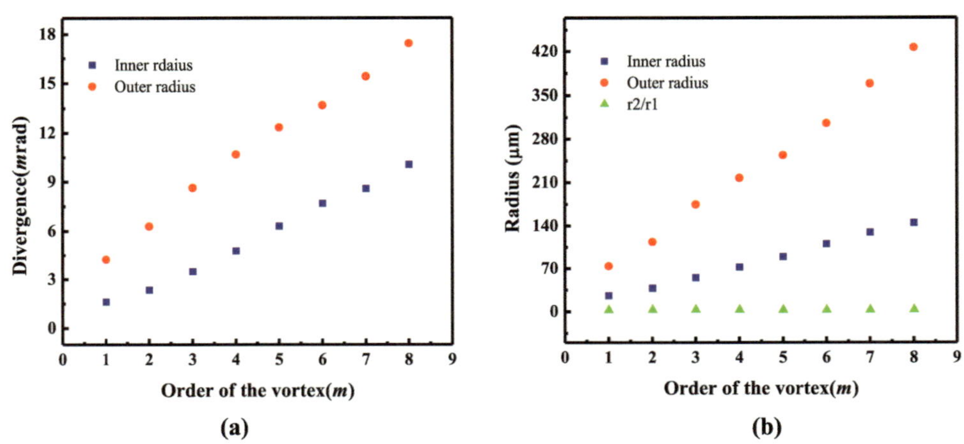

Figure 7.
Variation of inner and outer radii along with their divergence as a function of order of the coherence functions.

characterizing the order of a coherence vortex. One can also find the information about the incident spatial modes using these coherence vortices.

3. Autocorrelation studies for scattered optical vortices

For the applications in free-space optical communication using spatial modes, one needs to propagate these modes for longer distances. After propagating through the channel, the mode information gets disturbed, and one needs to find the mode information of these perturbed beams. Although there are many techniques to find the order of a higher-order coherent optical vortex [46], they are not suitable for partially coherent or incoherent vortices. A limited number of techniques are available for finding the order of a partially coherent vortex beam. In this section, we study the autocorrelation properties of scattered optical vortices for diagnosing the spatial mode information [83–85]. The number of zero points or dark rings present in 2D spatial correlation function provides the information about the spatial mode. The spatial autocorrelation function of a perturbed optical vortex is equivalent to the Fourier transform (FT) of its intensity in the source plane [86]. The number of dark rings presented in the spatial correlation function is equal to the topological charge of vortex beam which has also been verified by verifying the number of zero points present in Fourier transform of a coherent vortex beam which further will be discussed. Here, we show that the existence of the ring dislocations in the spatial correlation function corresponds to the scattering of the optical vortex field [87].

The theoretical background for the 2D autocorrelation function starts by assuming the field distribution of Laguerre-Gaussian beam with azimuthal index m and zero radial index in the source plane ($z = 0$) in cylindrical coordinates as:

$$E(\rho, \theta, 0) \propto \rho^{|m|} \exp\left(\frac{-\rho^2}{\omega_0{}^2}\right) \exp(im\theta) \tag{3}$$

where ω_0 is beam waist of the input beam and (ρ, θ) are the cylindrical coordinates in incident plane. The scattering of optical vortex (OV) beams through a ground glass plate (GGP) that can be well described by a random phase function $\exp(i\Phi)$ where Φ varies randomly from 0 to 2π. A particular way of obtaining this type of phase distribution Φ is by taking a 2D convolution between a random spatial function and a Gaussian correlation function [44]. The field $U(\rho, \theta)$ after the GGP can be obtained from the incident field $E(\rho, \theta)$ and can be written as:

$$U(\rho, \theta) \propto \exp(i\Phi) E(\rho, \theta) \tag{4}$$

where the autocorrelation of the phase exponential factor is a Dirac-delta function at plane (ρ, θ), which can be written mathematically as:

$$\langle \exp[i(\Phi(\rho_1, \quad \theta_1) - \Phi(\rho_2, \quad \theta_2))] \rangle = \delta(\rho_1 - \rho_2)\delta(\theta_1 - \theta_2) \tag{5}$$

where $\langle a \rangle$ denotes the ensemble average operation in a. The autocorrelation function between two speckle patterns of same order obtained by scattering of OV beams through GGP is given by:

$$\Gamma(r_1, \varphi_1; r_2, \varphi_2) = \langle U_1(r_1, \quad \varphi_1) U_2^*(r_2, \quad \varphi_2) \rangle \tag{6}$$

where (r, φ) are the coordinates at the detection plane. The filed at the detection plane in terms field at the incident plane can be evaluated using Fresnel's diffraction integral in cylindrical coordinates as [88, 89]:

$$U(r, \varphi, z) = \frac{e^{ikz}}{i\lambda z}\int \rho d\rho \int d\theta U(\rho, \theta) e^{\left\{\frac{ik}{2z}\left(\rho^2 + r^2 - 2\rho r\cos(\theta - \varphi)\right)\right\}} \tag{7}$$

Using Eq. (7) and Eq. (6), we have that

$$\begin{aligned}\Gamma(r_1, \varphi_1; r_2, \varphi_2) &= \left\langle U_1(r_1, \varphi_1) U_2^*(r_2, \varphi_2)\right\rangle \\ &= \left\langle \frac{e^{ikz}}{i\lambda z}\int \rho_1 d\rho_1 \int d\theta_1 U_1(\rho_1, \theta_1) e^{\left\{\frac{ik}{2z}\left({\rho_1}^2 + {r_1}^2 - 2\rho_1 r_1 \cos(\theta_1 - \varphi_1)\right)\right\}} \right. \\ &\quad \left. \times \frac{e^{-ikz}}{-i\lambda z}\int \rho_2 d\rho_2 \int d\theta_2 {U_2}^*(\rho_2, \theta_2) e^{\left\{\frac{-ik}{2z}\left({\rho_2}^2 + {r_2}^2 - 2\rho_2 r_2 \cos(\theta_2 - \varphi_2)\right)\right\}} \right\rangle \\ &= \frac{e^{\left\{\frac{ik}{2z}\left[r_1^2 - r_2^2\right]\right\}}}{\lambda^2 z^2}\int \rho_1 d\rho_1 \int d\theta_1 \int \rho_2 d\rho_2 \int d\theta_2 \langle U_1(\rho_1, \theta_1) {U_2}^*(\rho_2, \theta_2)\rangle \\ &\quad \times e^{\left\{\frac{ik}{2z}\left({\rho_1}^2 - {\rho_2}^2 - 2\rho_1 r_1 \cos(\theta_1 - \varphi_1) + 2\rho_2 r_2 \cos(\theta_2 - \varphi_2)\right)\right\}}\end{aligned} \tag{8}$$

which is a fourfold integral and includes cross-correlation of filed at the incident plane (ρ, θ) namely, $\left\langle U_1(\rho_1, \theta_1) U_2^*(\rho_2, \theta_2)\right\rangle$. Using Eq. (4) and Eq. (5), we get the cross-correlation function as

$$\begin{aligned}\langle U_1(\rho_1, \theta_1) {U_2}^*(\rho_2, \theta_2)\rangle &= \left\langle E_1(\rho_1, \theta_1) e^{i\Phi(\rho_1, \theta_1)} E_2^*(\rho_2, \theta_2) e^{i\Phi(\rho_2, \theta_2)}\right\rangle \\ &= E_1(\rho_1, \theta_1) E_2^*(\rho_2, \theta_2)\left\langle e^{i(\Phi(\rho_1, \theta_1) - \Phi(\rho_2, \theta_2))}\right\rangle \\ &= E_1(\rho_1, \theta_1) E_2^*(\rho_2, \theta_2) \times \delta(\rho_1 - \rho_2)\delta(\theta_1 - \theta_2)\end{aligned} \tag{9}$$

The ground glass plate (random phase screen) is modeled as a δ-correlated phase function. The autocorrelation function after using the same is

$$\langle U_1(\rho_1, \theta_1) {U_2}^*(\rho_2, \theta_2)\rangle = E(\rho_1, \theta_1) E^*(\rho_2, \theta_2) \tag{10}$$

Using Eq. (10) in Eq. (9) and the properties of the Dirac-delta function, the fourfold integral of the autocorrelation is reduced to the two-fold integral as:

$$\begin{aligned}\Gamma(r_1, \varphi_1; r_2, \varphi_2) &= \left\langle U_1(r_1, \varphi_1) U_2^*(r_2, \varphi_2)\right\rangle \\ &= \frac{e^{\left\{\frac{ik}{2z}\left[r_1^2 - r_2^2\right]\right\}}}{\lambda^2 z^2}\int \rho d\rho \int d\theta E(\rho, \theta) E^*(\rho, \theta) \\ &\quad \times e^{\left\{\frac{-ik}{z}\rho(r_1 \cos(\theta - \varphi_1) - r_2 \cos(\theta - \varphi_2))\right\}}\end{aligned} \tag{11}$$

$$\Gamma_{12}(\Delta r) = \frac{e^{\left\{\frac{ik}{2z}\left[r_1^2 - r_2^2\right]\right\}}}{\lambda^2 z^2}\iint |E(\rho, \theta)|^2 \exp\left[-\frac{ik}{z}(\rho \Delta r \cos(\varphi_s - \theta))\right]\rho d\rho d\theta \tag{12}$$

where $\Delta r \cos(\varphi_s - \theta) = [(r_1 \cos(\varphi_1) - r_2 \cos(\varphi_2)) \cos\theta] + [(r_1 \sin(\varphi_1) - r_2 \sin(\varphi_2)) \sin\theta]$ and $\Delta r^2 = r_1^2 + r_2^2 - 2r_1 r_2 \cos(\varphi_2 - \varphi_1)$.

Using Eq. (3), the absolute value of the field distribution is

$$|E(\rho,\ \phi,\ 0)|^2 = \rho^{2|m|} \exp\left(\frac{-2\rho^2}{\omega_0^2}\right) \tag{13}$$

Let us calculate the integral part of the correlation as

$$\begin{aligned}\Gamma_{12}(\Delta r) &= \frac{e^{\left\{\frac{ik}{2z}\left[r_1^2 - r_2^2\right]\right\}}}{\lambda^2 z^2} \iint \rho^{2|m|} \exp\left(\frac{-2\rho^2}{\omega_0^2}\right) \exp\left[-\frac{ik}{z}(\rho\Delta r \cos(\varphi_s - \theta))\right] \rho d\rho d\theta \\ &= \int \rho^{2|m|+1} \exp\left(\frac{-2\rho^2}{\omega_0^2}\right) d\rho \int \exp\left[-\frac{ik}{z}(\rho\Delta r \cos(\varphi_s - \theta))\right] d\theta\end{aligned} \tag{14}$$

and $I' = \int \exp\left[\frac{-ik}{z}(\rho\Delta r \cos(\varphi_s - \theta))\right] d\theta$ can be calculated by using Anger-Jacobi identity $e^{-iz\cos\theta} = \sum_{n=-\infty}^{\infty} (-1)^n i^n J_n(z) e^{in\theta}$, we get [90]

$$I' = \int \exp\left[\frac{-ik}{z}(\rho\Delta r \cos(\varphi_s - \theta))\right] d\theta = 2\pi J_0\left(\frac{k\rho}{z}\Delta r\right) \tag{15}$$

Substitute the aforementioned equation in Eq. (14), then the far-field autocorrelation function or the Fourier transform of the incident intensity in the source plane becomes

$$\Gamma_{12}(\Delta r) = \frac{2\pi e^{\left\{\frac{ik}{2z}\left[r_1^2 - r_2^2\right]\right\}}}{\lambda^2 z^2} \int_0^\infty \rho^{2|m|+1} \exp\left(\frac{-2\rho^2}{\omega_0^2}\right) J_0\left(\frac{k\rho}{z}\Delta r\right) d\rho \tag{16}$$

where $J_0\left(\frac{k\rho}{z}\Delta r\right)$ represents the zeroth-order Bessel function. Using the following integral,

$$\int_0^\infty x^\mu \exp(-\alpha x^2) J_\nu(xy) dx = \frac{y^\nu \Gamma\left(\frac{1}{2}(\mu - \nu + 1)\right)}{2^{\nu+1} \alpha^{\frac{1}{2}(\mu+\nu+1)}} \exp\left(\frac{-y^2}{4\alpha}\right) L^\nu_{\frac{1}{2}(\mu-\nu+1)}\left(\frac{y^2}{4\alpha}\right) \tag{17}$$

From Eqs. (16) and (17), the mutual coherence function will become

$$\Gamma_{12}(\Delta r) = \frac{\pi \omega_0^{2|m|+2} e^{\left\{\frac{ik}{2z}\left[r_1^2 - r_2^2\right]\right\}}}{2^{|m|+1} \lambda^2 z^2} \exp\left(\frac{-k^2 \omega_0^2 \Delta r^2}{8z^2}\right) L_{|m|}\left(\frac{k^2 \omega_0^2 \Delta r^2}{8z^2}\right) \tag{18}$$

where $\mu = 2|m| + 1, \alpha = \frac{2}{\omega_0^2}, y = \frac{k}{z}\Delta r, \nu = 0$ and $L_{|m|}\left(\frac{k^2 \omega_0^2 \Delta r^2}{8z^2}\right)$ represents the Laguerre polynomial of order m. The aforementioned equation represents the autocorrelation of scattered Laguerre-Gaussian (LG) beam, and it depends on the azimuthal index and propagation distance. In the spatial correlation filed, the number of dark rings or number of zero points in the Laguerre polynomial gives the information about the order or azimuthal index of the vortex beam. We verify these theoretical findings experimentally, and the details are given as follows.

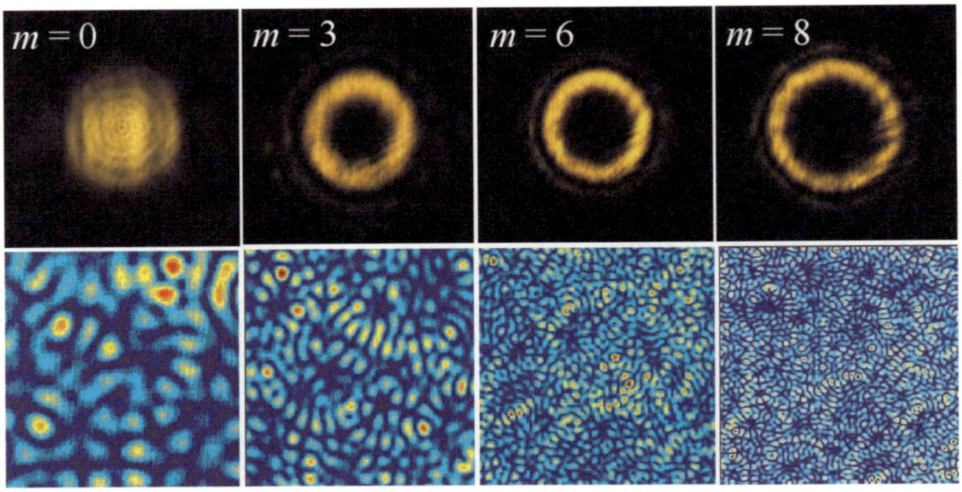

Figure 8.
Intensity distributions of optical vortices and their corresponding speckle patterns.

We have generated the optical vortices of orders m = 0–8 by displaying a computer-generated hologram on a spatial light modulator and scatter them through a GGP. We have shown the intensity distributions of optical vortices at the plane of GGP and the corresponding speckle patterns in **Figure 8** which have been recorded using a CCD camera. It is clear from the figure that the size of the speckles decreases with the increase in order, and we observe the structures in speckle distributions corresponding to higher orders.

Further, we have processed the speckle patterns for finding the autocorrelation function using MATLAB software. We found that the order or topological charge of a given spatial mode is given by the number of dark rings present in the autocorrelation function. This method is suitable for vortices with low topological charges. However, as we increase the order, we must identify the number of dark rings carefully because the adjacent two dark rings are very close to each other, and it is very difficult to distinguish them. This technique is alignment free as the autocorrelation function does not depend on the alignment. **Figure 9** shows the experimentally obtained spatial autocorrelation functions (top) for the speckle patterns corresponding to the vortices of orders m = 0–3 from left and right. The results are in good agreement with the theoretically obtained correlation function as shown in bottom row of the figure. It is clear from the figure that the order of vortex is equal to the number of dark rings present in the spatial autocorrelation field. One can also utilize the propagation characteristics for the better diagnosis of the information of a given spatial mode.

The generalized theory for autocorrelation functions of LG beams with nonzero radial index is provided and experimentally verified as well. The number of dark rings is equal to the sum of twice the radial index and azimuthal index [86]. The autocorrelation function of a scattered LG beam with nonzero radial and azimuthal indices is given by [91]:

$$\chi(\xi) = \frac{\pi\omega_0^{2|m|+2}}{2^{|m|+1}} \frac{(p+|m|)!}{p!} \exp\left(-\frac{\pi^2\omega_0^2\xi^2}{2}\right) L_p\left(-\frac{\pi^2\omega_0^2\xi^2}{2}\right) L_{p+|m|}\left(-\frac{\pi^2\omega_0^2\xi^2}{2}\right) \quad (19)$$

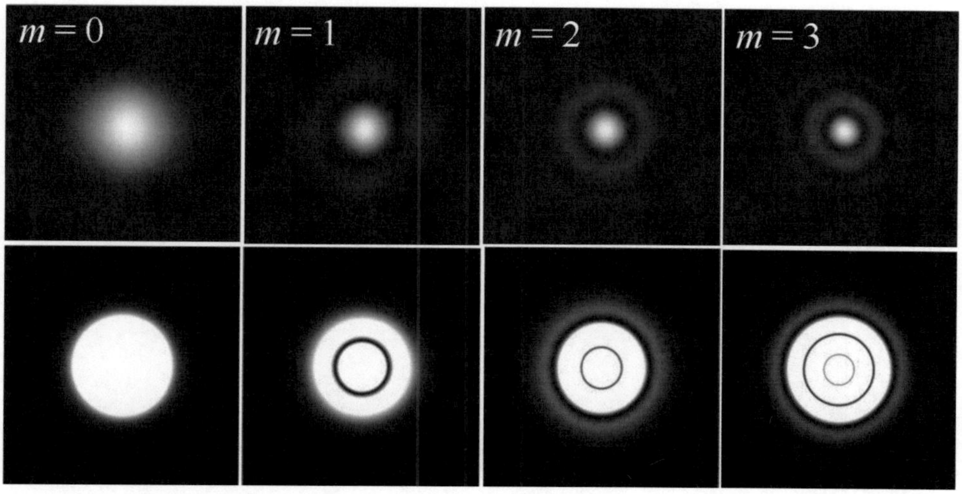

Figure 9.
Experimental (top) and theoretical (bottom) 2D spatial autocorrelation function for a speckle pattern generated by scattering a vortex beams of orders m = 0-3 *from left to right.*

where p is the radial index and m is the azimuthal index of a LG beam. We need to study the following subcases from the aforementioned expression for the better understanding of correlation function:

i. If $p = 0$, then the correlation function corresponds to the optical vortices that carry OAM and is given by:

$$\chi(\xi) = \frac{\pi\omega_0^{2|m|+2}}{2^{|m|+1}}|m|!\exp\left(-\frac{\pi^2\omega_0^2\xi^2}{2}\right)L_{|m|}\left(-\frac{\pi^2\omega_0^2\xi^2}{2}\right) \tag{20}$$

The aforementioned expression is exactly matching with our equation obtained for LG beams with zero radial index.

ii. If $m = 0$ (non-vortex beams), then the autocorrelation function is given by:

$$\chi(\xi) = \frac{\pi\omega_0^2}{2}\exp\left(-\frac{\pi^2\omega_0^2\xi^2}{2}\right)\left(L_p\left(-\frac{\pi^2\omega_0^2\xi^2}{2}\right)\right)^2 \tag{21}$$

From Eqs. (19), (20), and (21), one can obtain the relation between number of dark rings and radial and azimuthal indices as:

$$\begin{aligned} N &= 2p + |m| \quad \text{when} \quad m \neq 0 \\ &= p \quad \text{when} \quad m = 0. \end{aligned} \tag{22}$$

where N is the number of dark rings present in the autocorrelation function.

The numerical results for the LG beams of nonzero radial index have been shown in **Figure 10**. We can observe the number of dark rings in the far-field autocorrelation function which depends on both radial and azimuthal indices. The radial and azimuthal indices for the contour plots are (a) m = 1, p = 1; (b) m = 2, p = 1; (c) m = 3, p = 1; (d) m = 1, p = 2; (e) m = 2, p = 2; and (f) m = 3, p = 2.

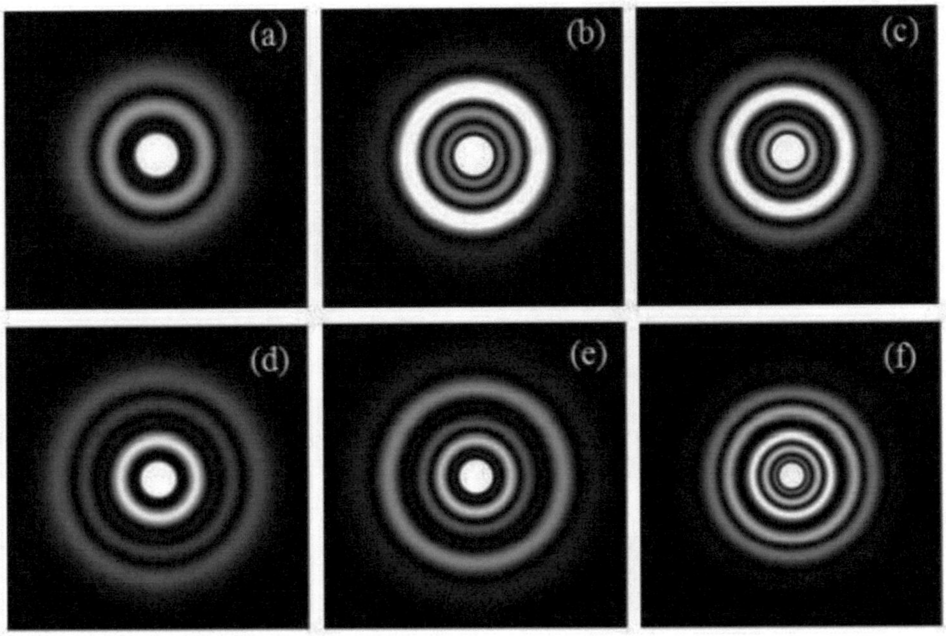

Figure 10.
Theoretical far-field auto-correlation function with different combinations of radial and azimuthal indices: (a) $m = 1, p = 1$; (b) $m = 2, p = 1$; (c) $m = 3, p = 1$; (d) $m = 1, p = 2$; (e) $m = 2, p = 2$; (f) $m = 3, p = 2$.

Now, we verify Van Cittert-Zernike theorem states that the autocorrelation function of a scattered light beam is same as the Fourier transform (FT) of intensity distribution incident on the rough surface, i.e. source plane. Here, we present the results obtained for the FT of intensity distribution of a LG beam in which the number of dark rings is equal to the order of the vortex [92] as shown in **Figure 11**.

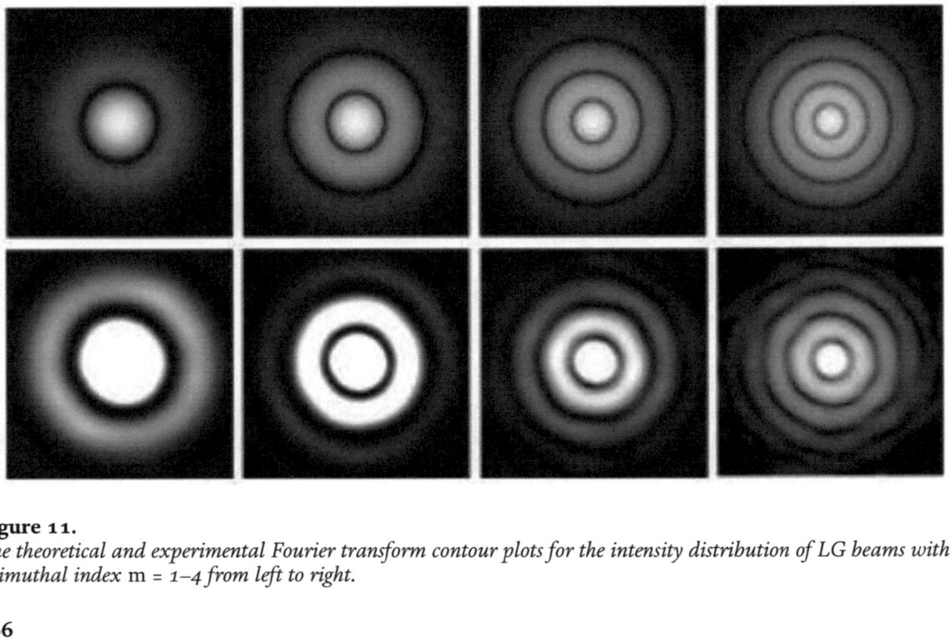

Figure 11.
The theoretical and experimental Fourier transform contour plots for the intensity distribution of LG beams with azimuthal index m = 1–4 *from left to right.*

3.1 Correlations in scattered perfect optical vortices

From the aforementioned discussions, it is known that the field and intensity distributions of an optical vortex are strongly influenced by its topological charge which may be a drawback for controlling them while they propagate through optical channels [93]. To overcome this, another class of vortices has been introduced, known as perfect optical vortices (POVs) with order-independent intensity distribution [94]. The POV beams are known for very thin annular rings whose width and radius are independent of topological charge [95–98]. The POV beams can be generated experimentally by Fourier transforming Bessel-Gauss beams which will be generated by passing optical vortex beams through an axicon [99–101]. The radius and width of the ring of a POV beam can easily be controlled by changing the apex angle or axicon parameter [100, 102]. We study the correlations present in scattered POV beams and utilize for generating the non-diffracting optical random fields [103]. The theoretical analysis for the cross-correlation between two speckle patterns obtained by scattering POV beams of different orders is as follows:

The field distribution of a perfect optical vortex (POV) beam, described by a thin annular ring of order m, which can be represented mathematically as:

$$E(\rho, \theta) = \delta(\rho - \rho_0)e^{im\theta} \tag{23}$$

where ρ_0 is the radius of the POV beam and $\boldsymbol{\delta}$ represents the Dirac-delta function. In practical, the POV beams can be realized with finite ring width which can be represented mathematically as:

$$E(\rho, \theta) = g(\rho - \rho_0; \varepsilon)e^{im\theta} \tag{24}$$

where $g(\rho - \rho_0; \varepsilon)$ is a narrow function in the radial direction with a finite "width" ε, such as Gaussian and height proportional to $1/\varepsilon$. The scattered field $U(\rho, \theta)$ after the GGP can be obtained from the incident field $E(\rho, \theta)$ on the GGP as

$$U(\rho, \theta) = e^{i\Phi(\rho, \theta)}E(\rho, \theta) \tag{25}$$

where the cross-correlation of the phase exponential factor is a Dirac-delta function at plane (ρ, θ), which implies the mutually independent inhomogeneities that can be expressed mathematically as:

$$\left\langle e^{i[\Phi(\rho_1, \theta_1) - \Phi(\rho_2, \theta_2)]} \right\rangle = \delta(\rho_1 - \rho_2)\delta(\theta_1 - \theta_2) \tag{26}$$

where $\langle a \rangle$ denotes the ensemble average operation in a [9]. The mutual coherence function between the two scattered POV fields at a distance of z from the GGP is given by:

$$\Gamma(r_1, \varphi_1; r_2, \varphi_2) = \langle U_1(r_1, \varphi_1)U_2^*(r_2, \varphi_2) \rangle \tag{27}$$

where ρ and θ are source plane coordinates, and r and φ are detection plane coordinates. The field at detection plane can be obtained using Fresnel diffraction integral in cylindrical coordinates as: [88, 89]

$$U(r, \varphi, z) = \frac{e^{ikz}}{i\lambda z}\int \rho d\rho \int d\theta U(\rho, \theta)e^{\left\{\frac{ik}{2z}\left(\rho^2 + r^2 - 2\rho r\cos(\theta - \varphi)\right)\right\}} \tag{28}$$

From Eq. (27) and Eq. (28), we have

$$\begin{aligned}\Gamma(r_1, \varphi_1; r_2, \varphi_2) &= \langle U_1(r_1, \varphi_1)U_2^*(r_2, \varphi_2)\rangle \\ &= \left\langle \frac{e^{ikz}}{i\lambda z}\int \rho_1 d\rho_1 \int d\theta_1 U_1(\rho_1, \theta_1) e^{\left\{\frac{ik}{2z}\left({\rho_1}^2+{r_1}^2-2\rho_1 r_1\cos(\theta_1-\varphi_1)\right)\right\}}\right. \\ &\quad \left. \times \frac{e^{-ikz}}{-i\lambda z}\int \rho_2 d\rho_2 \int d\theta_2 U_2^*(\rho_2, \theta_2) e^{\left\{\frac{-ik}{2z}\left({\rho_2}^2+{r_2}^2-2\rho_2 r_2\cos(\theta_2-\varphi_2)\right)\right\}}\right\rangle \\ &= \frac{e^{\left\{\frac{ik}{2z}\left[r_1^2-r_2^2\right]\right\}}}{\lambda^2 z^2}\int \rho_1 d\rho_1 \int d\theta_1 \int \rho_2 d\rho_2 \int d\theta_2 \langle U_1(\rho_1, \theta_1)U_2^*(\rho_2, \theta_2)\rangle \\ &\quad \times e^{\left\{\frac{ik}{2z}\left({\rho_1}^2-{\rho_2}^2-2\rho_1 r_1\cos(\theta_1-\varphi_1)+2\rho_2 r_2\cos(\theta_2-\varphi_2)\right)\right\}}\end{aligned} \tag{29}$$

which is a fourfold integral, and this integral includes the cross-correlation of the field at the incident plane (ρ, θ), namely, $\langle U_1(\rho_1, \theta_1)U_2^*(\rho_2, \theta_2)\rangle$. Using Eqs. (24) and (25), we can get the cross-correlation function at plane (ρ, θ) as:

$$\begin{aligned}\langle U_1(\rho_1, \theta_1)U_2^*(\rho_2, \theta_2)\rangle &= \left\langle E_1(\rho_1, \theta_1)e^{i\Phi(\rho_1, \theta_1)}E_2^*(\rho_2, \theta_2)e^{i\Phi(\rho_2, \theta_2)}\right\rangle \\ &= E_1(\rho_1, \theta_1)E_2^*(\rho_2, \theta_2)\left\langle e^{i(\Phi(\rho_1, \theta_1)-\Phi(\rho_2, \theta_2))}\right\rangle \\ &= E_1(\rho_1, \theta_1)E_2^*(\rho_2, \theta_2)\times\delta(\rho_1-\rho_2)\delta(\theta_1-\theta_2)\end{aligned} \tag{30}$$

Using Eq. (30) in Eq. (29) and the properties of the Dirac-delta function, the fourfold integral of the cross-correlation is reduced to the twofold integral as:

$$\begin{aligned}\Gamma(r_1, \varphi_1; r_2, \varphi_2) &= \langle U_1(r_1, \varphi_1)U_2^*(r_2, \varphi_2)\rangle \\ &= \frac{e^{\left\{\frac{ik}{2z}\left[r_1^2-r_2^2\right]\right\}}}{\lambda^2 z^2}\int \rho_1 d\rho_1 \int d\theta_1 E_1(\rho_1, \theta_1)E_2^*(\rho_1, \theta_1) \\ &\quad \times e^{\left\{\frac{-ik}{z}\rho\left(r_1\cos(\theta_1-\varphi_1)-r_2\cos(\theta_1-\varphi_2)\right)\right\}}\end{aligned} \tag{31}$$

In the special case of incident POV beams, we can use Eq. (24) to write

$$E_1(\rho_1, \theta_1)E_2^*(\rho_1, \theta_1) = g(\rho_1-\rho_{01}; \varepsilon)g(\rho_1-\rho_{02}; \varepsilon)e^{i(m_1\theta_1-m_2\theta_1)} \tag{32}$$

As we know that the radius of POV beams is independent of order, i.e. $\rho_{01} = \rho_{02} = \rho_0$, therefore

$$E_1(\rho_1, \theta_1)E_2^*(\rho_1, \theta_1) = g^2(\rho_1-\rho_0; \varepsilon)e^{i(m_1-m_2)\theta_1} \tag{33}$$

Under the condition $\varepsilon \to 0$, one can replace $g^2(\rho-\rho_0; \varepsilon)$ with a single Dirac-delta function $\delta(\rho-\rho_0)$ and the aforementioned expression becomes

$$E_1(\rho_1, \theta_1)E_2^*(\rho_1, \theta_1) = \delta(\rho_1-\rho_0)e^{i(m_1-m_2)\theta_1} \tag{34}$$

After substituting Eq. (34) in Eq. (31), we get

$$\Gamma(r_1,\varphi_1;r_2,\varphi_2)=\frac{e^{\left\{\frac{ik}{2z}\left[r_1^2-r_2^2\right]\right\}}}{\lambda^2 z^2}\int\rho_1 d\rho_1\int d\theta_1\delta(\rho_1-\rho_0)e^{i(m_1-m_2)\theta_1}e^{\left\{\frac{-ik}{z}\rho(r_1\cos(\theta_1-\varphi_1)-r_2\cos(\theta_1-\varphi_2))\right\}} \tag{35}$$

The integral in the aforementioned equation can be evaluated as:

$$\begin{aligned} I&=\int\delta(\rho_1-\rho_0)\rho_1 d\rho_1\int e^{i(m_1-m_2)\theta_1}e^{\frac{-ik}{z}\rho(r_1\cos(\theta_1-\varphi_1)-r_2\cos(\theta_2-\varphi_2))}d\theta_1\\ &=\int\delta(\rho_1-\rho_0)\rho_1 d\rho_1\int e^{i(m_1-m_2)\theta_1}e^{\frac{-ik}{z}(\rho\Delta r\cos(\varphi_s-\theta_1))}d\theta_1 \end{aligned} \tag{36}$$

For solving the integral of θ_1, we assume that $\theta'=\phi_s-\theta_1$ then we get the integral as:

$$I=e^{i(m_1-m_2)\varphi_s}\int\delta(\rho_1-\rho_0)\rho_1 d\rho_1\int e^{i(m_1-m_2)\theta'}e^{\frac{-ik}{z}(\rho\Delta r\cos\theta')}d\theta' \tag{37}$$

where θ' varies from $-\phi_s$ to $2\pi-\phi_s$and using Anger-Jacobi identity $e^{-iz\cos\theta}=\sum_{n=-\infty}^{\infty}(-1)^n i^n J_n(z)e^{in\theta}$, we get [90]

$$I=e^{i(m_1-m_2)\varphi_s}\int\delta(\rho_1-\rho_0)\rho_1 d\rho_1\sum_{n=-\infty}^{\infty}(-1)^n i^n J_n\left(\frac{k\rho}{z}\Delta r\right)\int e^{in\theta'}e^{i(m_1-m_2)\theta'}d\theta' \tag{38}$$

The aforementioned integral has nonzero value only when $n=m_2-m_1$ and the integral becomes

$$I=2\pi e^{i(m_1-m_2)\varphi_s}\int\delta(\rho_1-\rho_0)\rho_1 d\rho_1(-1)^{m_2-m_1}i^{m_2-m_1}J_{m_2-m_1}\left(\frac{k\rho}{z}\Delta r\right) \tag{39}$$

By using the integral properties of Dirac-delta function [90], we get that

$$I=2\pi\rho_0(-i)^{m_2-m_1}e^{i(m_1-m_2)\varphi_s}J_{m_2-m_1}\left(\frac{k\rho_0}{z}\Delta r\right) \tag{40}$$

Now, the cross-correlation function as defined in Eq. (35) becomes

$$\Gamma_{12}(\Delta r)=\frac{2\pi\rho_0(-i)^{m_2-m_1}e^{\frac{ik}{2z}\left(r_1^2-r_2^2\right)}}{\lambda^2 z^2}e^{i(m_1-m_2)\varphi_s}J_{m_2-m_1}\left(\frac{k\rho_0}{z}\Delta r\right) \tag{41}$$

The aforementioned equation representing the mutual coherence function of two speckle patterns is described well by the Bessel function of order $m=m_2-m_1$. The corresponding cross-correlation function of two speckle patterns is given by:

$$C(\Delta r)=(-i)^{m_2-m_1}e^{i(m_1-m_2)\varphi_s}J_{m_2-m_1}\left(\frac{k\rho_0}{z}\Delta r\right) \tag{42}$$

Normalized intensity distribution of the coherence function can be evaluated in terms of time-averaged intensity I_0 as:

$$I(\Delta r)=I_0^2\left(1+|C(\Delta r)|^2\right)=I_0^2\left(1+J_{m_2-m_1}^2\left(\frac{k\rho_0}{z}\Delta r\right)\right) \tag{43}$$

If two speckle patterns correspond to the same order, the cross-correlation function is converted into an autocorrelation function, which can be obtained by keeping $m_1 = m_2$ in the aforementioned equation. We obtain the autocorrelation function as:

$$I(\Delta r) = I_0^2\left(1 + J_0^2\left(\frac{k\rho_0}{z}\Delta r\right)\right) \tag{44}$$

It is clear from the aforementioned analysis that the autocorrelation functions can be described with Bessel functions of order zero and cross-correlation functions can be described with Bessel functions of nonzero orders $(m = m_2 - m_1)$.

The experimental validation of aforementioned theoretical findings has been done and the details are as follows: **Figure 12** shows the speckle patterns generated by the scattering of POV beams and the corresponding cross-correlation functions. From the figure, we confirm the Bessel-Gauss nature of coherence functions with order $m = m_2 - m_1$.

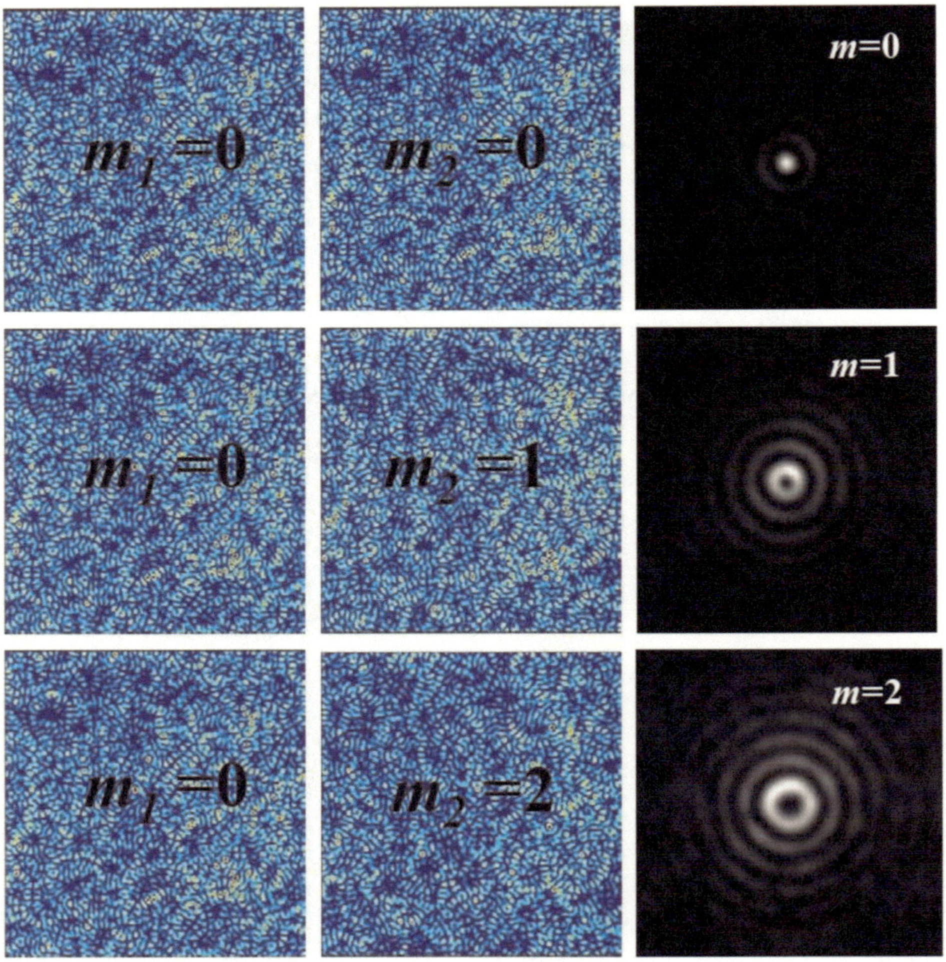

Figure 12.
The recorded speckle patterns and the corresponding cross-correlation functions, Bessel coherence functions (here $m = m_2 - m_1$).

From Eq. (43), we analyze the size of the speckles under the condition m_1=m_2, i.e. by considering the width of the autocorrelation function. The speckle size is defined as the spatial length up to which the correlations exist in the field [103, 104]. From Eq. (44), the first zero of zeroth-order Bessel function $J_0(x) = 0$ can happen at x = 2.4, and the correlation length or speckle size can be obtained as:

$$\Delta r = \frac{xz}{k\rho_0} = \frac{2.4z}{k\rho_0} \tag{45}$$

It is clear from the aforementioned equation is that the size of near-field speckles varies linearly with propagation distance z, independent of order m and inversely proportional to the ring radius ρ_0. We have observed that the speckle size and distribution are independent of the order, and they vary with propagation distance. The recorded speckles have been shown in **Figure 13**, and the linear variation of speckle size with propagation distance has been verified in **Figure 14**.

Figure 13.
The speckle patterns obtained by the scattering of POV beam of order m = 0 *at different propagation distances* z = 20 cm, z = 45 cm, *and* z = 70 cm *in the near field.*

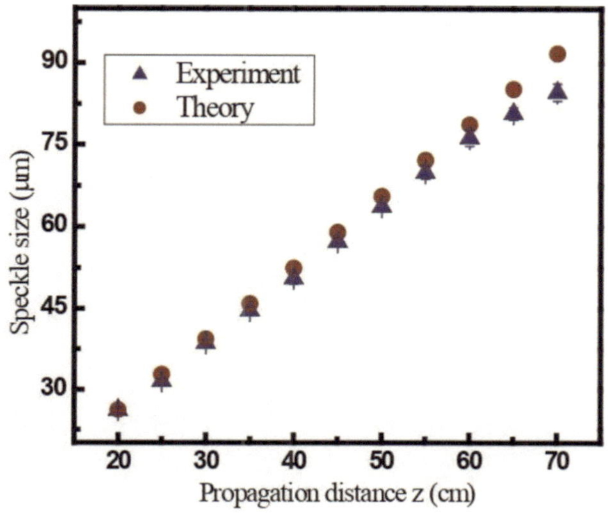

Figure 14.
Experimental (blue) and theoretical (red) results for the variation of near-field or diverging speckle size with propagation distance.

Further, we consider the Fourier transform of near-field speckles that produce the spatially invariant optical random fields. The Fourier transform can be realized with the help of a simple convex lens (here, we consider its focal length as f_2). The far-field autocorrelation function $\Gamma_{12}'(\Delta r')$ of scattered POV beams is given by [89]:

$$\Gamma_{12}'(\Delta r') = \frac{1}{\lambda^2 f^2} \iint |U_1(\rho,\ \theta)|^2 e^{\frac{-ik}{f_2}(\rho \Delta r' \cos(\varphi_{s1} - \theta))} \rho d\rho d\theta \tag{46}$$

Substituting Eq. (25) in Eq. (46) and following the same procedure, we get

$$\Gamma_{12}'(\Delta r') = \frac{2\pi \rho_0}{\lambda^2 f_2{}^2} J_0\left(\frac{k\rho_0}{f_2} \Delta r'\right) \tag{47}$$

From the aforementioned equation, one can easily observe that the correlation function is independent of order m as well as propagation distance z. As compared to the near-field diffraction, the spatial coherence function does not increase anymore with the propagation which can be utilized for communication and encryption applications.

From Eq. (47), we get the size of non-diffracting random fields as:

$$\Delta r' = \frac{x f_2}{k\rho_0} = \frac{2.4 f_2}{k\rho_0} \tag{48}$$

The speckle size is independent of propagation distance z and directly proportional to focal length f_2 and inversely proportional to ring radius ρ_0. We can control the size by just varying the focal length and ring radius that can be controlled by axicon parameter [103].

Figure 15 shows the speckle patterns recorded for different propagation distances and clear that are independent of propagation distances. **Figure 16** shows the variation of speckle for order 2 with propagation distance for different axicon parameters as mentioned. The size of the speckles decreases with the increase in axicon parameter which we attribute to the increase in area of illumination on the GGP. It is also shown that the speckle size is independent of the order.

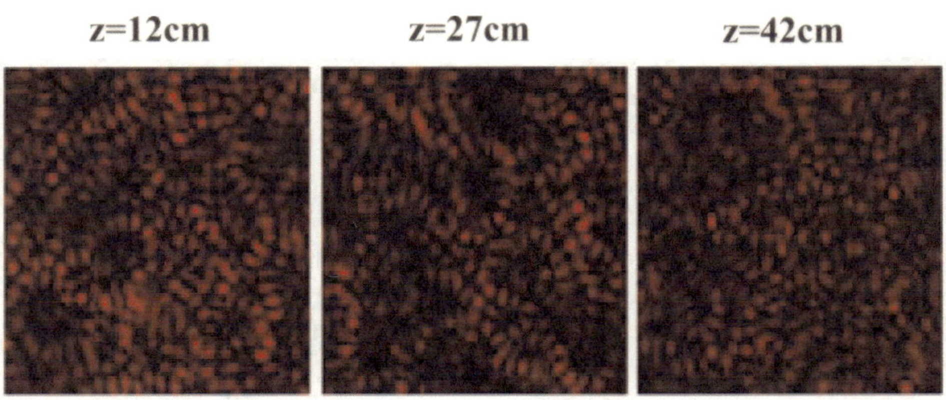

Figure 15.
The speckle patterns obtained by the scattering of POV beam of order m = 0 *at different propagation distances in the far field.*

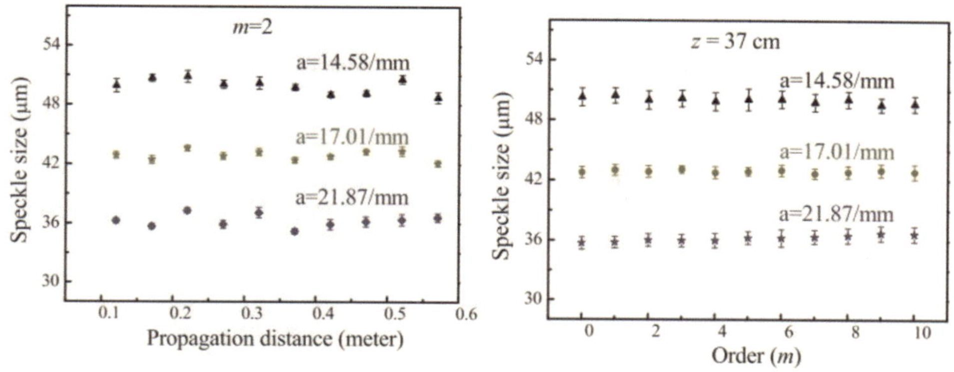

Figure 16.
The variation of speckle size with the propagation distance (left) and order (right) for different axicon parameters.

4. Physical unclonable functions using the correlations of scattered POV beams

Nowadays, securing private data, i.e. authenticating the authorized users to access the sensitive (personal) information, becomes mandatory. In the cryptographic algorithms, information that needs to be sent from a sender end is encrypted (i.e. input data are converted into an unreadable format) using secret keys. At the receiver end, by appropriately using the keys, encoded information can be retrieved (without loss), and this process is known as decryption. It is known that, depends on the cryptographic algorithm used, the keys for both the encryption and decryption process can be same or different [58, 105, 106]. Due to this reason, cryptographic algorithms are widely used in various fields, such as banking, healthcare, social medias, emails, and military communication, to name a few. However, recent developments in high-performance computers increased the vulnerability of cryptographic techniques for a number of different reasons [106]. To prevent from these attacks, a physical one-way function has been introduced in cryptographic systems which can be (physically) realized using the scattering of light beams [107]. These functions are, in general, known as physical unclonable functions (PUFs) and can be embedded into any optical systems for data authentication as this involves a scattering of light beams which results a random output, i.e. speckles [108]. Some of the advantages of PUF include (i) low cost (ii) high output complexity (iii) difficult to replicate, and (iv) high security against attacks [76, 109, 110]. Therefore, in this work, for the first time, we demonstrate an encryption system (i.e. linear canonical transform-based double random phase encoding (LCT-DRPE)) using PUFs that are generated by taking a correlation function between two speckle patterns obtained after scattering the POV beams through a ground glass plate. We wish to take the extra advantage of order-dependent correlation functions generated by the scattering of POV beams for producing the keys for encryption. Here, one should note that the speckle size and their distribution are order-independent, but the correlation between them is order-dependent [94]. We briefly describe the usage of the correlation functions as keys for encryption along with the decryption process as follows:

The LCT is a three-parameter class of linear integral transform and defined as [111]:

$$\Psi_{\alpha,\beta,\gamma}\{f(x,\ y)\} = C_1 \int\int_{-\infty}^{\infty} f(x,y)\, exp\left\{i\pi\left[\alpha\left(x^2+y^2\right) - 2\beta(ux+vy) + \gamma\left(u^2+v^2\right)\right]\right\} dxdy \quad (49)$$

where α,β,γ are the real-valued parameters that are independent of the coordinates that are applied symmetrically in both horizontally (x) and (y), i.e. 2D separable LCT. The encrypted (output) image $E(\omega,\ \varphi)$ can be expressed as [112, 113]:

$$E(\omega,\ \varphi) = LCT\{LCT\{f(x,\ \ y) \times O_1(x,\ \ y)\} \times O_2(x,\ y)\} \quad (50)$$

where $f(x,y)$ is the 2D input image, $O_1(x,y)$ and $O_2(x,y)$ are two random phase masks (RPMs) considered as secret keys which are generated using a correlation function obtained from two scattered POV light beams, i.e. speckles. The schematic for LCT-DRPE is shown in **Figure 17**.

The resultant encrypted image resembles a white noise, i.e. speckle image. Therefore, it does not reveal any of the input information. It is therefore possible to reverse this process called decryption and get the original image back without loss. This process is given mathematically as:

$$f(x,y) = ILCT\{ILCT\{E(\omega,\ \ \varphi)\} \times O_2^*(x,\ y)\} \times O_1^*(x,y) \quad (51)$$

where ILCT refers to inverse linear canonical transform and * denotes the complex conjugate operation. The LCT parameters alpha, beta, and gamma are set as 10,100,1, respectively.

Figure 18a shows the input image (i.e. reconstructed hologram of a 3D object) [113, 114], and **Figure 18b** is the amplitude of the complex encrypted image, and information contained in it is very difficult to be observed. **Figure 18c** shows the decrypted image using appropriate secret keys. The decrypted image quality is the classical mean squared error (MSE) which is calculated between the input image and decrypted image.

In **Figure 19**, changes in the LCT parameter yield the fruitful results, i.e. not able to get proper decrypted images for the corresponding input data.

Encryption using Fourier domain: This is the one of the methods that allows to encode a primary image into a stationary white noise. We demonstrate how

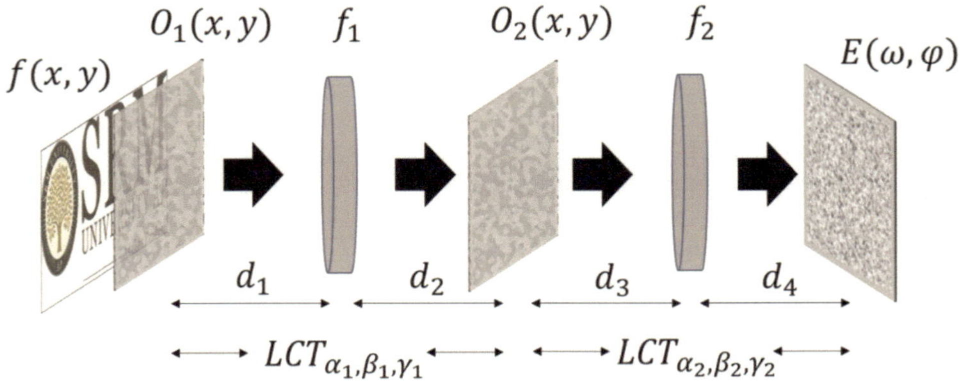

Figure 17.
The schematic for LCT-based DRPE system.

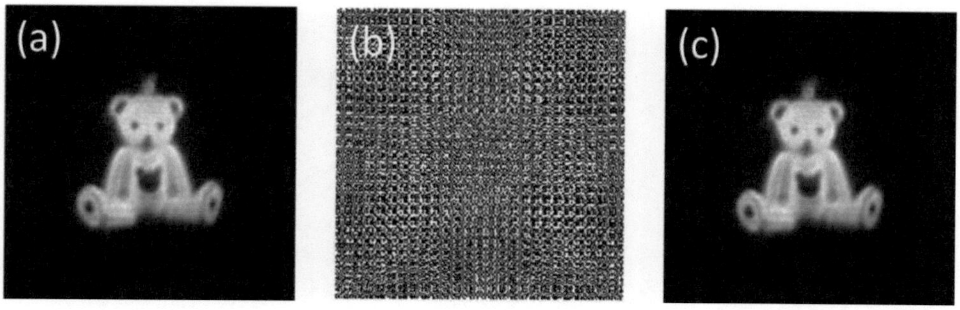

Figure 18.
Simulation results: (a) input grayscale image, (b) encrypted image, and (c) decrypted image (MSE = 1.3685e–27).

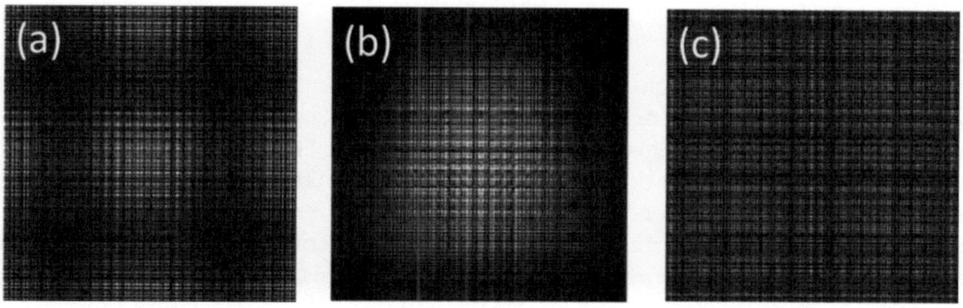

Figure 19.
Decryption with wrong LCT parameters: (a) alpha is wrong, (b) beta is wrong, and (c) gamma is wrong.

straightforward and reliable it is to rebuild the original image using the encoded image [115]. In fact, it is critical to have the ability to encrypt data in a way that makes it challenging to decode without a key yet simple to do so with a key but easy if one knows that key [116]. Let us consider the input signal to be encoded is a face Images, that is, since the image is a positive function and is two-dimensional, it is well known that it is possible to reconstruct an image from its Fourier magnitude [91, 117, 118]. The encoded image can be expressed as:

$$\psi(x) = \{f(x)\exp(i2\pi n(x))\} * h(x) \tag{52}$$

where $f(x)$ is the input function and $\exp(i2\pi n(x))$ is the random phase mask. Then, we convolve this image by the impulse response $h(x)$ is the Fourier transform of the $\exp(i2\pi b(\nu))$ (**Figure 20**).

The encrypted image $\psi(x)$ is optically Fourier-transformed and multiplied by the phase mask $\exp(-i2\pi b(\nu))$ and then inverse Fourier-transformed to produce decrypted image. The decrypted image is expected for the input image with the addition of some noise $u(x)$.

5. Conclusion

In conclusion, we have briefly explained about the correlations present in scattered phase singular beams and their applications toward communication and encryption.

Figure 20.
Simulation results: (a) input grayscale image, (b) encrypted image, and (c) decrypted image.

We have shown that the number of dark rings present in the autocorrelation function of speckles provides the information about the incident spatial mode, and one can utilize these results in free-space optical communication. Further we have utilized, the cross-correlations present in speckle patterns corresponding to vortices of different orders for generating the coherence vortices. We have discussed the geometry of coherence vortices along with their propagation characteristics. We further discussed about the correlations present in the scattered POV beams which produce the order-independent speckle patterns. Finally, we utilized these cross-correlation functions for encryption applications and discussed in detail.

Author details

Vanitha Patnala[1], Gangi Reddy Salla[1*] and Ravindra Pratap Singh[2]

1 Department of Physics, SRM University-AP, Amaravati, India

2 Physical Research Laboratory, Ahmedabad, India

*Address all correspondence to: gangireddy.s@srmap.edu.in

References

[1] Wu Y, Xu C, Qiu H, Xu D, Deng D. Guiding the optical vortex along predesigned parabolic trajectories from circular symmetric Airy-like beams. Applied Optics. 2022;**61**(8):1906-1911

[2] Song D, Wang H, Ma J, Tang L, Zheng X, Hu Y, et al. Synthetic optical vortex beams from the analogous trajectory change of an artificial satellite. Photonics Research. 2019;7(9):1101-1105

[3] Rode AV, Desyatnikov AS, Shvedov VG, Krolikowski W, Kivshar YS. Optical guiding of absorbing nanoclusters in air. Optics Express. 2009;**17**(7):5743-5757

[4] Liu F, Zhang Q, Cheng T, Wu X, Wei Y, Zhang Z. Photophoretic trapping of multiple particles in tapered-ring optical field. Optics Express. 2014; **22**(19):23716-23723

[5] Zhao J, Chremmos ID, Song D, Christodoulides DN, Efremidis NK, Chen Z. Curved singular beams for three-dimensional particle manipulation. Scientific Reports. 2015;**5**:12086

[6] Gong L, Lu R-D, Chen Y, Ren Y-X, Fang Z-X. Generation and characterization of a perfect vortex beam with a large topological charge through a digital micromirror device. Applied Optics. 2015;**54**(27):8030-8035

[7] Stoyanov L, Maleshkov G, Stefanov I, Paulus GG, Dreischuh A. Focal beam structuring by triple mixing of optical vortex lattices. Optical and Quantum Electronics. 2022;**54**:34

[8] Mao D, Zheng Y, Zeng C, Lu H, Wang C, Zhang H, et al. Generation of polarization and phase singular beams in fibers and fiber lasers. Advanced Photonics. 2021;**5**:014002

[9] Bhattacharya R. Generation of phase singular optical beams in microstructure optical fibers. Optics Communications. 2018;**428**:15-21

[10] Wang W, Yokozeki T, Ishijima R, Wada A, Miyamoto Y, Takeda M, et al. Optical vortex metrology for nanometric speckle displacement measurement. Optics Express. 2006;**14**(1):120-127

[11] Shen Y, Wang X, Xie Z, Min C, Fu X, Liu Q, et al. Optical vortices 30 years on: OAM manipulation from topological charge to multiple singularities. Light: Science & Applications. 2019;**8**:90

[12] Kovalev AA, Porfirev AP, Kotlyar VV. Asymmetric Gaussian optical vortex. Optics Letters. 2017; **42**(1):139-142

[13] Curtis JE, Grier DG. Structure of optical vortices. Physical Review Letters. 2003;**90**(13):133901-133904

[14] Allen L, Beijersbergen MW, Spreeuw RJC, Woerdman JP. Orbital angular momentum of light and the transformation of Laguerre-Gaussian laser modes. Physical Review A. 1992;**45**: 8185-8189

[15] Berry MV. Optical vortices evolving from helicoidal integer and fractional phase steps. Journal of Optics A: Pure and Applied Optics. 2004;**6**:259-268

[16] Franke-Arnold S, Allen L, Padgett M. Advances in optical angular momentum. Laser And Photonics Reviews. 2008;**2**(4):299-313

[17] Wang WB, Gozali R, Nguyen TA, Alfano RR. Propagation and transmission of optical vortex beams through turbid scattering wall with

orbital angular momentums. Proceedings of SPIE. 2015;**9318**:931805

[18] Shi L, Lindwasser L, Alfano RR, Gozali R, Wang WB. Deep transmission of Laguerre–Gaussian vortex beams through turbid scattering media. Optics Letters. 2016;**41**(9):2069-2072

[19] Zhu K, Li S, Tang Y, Yu Y, Tang H. Study on the propagation parameters of Bessel-Gaussian beams carrying optical vortices through atmospheric turbulence. Journal of the Optical Society of America A. 2012;**29**(3):251-257

[20] Zhou G, Tang H, Zhu K, Zheng X, Li X. Propagation of Bessel-Gaussian beams with optical vortices in turbulent atmosphere. Optics Express. 2008; **16**(26):21315-21320

[21] Hufnagel F, Sit A, Grenapin F, Bouchard F, Heshami K, Heshami K, et al. Characterization of an underwater channel for quantum communications in the Ottawa River. Optics Express. 2019; **27**(19):26346-26354

[22] Mobashery A, Parmoon B, Saghafifar H, Karahroudi MK, Moosavi SA. Performance evaluation of perfect optical vortices transmission in an underwater optical communication system. Applied Optics. 2018;**57**(30): 9148-9154

[23] Singh M, Atieh A, Grover A, Barukab O. Performance analysis of 40 Gb/s free space optics transmission based on orbital angular momentum multiplexed beams. Alexandria Engineering Journal. 2022;**61**(7): 5203-5212

[24] Li L, Zhang R, Zhao Z, Xie G, Liao P, Pang K, et al. High-capacity free-space optical communications between a ground transmitter and a ground receiver via a UAV using multiplexing of multiple orbital-angular-momentum beams. Scientific Reports. 2017;**7**:17427

[25] Gibson G, Courtial J, Vasnetsov M, Padgett MJ, Franke-Arnold S, Barnett SM, et al. Free-space information transfer using light beams carrying orbital angular momentum. Optics Express. 2004;**12**(22):5448-5456

[26] Qu Z, Djordjevic IB. Orbital angular momentum multiplexed free-space optical communication systems based on coded modulation. Applied Sciences. 2018;**8**(12):2179

[27] Willner AE, Pang K, Song H, Zou K, Zhou H. Orbital angular momentum of light for communications. Applied Physics Reviews. 2021;**8**:041312

[28] Eyyuboğlu HT. Optical communication system using Gaussian vortex beams. Journal of the Optical Society of America A. 2020;**37**(**10**): 1531-1538

[29] Wang J, Liu J, Li S, Zhao Y, Du J, Zhu L. Orbital angular momentum and beyond in free-space optical communications. Nano. 2022;**11**(4): 645-680

[30] Chen R, Zhou H, Moretti M, Wang X, Li J, Member S. Orbital angular momentum waves: Generation. Detection and Emerging Applications, IEEE Communications Surveys & Tutorials. 2020;**22**(2):840-868

[31] White AG, Smith CP, Heckenberg NR, McDuff R. Generation of optical phase singularities by computer-generated holograms. Optics Letters. 1992;**17**(3):221-223

[32] Li S, Wang Z. Generation of optical vortex based on computer-generated holographic gratings by

photolithography. Applied Physics Letters. 2013;**103**:141110

[33] Li D, Xuan L, Hu L, Mu Q, Liu Y, Cao Z. Phase-only liquid-crystal spatial light modulator for wave-front correction with high precision. Optics Express. 2004;**12**(26):6403-6409

[34] Efron U. Spatial Light Modulator Technology: Materials, Devices, and Applications. New York: Marcel Dekker; 1995

[35] Kovalev AA, Moiseev OY, Skidanov RV, Khonina SN, Soĭfer VA, Kotlyar VV. Simple optical vortices formed by a spiral phase plate. Journal of Optical Technology. 2007;**74**(10): 686-693

[36] Khonina SN, Ustinov AV, Logachev VI, Porfirev AP. Properties of vortex light fields generated by generalized spiral phase plates. Physical Review A. 2020;**101**:043829

[37] Beijersbergen MW, Allen L, van der Veen HELO, Woerdman JP. Astigmatic laser mode converters and transfer of orbital angular momentum. Optics Communications. 1993;**96**(1-3):123-132

[38] Kano K, Kozawa Y, Sato S. Generation of a purely single transverse mode vortex beam from a He-Ne laser cavity with a spot-defect mirror. International Journal of Optics. 2012; **2012**:359141

[39] Guo Y, Pu M, Zhao Z, Wang Y, Jin J, Gao P, et al. Merging geometric phase and plasmon retardation phase in continuously shaped metasurfaces for arbitrary orbital angular momentum generation. ACS Photonics. 2016;**3**(11): 2022-2029

[40] Liu C. Vortex beam and its application in optical tweezers. Journal of Physics: Conference Series. 2020; **1549**:032012

[41] D'Ambrosio V, Nagali E, Walborn SP, Aolita L, Slussarenko S, Marrucci L, et al. Complete experimental toolbox for alignment-free quantum communication. Nature Communications. 2012;**3**:961

[42] Yu S, Pang F, Liu H, Li X, Yang J, Wang T. Compositing orbital angular momentum beams in Bi4Ge3O12 crystal for magnetic field sensing. Applied Physics Letters. 2017;**111**:091107

[43] Nye J, Berry M. Dislocations in wave trains. Proceedings of the Royal Society A Mathematical Physical and Engineering Sciences. 1974;**336**(1605): 165-190

[44] Goodman JW. Speckle Phenomena in Optics: Theory and Applications. Second Edition. Washington, USA: SPIE Press; 2013

[45] Dainty JC. Laser Speckle and Related Phenomena. London: Springer; 1976

[46] Reddy SG, Prabhakar S, Kumar A, Banerji J, Singh RP. Higher order optical vortices and formation of speckles. Optics Letters. 2014;**39**(15):4364-4367

[47] Wang W, Hanson SG, Miyamoto Y, Takeda M. Experimental investigation of local properties and statistics of optical vortices in random wave fields. Physical Review Letters. 2005;**94**:103902

[48] Vinu RV. Shaping and analysis of laser speckle for imaging applications, 308210515 [Accessed: June 30, 2022]

[49] Hu XB, Dong MX, Zhu ZH, Gao W, Rosales-Guzmán C. Does the structure of light influence the speckle size? Scientific Reports. 2020;**10**:199

[50] Facchin M, Dholakia K, Bruce GD. Wavelength sensitivity of the speckle patterns produced by an integrating sphere. Journal of Physics: Photonics. 2021;**3**:035005

[51] Wang S, Fan X, Wan Y, Zhang Z, He Z. High-resolution wavemeter using Rayleigh speckle obtained by optical time domain reflectometry. Optics Letters. 2020;**45**(4):799-802

[52] Bruce GD, Dholakia K, O'Donnell L, Chen M, Facchin M. Femtometer-resolved simultaneous measurement of multiple laser wavelengths in a speckle wavemeter. Optics Letters. 2020;**45**(7):1926-1929

[53] Gbur G, Visser TD. Phase singularities and coherence vortices in linear optical systems. Optics Communications. 2006;**259**(2):428-435

[54] Wang W, Duan Z, Hanson SG, Miyamoto Y, Takeda M. Experimental study of coherence vortices: Local properties of phase singularities in a spatial coherence function. Physical Review Letters. 2006;**96**:073902

[55] Jesus-Silva AJ, Alves CR, Fonseca EJS. Characterizing coherence vortices through geometry. Optics Letters. 2015;**40**(12):2747-2750

[56] Heeman W, Steenbergen W, van Dam GM, Boerma EC, Heeman W. Clinical applications of laser speckle contrast imaging: A review. Journal of Biomedical Optics. 2019;**24**(8):080901

[57] Leibov L, Ismagilov A, Zalipaev V, Nasedkin B, Grachev Y, Petrov N, et al. Speckle patterns formed by broadband terahertz radiation and their applications for ghost imaging. Scientific Reports. 2021;**11**:20071

[58] Stallings W. Cryptography and Network Security Principles and Practice. New York: Prentice Hall; 2022

[59] Muniraj I, Sheridan JT. Optical Encryption and Decryption. SPIE Press Book; 2019

[60] Vanitha P, Manupati B, Reddy SG, Singh RP, Muniraj I, Anamalamudi S. Augmenting data security: Physical Unclonable Functions for digital holography based quadratic phase cryptography. 2022. DOI: 10.21203/rs.3.rs-1509081/v1

[61] Léger D, Perrin JC. Real-time measurement of surface roughness by correlation of speckle patterns. Journal of the Optical Society of America. 1976;**66**(11):1210-1217

[62] Chen C, Yang H. Correlation between turbulence-impacted optical signals collected via a pair of adjacent spatial-mode receivers. Optics Express. 2020;**28**(10):14280-14299

[63] Dong K, Cheng M, Lavery MPJ, Geng S, Wang P, Guo L. Scattering of partially coherent vortex beam by rough surface in atmospheric turbulence. Optics Express. 2022;**30**(3):4165-4178

[64] Liu X, Liu L, Chen Y, Cai Y. Partially coherent vortex beam: From theory to experiment, vortex dynamics and optical vortices. London: Springer; 2017

[65] Salem M, Shirai T, Dogariu A, Wolf E. Long-distance propagation of partially coherent beams through atmospheric turbulence. Optics Communications. 2003;**216**(4-6):261-265

[66] Dong M, Yang Y. Coherent vortices properties of partially coherent Elegant Laguerre-Gaussian beams in the free space. Optics and Photonics Journal. 2020;**10**(6):159-166

[67] Gbur G, Visser TD. Coherence vortices in partially coherent beams.

Optics Communications. 2003;**222**(1-6): 117-125

[68] Yadav BK, Kandpal HC, Joshi S. Experimental observation of the effect of generic singularities in polychromatic dark hollow beams. Optics Letters. 2014; **39**(16):4966-4969

[69] Fel'de CV, Wolf E, Bogatyryova GV, Soskin MS, Polyanskii PV, Ponomarenko SA. Partially coherent vortex beams with a separable phase. Optics Letters. 2003;**28**(11):878-880

[70] Jesus-Silva AJ, Neto APS, Alves CR, Amaral JP, Neto JGMN. Measuring the topological charge of coherence vortices through the geometry of the far-field cross-correlation function. Applied Optics. 2020;**59**(6):1553-1557

[71] Paterson C. Atmospheric turbulence and orbital angular momentum of single photons for optical communication. Physical Review Letters. 2005;**94**:153901

[72] Davidson FM, Ricklin JC. Atmospheric turbulence effects on a partially coherent Gaussian beam: Implications for free-space laser communication. Journal of the Optical Society of America A. 2002;**19**(9): 1794-1802

[73] Liu X, Peng X, Liu L, Wu G, Zhao C, Wang F, et al. Self-reconstruction of the degree of coherence of a partially coherent vortex beam obstructed by an opaque obstacle. Applied Physics Letters. 2017;**110**(18):181104

[74] Wu G, Cai Y. Detection of a semirough target in turbulent atmosphere by a partially coherent beam. Optics Letters. 2011;**36**(10): 1939-1941

[75] Kermisch D. Partially coherent image processing by laser scanning. Journal of the Optical Society of America. 1975;**65**(8):887-891

[76] Wang P, Chen F, Li D, Sun S, Huang F, Zhang T, et al. Authentication of optical physical unclonable functions based on single-pixel detection. Physical Review Applied. 2021;**16**:054025

[77] Mandel L, Wolf E. Optical Coherence and Quantum Optics. New York: Cambridge University Press; 1995

[78] Wolf E, Gbur G, Schouten HF, Visser TD. Phase singularities of the coherence functions in Young's interference pattern. Optics Letters. 2003;**28**(12):968-970

[79] Vanitha P, Lal N, Rani A, Das BK, Salla GR, Singh RP. Correlations in scattered perfect optical vortices. Journal of Optics. 2021;**23**:095601

[80] Jesus-Silva AJ, Hickmann JM, Fonseca EJS, Allen L, Beijersbergen MW, Spreeuw RJ, et al. Strong correlations between incoherent vortices. Optics Express. 2012;**20**(18):19708-19713

[81] Anwar A, Permangatt C, Banerji J, Singh RP, Reddy SG, Prabhakar S. Divergence of optical vortex beams. Applied Optics. 2015;**54**(22):6690-6693

[82] Vallone G, Parisi G, Spinello F, Mari E, Tamburini F, Villoresi P. General theorem on the divergence of vortex beams. Physical Review A. 2016;**94**: 023802

[83] Mazilu M, Mourka A, Vettenburg T, Wright EM, Dholakia K. Determination of the azimuthal and radial mode indices for light fields possessing orbital angular momentum. Frontiers in Optics/Laser Science XXVIII, Paper FW4A.4. 2012

[84] Mazilu M, Mourka A, Vettenburg T, Wright EM, Dholakia K. Simultaneous

determination of the constituent azimuthal and radial mode indices for light fields possessing orbital angular momentum. Applied Physics Letters. 2012;**100**(23):231115

[85] Yang Y, Chen M, Mazilu M, Mourka A, Liu YD, Dholakia K. Effect of the radial and azimuthal mode indices of a partially coherent vortex field upon a spatial correlation singularity. New Journal of Physics. 2013;**15**:113053

[86] Yang Y, Dong Y, Zhao C, Liu Y, Cai Y, McMorran BJ, et al. Autocorrelation properties of fully coherent beam with and without orbital angular momentum. Optics Express. 2014;**22**:2925

[87] Vanitha P, Manupati B, Reddy SG, Annapureddy V, Prabhakar S, Singh RP. Intensity correlations in perturbed optical vortices: Diagnosis of the topological charge. 2022. DOI: 10.48550/arxiv.2206.02365

[88] Peatross J, Ware M. Physics of light and optics: A free online textbook. Frontiers in Optics/ Laser Science XXVI, Paper JWA64. 2010

[89] Dogariu A, Acevedo CH. Non-evolving spatial coherence function. Optics Letters. 2018;**43**(23):5761-5764

[90] Gradshteyn IS, Ryzhik IM. Table of integrals, series, and products. Washington, USA: Academic Press; 2014

[91] Goodman JW. Introduction to Fourier optics. Colorado, USA: Robert and Company Publishers; 1996

[92] Kumar A, Banerji J, Singh RP, Prabhakar S. Revealing the order of a vortex through its intensity record. Optics Letters. 2011;**36**(22):4398-4400

[93] Yue Y, Yan Y, Ahmed N, Yang JY, Zhang L, Ren Y, et al. Mode and propagation effects of optical orbital angular momentum (OAM) modes in a ring fiber. IEEE Photonics Journal. 2012; **4**(2):535-543

[94] Ostrovsky AS, Rickenstorff-Parrao C, Arrizón V. Generation of the perfect optical vortex using a liquid-crystal spatial light modulator. Optics Letters. 2013;**38**(4):534-536

[95] Anaya Carvajal N, Acevedo CH, Torres Moreno Y. Generation of perfect optical vortices by using a transmission liquid crystal spatial light modulator. International Journal of Optics. 2017; **2017**:6852019

[96] Yuan W, Xu Y, Xu Y, Zheng K, Fu S, Fu S, et al. Experimental generation of perfect optical vortices through strongly scattering media. Optics Letters. 2021; **46**(17):4156-4159

[97] Kotlyar VV, Kovalev AA, Porfirev AP. Elliptic perfect optical vortices. Optik. 2018;**156**:49-59

[98] Chen M, Mazilu M, Arita Y, Wright EM, Dholakia K. Dynamics of microparticles trapped in a perfect vortex beam. Optics Letters. 2013; **38**(22):4919-4922

[99] Rusch L, Vaity P. Perfect vortex beam: Fourier transformation of a Bessel beam. Optics Letters. 2015;**40**(4): 597-600

[100] Kumar A, Singh RP, Reddy SG, Prabhakar S. Experimental generation of ring- shaped beams with random sources. Optics Letters. 2013;**38**(21): 4441-4444

[101] Gori F, Guattari G, Padovani C. Bessel-Gauss beams. Optics

Communications. 1987;**64**(6): 491-495

[102] McLeod JH. The Axicon: A new type of optical element. Journal of the Optical Society of America. 1954;**44**(8): 592-597

[103] Reddy SG, Chithrabhanu P, Vaity P, Aadhi A, Prabhakar S, Singh RP. Non- diffracting speckles of a perfect vortex beam. Journal of Optics. 2016;**18**: 055602

[104] Vanitha P, Rani A, Annapureddy V, Reddy SG, Singh RP. Diffracting and Non-diffracting random fields. 2021. https://doi.org/10.48550/arxiv.2111.12388.

[105] Vaudenay S. A Classical Introduction to Cryptography. New York: Springer; 2008

[106] Stallings W. Cryptography and Network Security Principles and Practice. New York: Prentice Hall; 2011

[107] Pappu R, Recht B, Taylor J, Gershenfeld N. Physical one-way functions. Science. 2002;**297**(5589): 2026-2030

[108] Bohm C, Hofer M, Physical Unclonable functions in theory and practice. New York: Springer Publishers; 2013

[109] Chen K, Huang F, Wang P, Wan Y, Li D, Yao Y. Fast random number generator based on optical physical unclonable functions. Optics Letters. 2021;**46**(19):4875-4878

[110] Shamsoshoara A, Korenda A, Afghah F, Zeadally S. A survey on physical unclonable function (PUF)-based security solutions for Internet of Things. Computer Networks. 2020;**183**: 107593

[111] Lee B-G, Guo C, Muniraj I, Ryle JP, Healy JJ, Sheridan JT, et al. Low photon count based digital holography for quadratic phase cryptography. Optics Letters. 2017;**42**(14):2774-2777

[112] Liu S, Guo C, Sheridan JT. A review of optical image encryption techniques. Optics & Laser Technology. 2014;**57**: 327-342

[113] Muniraj I, Ryle JP, Healy JJ, Sheridan JT, Wan M, Chen N, et al. Orthographic projection images-based photon-counted integral Fourier holography. Applied Optics. 2019; **58**(10):2656-2661

[114] Lam EY, Chen N, Ren Z. High-resolution Fourier hologram synthesis from photographic images through computing the light field. Applied Optics. 2016;**55**(7):1751-1756

[115] Javidi B, Refregier P. Optical image encryption based on input plane and Fourier plane random encoding. Optics Letters. 1995;**20**(7):767-769

[116] Jumarie G. Relative Information. Berlin: Springer-Verlag; 1990

[117] Hayes MH. In: Stark H, editor. Image Recovery: Theory and Application. San Diego, CA: Academic; 1987

[118] Dainty JC, Fienup JR. In: Stark H, editor. Image Recovery: Theory and Application. San Diego, CA: Academic; 1987